T0074945

Membrane Technology and Applications

Membrane Technology and Applications

Fourth Edition

RICHARD W. BAKER

Membrane Technology and Research, Inc
CA, USA

Registered Offices
John Wiley & Sons, Inc., 111 River Street, Hoboken, NJ 07030, USA
John Wiley & Sons Ltd, The Atrium, Southern Gate, Chichester, West Sussex, PO19 8SQ, UK

For details of our global editorial offices, customer services, and more information about Wiley products visit us at www.wiley.com.

Wiley also publishes its books in a variety of electronic formats and by print-on-demand. Some content that appears in standard print versions of this book may not be available in other formats.

Library of Congress Cataloging-in-Publication

Names: Baker, Richard W. (Richard William), 1941– author.
Title: Membrane technology and applications / Richard W. Baker.
Description: Fourth edition. | Hoboken, NJ : Wiley, 2024. | Includes index.
Identifiers: LCCN 2023032341 (print) | LCCN 2023032342 (ebook) | ISBN
    9781119685982 (hardback) | ISBN 9781119686002 (adobe pdf) | ISBN
    9781119685999 (epub)
Subjects: LCSH: Membranes (Technology) | Membrane separation.
Classification: LCC TP159.M4 B35 2024 (print) | LCC TP159.M4 (ebook) |
    DDC 660/.28424–dc23/eng/20230812
LC record available at https://lccn.loc.gov/2023032341
LC ebook record available at https://lccn.loc.gov/2023032342

Cover Design: Wiley
Cover Images: Membrane Image: © Richard Baker
Image on the bottom left and background: © Roplant/Shutterstock

Set in 10/12pt Times-Roman by Straive, Pondicherry, India
SKY10059195_110223

Contents

Preface

My introduction to membranes was as a graduate student in 1963. At that time membrane permeation was a sub-study of materials science. What is now called membrane technology did not exist, nor did any large industrial applications of membranes. Since then, sales of membranes and membrane equipment have increased more than 1000-fold and tens of millions of square meters of membrane are produced each year – a membrane industry has been created.

This membrane industry is very fragmented. Industrial applications are divided into six main subgroups: reverse osmosis, ultrafiltration, microfiltration, gas separation, pervaporation, and electrodialysis. Medical applications are divided into three more: artificial kidneys, blood oxygenators, and controlled release pharmaceuticals. Few companies are involved in more than one subgroup of the industry. Because of these divisions, it is difficult to obtain an overview of membrane science and technology; this book is an attempt to give such an overview.

The book starts with a series of general chapters on membrane preparation, transport theory, and concentration polarization. Thereafter, each major membrane application is treated in a single 20- to 50-page chapter. In a book of this size, it is impossible to describe every membrane process in detail, but the major processes are covered. However, medical applications are shortchanged and some applications – battery separators and membrane sensors, for example – are not covered at all.

Each application chapter starts with a short historical background to acknowledge the developers of the technology. I am conscious that my views of what was important in the past differ from those of many of my academic colleagues. In this book, I have given more credit than is usual to the engineers who actually made the processes work.

Membrane technology continues to expand, and some change has been made to every chapter of this edition to reflect new developments. Also, two new chapters have been added. The first consolidates recent developments using membrane contactors into a new chapter describing the principles linking these developments. The second new chapter covers finely microporous membranes made from zeolites, microporous carbons, polymers of intrinsic microporosity, and metal organic frameworks. Although these membranes are still struggling to leave the laboratory, they sometimes have exceptional permeation properties and are areas of considerable current academic interest.

Readers of the Theory sections (Chapters 2 and 3) and elsewhere in the book will see that membrane permeation is described using simple phenomenological equations, most commonly, Fick's law. There is no mention of irreversible thermodynamics or Stefan–Maxwell formalism. The irreversible thermodynamic approach to permeation was very fashionable when I began to work with membranes in the 1960s. This approach has the appearance of rigor but hides the physical reality of even simple processes behind a fog of tough equations. As a student and young researcher, I struggled with irreversible thermodynamics for more than 15 years before finally giving up in the 1970s. I have lived happily ever after.

Finally, a few words on units. Unlike the creators of the Pascal, I am not a worshipper of mindless uniformity. Nonetheless, in this edition, I have used metric units to describe most of the processes covered in this book. British/US units are now only used when they are the industry standard, and metric units would lead to confusion.

Acknowledgments

Acknowledgments for the Fourth Edition

This is the fourth and, I suspect, the final edition of this book. Membrane technology has now expanded to a point when it is difficult for a single author to write a comprehensive text covering the whole industry. I originally told Jenny Cossham, my contact at John Wiley, that I needed about a year to do the revision – it took three. What made even this possible was the support I received from Janet Farrant, who agreed to be the editor for this edition. Janet worked for 30 years as the head of our company's patent and contracts group. In this role, she acquired a firm understanding of current membrane technology. This meant she was not only able to fix my grammar but also to critique, challenge, and rework my presentation of the subject. Working with her was sometimes humbling but always a pleasure. I am appreciative of the many hours she spent improving this book.

As in earlier versions of this book, Jenny Valcov converted my handwritten yellow notepads into manuscript form. She also handled all the copyright permissions needed and, working with David Lehmann, added to and revised the 400 figures in this book. I am grateful to these colleagues for their help.

Acknowledgments for the Third Edition

As with the earlier editions of this book, I would not have been able to produce this manuscript without the support of my co-workers at Membrane Technology and Research, Inc. I work with a group of scientist-engineers interested in many aspects of membrane technology. This has kept me informed on new developments affecting our own company's interests and on developments across the membrane field. I also had the assistance of Sara Soder, our company's technical editor, who by mastering my spelling and handwriting, was able to provide me with a polished manuscript draft and who then had the patience to allow me to change and re-change the draft as I clarified my thoughts. Crystal Min and David Lehmann revised and added the nearly 400 figures, and Beth Godfrey, Jenny Valcov, and Linda Szkoropad pitched in to assist with the figure permissions and final manuscript preparation. I am grateful to all of these colleagues for their help.

Acknowledgments for the Second Edition

Eighteen months after the first edition of this book appeared, it was out of print. Fortunately, John Wiley and Sons agreed to publish a second edition, and I have taken the opportunity to update and revise a number of sections. Tessa Ennals, long-time editor at Membrane Technology and Research, postponed her retirement to help me finish the new edition. Tessa has the standards of an earlier time, and here, as in the past, she gave the task nothing but her best effort. I am indebted to her and wish her a long and happy retirement. Marcia Patten, Eric Peterson, David Lehmann, Cindy Dunnegan, and Janet Farrant assisted Tessa by typing new sections, revising and adding figures, and checking references, as well as helping with proofing the manuscript. I am grateful to all of these colleagues for their help.

Acknowledgments for the First Edition

As a schoolboy I once received a mark of ½ out of a possible 20 in an end-of-term spelling test. My spelling is still weak, and the only punctuation I ever really mastered was the period. This made the preparation of a polished final book draft from my yellow notepads a major undertaking. This effort was headed by Tessa Ennals and Cindi Wieselman. Cindi typed and retyped the manuscript with amazing speed, through its numerous revisions, without complaint. Tessa corrected my English, clarified my language, unsplit my infinitives, and added every semicolon found in this book. She also chased down a source for all of the illustrations used and worked with David Lehmann, our graphics artist, to prepare the figures. It is a pleasure to acknowledge my debt to these people. This book would have been far weaker without the many hours they spent working on it. I also received help from other friends and colleagues at MTR. Hans Wijmans read, corrected, and made numerous suggestions on the theoretical section of the book (Chapter 3). Ingo Pinnau also provided data, references, and many valuable suggestions in the area of membrane preparation and membrane material sciences. I am also grateful to Kenji Matsumoto, who read the section on Reverse Osmosis and made corrections, and to Heiner Strathmann, who did the same for Electrodialysis. The assistance of Marcia Patten, who proofed the manuscript, and Vivian Tran, who checked many of the references, is also appreciated.

1

Overview of Membrane Science and Technology

Membrane Technology and Applications, Fourth Edition. Richard W. Baker.
© 2024 John Wiley & Sons Ltd. Published 2024 by John Wiley & Sons Ltd.

1.1 Introduction

In the last 50 years, synthetic membranes have changed our lives. Two million people worldwide are kept alive by hemodialysis, 10 trillion gallons of drinking water are produced annually by reverse osmosis (also sometimes known as RO), and there has been a revolution in municipal sewage treatment thanks to modern microfiltration and ultrafiltration systems. Transdermal patches can help you quit smoking, and avoid getting seasick or pregnant. Membranes are on the brink of being able to capture 90% of the carbon dioxide emitted by power plants. Membranes have become big business.

The key property of a membrane is that it has limited permeability. A membrane is simply a barrier through which different chemical species pass at different rates. In controlled drug delivery, this mechanism is used to moderate the rate at which a drug is dispensed from a reservoir to the body. In industrial applications, the membrane is chosen such that one component of a mixture permeates much faster than another, allowing separation of the components.

This book provides a general introduction to membrane science and technology. Chapters 2–5 cover membrane science, that is, topics that are basic to all membrane processes, such as transport mechanisms, membrane preparation, and boundary layer effects. The next six chapters cover the industrial membrane separation processes that represent the heart of current membrane technology. Carrier transport is covered next, followed by a chapter on membrane contactors and one reviewing the medical applications of membranes. The book closes with a chapter that describes some intriguing membranes and processes that are worthy of inclusion by virtue of their technical ingenuity and level of ongoing research, although they may never make it to the big leagues.

Chapters 2 and 3 on the theory of membrane permeation are not an easy read. The chapter on the solution-diffusion model has more than 100 equations. Readers interested in membrane technology as a unit operation can skip these theory chapters and move directly to the process chapters. Each of these starts with a short theoretical introduction and the basic equations, and then it is on to process design and applications. Readers who are using the book as an introduction to research careers will find the theory chapters of more interest.

1.2 Historical Development of Membranes

Systematic studies of membrane phenomena can be traced to the eighteenth-century philosopher scientists. In 1748, Abbé Nolet, better known for his work on electricity, but a man of many interests, coined the word 'osmosis' to describe permeation of water through a diaphragm. Through the nineteenth and early twentieth centuries, membranes had no industrial or commercial uses, but were used as laboratory tools to develop physical/chemical theories. For example, the measurements of solution osmotic pressure made with membranes by Traube and Pfeffer were used by van't Hoff in 1887 to develop his limit law, which explains the behavior of ideal dilute solutions; this work led directly to the van't Hoff equation. At about the same time, the concept of a perfectly selective semipermeable membrane was used by Maxwell and others in developing the kinetic theory of gases.

Early membrane investigators experimented with every type of diaphragm available to them, such as bladders of pigs, cattle, or fish, and sausage casings made of animal gut. Later, collodion (nitrocellulose) membranes were preferred, because they could be made reproducibly. In 1907, Bechhold devised a technique to prepare nitrocellulose membranes of graded pore size, which he determined by a bubble test [1]. Other early workers improved on Bechhold's technique, and by the early 1930s microporous nitrocellulose membranes were commercially available for laboratory use. During the next 20 years, this early microfiltration membrane technology was expanded to other polymers, notably cellulose acetate.

Membranes found their first substantial application in the testing of drinking water at the end of World War II. Drinking water supplies serving large communities in Germany and elsewhere in Europe had broken down, and laboratory filters to test for water safety were needed urgently. The research effort to develop these filters, sponsored by the US Army, was later exploited by the Millipore Corporation, the first and still the largest US microfiltration membrane producer.

By 1960, the elements of modern membrane science had been developed, but membranes were used only in laboratories and for a few small, specialized industrial applications. No significant membrane industry existed, and total annual sales probably did not exceed US$20 million in today's dollars. Membranes suffered from four problems that prohibited their widespread use: they were too unreliable, too slow, too unselective, and too expensive. Solutions to each of these problems have been developed, and membrane-based separation processes are now commonplace.

The seminal discovery that transformed membrane separation from a laboratory to an industrial process was the development, in the early 1960s, of the Loeb–Sourirajan process for making defect-free, high-flux membranes [2]. These membranes were, and still are, made from a polymer solution by a casting process that produces a structure that is asymmetric or anisotropic (both words are commonly used) in cross section. The membrane consists of an ultrathin, selective surface film, which performs the separation, supported on a much thicker but much more permeable microporous layer, which provides the mechanical strength. The process Loeb and Sourirajan were interested in was reverse osmosis, a way to make drinking water from salt water by permeating the water and retaining the salt. Because the selective layer was so thin, the flux of the first Loeb–Sourirajan membrane was 20 times higher than that of any membrane then available and made RO a feasible method of desalting water. The work of Loeb and Sourirajan, and the timely infusion of large sums of research and development dollars from the US Department of Interior, Office of Saline Water (OSW), resulted in the commercialization of reverse osmosis.

Concurrent with the development of industrial applications was the independent development of membranes for medical separation processes, in particular, the artificial kidney. Kolff and Berk [3] had demonstrated the first successful artificial kidney in the Netherlands in 1943. Their device consisted of a coil of cellophane membrane through which the patient's blood was pumped, while a solution of isotonic saline circulated on the other side. Toxic components permeated from the blood to the saline solution. It took almost 20 years to refine the technology for use on a large scale, but these developments were complete by the early 1960s. Since then, the use of membranes in artificial organs has become a major life-saving procedure. About two million people are now sustained by artificial kidneys. A further million undergo open-heart surgery each year, a procedure made possible by the membrane blood oxygenator, in which carbon dioxide in the patient's blood is removed and replaced with oxygen, a task normally performed by the lungs. The sales of these medical devices exceed the total industrial membrane separation market.

The creation of the modern membrane industry can be divided into the four phases shown in Figure 1.1. In the first phase, building on the original Loeb–Sourirajan technique, other membrane formation processes, including interfacial polymerization and multilayer composite casting and coating, were developed. Using these processes, high-performance membranes with selective surface layers as thin as 0.1 μm or less are now produced commercially by many companies. Methods of packaging membranes into large-membrane-area spiral-wound, hollow fine fiber, capillary, and plate-and-frame modules were also developed, and advances were made in improving membrane stability. The support of the OSW was key to these developments.

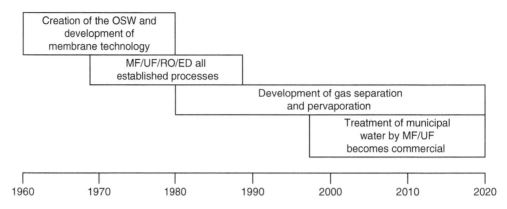

Figure 1.1 *The development of the membrane separation industry, 1960–2020.*

In the second phase, beginning in the early 1970s, the developments that came out of the OSW program began to appear in commercial membrane units; by the 1980s microfiltration (to remove bacteria and viruses from water), ultrafiltration (to separate proteins and other dissolved macromolecules), and reverse osmosis and electrodialysis (both for desalting water) were all established processes.

The third phase, which began in the 1980s, was the emergence of industrial gas separation processes. The first major product was the Monsanto Prism® membrane for hydrogen separation, introduced in 1980 [4]. Within a few years, Dow was producing systems to separate nitrogen from air, and Cynara and Separex were producing systems to separate carbon dioxide from natural gas. Gas separation technology is continuing to evolve and expand; further growth will be seen in the coming years.

The final development phase, which began in earnest in the late 1980s and came to fruition in the mid-1990s, was the development of ultrafiltration systems for the treatment of municipal drinking water and municipal sewage. These applications had been targets for membrane developers for more than 20 years, but their commercialization was impeded by low fluxes, caused by intractable membrane fouling. In the late 1980s, Dr. Kazuo Yamamoto began to develop low-pressure, submerged, air-sparged membranes [5]. It took another 10 years for companies like Kubota, Mitsubishi Rayon, and Zenon to bring this breakthrough concept to the commercial stage, but by the late 1990s, systems with effective fouling control began to be installed. Since then, treatment of municipal water has become one of the most rapidly growing areas of membrane technology. Membrane systems are cost competitive with conventional biological treatment and produce a far superior treated water product. Membrane plants that contain more than one million square meters (250 acres) of membrane, and able to treat the wastewater from a city, are in operation.

1.3 Membrane Transport Theory

The most important practical property of membranes is their ability to sort sheep from goats. In other words, they can discriminate between closely related species. Throughout the twentieth century, two models – the pore-flow model and the solution-diffusion model – were used to describe the mechanisms of permeation. These models are illustrated in Figure 1.2a, b. Over

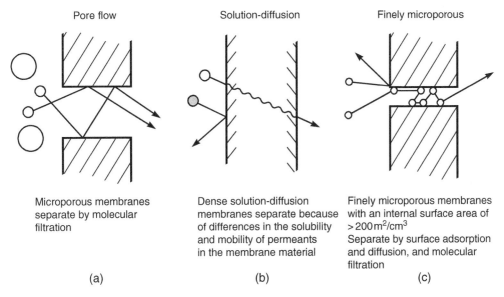

Pore flow   Solution-diffusion   Finely microporous

Microporous membranes separate by molecular filtration	Dense solution-diffusion membranes separate because of differences in the solubility and mobility of permeants in the membrane material	Finely microporous membranes with an internal surface area of >200 m²/cm³ Separate by surface adsorption and diffusion, and molecular filtration
(a)	(b)	(c)

Figure 1.2 *In traditional membranes, molecular transport can be described by a flow through permanent pores (as in (a), for porous membranes), or by the solution-diffusion mechanism (as in (b) for dense membranes). Classes of very finely microporous membranes also exist, as shown in (c). In such membranes, the mechanism of permeation is complex, conforming to neither model.*

the last 20 years, membranes with extremely fine pores, some as small as 5 Å in diameter, have been developed; their mechanism of permeation is complicated and conforms to neither traditional model. These membranes are shown in Figure 1.2c.

According to the pore-flow model of Figure 1.2a, permeants are transported by flow through tiny pores. The flow is convective and occurs because a driving force, usually a pressure gradient, exists between the feed and permeate sides of the membrane. Separation occurs by filtration, that is, the selective retention on the feed side of species that are too large to pass through the pores. Care must be taken in characterizing pore size because most pores are not the perpendicular, straight cylinders shown in the figure, nor are they uniform in size. Nevertheless, this familiar model describes permeation in ultrafiltration and microfiltration processes, where the membrane pores are more than 20 Å in diameter.

The solution-diffusion model, Figure 1.2b, describes transport in membranes that lack permanent pores. In this case, permeants dissolve in the membrane material as in a liquid, and then diffuse through the membrane down a concentration gradient. Transport depends both on the solubility of the permeant in the polymer of which the membrane is made and the diffusion rate once dissolved. Depending on the polymer and the permeant, either solution or diffusion may be the dominant controlling phenomenon. The product of these terms, known as the permeability, is characteristic for a given permeant, and permeants are separated because of differences in their permeabilities. The solution-diffusion model, once controversial, has a secure mathematical basis and is now firmly accepted as describing permeation through the dense polymer membranes used in dialysis, gas separation, reverse osmosis, and pervaporation.

Work in the last 20 years has given rise to a spectrum of finely microporous membrane types that fall between the clearly nonporous solution-diffusion membranes of Figure 1.2b and the clearly microporous filtration membranes of Figure 1.2a. Shown in Figure 1.2c, these intermediate types, which may be made from a variety of materials, including zeolites, metal/organic hybrids, and carbonized polymers, are characterized by pores of diameter in the 5–25 Å range. These pores may change in size and shape over extended periods of time, but in the time frame of permeation, they are permanent.

Because the pores are so small, the ratio of the pore surface area to the pore volume is very large, typically of the order 100–1000 m^2/cm^3 of membrane material. Transport through these membranes is controlled not only by molecular sieving, as in ultrafiltration, but also by adsorption effects on the huge internal surface area of the pore walls; modeling of their transport behavior is more complicated, therefore, than for either of the traditional membrane types. Though their transport properties defy a simple theoretical model, these very finely microporous materials can manifest exceptional permeances and selectivities and are the subject of much current research.

1.4   Types of Membranes

This book is limited to synthetic membranes, excluding all biological structures, but the topic is still large enough to include a wide variety of membranes that differ in chemical and physical composition and in the way they operate. In essence, a membrane is nothing more than a discrete, thin interface that moderates the permeation of chemical species in contact with it. A normal filter meets this definition of a membrane, but, by convention, the term filter is usually used for structures that retain particles larger than about 1–10 μm in diameter. A polymeric membrane may be referred to as isotropic or symmetrical, terms that are used interchangeably for membranes that are homogeneous (uniform in composition and structure). In the alternative, it may be anisotropic or asymmetric, that is, chemically or physically heterogeneous (containing holes or pores of varying dimensions or consisting of some form of layered structure). Other special types of membrane that may themselves be isotropic or anisotropic, but that are generally referred to by their distinguishing characteristic, include charged membranes, supported liquid membranes, and mixed matrix membranes. The principal types of membrane are shown schematically in Figures 1.3–1.5 and are described briefly below.

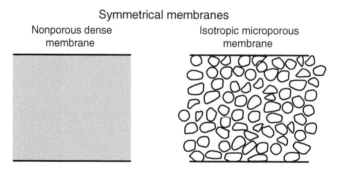

Figure 1.3   *Isotropic membranes. The membrane may be microporous or dense but has a uniform structure throughout.*

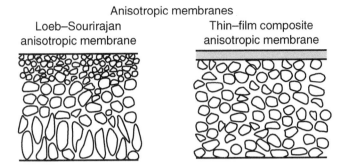

Figure 1.4 *Anisotropic membranes, characterized by a relatively open support structure underlying the selective layer.*

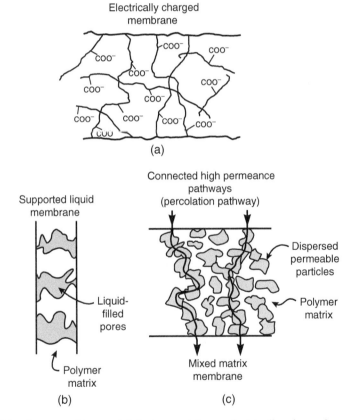

Figure 1.5 *Membranes with special features: (a) an electrically charged membrane; (b) a supported liquid membrane; and (c) a mixed matrix membrane.*

1.4.1   Isotropic Membranes

1.4.1.1   Nonporous, Dense Films

Nonporous, dense polymer films transport permeants by diffusion under the driving force of a pressure, concentration, or electrical potential gradient. There are no permanent pores; the separation of various components of a mixture is related to their relative transport rate within the polymer material, which is determined by a combination of their diffusivity and solubility. Thus, nonporous, dense membranes can separate permeants of similar size (and hence diffusivity) if the permeant concentrations in the membrane material (that is, their solubilities in the polymer) differ significantly. On the other hand, if the permeants have similar solubilities but differ in size, diffusion determines the separation that can be achieved, because small molecules diffuse faster than big ones.

Unsupported dense films of the type shown in Figure 1.3 need to be at least a few tens of microns thick to have any mechanical strength at all. Hence their fluxes are generally much too low for cost-effective industrial separations. They are used for research purposes to determine the intrinsic permeation properties of polymer materials and as drug delivery mediators in transdermal devices.

1.4.1.2   Microporous Membranes

A microporous isotropic membrane is similar in structure and function to a conventional filter. It has a rigid, highly voided structure with interconnected pores. However, these pores differ from those in a conventional filter by being extremely small, on the order of 0.01–10 µm in diameter. All particles larger than the pore diameter are completely rejected by the membrane. Particles smaller than the pore diameter can enter the membrane but may then be captured by adsorption onto the pore walls or entanglement at pore constrictions in the interior of the membrane. As a result, microporous membranes can be impermeable to particles much smaller than their geometric pore diameter would suggest. For this reason, microporous membranes are generally characterized by their *effective* pore diameter, that is, the size of the smallest particle that is completely retained by the membrane. This type of isotropic membrane is mostly used for microfiltration.

1.4.2   Anisotropic Membranes

The transport rate of a species through a membrane is inversely proportional to the membrane thickness; the thicker the membrane, the lower the flux. High transport rates are desirable in membrane separation processes for economic reasons; therefore, the membrane should be as thin as possible. Conventional film fabrication technology limits manufacture of mechanically strong, defect-free films to thicknesses of about 20 µm or above. The development of novel fabrication techniques to produce anisotropic membrane structures was one of the major breakthroughs of membrane technology. Anisotropic membranes consist of an extremely thin surface layer supported on a much thicker, porous substructure, as shown in Figure 1.4. The surface layer and its substructure may be formed in a single operation, as in the Loeb–Sourirajan technique. Alternatively, a composite membrane, in which the layers are made from different polymers, can be made by forming the support membrane, then applying a very thin dense coating. In either case, the separation properties and permeation rates of the membrane are determined mainly by the surface skin or applied coating layer; the substructure offers little resistance to permeation of the species to be separated, and functions solely as a mechanical support. The advantages of the higher fluxes provided by anisotropic and composite polymer membranes are so great that almost all reverse osmosis, gas separation, pervaporation, and ultrafiltration processes use such membranes.

1.4.3   Membranes with Special Features

1.4.3.1   Electrically Charged Membranes

Electrically charged membranes can be dense or microporous, but are most commonly very finely microporous, with the pore walls carrying fixed positively or negatively charged ions. A membrane with fixed positive ions is referred to as an anion exchange membrane, because it attracts and transports negatively charged anions from the surrounding fluid. Similarly, a membrane containing fixed negative ions is called a cation exchange membrane. Separation with charged membranes is achieved mainly by exclusion of ions of the same charge as the fixed ions of the membrane structure. The separation is affected by the charge and concentration of the ions in solution. Electrically charged membranes are used for processing electrolyte solutions in electrodialysis.

1.4.3.2   Mixed Matrix Membranes

In recent years, a great deal of work has been spent on mixed matrix membranes. These membranes consist of a dispersed phase of selective, highly permeable fine particles, for example zeolite particles, in a polymer matrix that holds everything together. The idea is to combine the selectivity of the molecular level pores in the zeolite particles with the ease of fabrication and stability of polymeric membranes. When the volume fraction of zeolite particles in the polymer is above about 30%, there is enough particle-to-particle contact to form a continuous pathway through the zeolite phase of the membrane; this is called its percolation threshold. Despite many years of research, problems with stability and difficulty of scale-up have kept mixed matrix membranes in the laboratory.

1.4.3.3   Supported Liquid Membranes

Supported liquid membranes are microporous polymer membranes in which the pores are filled with a liquid. As with mixed matrix membranes, the microporous polymer structure serves simply as a scaffold; the liquid, which is constrained inside the pores by capillary forces, provides the separation capability. To do this, the liquid incorporates an agent that can bind reversibly with one of the components in the mixture to be treated, enabling that component to be transported preferentially across the membrane. Though they have been a subject of research interest for many years, these membranes remain undeveloped because of unsolved stability problems.

1.4.3.4   Ceramic and Metal Membranes

The discussion so far implies that all membrane materials are organic polymers and, in fact, the vast majority of membranes used commercially are polymer-based. However, there has always been interest in using other materials. Dense metal membranes, particularly palladium membranes for separating hydrogen from gas mixtures, were tried out in a refinery over 60 years ago. Ceramic membranes, a special class of microporous membranes made from alumina, silica, and other refractory materials, are characterized by their exceptional chemical and heat resistance. They are used in ultrafiltration and microfiltration applications for which solvent resistance and thermal stability are required.

1.5   Membrane Processes

Five developed, and a number of developing and yet-to-be-developed, industrial membrane processes are discussed in this book. In addition, sections are included describing the use of membranes in medical applications such as kidney dialysis, blood oxygenation, and controlled drug delivery. The status of these processes is summarized in Table 1.1.

Table 1.1 *Membrane technologies characterized by their stage of development and use.*

Category	Process	Status
Developed industrial membrane separation technologies	– Microfiltration – Ultrafiltration – RO – Electrodialysis – Gas separation	Well-established unit operations
Developing industrial membrane separation technologies	– Pervaporation – Organic/organic (hyperfiltration) – Membrane contactors	A number of plants have been installed. Market size and number of applications are small but expanding
To-be-developed industrial membrane separation technologies	– Carrier transport – Piezodialysis – PRO, FO	Major problems remain to be solved before industrial systems will be installed on a large scale
Medical applications of membranes	– Artificial kidneys – Artificial lungs – Controlled drug delivery	Well-established processes. Still the focus of research to improve performance, for example, improving biocompatibility

The developed industrial membrane separation processes are microfiltration, ultrafiltration, reverse osmosis, electrodialysis, and gas separation. These processes are all well established, and their markets are served by a number of experienced companies.

1.5.1   Reverse Osmosis, Ultrafiltration, Microfiltration

The range of application of the three pressure-driven water separation processes – reverse osmosis, ultrafiltration, and microfiltration – is illustrated in Figure 1.6. Ultrafiltration (Chapter 7) and Microfiltration (Chapter 8) are similar in that the mode of separation includes molecular sieving through increasingly fine pores. The figure shows the distinction between the two as being determined by pore size, which is the simple conventional definition. It can be more convenient for practical purposes to differentiate the processes in terms of their modes of operation, but discussion of this is deferred until the two relevant chapters. Ultrafiltration and microfiltration membranes contain permanent pores usually ranging from 20 Å to 10 μm in diameter. Ultrafiltration membranes mostly operate by capturing the retained material on the surface of the membrane by molecular sieving or other effects. Microfiltration membranes operate by a combination of molecular sieving at the membrane surface and adsorption of particulates in the interior of the membrane. The uses of these filtration processes are extremely diverse and include preparation of pharmaceuticals, drinking water sterilization, and wastewater and sewage treatment.

The mechanism of separation by reverse osmosis (Chapter 6) is quite different from that of ultrafiltration and microfiltration. The membrane pores are so small, from 3 to 5 Å in diameter, that they are within the range of thermal motion of the polymer chains that form the membrane. This means there are no permanent pores in a reverse osmosis membrane, and the accepted mechanism of permeant transport is the solution-diffusion model. According to this model, solutes permeate the membrane by dissolving in the membrane material and diffusing down a concentration gradient. Separation occurs because of the difference in solubilities and mobilities of different solutes in the membrane. The major applications of reverse osmosis are desalination of seawater and brackish groundwater.

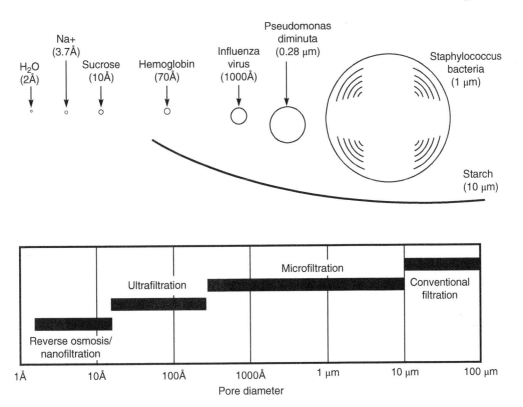

Figure 1.6   *Reverse osmosis, ultrafiltration, microfiltration, and conventional filtration are related processes differing principally in the average pore diameter of the membrane filter. RO membranes are so dense that discrete pores do not exist; transport occurs via statistically distributed free volume areas. The relative size of different solutes removed by each class of membrane is illustrated in this schematic.*

Reverse osmosis, ultrafiltration, and microfiltration are all directed to removing an undesired substance from water. From Figure 1.6, the typical pore diameter of a microfiltration membrane is about 1 μm. This is 100-fold larger than the average ultrafiltration pore and 1000-fold larger than the (nominal) diameter of pores in reverse osmosis membranes. Because flux is proportional to the square of the pore diameter, the permeance (that is, flux per unit transmembrane pressure difference ($J/\Delta p$)) of microfiltration membranes is enormously higher than that of ultrafiltration membranes, which in turn is much higher than that of reverse osmosis membranes. These differences significantly impact the operating pressure and the way that these membranes are used industrially. Reverse osmosis membranes typically operate at transmembrane pressures of 10–100 bar, ultrafiltration membranes at 0.2–5 bar, and microfiltration membranes at below 0.5 bar.

1.5.2   Electrodialysis

The fourth mature membrane process is electrodialysis (Chapter 11), in which charged membranes are used to separate ions from aqueous solutions under the driving force of an electrical potential difference. The process utilizes an electrodialysis stack, built on the filter-press principle and containing several hundred individual cells, each bounded by an

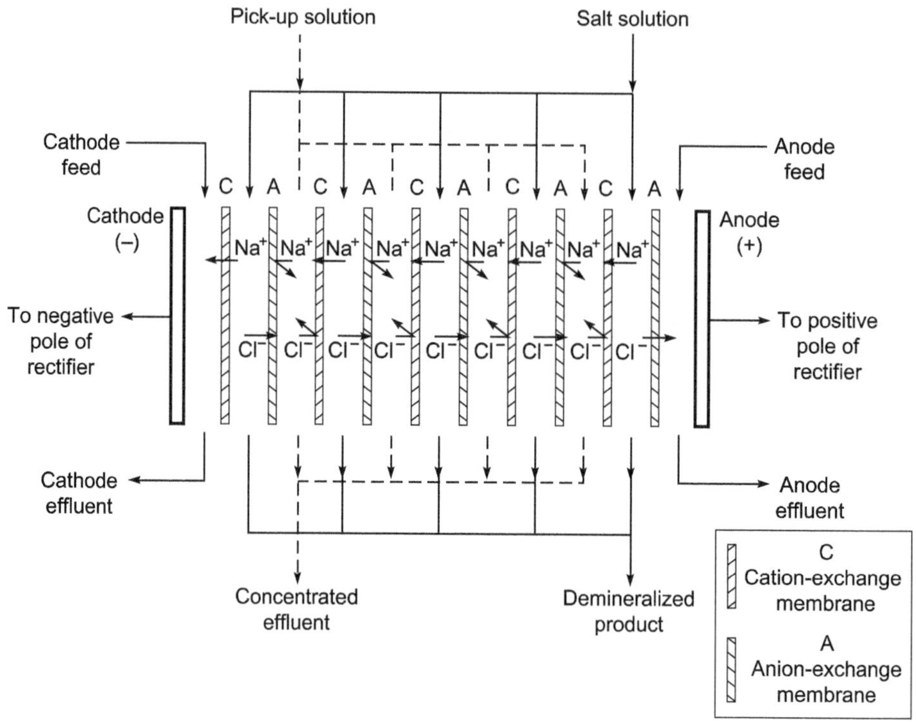

Figure 1.7 *Schematic diagram of an electrodialysis process.*

anion exchange membrane on one side and a cation exchange membrane on the other. Salt solution flows through every other cell; a much slower flow of pick-up solution flows through the adjacent cells. When a current is applied across the stack, anions and cations in the solutions carry the current toward their respective counter-charged electrodes. As a consequence, salt is removed from the salty feed solution cells, forming a demineralized product solution. The removed salt is concentrated into the pick-up cells and is removed as a concentrated brine. A schematic of the process is shown in Figure 1.7. The principal application of electrodialysis is desalting of brackish groundwater.

1.5.3   Gas Separation

The final established process, with annual sales in the US$1–2 billion/year range, is gas separation. At least 20 companies worldwide offer industrial systems for a variety of applications. In gas separation, a gas mixture at an elevated pressure is passed across the surface of a membrane that is selectively permeable to one component of the feed mixture; the membrane permeate is enriched in this species. The basic process is illustrated in Figure 1.8. Major current applications are the separation of hydrogen from nitrogen, argon and methane, used in refineries, petrochemical and ammonia plants; the production of nitrogen from air; the separation of carbon dioxide from methane in natural gas operations; the recovery of organic feedstocks in petrochemical plants; and the removal of C_{3+} hydrocarbons from raw natural gas. Membrane gas separation is an area of research interest, and the number of applications may increase.

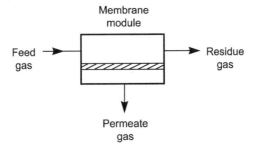

Figure 1.8   Schematic diagram of the basic membrane gas separation process.

In particular, much effort is being expended on the development of carbon dioxide/nitrogen processing schemes that could potentially handle the enormous streams of carbon-dioxide-laden flue gas that emanate from power plants, cement plants, steel mills, and the like.

1.5.4   Pervaporation

In pervaporation, a warm liquid mixture contacts one side of a membrane, and the permeate is removed as a vapor from the other. The driving force for the process is the low vapor pressure on the permeate side of the membrane, generated by cooling and condensing the permeate vapor. The attraction of pervaporation is that the separation obtained is proportional to the rate of permeation of the components of the liquid mixture through the selective membrane. Therefore, pervaporation offers the possibility of separating closely boiling mixtures and azeotropes, which are difficult to separate by distillation or other means.

A schematic of a pervaporation process in its simplest form, using a condenser to generate the permeate vacuum, is shown in Figure 1.9. Currently, the main industrial application of

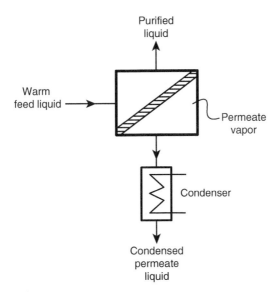

Figure 1.9   Schematic diagram of the basic pervaporation process.

pervaporation is the dehydration of organic solvents, in particular, the dehydration of ethanol solutions, a difficult separation problem because of the ethanol/water azeotrope at 95% ethanol. Pervaporation membranes that selectively permeate water can produce 99.9% ethanol from these solutions. Pervaporation processes are also being developed for the removal of dissolved organics from water and for the separation of organic mixtures, for example of aromatics from aliphatics in refineries. Despite its apparent advantages, pervaporation has been in the 'developing' category for many years and has been slow to take off. Two related problems that have hampered its industrial success are scale-up and stability. A successful laboratory experiment requires only $100\,cm^2$ of membrane that can hold up under test conditions for a few days. An industrial process needs thousands of square meters of membrane stable enough to last a few years in a harsh organic environment. Organic polymer membranes do not fare well under this challenge. For this reason, many pervaporation membranes that have made it to industrial usage are ceramic based. Though stable when confronted with most organic components, these membranes are very expensive. In light of these issues, current annual worldwide sales of pervaporation units are probably no more than about US$50 million.

1.5.5   Hyperfiltration

Development of reverse osmosis to separate water from salt solutions using a pressure driving force has promoted the development of related polymeric membranes to separate organic solvent mixtures. The process is called hyperfiltration. The membranes used can be dense polymer films that separate permeants by the solution-diffusion process or they can be finely microporous membranes containing interconnected pores with diameters in the 5–25 Å range.

Today, hyperfiltration has only a few applications. One problem is the limited stability of polymeric membranes in organic solvent mixtures. A second problem is the high osmotic pressure that must be overcome to generate a solvent flow across the membrane. Nonetheless, a number of small, high value applications in the pharmaceutical and fine chemicals industries have developed [6]. Typical applications are the separation of catalysts or pharmaceutical products in the 200–300 MW range from solvents such as toluene or hexane.

1.5.6   Membrane Contactors

The most common use of membranes is as a filter in which a driving force of pressure across the membrane produces a flow of components from feed to permeate. There are also a number of applications where the function of the membrane is to provide a permeable barrier separating and preventing mixing of different fluids on either side of the membrane. The membrane may be selective for components in the fluids but this is not required. These devices, called membrane contactors, allow interchange of components between the fluids flowing on either side of the membrane. Permeation is usually produced by a difference in temperature or concentration. Membrane contactors are discussed in Chapter 13.

Membrane contactors are being used in a number of applications. One of the largest is in building air conditioners to allow water vapor to exchange between the dry ventilation air leaving the building and the humid hot fresh air entering. Another common application is to use a flow of nitrogen to remove dissolved oxygen from boiler feed water. A hollow fiber membrane contactor provides a large interchange area in a simple compact device. Applications of membrane contactors have grown in the last decades and this trend is likely to continue.

1.5.7 Carrier Transport

A number of processes are in the to-be-developed category in Table 1.1. Perhaps the most important of these is carrier transport (Chapter 12), which often employs liquid membranes containing a carrier agent. The carrier agent reacts with one component of a mixture on the feed side of the membrane and then diffuses across the membrane to release the component on the permeant side. The reformed carrier agent then diffuses back to the feed side. Thus, the carrier agent acts as a shuttle to selectively transport one component through the membrane.

Carrier transport can be used to separate gases, in which case transport is driven by a difference in the gas partial pressure across the membrane. Metal ions can also be selectively transported across a membrane, driven by a flow of hydrogen or hydroxyl ions in the other direction. This process is sometimes called coupled transport. Examples of carrier transport processes for gas and ion transport are shown in Figure 1.10.

Because carrier transport employs a reactive carrier species, very high membrane selectivities can be achieved. These selectivities are often far larger than those achieved by other membrane processes. This one fact has maintained interest in the technology for the past 40 years, but no commercial applications have developed. The principal problems are the physical instability of the liquid membrane and the chemical instability of the carrier agent, both of which have yet to be satisfactorily solved.

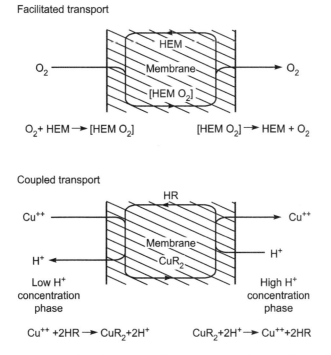

Figure 1.10 *Schematic examples of carrier-facilitated transport of gas and ions. The gas-transport example shows the transport of oxygen across a membrane using hemoglobin dissolved in water as the carrier agent. The ion-transport example shows the transport of copper ions across a membrane using a liquid ion-exchange reagent dissolved in a water immiscible solvent as the carrier agent.*

1.5.8 Medical Applications

The largest application of membranes in medicine is to remove toxic metabolites from blood in patients suffering from kidney failure. The first successful artificial kidney was based on cellophane (regenerated cellulose) dialysis membranes and was developed in 1943–1944. Many changes have been made since then. Currently, most artificial kidneys are based on hollow fiber membranes formed into modules having a membrane area of about 1 m²; their use is illustrated in Figure 1.11. Blood is circulated through the center of the fiber; isotonic saline, the dialysate, is pumped countercurrently around the outside of the fibers. Urea, creatinine, and other low-molecular-weight metabolites in the blood diffuse across the fiber wall and are removed with the saline solution. The process is slow, usually requiring several hours to remove the required amount of the metabolite from the patient, and must be repeated a couple of times per week. In terms of membrane area used and dollar value of the product, artificial kidneys are the single largest application of membranes.

Following the success of the artificial kidney, similar devices were developed to remove carbon dioxide from, and deliver oxygen to, the blood. These so-called artificial lungs are used in surgical procedures during which the patient's lungs cannot function. The dialysate fluid shown in Figure 1.11 is replaced with a carefully controlled sweep gas containing oxygen, which is delivered to the blood, and carbon dioxide, which is removed. These two medical applications of membranes are described in Chapter 14.

Another major medical use of membranes is in controlled drug delivery, also described in Chapter 14. Controlled drug delivery can be achieved by a wide range of techniques, many of which involve membranes; a simple example is the transdermal patch illustrated in Figure 1.12. In this device, designed to adhere to and deliver drugs through the skin, drug is contained in a reservoir that is separated from the skin by a membrane. With such a system, the release of drug through the membrane is constant as long as a constant concentration of drug is maintained within the device. A constant concentration is maintained if the reservoir contains a saturated solution and sufficient excess of solid drug. Systems that operate on this principle are used to moderate delivery of drugs such as nitroglycerine (for angina), nicotine (for smoking cessation), fentanyl (for pain), and estradiol (for hormone replacement therapy) through the skin. Other devices using osmosis or biodegradation as the rate-controlling mechanism are produced as implants and tablets.

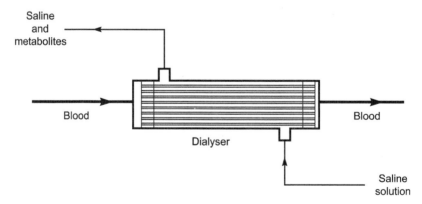

Figure 1.11 *Schematic of a hollow fiber artificial kidney dialyzer used to remove urea and other toxic metabolites from blood. About 100 million of these devices are used every year.*

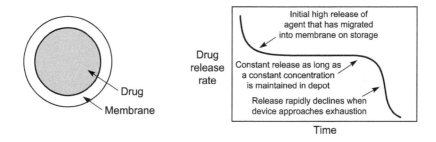

Diagram and release curve for a simple reservoir system

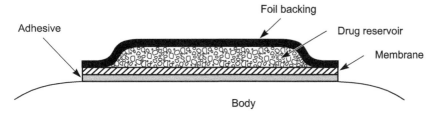

Figure 1.12   *Schematic of transdermal patch in which the rate of delivery of drug to the body is controlled by a polymer membrane. Such patches are used to deliver many drugs including nitroglycerine, estradiol, nicotine, and scopolamine.*

References

1. Bechhold, H. (1907). Kolloidstudien mit der Filtrationsmethode. *Z. Physik Chem.* **60**: 257.
2. Loeb, S. and Sourirajan, S. (1963). Sea water demineralization by means of an osmotic membrane. In: *Saline Water Conversion II*, Advances in Chemistry Series Number, vol. **38** (ed. R.F. Gould), 117–132. Washington, DC: American Chemical Society.
3. Kolff, W.J., Berk, H.T.J., ter Welle, M. et al. (1944). The artificial kidney: a dialyzer with great area. *Acta Med. Scand.* **117**: 121.
4. Henis, J.M.S. and Tripodi, M.K. (1980). A novel approach to gas separation using composite hollow fiber membranes. *Sep. Sci. Technol.* **15**: 1059.
5. Yamamoto, K., Hiasa, M., and Mahmood, T. (1989). Direct solid-liquid separation using hollow fiber membrane in an activated sludge aeration tank. *Water Sci. Technol.* **21**: 43.
6. Marchetti, P., Jimenez Solomon, M.F., Szkely, G., and Livingston, A.G. (2014). Molecular separation with organic solvent nanofiltration: a critical review. *Chem. Rev.* **114**: 10735.

2

Membrane Transport Theory – Solution-Diffusion

2.1 Introduction

The most important property of membranes is their ability to control the rate of permeation of different species. The two models most commonly used to describe the mechanism of permeation are illustrated in Figure 2.1.

Membrane Technology and Applications, Fourth Edition. Richard W. Baker.
© 2024 John Wiley & Sons Ltd. Published 2024 by John Wiley & Sons Ltd.

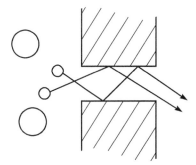

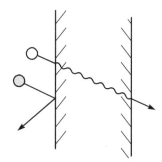

Microporous membranes
separate by molecular
filtration

Dense solution–diffusion
membranes separate because
of differences in the solubility
and mobility of permeants dissolved
in the membrane material

Figure 2.1 *Molecular transport through membranes can be described by flow through permanent pores or by the solution-diffusion mechanism.*

One is the pore-flow model, in which permeants are transported by pressure-driven convective flow through tiny pores. Separation occurs because one of the permeants is excluded (filtered) from some of the pores in the membrane through which other permeants move. The other model is the solution-diffusion model, in which permeants dissolve in the membrane material and then diffuse through the membrane down a concentration gradient. Permeants are separated because of differences in their solubilities in the membrane and differences in the rates at which they diffuse through it. Both models were proposed in the nineteenth century, but the pore-flow model, because it was closer to common experience of how filters behave, dominated the thinking until the mid-1940s (and remains the model for understanding microfiltration and ultrafiltration). In the decades following World War II, however, the solution-diffusion model was used to explain transport of gases through dense polymeric films (films without permanent pores). Use of the solution-diffusion model for gas transport fitted the facts and was soon accepted as uncontroversial. The transport mechanism in reverse osmosis (RO), however, remained hotly debated through the 1960s and 1970s, even after the process had taken off commercially [1–6]. By 1980, the proponents of solution-diffusion had carried the day; solution-diffusion is now the definitive model for understanding reverse osmosis and gas separation through dense membranes; pore-flow is the definitive model for ultrafiltration or microfiltration membranes that separate large solutes, such as proteins with diameters in the 10–100 Å range.

In the last 20 years, newer membranes, made from materials whose permeation properties fit neatly into neither model, have begun to be used. These membranes are made from polymers with intrinsic microporosity (PIMs), thermally rearranged (TR) polymers, metal organic frameworks (MOFs), and zeolites. Taken as a group, they appear to have a network of interconnected pores with diameters in the 5–25 Å range and are called finely microporous membranes. In these membranes, all or almost all of the permeating component is contained in the open space of the pores, but most is adsorbed onto the surface of the pore walls. Diffusion of different components takes place within the open space of the pores and on the pore surface. Separation of different components occurs because of mechanical filtering at pore constrictions, but also because of different rates of surface diffusion.

This chapter will cover permeation through dense polymeric membranes using the solution-diffusion model. Transport through finely microporous membranes and simple microporous ultrafiltration membranes will be covered in the following chapter after a very brief introduction here.

Diffusion, the basis of the solution-diffusion model, is the process by which permeants are transported from one part of a system to another by a concentration gradient. The individual permeant molecules in a membrane medium are in constant random molecular motion, but, in an isotropic medium, individual molecules have no preferred direction of motion. Although the average displacement of a molecule from its starting point can be calculated after a period of time, nothing can be said about the direction in which any individual molecule will move. However, if a concentration gradient of permeant molecules is formed in the medium, simple statistics show that a net transport of matter will occur from the high concentration region to the low concentration region. When two adjacent volume elements with slightly different permeant concentrations are separated by an interface, then simply because of the difference in the number of molecules in each volume element, more molecules will move from the concentrated side to the less concentrated side of the interface than will move in the other direction. This concept was first recognized by Fick theoretically and experimentally in 1855 [7]. Fick formulated his results as the equation now called Fick's law of diffusion, which states

$$J_i = -D_i \frac{dc_i}{dx} \tag{2.1}$$

where J_i is the rate of transfer of component i or flux (g/cm$^2 \cdot$ s) and dc_i/dx is the concentration gradient of component i (g/cm$^3 \cdot$ cm). The term D_i is called the diffusion coefficient (cm$^2 \cdot$ s) and is a measure of the mobility of the individual molecules. The minus sign shows that the direction of diffusion is down the concentration gradient. Diffusion is an inherently slow process. In practical diffusion-controlled separation processes, useful fluxes across the membrane are achieved by making the membranes very thin and creating large concentration gradients.

Pressure-driven convective flow, the basis of the pore-flow model, is most commonly used to describe flow in a capillary or porous medium. The basic equation covering this type of transport is Darcy's law, which can be written as

$$J_i = K' c_i \frac{dp}{dx} \tag{2.2}$$

where dp/dx is the pressure gradient existing in the porous medium, c_i is the concentration of component i in the medium, and K' is a coefficient reflecting the nature of the medium. In general, pore-flow pressure-driven membrane fluxes are high compared with those obtained by simple diffusion.

The difference between the solution-diffusion and pore-flow mechanisms lies in the relative size and permanence of the pores. For membranes in which transport is best described by the solution-diffusion model and Fick's law, the free volume elements (pores) in the membrane are tiny spaces between polymer chains caused by thermal motion of the polymer molecules. These volume elements appear and disappear on about the same time scale as the motions of the permeants traversing the membrane. On the other hand, for a membrane in which transport is best described by a pore-flow model and Darcy's law, the free volume elements (pores) are relatively large and fixed and do not fluctuate in position or volume on the time scale of permeant motion. These pores are usually connected to one another. The larger the individual free volume elements (pores), the more likely they are to be present long enough to produce pore-flow characteristics in the membrane. As a rough rule of thumb, the transition between transient (solution-diffusion) and permanent (pore-flow) pores is a diameter in the range of 5–6 Å.

The average pore diameter in a membrane is difficult to measure directly and must often be inferred from the size of the molecules that permeate the membrane or by some other indirect

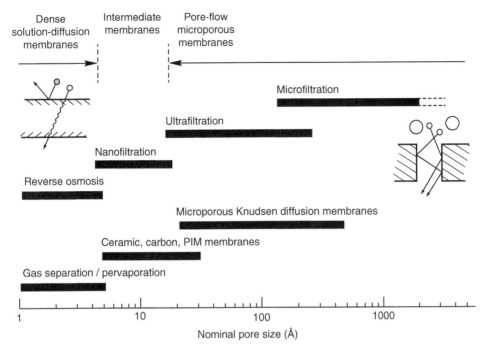

Figure 2.2   *Schematic representation of the nominal pore size and best theoretical model for the principal membrane separation processes.*

technique. With this caveat in mind, membranes can be organized into the three general groups shown in Figure 2.2:

- *Ultrafiltration, microfiltration, and microporous gas separation membranes.* These membranes contain pores larger than about 20 Å in diameter and transport occurs by pore flow.
- *Reverse osmosis, pervaporation, and polymeric gas separation membranes.* These membranes contain a dense selective polymer layer with no visible pores and show selective transport for molecules as small as 2–3 Å in diameter. Transport occurs by solution-diffusion, taking advantage of transient free volume elements, and fluxes are much lower than through microporous membranes.
- *Membranes in the third group.* These contain pores with diameters between 5 and 25 Å and are intermediate between truly microporous and truly solution-diffusion membranes. For example, nanofiltration membranes have properties that fall between ultrafiltration membranes – clearly microporous – and reverse osmosis membranes – clearly dense films. Some gas separation membranes that use finely microporous materials also fall into this category.

2.2   The Solution-Diffusion Model

2.2.1   Molecular Dynamics Simulations

The solution-diffusion model applies to reverse osmosis, pervaporation, and gas permeation in polymer films. At first glance, these processes appear very different. Reverse osmosis uses a large pressure difference across the membrane to separate water from salt solutions. In pervaporation, the membrane separates a liquid feed solution from a permeate vapor.

The pressure difference across the membrane is small, and the process is driven by the vapor pressure difference between the feed liquid and the low partial pressure of the permeate vapor. Gas permeation involves transport of gases down a pressure or concentration gradient. However, all of these processes involve diffusion of molecules in a dense polymer. The pressure, temperature, and composition of the fluids on either side of the membrane determine the concentration of the diffusing species at the membrane surface in equilibrium with the fluid. Once dissolved in the membrane, individual permeating molecules move by the same process of random molecular diffusion, no matter whether the membrane is being used in reverse osmosis, pervaporation, or gas permeation. Often, similar membrane materials are used in very different processes. For example, cellulose acetate membranes were developed for desalination of water by reverse osmosis, but essentially identical membranes have been used in pervaporation to dehydrate alcohol and are widely used in gas permeation to separate carbon dioxide from natural gas. Similarly, silicone rubber membranes used in hyperfiltration processes to separate polyaromatics and dissolved color components from diesel can also be used to separate volatile organics from water by pervaporation and organic vapors from air by gas permeation.

Fluctuations in the volumes between polymer chains due to thermal motion can be modeled by computer. Figure 2.3 shows the results of a molecular dynamics simulation for a small volume element of a polymer, in this case a polyimide. The change in position of individual polymer molecules in the small volume element can be calculated at short enough time intervals to represent the normal thermal motion occurring in a polymeric matrix. If a penetrant molecule

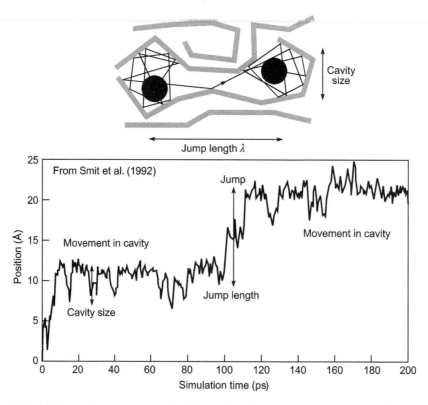

Figure 2.3 *Motion of a carbon dioxide molecule in a 6FDA-4PDA polymer matrix. Source: Reproduced with permission from [8]. Copyright (1992) Elsevier.*

is placed in one of the small free volume microcavities between polymer chains, its motion can also be calculated. The simulated motion of a carbon dioxide molecule in a 6FDA-4PDA polyimide matrix is shown in Figure 2.3 [8]. From kinetic theory, the average velocity of the carbon dioxide molecule at ambient temperatures is about 5 Å per picosecond (1 ps = 10^{-12} s).

During the first 100 ps of the simulation, the carbon dioxide molecule bounces around in the cavity where it has been placed, never moving more than about 5 Å, the diameter of the microcavity. After 100 ps, however, a chance thermal motion moves a segment of the polymer chain sufficiently for the carbon dioxide molecule to jump approximately 10 Å to an adjacent space, where it remains until another movement of the polymer chain allows it to make another jump. By repeating these calculations many times and averaging the distance moved by the gas molecule, its diffusion coefficient can be calculated.

An alternative method of representing the movement of an individual molecule by computational techniques is shown in Figure 2.4 [9]. This figure shows the movements of three different permeant molecules over a period of 200 ps in a silicone rubber polymer matrix. The small helium molecule moves more frequently and makes larger jumps than the larger methane molecule. This is because helium, with a molecular diameter of 2.6 Å, has many more opportunities to move from one position to another than methane, which has a molecular diameter of 3.7 Å. Oxygen, with a molecular diameter of 3.5 Å, has intermediate mobility.

The effect of polymer structure on diffusion can be seen by comparing the distance moved by the gas molecules in the same 200 ps period in Figures 2.3 and 2.4. Figure 2.3 simulates diffusion in a glassy rigid backbone polyimide. In 200 ps, the permeant molecule has made only one large jump. Figure 2.4 simulates diffusion in silicone rubber, a material with a flexible polymer backbone. In 200 ps, all the permeants in silicone rubber have made a number of large jumps from one space to another.

This type of calculation also explains the anomalously high diffusion coefficient of carbon dioxide compared to methane. Carbon dioxide has a molar volume of 18.7 cm^3/mol. The molar volume of methane is a little lower, at 17.1 cm^3/mol, so methane is a slightly smaller

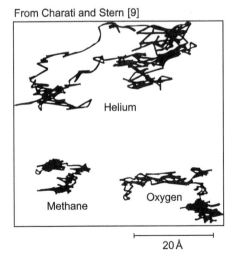

From Charati and Stern [9]

Helium

Methane

Oxygen

20 Å

Figure 2.4 *Simulated trajectories of helium, oxygen, and methane molecules during a 200 ps time period in a poly(dimethylsiloxane) matrix. Source: Reproduced with permission from [9]. Copyright (1998) American Chemical Society.*

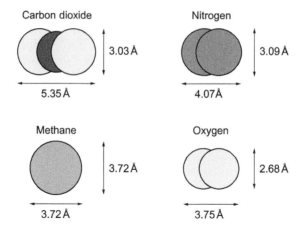

Figure 2.5 *Illustration showing the approximate molecular dimensions of CO_2, CH_4, N_2, and O_2. Shape and size both influence permeant diffusion.*

molecule. However, as shown in Figure 2.5, methane is spherical, with a kinetic diameter of 3.72 Å; carbon dioxide is an oblate ellipsoid. Viewed from the top, the carbon dioxide molecule is 5.25 Å long, but seen end on, the diameter is only 3.03 Å. As carbon dioxide molecules bounce around in their microcavities, they occasionally become oriented so they can slide through transient gaps between the polymer chains that are as small as 3.03 Å, a gap small enough to block the methane molecules no matter how they might rotate. The result is that the diffusion coefficient of carbon dioxide, as calculated by molecular dynamics simulation of glassy polymers, is two to six times larger than methane. This result is also obtained experimentally.

Similarly, nitrogen, another oblate spheroid, although it has almost the same molar volume as methane, has a diffusion coefficient two to four times larger than methane in most polymers.

Molecular dynamics simulations also allow the transition from solution-diffusion to finely microporous pore-flow transport to be seen. As the microcavities become larger, the transport mechanism changes from the diffusion process simulated in Figures 2.3 and 2.4 to a pore-flow mechanism. Pores that are permanent on the time scale of 1000 ps begin to form when the microcavities are larger than about 5 Å in diameter.

Although molecular dynamics calculations are able to rationalize many membrane permeation properties in a qualitative way, they are seldom able to predict them quantitatively. Current estimates of diffusion coefficients from these simulations generally do not match the experimental values, and a better understanding of the interactions between the molecules of polymer chains will be required to produce accurate predictions. Moreover, these simulations require enormous computing capacity, and the longest times that can be easily simulated are a few hundred picoseconds. Extrapolation techniques can be used, but the results are questionable. Several reviews of the development of molecular simulation techniques are available [10]. Despite the lack of quantitative success, molecular dynamic calculations demonstrate the qualitative basis of the solution-diffusion model in a very graphic way. However, the best quantitative description of permeation uses phenomenological equations, particularly Fick's law. This description is given in the section that follows, which outlines the mathematical basis of the solution-diffusion model. Much of this section is adapted from two papers written with my colleague, Hans Wijmans [11, 12].

2.2.2   Concentration and Pressure Gradients in Membranes

The starting point for the mathematical description of diffusion in membranes is the proposition, solidly based in thermodynamics, that the driving forces of pressure, temperature, concentration, and electrical potential are interrelated and that the overall driving force producing movement of a permeant is a gradient in its chemical potential. Thus, the flux, J_i (g/cm^2 · s), of a component, i, is described by the simple equation

$$J_i = -L_i \frac{d\mu_i}{dx} \tag{2.3}$$

where $d\mu_i/dx$ is the chemical potential gradient of component i and L_i is a coefficient of proportionality (not necessarily constant) linking this chemical potential driving force to flux. Driving forces, such as gradients in concentration, pressure, temperature, and electrical potential, can be expressed as chemical potential gradients, and their effect on flux expressed by this equation. This approach is extremely useful, because many processes involve more than one driving force, for example, both pressure and concentration in RO. Restricting the approach to driving forces generated by concentration and pressure gradients, the chemical potential is written as

$$d\mu_i = RTd \, \ln(\gamma_i n_i) + v_i dp \tag{2.4}$$

where n_i is the mole fraction(mol/mol) of component i, γ_i is the activity coefficient (mol/mol) linking mole fraction with activity, p is the pressure, and v_i is the molar volume of component i.

In incompressible phases, such as in a liquid or a solid membrane, volume does not change with pressure. In this case, integrating Eq. (2.4) with respect to concentration and pressure gives

$$\mu_i = \mu_i^o + RT \, \ln(\gamma_i n_i) + v_i\left(p - p_i^o\right) \tag{2.5}$$

where μ_i^o is the chemical potential of pure i at a reference pressure, p_i^o.

In compressible gases, the molar volume changes with pressure. Using the ideal gas laws in integrating Eq. (2.4) gives

$$\mu_i = \mu_i^o + RT \, \ln(\gamma_i n_i) + RT \, \ln \frac{p}{p_i^o} \tag{2.6}$$

To ensure that the reference chemical potential μ_i^o is identical in Eqs. (2.5) and (2.6), the reference pressure p_i^o is defined as the saturation vapor pressure of i, $p_{i_{sat}}$. Equations (2.5) and (2.6) can then be rewritten as

$$\mu_i = \mu_i^o + RT \, \ln(\gamma_i n_i) + v_i\left(p - p_{i_{sat}}\right) \tag{2.7}$$

for incompressible liquids and solid membranes, and as

$$\mu_i = \mu_i^o + RT \, \ln(\gamma_i n_i) + RT \, \ln \frac{p}{p_{i_{sat}}} \tag{2.8}$$

for compressible gases.

Several assumptions must be made to define any permeation model. Usually, the first assumption governing transport through membranes is that the fluids on either side of the membrane are in equilibrium with the membrane material at the interface. This assumption means that the gradient in chemical potential from one side of the membrane to the other is continuous. Implicit in this assumption is that the rates of absorption and desorption at the membrane interface are much higher than the rate of diffusion through the membrane. This appears to be the case in almost all membrane processes, but may fail in transport processes involving chemical reactions, such as facilitated transport, or in diffusion of gases through ion-conducting membranes, where interfacial absorption can be slow.

The second assumption concerns the way the chemical potential gradient across the membrane is expressed within the membrane:

- The solution-diffusion model assumes that the pressure within a membrane is uniform and that the chemical potential gradient of a permeant across the membrane is represented only as a concentration gradient.
- The simple pore-flow model assumes that the permeant concentration within a membrane is uniform and that the chemical potential gradient across the membrane is represented only as a pressure gradient.

The consequences of these two assumptions are illustrated in Figure 2.6, which compares pressure-driven permeation of a one-component solution by solution-diffusion and by pore flow.

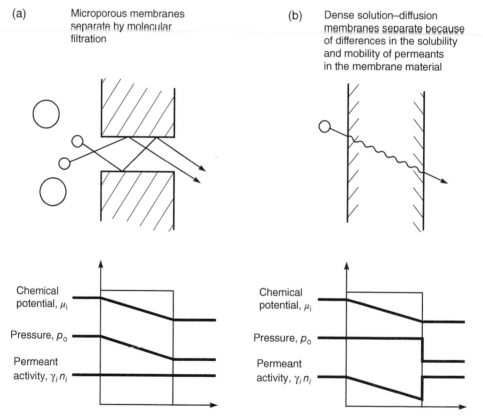

Figure 2.6   *A comparison of the driving force gradients acting on a one-component solution permeating (a) a pore-flow and (b) a solution-diffusion membrane.*

In both models, the difference in pressure across the membrane ($p_o - p_\ell$) produces a gradient in chemical potential (Eqs. (2.7) and (2.8)). In the pore-flow model, the pressure difference produces a smooth gradient in pressure through the membrane, but the solvent concentration remains constant within the membrane. The solution-diffusion model on the other hand assumes that when a pressure is applied across a dense membrane, the pressure everywhere within the membrane is constant at the high-pressure value. This assumes, in effect, that solution-diffusion membranes transmit pressure in the same way as liquids. Consequently, the chemical potential difference across the membranes is represented as a concentration gradient within the membrane, with Eqs. (2.7) and (2.8) providing the mathematical link between pressure and concentration.

In this chapter, we are only concerned with the solution-diffusion model, Figure 2.6b. According to this model, the pressure within the membrane is constant at the high pressure value (p_o), and the gradient in chemical potential across the membrane is expressed as a smooth gradient in solvent activity ($\gamma_i n_i$). The flow that occurs down this gradient is expressed by Eq. (2.3), but because no pressure gradient exists within the membrane, Eq. (2.3) can be rewritten by combining Eqs. (2.3) and (2.4) as

$$J_i = -\frac{RTL_i}{n_i} \cdot \frac{dn_i}{dx}$$
(2.9)

In Eq. (2.9), the gradient of component i across the membrane is expressed as a gradient in mole fraction n_i of the component i. Using the more practical term concentration (g/cm^3) defined as

$$c_i = m_i \rho n_i$$
(2.10)

where m_i is the molecular weight of i (g/mol) and ρ is the molar density (mol/cm^3), Eq. (2.9) can be written as

$$J_i = -\frac{RTL_i}{c_i} \cdot \frac{dc_i}{dx}$$
(2.11)

Equation (2.11) has the same form as Fick's law, in which the term RTL_i/c_i can be replaced by the diffusion coefficient D_i. Thus,

$$J_i = -\frac{D_i dc_i}{dx}$$
(2.12)

Integrating over the thickness of the membrane then gives[1]

$$J_i = \frac{D_i \left(c_{i_{o(m)}} - c_{i_{\ell(m)}} \right)}{\ell}$$
(2.13)

---

[1] In the equations that follow, the terms i and j represent components of a solution, and the terms o and ℓ represent the positions of the feed and permeate interfaces, respectively, of the membrane. Thus the term c_{i_o} represents the concentration of component i in the fluid (gas or liquid) in contact with the membrane at the feed interface. The subscript m is used to represent the membrane phase. Thus, $c_{i_{o(m)}}$ is the concentration of component i in the membrane at the feed interface (point o).

We start by using reverse osmosis as an example. In simple terms, if a selective membrane (that is, a membrane freely permeable to water, but much less permeable to salt) separates a salt solution from pure water, water will pass through the membrane from the pure water side of the membrane into the side less concentrated in water (salt side) as shown in Figure 2.7. This process is normal osmosis. If a hydrostatic pressure is applied to the salt side of the membrane, the flow of water can be retarded and, when the applied pressure is sufficient, the flow ceases. The hydrostatic pressure (p) required to stop the water flow is called the osmotic pressure

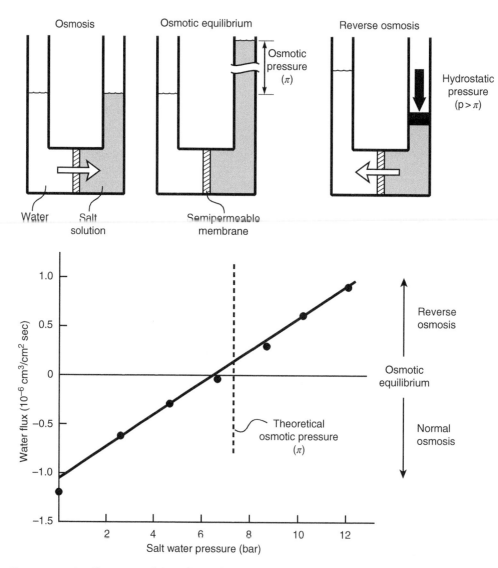

Figure 2.7 *An illustration of the relationship between osmosis, osmotic equilibrium, and reverse osmosis. The data shown are from a paper by Thorsen and Holt [13] using a cellulose acetate membrane. Because the membrane is not perfectly selective for salt, the point of osmotic equilibrium (no flow across the membrane) is slightly below the theoretical osmotic pressure.*

(π). If pressures greater than the osmotic pressure are applied to the salt side of the membrane ($p > \pi$), then the flow of water is reversed, and water begins to flow from the salt solution to the pure water side of the membrane. This process is called RO, which is an important method of producing pure water from salt solutions.

The Fick's law Eqs. (2.12) and (2.13) are approximations. The diffusion coefficient is a constant independent of concentration and the concentration (sorption) term is calculated from Henry's law, again with constant coefficients. Both of these assumptions are flawed. The ability of Fick's law to represent experimental measurements could be improved by using Flory–Huggins methodology or similar for the sorption term, and the Stefan–Maxwell approach to represent the diffusion coefficient term [14, 15]. The resulting derivations might then more accurately represent experiments. They would also be much more complicated, and the nature of the physical processes described would be hidden behind a fog of equations. For this reason, we stay with Fick's law in this chapter, which gives a clear, even if quantitatively shakier, description of permeation.

The chemical potential, pressure, and solvent concentration gradients within and across a semipermeable membrane separating a salt solution from pure solvent (water) are shown in Figure 2.8. In Figure 2.8a, the pressure is the same on both sides of the membrane. For simplicity, the gradient of salt (component j) is not shown in this figure, but the membrane is assumed to be very selective, so the concentration of salt within the membrane is small. The difference in concentration across the membrane results in a continuous, smooth gradient in the chemical potential of the water (component i) across the membrane, from μ_{i_o} on the water side to a slightly lower value μ_{i_o} on the salt side. The pressure within and across the membrane is constant (that is, $p_o = p_m = p_\ell$) and the solvent activity gradient ($\gamma_{i(m)} n_{i(m)}$) falls continuously from the pure water (solvent) side to the saline (solution) side of the membrane. Consequently, water passes across the membrane from right to left.

Figure 2.8b shows the situation when the external pressure applied on the saline side is raised sufficiently high to balance the then existing osmotic pressure difference $\Delta\pi$ between the pure water and the saline solution. This brings the flow across the membrane to zero. As shown in Figure 2.8b, the pressure within the membrane is assumed to be constant at the high-pressure value (p_o). There is a discontinuity in pressure at the permeate side of the membrane, where the pressure falls abruptly from p_o to p_ℓ, the pressure on the solvent side of the membrane. This pressure difference ($p_o - p_\ell$) can be expressed in terms of the concentration difference between the feed and permeate solutions.

The membrane in contact with the permeate side solution is in equilibrium with this solution. That is, the chemical potentials are equal

$$\mu_{i_{\ell(m)}} = \mu_{i_\ell} \tag{2.14}$$

and Eq. (2.7) can be used to link the two phases in terms of their chemical potentials, thus

$$RT \ln\left(\gamma_{i_{\ell(m)}} n_{i_{\ell(m)}}\right) + v_i p_o = RT \ln\left(\gamma_{i_\ell} n_{i_\ell}\right) + v_i p_\ell \tag{2.15}$$

On rearranging, this gives

$$RT \ln\left(\gamma_{i_{\ell(m)}} n_{i_{\ell(m)}}\right) - RT \ln\left(\gamma_{i_\ell} n_{i_\ell}\right) = -v_i(p_o - p_\ell) \tag{2.16}$$

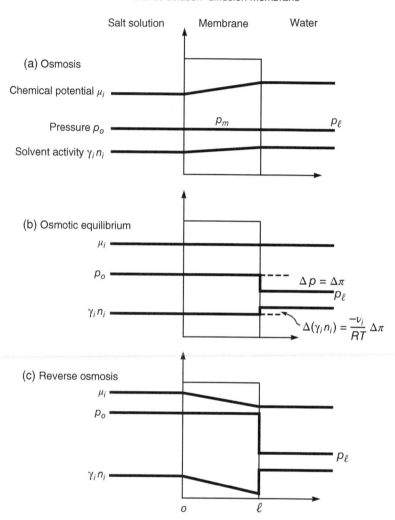

Figure 2.8 *Chemical potential, pressure, and solvent gradients in an osmotic membrane following the solution-diffusion model during osmosis (a), reverse osmosis (c), and under osmotic equilibrium (b). In all cases, the pressure in the membrane is uniform and equal to the high-pressure value, so the chemical potential gradient within the membrane is expressed as a concentration gradient.*

At osmotic equilibrium, $\Delta(\gamma_i n_i)$ can also be defined by

$$\Delta(\gamma_i n_i) = \gamma_{i_\ell} n_{i_\ell} - \gamma_{i_{\ell(m)}} n_{i_{\ell(m)}} \tag{2.17}$$

Since the permeate fluid is pure water, $\gamma_{i_\ell} n_{i_\ell} \approx 1$, and it follows, on substituting Eq. (2.17) into (2.16), that

$$RT \ln[1 - \Delta(\gamma_i n_i)] = -v_i(p_o - p_\ell) \tag{2.18}$$

Since $\Delta(\gamma_i n_i)$ is small, $\ln[1 - \Delta(\gamma_i n_i)] \approx \Delta(\gamma_i n_i)$. Also, since $\Delta p = \Delta\pi$, Eq. (2.18) reduces to

$$\Delta(\gamma_i n_i) = \frac{-\upsilon_i(p_o - p_\ell)}{RT} = \frac{-\upsilon_i \Delta\pi}{RT} \tag{2.19}$$

Thus, the pressure difference across the membrane balances the difference in solvent activity, $\Delta(\gamma_i n_i)$, and the flow is zero.

If a pressure higher than the osmotic pressure is applied to the feed side of the membrane, as shown in Figure 2.8c, then the solvent activity difference across the membrane increases further, resulting in a flow from left to right. This is the process of RO.

The important conclusion illustrated by Figure 2.8 is that, although the fluids on either side of a membrane may be at different pressures and concentrations, within a perfect solution-diffusion membrane, there is no pressure gradient – only a concentration gradient. Flow through this type of membrane is expressed by Fick's law, Eq. (2.13).

2.2.3   Application of the Solution-Diffusion Model to Specific Processes

In this section, the solution-diffusion model is used to describe transport in dialysis, reverse osmosis, gas permeation, and pervaporation processes. The resulting equations, linking the driving forces of pressure and concentration with flow, are then shown to be consistent with experimental observations.

The general approach is to use the first assumption of the solution-diffusion model, namely, that the chemical potentials of the feed and permeate fluids are in equilibrium with the adjacent membrane surfaces. From this assumption, the chemical potential in the fluid and membrane phases can be equated using the appropriate expressions for chemical potential given in Eqs. (2.7) and (2.8). By rearranging these equations, the concentrations of the different species in the membrane at the fluids interface ($c_{i_{o(m)}}$ and $c_{i_{\ell(m)}}$) can be obtained in terms of the pressure and composition of the feed and permeate fluids. These values for $c_{i_{o(m)}}$ and $c_{i_{\ell(m)}}$ can then be substituted into the Fick's law expression, Eq. (2.13), to give the transport equation for the particular process.

2.2.3.1   *Dialysis*

Dialysis is the simplest application of the solution-diffusion model because only concentration gradients are involved. In dialysis, a membrane separates two solutions of different compositions. The concentration gradient across the membrane causes a flow of solute and solvent from one side of the membrane to the other.

Following the general procedure described above, equating the chemical potentials in the solution and membrane phase at the feed side interface of the membrane gives

$$\mu_{i_o} = \mu_{i_{o(m)}} \tag{2.20}$$

Substituting the expression for the chemical potential of incompressible fluids from Eq. (2.7) gives

$$\mu_i^o + RT \ln\left(\gamma_{i_o}^L n_{i_o}\right) + \upsilon_i\left(p_o - p_{i_{sat}}\right) = \mu_i^o + RT \ln\left(\gamma_{i_{o(m)}} n_{i_{o(m)}}\right) + \upsilon_i\left(p_o - p_{i_{sat}}\right) \tag{2.21}$$

which leads to[2]

---

[2] The superscripts G and L are used here and later to distinguish between gas-phase and liquid-phase activity coefficients, sorption coefficients, and permeability coefficients.

$$\ln\left(\gamma_{i_o}^L n_{i_o}\right) = \ln\left(\gamma_{i_{o(m)}} n_{i_{o(m)}}\right) \tag{2.22}$$

and thus

$$n_{i_{o(m)}} = \frac{\gamma_{i_o}^L}{\gamma_{i_{o(m)}}} \cdot n_{i_o} \tag{2.23}$$

The mole fraction terms from Eq. (2.10) ($n_{i_{o(m)}}$ and n_{i_o}) can then be converted to concentrations using

$$c_{i_{o(m)}} = \frac{\gamma_{i_o}^L \rho_m}{\gamma_{i_{o(m)}} \rho_o} \cdot c_{i_o} \tag{2.24}$$

It follows that the sorption coefficient K_i^L is

$$K_i^L = \frac{\gamma_{i_o}^L \rho_m}{\gamma_{i_{o(m)}} \rho_o} \tag{2.25}$$

and Eq. (2.24) becomes

$$c_{i_{o(m)}} = K_i^L \cdot c_{i_o} \tag{2.26}$$

That is, the concentration in the membrane at the feed side surface $c_{i_{o(m)}}$ is proportional to the concentration of the feed solution, c_{i_o}.

On the permeate side of the membrane, the same procedure can be followed, leading to an equivalent expression

$$c_{i_{\ell(m)}} = K_i^L \cdot c_{i_\ell} \tag{2.27}$$

The concentrations of permeant within the membrane phase at the two interfaces can then be substituted from Eqs. (2.28) and (2.29) into the Fick's law expression, Eq. (2.12), to give the familiar expression describing permeation through dialysis membranes

$$J_i = \frac{D_i K_i^L}{\ell}\left(c_{i_o} - c_{i_\ell}\right) = \frac{P_i^L}{\ell}\left(c_{i_o} - c_{i_\ell}\right) \tag{2.28}$$

The product $D_i K_i^L$ is normally referred to as the permeability coefficient, P_i^L, which has units cm^2/s. For many systems, K_i^L and hence P_i^L are concentration-dependent. Thus, Eq. (2.28) implies the use of values for K_i^L, and P_i^L that are averaged over the membrane thickness.

The permeability coefficient P_i^L is often treated as a pure materials constant, depending only on the permeant and the membrane material, but the nature of the solvent used in the liquid phase is also important. From Eqs. (2.28) and (2.25), P_i^L can be written as

$$P_i^L = D_i \cdot \gamma_i^L / \gamma_{i_{(m)}} \cdot \frac{\rho_m}{\rho_o} \tag{2.29}$$

The presence of the term γ_i^L makes the permeability coefficient a function of the solvent in which the permeant is dissolved; change the solvent and the activity coefficient γ_i^L changes, and so does the permeability. Some measurements of the flux of the drug progesterone through the same membrane when dissolved in different solvents illustrate this effect, as shown in Figure 2.9 [16]. The figure is a plot of the product, $J_i\ell$ (progesterone flux times membrane thickness), against the concentration difference across the membrane, $(c_{i_o} - c_{i_\ell})$. From Eq. (2.28), the slope of this line is the permeability, P_i^L. Three sets of experiments are reported, with water, silicone oil, and polyethylene glycol MW 600 (PEG 600) as solvents. The permeability calculated from these plots varies from 9.5×10^{-7} cm^2/s for water to 6.5×10^{-10} cm^2/s for PEG 600. This difference reflects the activity term γ_i^L in Eq. (2.29). However, when the driving force across the membrane is represented not as a difference in concentration but as a difference in fractional saturation between the feed and permeate solution, all the data fall on a single line as shown in Figure 2.9d. The slope of this line is the term $P_i^L c_{i_{sat}}$. This result is also in agreement with Eq. (2.29); when combined with the approximation that, for dilute solutions, the activity of component i can be written as

$$\gamma_i^L = \frac{1}{n_{i_{sat}}} = \frac{m_i \rho_o}{c_{i_{sat}}} \tag{2.30}$$

the result is

$$P_i^L c_{i_{sat}} = \frac{D_i m_i \rho_m}{\gamma_{i_{(m)}}} \tag{2.31}$$

The terms $D_i m_i \rho_m / \gamma_{i_{(m)}}$ and, therefore, $P_i^L c_{i_{sat}}$ are determined solely by the permeant and the membrane material and are thus independent of the liquid phase surrounding the membrane.

2.2.3.2   Reverse Osmosis

RO usually involves two components, water (i) and salt (j). Following the general procedure for application of the solution-diffusion model, the chemical potentials at both sides of the membrane are first equated. At the feed interface, the pressures in the feed solution and within the membrane are identical (as shown in Figure 2.8c). Equating the chemical potentials at this interface gives the same expression as in dialysis (cf. Eq. (2.26))

$$c_{i_{o(m)}} = K_i^L \cdot c_{i_o} \tag{2.32}$$

At the permeate interface, a pressure difference exists from p_o within the membrane to p_ℓ in the permeate solution (as shown in Figure 2.8c). Equating the chemical potentials across this interface gives

$$\mu_{i_\ell} = \mu_{i_{\ell(m)}} \tag{2.33}$$

Substituting the appropriate expression for the chemical potential of an incompressible fluid to the liquid and membrane phases (Eq. (2.7)) yields

$$\mu_i^o + RT \ln\left(\gamma_{i_\ell}^L n_{i_\ell}\right) + \upsilon_i\left(p_\ell - p_{i_{sat}}\right) = \mu_i^o + RT \ln\left(\gamma_{i_{\ell(m)}} n_{i_{\ell(m)}}\right) + \upsilon_i\left(p_o - p_{i_{sat}}\right) \tag{2.34}$$

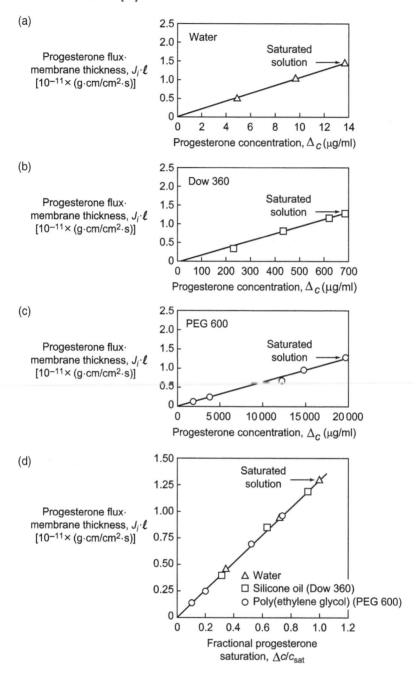

Figure 2.9 *Permeation of progesterone through polyethylene vinyl acetate films. The thickness-normalized progesterone flux $J_i\ell$ is plotted against the fractional progesterone concentration across the membrane. The solvents for the progesterone are (a) water, (b) silicone oil (Dow 360), and (c) polyethylene glycol (PEG 600). Because of the different solubilities of progesterone in these solvents, the permeabilities calculated from these data using Eq. (2.28) vary 1000-fold. All the data can be rationalized onto a single curve by plotting the thickness-normalized flux against fractional progesterone saturation as shown in (d). The slope of this line ($P_i^L c_{i_{sat}}$ or $D_i m_i \rho_m / \gamma_{i_{(m)}}$) is a true materials property, dependent only on the membrane material and the permeant, and independent of the solvent. Source: Reproduced with permission from [16]. Copyright (1976) Elsevier.*

which leads to

$$\ln\left(\gamma_{i_\ell}^L n_{i_\ell}\right) = \ln\left(\gamma_{i_{(m)}}^L n_{i_{\ell(m)}}\right) + \frac{\upsilon_i(p_o - p_\ell)}{RT} \tag{2.35}$$

Rearranging and substituting concentration c_i for mole fraction n_i (Eq. (2.10)) and using the expression for the sorption coefficient, K_i^L (Eq. (2.25)), gives the expression

$$c_{i_{\ell(m)}} = K_i^L \cdot c_{i_\ell} \cdot \exp\left[\frac{-\upsilon_i(p_o - p_\ell)}{RT}\right] \tag{2.36}$$

The expressions for the concentrations within the membrane at the interface in Eqs. (2.32) and (2.36) can now be substituted into Eq. (2.13) to yield

$$J_i = \frac{D_i K_i^L}{\ell}\left\{c_{i_o} - c_{i_\ell}\exp\left[\frac{-\upsilon_i(p_o - p_\ell)}{RT}\right]\right\} \tag{2.37}$$

The term $D_i K_i^L$ can also be written as a permeability P_i^L with units cm²/s, and so Eq. (2.37) becomes

$$J_i = \frac{P_i^L}{\ell}\left\{c_{i_o} - c_{i_\ell}\exp\left[\frac{-\upsilon_i(p_o - p_\ell)}{RT}\right]\right\} \tag{2.38}$$

Equation (2.38) and the equivalent expression for component j give the water flux and the salt flux across the RO membrane in terms of the pressure and concentration differences across the membrane.

The membrane permeability term P_i^L in Eq. (2.38) is a fundamental property of the membrane material, but determining P_i^L from permeation experiments requires a knowledge of the membrane thickness. Industrial membranes often comprise a layer of the selective material less than 1 µm thick coated onto a much thicker support, and the boundary between the layers is often blurred. This makes it difficult to measure the thickness of the layer responsible for the separation properties with any accuracy. To avoid this problem, the term permeance, (P_i^L/ℓ), (a membrane specific rather than a materials property) is widely used to characterize the membrane performance. From Eq. (2.38), the permeance can be calculated by measuring the flux and the feed and permeate pressures and concentrations, all of which can be easily determined with good accuracy. The resulting permeance number can be used reliably for the engineering calculations needed to design and size useful systems.

One result of Eq. (2.38) and the solution-diffusion model illustrated in Figure 2.8c is that the action of an applied pressure on the feed side of the membrane is to *decrease* the concentration of the permeant on the *low-pressure* side of the membrane. A number of workers have verified this intuitively odd prediction experimentally with a variety of membrane/permeant combinations, ranging from diffusion of water in glassy cellulose acetate membranes to diffusion of organics in swollen rubbers [17–19]. Convincing examples include the results of Rosenbaum and Cotton shown in Figure 2.10 [17]. In these experiments, four thin cellulose acetate films were laminated together, placed in a high-pressure RO cell, and subjected to feed pressures of 69 bar or 138 bar.

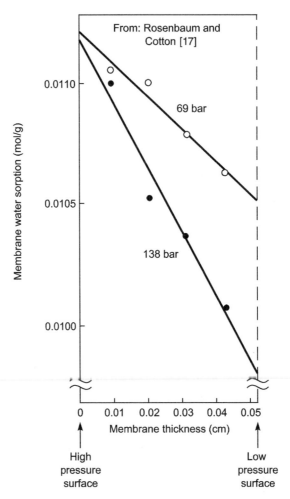

Figure 2.10 *Measurements of Rosenbaum and Cotton of the water concentration gradients in a laminated RO cellulose acetate membrane under applied pressures of 69 and 138 bar. Source: Reproduced with permission from [17]. Copyright (1969) John Wiley & Sons, Inc.*

The permeate was maintained at atmospheric pressure. After the membrane laminate had reached a steady state, the membrane was quickly removed from the cell, and the water concentration in each laminate measured. As predicted by the solution-diffusion model, the applied pressure decreases the concentration of water on the permeate side of the membrane. Also, the concentration difference across the membrane at 138 bar applied pressure is about twice that observed at 69 bar, and the measured concentration on the permeate side is within 20% of the expected value calculated from Eq. (2.36).

Equation (2.38) predicts that as the pressure difference across the membrane increases, the concentration gradient formed in the membrane will also increase, causing the membrane flux to increase. At small pressure differences, the concentration gradient and flux will increase almost linearly, but at very high-pressure differences, Eq. (2.38) predicts the concentration gradient and hence flux will plateau and approach a limiting value. The concentration on the permeate side of the membrane, $c_{i_{\ell(m)}}$, will approach zero, and the flux will approach a limiting value $P_i^L c_{i_o}/\ell$. As Figure 2.10 shows, even at pressure differences of 69–138 bar, water sorption into the membrane, and hence water flux, is still in the linear part of the curve predicted by Eq. (2.38). This result is because the molar volume (v_i) of water is small – 18 cm^3/mol – and this reduces the effect of increasing pressure in Eq. (2.38). Solvents of larger molar volume, for example, isooctane (molar volume 162 cm^3/mol) have proportionally larger concentration gradients at comparable pressures. Figure 2.11 shows the flux of isooctane and methyl ethyl ketone as a function of applied pressure, calculated using Eq. (2.38). At transmembrane pressure differences of 30 bar and above, the concentration on the permeate side of the membrane tends to zero and flux has approached a limiting value, $J_{i_{\max}}$. This plateauing of flux with applied pressure takes place at very high pressures in reverse osmosis and so is not usually noticed, but plateauing has been observed in hyperfiltration of organic liquids [4].

The above derivation of Eq. (2.38) relies on the simplifying assumption that the molar volumes of the permeant in the membrane phase and in the liquid phase in contact with the membrane are equal. This assumption is not always valid. Transport equations can be derived for the case when the molar volumes in the two phases are different; the resulting equations differ by a term called the 'molar volume correction factor'. This correction factor is absent in the gas transport equation, and is insignificant for dialysis and pervaporation. For high-pressure hyperfiltration of organic mixtures containing relatively large molecules, the molar volume correction factor can be large enough to affect the dependence of flux on pressure. This correction is explained in a paper by Wijmans [20].

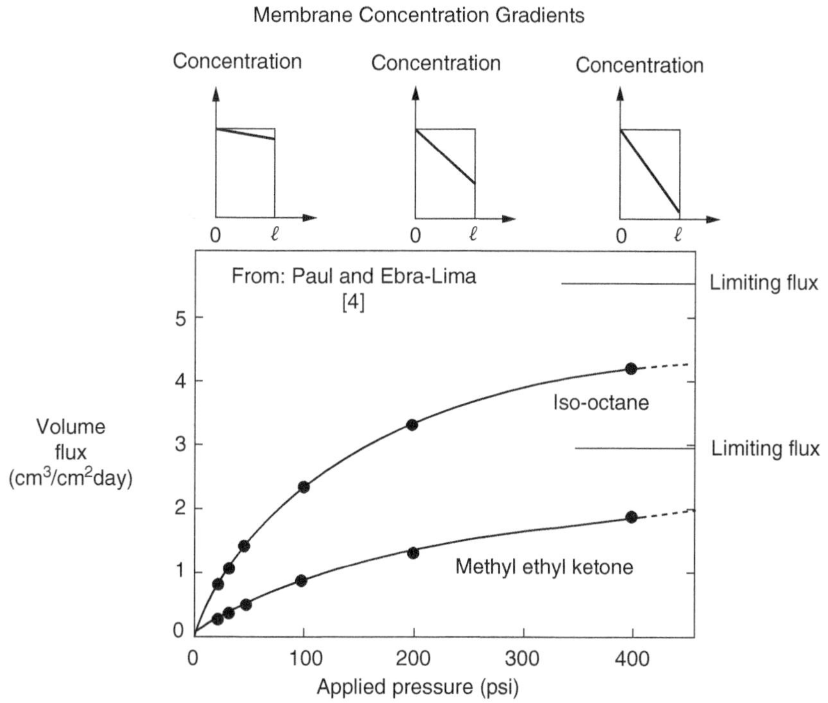

Figure 2.11 *Permeation of iso-octane and methyl ethyl ketone through crosslinked 265 μm thick natural rubber membranes by RO. The change in the concentration gradient in the membrane as the applied pressure is increased is illustrated by the inserts. At high applied pressures, the concentration gradient and the permeation fluxes approach their limiting values. Source: Adapted from [4].*

2.2.3.3   Characterization of Reverse Osmosis and Hyperfiltration Membranes

Equation (2.37) allows the performance of a membrane to be calculated for known permeances, $D_i K_i^L/\ell$ and $D_j K_j^L/\ell$, feed concentrations, c_{i_o} and c_{j_o}, and membrane thickness, ℓ. More commonly, however, Eq. (2.37) is simplified by assuming that the membrane selectivity is high, and that the water permeance is much higher than the salt permeance. This is a good assumption for most RO membranes. Consider the water flux first. When the applied hydrostatic pressure balances the water activity gradient, that is, the point of osmotic equilibrium in Figure 2.8b, the flux of water across the membrane is zero. Equation (2.37) then becomes

$$J_i = 0 = \frac{D i K_i^L}{\ell}\left\{c_{i_o} - c_{i_\ell}\exp\left[\frac{-vi(\Delta\pi)}{RT}\right]\right\} \tag{2.39}$$

and, on rearranging

$$c_{i_o} = c_{i_\ell}\exp\left[\frac{-v_i(\Delta\pi)}{RT}\right] \tag{2.40}$$

At hydrostatic pressures higher than $\Delta\pi$, Eqs. (2.37) and (2.40) can be combined to yield

$$J_i = \frac{D_i K_i^L c_{i_o}}{\ell}\left(1 - \exp\left\{\frac{-n_i[(p_o - p_\ell) - \Delta\pi]}{RT}\right\}\right) \tag{2.41}$$

or

$$J_i = \frac{D_i K_i^L c_{i_o}}{\ell}\left\{1 - \exp\left[\frac{-v_i(\Delta p - \Delta\pi)}{RT}\right]\right\} \tag{2.42}$$

where, as in Figure 2.8, Δp is the difference in hydrostatic pressure across the membrane $(p_o - p_\ell)$. A trial calculation shows that the term $-v_i(\Delta p - \Delta\pi)/RT$ is small under the normal conditions of RO. For example, in water desalination, when $\Delta p = 100$ bar, $\Delta\pi = 10$ bar, and $v_i = 18$ cm^3/mol, the term $vi(\Delta p - \Delta\pi)/RT$ is about 0.06.

Under these conditions, the simplification $1 - \exp(x) \to x$ as $x \to 0$ can be used, and Eq. (2.42) can be written to a very good approximation as

$$J_i = \frac{D_i K_i^L c_{i_o} v_i(\Delta p - \Delta\pi)}{\ell RT} \tag{2.43}$$

This equation can be simplified to

$$J_i = A(\Delta p - \Delta\pi) \tag{2.44}$$

where A is a constant equal to the term $D_i K_i^L c_{i_o} v_i/\ell RT$. In the RO literature, the constant A is usually called the *water permeability constant*.

Similarly, a simplified expression for the salt flux, J_j, through the membrane can be derived, starting with the salt equivalent to Eq. (2.37)

$$J_j = \frac{D_j K_j^L}{\ell} \left\{ c_{j_o} - c_{j_\ell} \exp\left[\frac{-v_j(p_o - p_\ell)}{RT}\right] \right\} \tag{2.45}$$

Because the term $-v_j(p_o - p_\ell)/RT$ is small, the exponential term in Eq. (2.45) is close to one, and Eq. (2.45) can then be written as

$$J_j = \frac{D_j K_j^L}{\ell} \left(c_{j_o} - c_{j_\ell} \right) \tag{2.46}$$

or

$$J_j = B\left(c_{j_o} - c_{j_\ell} \right) \tag{2.47}$$

where B is usually called the *salt permeability constant* and has the value

$$B = \frac{D_j K_j^L}{\ell} \tag{2.48}$$

Predictions of salt and water transport can be made from this application of the solution-diffusion model to reverse osmosis (first derived by Merten and coworkers) [21, 22]. According to Eq. (2.44), the water flux through a reverse osmosis membrane remains small up to the osmotic pressure of the salt solution and then increases with applied pressure, whereas according to Eq. (2.47), the salt flux is essentially independent of pressure. Some typical results are shown in Figure 2.12. This ability of reverse osmosis membranes to separate permeants can be expressed in several ways. Industrially, a term called the rejection coefficient, $\mathbb{R}$, is used, which is defined as

$$\mathbb{R} = \left(1 - \frac{c_{j_\ell}}{c_{j_o}} \right) \times 100\% \tag{2.49}$$

The rejection coefficient is a measure of the ability of the membrane to separate salt from the feed solution.

For a perfectly selective membrane, the permeate salt concentration $c_{j_\ell} = 0$ and $\mathbb{R} = 100\%$; for a completely unselective membrane, the permeate salt concentration is the same as the feed salt concentration, $c_{j_\ell} = c_{j_o}$ and $\mathbb{R} = 0\%$. The rejection coefficient increases with applied pressure as shown Figure 2.12, because the water flux increases with pressure, but the salt flux does not.

Characterizing membrane properties in terms of the salt and water permeability constants, A and B, is widely used in the reverse osmosis industry. This is because these membranes have been developed for one application – desalination of sea water – and so the benefits of more fundamental measures of performance do not outweigh their greater complexity. As reverse osmosis begins to be applied to a wider range of application, especially in the related process of hyperfiltration of organic mixtures, the use of permeabilities (P_i^L, P_j^L), permeances ($P_i^L/\ell, P_j^L/\ell$), and selectivities (α_{ij}) is becoming more common.

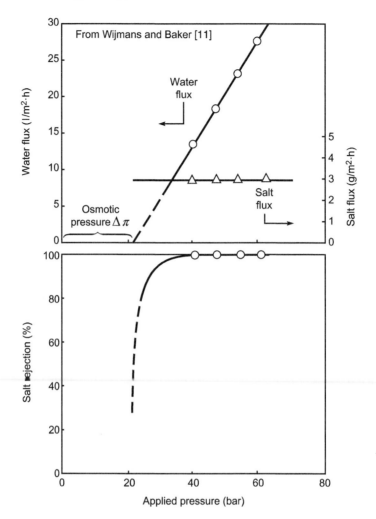

Figure 2.12   *Flux and rejection data for a model seawater solution (3.5% sodium chloride) in a good quality RO membrane (FilmTec Corp. FT 30 membrane) as a function of pressure. The salt flux, in accordance with Eq. (2.47), is essentially constant and independent of pressure. The water flux, in accordance with Eq. (2.44), increases with pressure, and, at zero flux, meets the pressure axis at the osmotic pressure of seawater (~23 bar).*

Membrane selectivity is best defined as the ratio of the permeability or permeances of components through the membrane. Thus

$$\alpha_{ij}^L = \frac{P_i^L}{P_j^L} \tag{2.50}$$

This can be written as

$$\alpha_{ij}^L = \left(\frac{D_i}{D_j}\right)\left(\frac{K_i^L}{K_j^L}\right) \tag{2.51}$$

This equation is useful because it illustrates the factors that determine membrane selectivity. The ratio (D_i/D_j) is the ratio of the diffusion coefficients and can be viewed as mobility selectivity, reflecting the relative size of the two permeants. Diffusion coefficients decrease with increasing molecular size because large molecules (or ions) interact with more segments of the polymer than small molecules. Hence, the ratio D_i/D_j always favors the permeation of water (i) over large hydrated ions such as Na^+ or Cl^- (j).

The ratio K_i^L/K_j^L is the ratio of the sorption coefficients. The magnitude of this term will depend on the nature of the permeants and the membrane. For water and salt diffusing in hydrophilic neutral polymers, this ratio also favors sorption of water over salt.

The membrane selectivity can be calculated from Eqs. (2.42) and (2.45), but for very selective membranes, the approximate forms (Eqs. (2.44) and (2.47)) can be used without significant error. The membrane selectivity can then be written in terms of the water and salt permeability coefficients as

$$\alpha^L = \frac{A}{B} \cdot \frac{RT}{c_{i_o} v_i} \tag{2.52}$$

The data in Figure 2.12 show that with this type of good quality seawater desalination membrane the selectivity for water over salt is more than 10 000.

For the past 20 years, the bulk of RO membrane data have been measured with interfacial composite membranes, where measurement of diffusion and sorption coefficients and membrane thickness is not possible. However, the first-generation membrane material cellulose acetate can be cast into 100 μm films, allowing such measurements to be made. These data are shown in Table 2.1 [21].

Some trends are immediately clear. First, the bulk of the selectivity is due to a very large mobility selectivity in favor of water. The sorption selectivity, although it also favors permeation of water, is small in comparison. Second, as the acetate content increases, making the membrane more hydrophobic, the water sorption term (K_w) decreases and the sorption selectivity and mobility selectivity both increase. As a consequence, the overall selectivity $\alpha_{w/s}^L$ increases almost 100-fold. Concurrently, the water permeability ($D_w K_w^L$) decreases 10-fold. This trade-off between permeation and selectivity is observed with other separation processes, gas separation and pervaporation for example.

Table 2.1 *Water (w) and sodium chloride (s) diffusion and sorption data measured for a series of cellulose acetate films [21].*

Properties			Membrane acetate content (wt%)			
			33.6	37.6	39.8	43.2
Water	Permeability	$D_w K_w^L (10^{-7} cm^2/s)$	16	5.7	2.6	1.5
	Diffusion coefficient	$D_w (10^{-6} cm^2/s)$	5.7	2.9	1.6	1.3
	Sorption coefficient	$K_w^L (-)$	0.29	0.20	0.16	0.12
Salt	Permeability	$D_s K_s^L (10^{-11} cm^2/s)$	500	27	0.33	0.059
	Diffusion coefficient	$D_s (10^{-10} cm^2/s)$	290	43	9.4	0.39
	Sorption coefficient	$K_s^L (-)$	0.17	0.062	0.035	0.015
Mobility selectivity		D_w/D_s	200	670	1700	3300
Solubility selectivity		K_w^L/K_s^L	1.6	3.2	4.6	8.0
Overall selectivity		$\alpha_{w/s}^L$	320	2100	7900	25 000

A membrane acetate content of 33.6% acetate is close to cellulose diacetate; 43.2% acetate is close to cellulose triacetate.

The very high overall selectivities shown in Table 2.1 suggest that membranes with rejection coefficients of 99.9% or more could be made from these materials. Commercial RO membranes almost always contain small imperfections, however, so the theoretical rejection can only be obtained with small membrane samples prepared in an ultraclean environment [23]. Typical water/NaCl selectivities of RO membranes today are about 10 000.

By convention, the term reverse osmosis is used to describe the pressure-driven separation of an aqueous salt solution. The same type of process has been applied to the separation of organic mixtures. For example, Mobil Oil installed a large plant to separate methyl ethyl ketone (MEK) from MEK–oil mixtures created in the production of lubricating oil [24]. Membranes can also be considered for the separation of aromatics, such as toluene, from aliphatics, such as octane, as described in Chapter 6. Though the transport mechanism is the same as for traditional RO, separation of such mixtures is best called hyperfiltration to avoid confusion.

The mathematical description of hyperfiltration is identical to that for RO given in Eq. (2.37) and leads to the same expressions for the solute and solvent fluxes

$$J_i = \frac{D_i K_i^L}{\ell} \left\{ c_{i_o} - c_{i_\ell} \exp\left[\frac{-v_i(p_o - p_\ell)}{RT} \right] \right\}$$ (2.53)

and

$$J_j = \frac{D_j K_j^L}{\ell} \left\{ c_{j_o} - c_{j_\ell} \exp\left[\frac{-v_j(p_o - p_\ell)}{RT} \right] \right\}$$ (2.54)

The numerical solution to these equations is straightforward, even for multi-component mixtures. Figure 2.13 shows a calculation for the separation of a 20 wt% solution of n-decane in methyl ethyl ketone (MEK). The selectivity for MEK over n-decane is set at 10. The curves have essentially the same form as the salt solution flux data in Figure 2.12. At high pressures, the rejection approaches a limiting value of 90%, and the limiting flux for the solvent flux (MEK) and solute flux (n-decane) given in the discussion of Eq. (2.38) apply.

2.2.3.4   Gas Separation

In gas separation, a high-pressure gas mixture at a pressure p_o is applied to the feed side of the membrane, and the permeate gas at a lower pressure p_ℓ is removed from the downstream side.

The concentration and pressure gradients through a dense polymer membrane performing gas separation are shown graphically in Figure 2.14. The figure has the features previously illustrated for RO in Figure 2.8. The pressure within the membrane is constant at the feed pressure and the chemical potential gradient is expressed as a concentration gradient, which can be changed by changing the feed or permeate pressure. As the pressure is increased on the feed side of the membrane, the concentration of component i in the membrane at the feed interface increases, reaching a maximum when the vapor pressure reaches the saturation vapor pressure, $p_{i_{sat}}$. Similarly, the concentration in the membrane at the permeate side interface decreases with decreasing permeate pressure, reaching zero when a hard vacuum is created on the permeate side. In gas separation therefore, the partial pressures of a component on either side of the membrane can be linked by the expression

$$p_{i_{sat}} \geq p_{i_o} \geq p_{i_\ell}$$ (2.55)

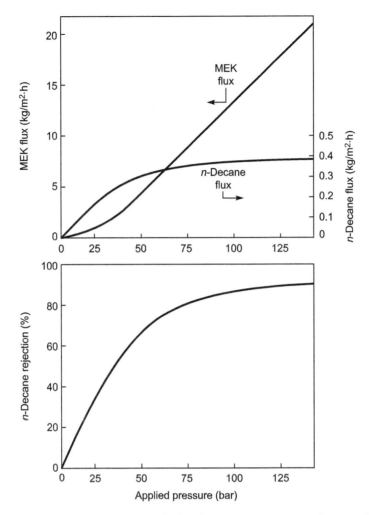

Figure 2.13 Flux and rejection curves calculated using Eqs. (2.53) and (2.54) for a 20 wt% n-decane solution in methyl ethyl ketone (MEK). MEK is assumed to be 10 times more permeable than n-decane.

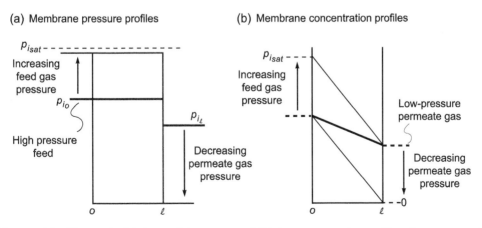

(a) Membrane pressure profiles

(b) Membrane concentration profiles

Figure 2.14 Changes in (a) the partial pressure and (b) the concentration profiles of a component i through a membrane operating in gas separation mode, according to the solution-diffusion model, as the feed pressure increases and permeate pressure decreases. The feed partial pressure of component i (p_{i_o}) cannot exceed its saturation vapor pressure ($p_{i_{sat}}$); the permeate pressure (p_{i_ℓ}) is always greater than zero.

As before, the starting point for the derivation of the gas separation transport equation is to equate the chemical potentials on either side of the gas/membrane interface. This time, however, the chemical potential for the gas phase is given by Eq. (2.8) for a compressible gas, and Eq. (2.7) for an incompressible medium is applied to the membrane phase. Equating the potentials at the feed gas/membrane interface then yields[3]

$$\mu_i^o + RT \ln\left(\gamma_{i_o}^G n_{i_o}\right) + RT \ln \frac{p_o}{p_{i_{sat}}} = \mu_i^o + RT \ln\left(\gamma_{i_{o(m)}} n_{i_{o(m)}}\right) + \upsilon_i\left(p_o - p_{i_{sat}}\right) \tag{2.56}$$

which rearranges to

$$n_{i_{o(m)}} = \frac{\gamma_{i_o}^G}{\gamma_{i_{o(m)}}} \cdot \frac{p_o}{p_{i_{sat}}} \cdot n_{i_o} \exp\left[\frac{-\upsilon_i\left(p_o - p_{i_{sat}}\right)}{RT}\right] \tag{2.57}$$

Substituting concentration $c_{i_{o(m)}}$ for mole fraction $n_{i_{o(m)}}$ using Eq. (2.10) and expressing the term $p_o n_{i_o}$ as the partial pressure p_{i_o}, Eq. (2.54) can be rewritten as

$$c_{i_{o(m)}} = \frac{m_i \rho_m \gamma_{i_o}^G}{\gamma_{i_{o(m)}} p_{i_{sat}}} \cdot p_{i_o} \exp\left[\frac{-\upsilon_j\left(p_o - p_{i_{sat}}\right)}{RT}\right] \tag{2.58}$$

Defining a gas phase sorption coefficient K_i^G in a similar way to the liquid phase coefficient K_i^L Eq. (2.25) gives

$$K_i^G = \frac{m_i \rho_m \gamma_{i_o}^G}{\gamma_{i_{o(m)}} p_{i_{sat}}} \tag{2.59}$$

and so Eq. (2.58) becomes

$$c_{i_{o(m)}} = K_i^G \cdot p_{i_o} \cdot \exp\left[\frac{-\upsilon_i\left(p_o - p_{i_{sat}}\right)}{RT}\right] \tag{2.60}$$

In exactly the same way, the process represented by Eqs. (2.56)–(2.58) can be repeated at the membrane/permeate interface and the concentration of component i in the membrane at the membrane/permeate gas interface can be shown to be

$$c_{i_{\ell(m)}} = K_i^G \cdot p_{i_\ell} \cdot \exp\left[\frac{-\upsilon_i\left(p_o - p_{i_{sat}}\right)}{RT}\right] \tag{2.61}$$

---

[3] At this point the superscript G is introduced to denote the gas phase. For example, γ_i^G is the activity of component i in the gas phase, and K_i^G is the sorption coefficient of component i between the gas and membrane phases (Eq. (2.59)). Cf footnote 2 on page 23.

Combining Eqs. (2.60) and (2.61) with Eq. (2.13) then gives

$$J_i = \frac{D_i K_i^G \left(p_{i_o} - p_{i_\ell}\right)}{\ell} \cdot \exp\left[\frac{-v_i\left(p_o - p_{i_{sat}}\right)}{RT}\right] \tag{2.62}$$

Equation (2.62) shows that the permeation of gas through a membrane is the product of two terms. The first term contains the partial pressure difference of component i across the membrane. The second exponential term contains the total gas pressure p_o of all gas components $p_o = (p_i + p_j + \cdots)$ on the feed side of the membrane. Gas mixtures can have the same partial pressures for one of the components (i), but very different total pressures, so the exponential term is a measure of how much the total gas pressure, produced by the presence of other components, affects the partial pressure driving force for component i. It should be noted that the term v_i in Eq. (2.62) is not the molar volume of i in the gas phase, but the partial molar volume of i dissolved in the membrane material, which is approximately equal to the molar volume of liquid i. As a result, the exponential term (known as the Poynting correction) in Eq. (2.62) is usually very close to 1 for permanent gases and only becomes significant for vapors with larger molar volumes at high pressures. For most gas permeation processes, Eq. (2.62) reduces to

$$J_i = \frac{D_i K_i^G \left(p_{i_o} - p_{i_\ell}\right)}{\ell} \tag{2.63}$$

The product $D_i K_i^G$ is often abbreviated to a permeability coefficient P_i^G, leading to the familiar expression

$$J_i = \frac{P_i^G \left(p_{i_o} - p_{i_\ell}\right)}{\ell} \tag{2.64}$$

Equation (2.64) is widely used to rationalize and predict the properties of gas permeation membranes with good accuracy.

The derivation of Eq. (2.64) might be seen as a long-winded way of arriving at an obvious result. However, this derivation explicitly clarifies the assumptions behind the equation. First, a gradient in concentration occurs within the dense polymer membrane, but there is no gradient in pressure. Second, absorption of a component into the membrane is proportional to its activity (partial pressure) in the adjacent gas, but is independent of the total gas pressure. This is related to the approximation made following Eq. (2.62), in which the Poynting correction was assumed to be 1.

The permeability coefficient, P_i, equal to the product $D_i K_i^G$, can be expressed from the definition of K_i^G in Eq. (2.59) as

$$P_i^G = \frac{\gamma_i^G D_i m_i \rho_m}{\gamma_{i(m)} p_{i_{sat}}} \tag{2.65}$$

In Eq. (2.64) the gas flux, J_i, is a mass flux (g/cm$^2 \cdot$ s). The equivalent expression used to define P_i^L in reverse osmosis (Eq. (2.38)) also uses J_i as a mass flux. P_i^G and P_i^L are thus both 'mass' permeabilities. However, the gas separation literature predominantly uses a molar flux, typically expressed in the units cm³(STP)/cm² · s. The molar flux, j_i, can be linked to the mass flux, J_i, by the expression

$$j_i = J_i \frac{v_i^G}{m_i} \tag{2.66}$$

where v_i^G is the molar volume of gas i (22.4 l(STP)/mol). Similarly, the mass permeability unit P_i^G, defined in Eq. (2.64), can be linked to the molar gas permeability $\mathcal{P}_i^G$, usually in the units cm³(STP) · cm/cm² · s · cmHg, as

$$\mathcal{P}_i^G = \frac{P_i^G v_i^G}{m_i} \tag{2.67}$$

Equation (2.64) can then be written as

$$j_i = \frac{\mathcal{P}_i^G}{\ell} \left(p_{i_o} - p_{i_\ell} \right) \tag{2.68}$$

The ability of a membrane to separate two gases can then be measured by the membrane selectivity term

$$\alpha_{ij}^G = \frac{\mathcal{P}_i^G}{\mathcal{P}_j^G} = \left(\frac{D_i}{D_j} \right) \left(\frac{K_i^G}{K_j^G} \right) \tag{2.69}$$

The effect of polymer chemistry and morphology on membrane permeabilities and selectivities is covered in the Gas Separation chapter (Chapter 9).

2.2.3.5   Pervaporation

Pervaporation is a process intermediate between gas separation and RO/hyperfiltration. The feed membrane interface is contacted with a feed fluid in the liquid phase at a pressure greater than the saturation vapor pressure; the permeate interface is in contact with a permeate fluid in the gas phase at a pressure below the saturation vapor pressure.

The selective layer of almost all pervaporation membranes is a dense polymer film and the pressure and concentration profiles that form across the membrane in pervaporation are shown in Figure 2.15. The pressure within the membrane is the same as the feed pressure. At the permeate side interface, the pressure drops to a value below the saturation vapor pressure. The pressures on either side of the membrane for each component of the feed mixture can be linked by the expression

$$p_{i_o} \geq p_{i_{sat}} \geq p_{i_\ell} \tag{2.70}$$

As before, the flux through the membrane can be determined by calculating the concentration in the membrane at the two interfaces.

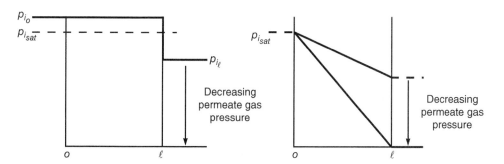

(a) Pressure profile

(b) Concentration profile

Figure 2.15   *Changes in (a) the pressure and (b) the concentration profiles through a pervaporation membrane as the permeate pressure changes, according to the solution-diffusion model. In pervaporation, the feed is a liquid; therefore, the feed pressure* p_{i_o} *exceeds the saturation vapor pressure* $p_{i_{sat}}$*.*

At the liquid solution/membrane feed interface, the chemical potential of the feed liquid is equilibrated with the chemical potential in the membrane at the same pressure. Equation (2.7) then gives

$$\mu_i^o + RT \ln\left(\gamma_{i_o}^L n_{i_o}\right) + v_i\left(p_o - p_{i_{sat}}\right) = \mu_i^o + RT \ln\left(\gamma_{i_{o(m)}} n_{i_{o(m)}}\right) + v_i\left(p_o - p_{i_{sat}}\right) \tag{2.71}$$

which leads to an expression for the concentration at the feed-side interface

$$c_{i_{o(m)}} = \frac{\gamma_{i_o}^L \rho_m}{\gamma_{i_{o(m)}} \rho_o} \cdot c_{i_o} = K_i^L \cdot c_{i_o} \tag{2.72}$$

where K_i^L is the liquid-phase sorption coefficient defined in Eq. (2.25).

At the permeate gas/membrane interface, the pressure drops from p_o in the membrane to p_ℓ in the permeate vapor. The equivalent expression for the chemical potentials in each phase is then

$$\mu_i^o + RT \ln\left(\gamma_{i_\ell}^G n_{i_o}\right) + RT \ln\left(\frac{p_\ell}{p_{i_{sat}}}\right) = \mu_i^o + RT \ln\left(\gamma_{i_{\ell(m)}} n_{i_{\ell(m)}}\right) + v_i\left(p_o - p_{i_{sat}}\right) \tag{2.73}$$

Rearranging Eq. (2.73) gives

$$n_{i_{\ell(m)}} = \frac{\gamma_{i_\ell}^G}{\gamma_{i_{\ell(m)}}} \cdot \frac{p_\ell}{p_{i_{sat}}} \cdot n_{i_\ell} \cdot \exp\left[\frac{-v_i\left(p_o - p_{i_{sat}}\right)}{RT}\right] \tag{2.74}$$

As before, and discussed under Eq. (2.62), the exponential term, the Poynting correction, is close to unity; thus, the concentration at the permeate side interface is

$$n_{i_{(m)}} = \frac{\gamma_{i_\ell}^G}{\gamma_{i_{\ell(m)}}} \cdot n_{i_\ell} \cdot \frac{p_\ell}{p_{i_{sat}}} \tag{2.75}$$

The product $n_{i_\ell} p_\ell$ can be replaced by the partial pressure term p_{i_ℓ}, thus

$$n_{i_{\ell(m)}} = \frac{\gamma_{i_\ell}^G}{\gamma_{i_{\ell(m)}}} \cdot \frac{p_{i_\ell}}{p_{i_{sat}}} \tag{2.76}$$

and substituting concentration for mole fraction from Eq. (2.10),

$$c_{i_{\ell(m)}} = m_i \rho_m \cdot \frac{\gamma_{i_\ell}^G p_{i_\ell}}{\gamma_{i_{\ell(m)}} p_{i_{sat}}} = K_i^G p_{i_\ell} \tag{2.77}$$

where K_i^G is the gas-phase sorption coefficient defined in Eq. (2.59).

The concentration terms in Eqs. (2.72) and (2.77) can be substituted into Eq. (2.13) to obtain an expression for the membrane flux.

$$J_i = \frac{D_i \left(K_i^L c_{i_o} - K_i^G p_{i_\ell} \right)}{\ell} \tag{2.78}$$

Equation (2.78) contains two different sorption coefficients, deriving from Eqs. (2.25) and (2.59). The sorption coefficient in Eq. (2.25) is a liquid-phase coefficient, whereas the sorption coefficient in Eq. (2.59) is a gas-phase coefficient. The interconversion of these two coefficients can be handled by considering a hypothetical vapor in equilibrium with the feed solution. The vapor–liquid equilibrium can then be written from Eqs. (2.7) and (2.8) as

$$\mu_i^o + RT \ln\left(\gamma_i^L n_i^L\right) + v_i\left(p - p_{i_{sat}}\right) = \mu_i^o + RT \ln\left(\gamma_i^G \cdot n_i^G\right) + RT \ln\left(\frac{p}{p_{i_{sat}}}\right) \tag{2.79}$$

Following the same steps as were taken from Eqs. (2.73) through (2.77), Eq. (2.79) becomes Eq. (2.80)

$$n_i^L = \frac{\gamma_i^G p_i}{\gamma_i^L p_{i_{sat}}} \tag{2.80}$$

Converting from mole fraction to concentration using Eq. (2.10) gives

$$c_i^L = m_i \rho \left(\frac{\gamma_i^G p_i}{\gamma_i^L p_{i_{sat}}} \right) \tag{2.81}$$

and so

$$c_i^L = \left(\frac{K_i^G}{K_i^L} \right) p_i \tag{2.82}$$

This expression links the concentration of component i in the liquid phase, c_i^L with p_i, the partial vapor pressure of i in equilibrium with the liquid. Substitution of Eq. (2.82) into Eq. (2.78) yields

$$J_i = \frac{D_i K_i^G \left(p_{i_o} - p_{i_\ell} \right)}{\ell} \tag{2.83}$$

where p_{i_o} and p_{i_ℓ} are the partial vapor pressures of component i on either side of the membrane. Equation (2.83) can also be written as

$$J_i = \left(\frac{P_i^G}{\ell}\right)\left(p_{i_o} - p_{i_\ell}\right) \tag{2.84}$$

where P_i^G is the gas permeability coefficient. Equation (2.84) explicitly expresses the driving force in pervaporation as the vapor pressure difference across the membrane, a form of the pervaporation transport equation derived first by Kataoka et al. [25] and later, independently by Wijmans and Baker [26]. This is the commonly used way to characterize the performance of pervaporation membranes. As in the derivation of Eq. (2.64) for gas separation, only partial pressure terms appear in Eq. (2.84). In other words, the flux is essentially independent of the hydrostatic pressure of the feed liquid.

In the derivations given above, Eq. (2.82) links the concentration of a vapor in the liquid phase $\left(c_i^L\right)$ with the equilibrium partial pressure of the vapor. This relationship is more familiarly known as Henry's law, written as

$$H_i \cdot c_i^L = p_i \tag{2.85}$$

where H_i is the Henry's law coefficient.

From Eqs. (2.81) and (2.85), it follows that H_i can be written as

$$H_i = \frac{K_i^L}{K_i^G} = \frac{\gamma_i^L p_{i_{sat}}}{m_i \rho \gamma_i^G} \tag{2.86}$$

These expressions can be used to rewrite Eq. (2.78) as

$$J_i = \frac{P_i^G}{\ell}\left(c_{i_o} H_i - p_{i_\ell}\right) \tag{2.87}$$

where P_i^G is the gas permeability coefficient, or

$$J_i = \frac{P_i^L}{\ell}\left(c_{i_o} - p_{i_\ell}/H_i\right) \tag{2.88}$$

where P_i^L is the liquid (hyperfiltration) permeability coefficient.

At low permeate side vapor pressure, the two alternative ways of describing the pervaporation flux reduce to

$$J_i = \frac{P_i^G}{\ell}p_{i_o} = \frac{P_i^L}{\ell}c_{i_o} \tag{2.89}$$

This equation shows that in pervaporation, the dependence of vapor pressure (driving force) on temperature, explicit in Eq. (2.84), is hidden in the term P_i^L of Eq. (2.88). P_i^L increases exponentially with temperature as the vapor pressure term imbedded in P_i^L increases. In contrast, the gas phase permeability coefficient P_i^G is only a modest function of temperature.

The benefit of using the gas permeability constant P_i^G and Eq. (2.84) to describe pervaporation has been amply demonstrated experimentally. For example, Figure 2.16 shows data for the pervaporation water flux through a silicone rubber membrane as a function of permeate pressure. As the permeate pressure (p_{i_ℓ}) increases, the water flux falls in accordance with Eq. (2.84), reaching zero flux when the permeate pressure is equal to the vapor pressure (p_{i_o}) of the feed liquid at the temperature of the experiment. The straight lines in Figure 2.16 indicate that the permeability coefficient of water in silicone rubber is constant. This can be expected in systems in which the membrane material is a rubbery polymer and the permeant swells the polymer only moderately.

Thompson and coworkers [27] have studied the effects of feed and permeate pressure on pervaporation flux in some detail. Some illustrative results are shown in Figure 2.17. As Figure 2.17a shows, the dependence of flux on permeate pressure is in accordance with

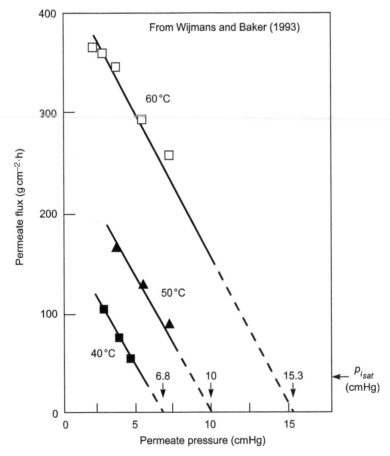

Figure 2.16 *The effect of permeate pressure on the water flux through a silicone rubber membrane during pervaporation. The arrows on the horizontal axis represent the saturation vapor pressure of the feed solution at the experiment temperatures. Source: Reproduced with permission from [11]. Copyright (1995) Elsevier.*

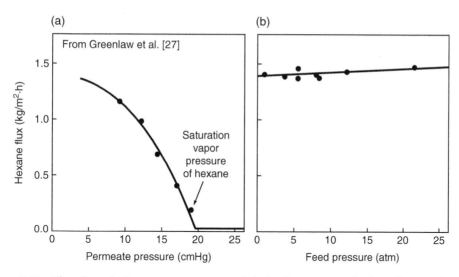

Figure 2.17   *The effect of (a) permeate pressure and (b) feed pressure on the flux of hexane through a rubbery pervaporation membrane. The flux is extremely sensitive to permeate pressure, but essentially independent of feed pressure up to 20 bar. Eq. (2.84) explains this behavior. Source: Reproduced with permission from [27]. Copyright (1977) Elsevier.*

Eq. (2.84). At very low permeate pressures, p_{i_ℓ} approaches zero and the membrane flux has its maximum value $J_{i_{max}}$, equal to $P_i^G p_{i_o}/\ell$. As the permeate pressure increases, the flux decreases, reaching a zero flux when the permeate pressure equals the saturation vapor pressure of the feed.

From Eq. (2.84), the slope of the curve at any point in Figure 2.17a is a measure of the membrane permeability at that point. At permeate pressures close to the saturation vapor pressure, the membrane is highly swollen throughout, and the steep slope of the curve indicates that the average permeability across the thickness of the membrane is correspondingly high. When the permeate pressure is lower, and thus farther away from the saturated vapor pressure, the membrane becomes less swollen toward the permeate side. As a result, the permeability of the membrane material decreases noticeably from the feed side to the permeate side, and the decline in average permeability is indicated by the flattening of the curve. This behavior is typical of membranes that are swollen significantly by the permeant. The pressure on the feed side, on the other hand, has almost no influence on flux, because the feed in pervaporation is already in the liquid phase, and the vapor pressure of a liquid increases very little with increased hydrostatic pressure. If the permeate pressure is fixed at a low value, the hydrostatic pressure of the feed liquid can be increased to as much as 20 bar without any significant change in the flux, as shown in Figure 2.17b. Equation (2.84) shows that the feed vapor pressure, not hydrostatic pressure, is the true measure of the driving force for transport through the membrane. Thus, the properties of pervaporation membranes illustrated in Figures 2.16 and 2.17 are easily rationalized by the solution-diffusion model. They are much more difficult to explain by a pore-flow mechanism, although this has been tried.

Many papers on pervaporation report membrane separation performance as raw data; that is, as fluxes J_i and J_j and separation factor β_{ij} defined as

$$\beta_{ij} = \frac{c_{i_\ell}/c_{j_\ell}}{c_{i_o}/c_{j_o}} \tag{2.90}$$

These values are a function of the intrinsic properties of the membrane and the operating conditions of the experiments (feed concentration, permeate pressure, feed temperatures); change the operating conditions and all the numbers change. A better way of measuring separation performance is to report data in a normalized form; that is, as permeabilities (P_i^G), permeances (P_i^G/ℓ), and selectivities (α_{ij}^G) [28]. The connection between these parameters and flux and separation factors is described in Chapter 10.

2.2.4   A Unified View

In the preceding sections, the solution-diffusion model was used to describe gas separation, pervaporation, and reverse osmosis. The equations describing the flux through the membranes contain the coefficients D_i, K_i, and P_i irrespective of the specific process. This happens because the driving force for permeation is the same in each case – a concentration gradient within the membrane. The fluids on either side of the membrane can change from a gas to a pressurized liquid, but the only effect of these changes on permeation within the membrane is to alter the concentration gradient driving force.

The pressure and concentration profiles within a solution-diffusion membrane for gas separation, pervaporation, and reverse osmosis/hyperfiltration are compared in Figure 2.18. Considering gas separation first, the feed and permeate sides of the membrane are both below the saturation vapor pressure and we can write

$$p_{i_{sat}} > p_{i_o} > p_{i_\ell} \qquad \text{(Gas separation)}$$

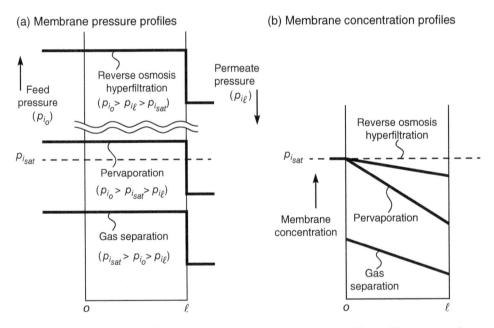

Figure 2.18   *(a) Pressure profiles in gas separation, pervaporation and hyperfiltration membranes, relative to the saturation vapor pressure ($p_{i_{sat}}$). (b) Concentration profile created within the same membranes created by the pressure profiles shown in (a).*

The concentration in the membrane at the interfaces is proportional to the adjacent gas phase pressure as described in Eqs. (2.60) and (2.61).

If the feed gas pressure is raised until it exceeds the saturation vapor pressure, then the membrane enters the pervaporation region, liquid forms on the feed side of the membrane and

$$p_{i_o} > p_{i_{sat}} > p_{i_\ell} \qquad \text{(Pervaporation)}$$

At this point, the concentrations in the membrane at the interfaces are described by Eqs. (2.72) and (2.77).

If the permeate side pressure is then increased, so that the saturation vapor pressure is exceeded on both sides of the membrane, liquid forms at both membrane interfaces. The membrane then enters the reverse osmosis region and

$$p_{i_o} > p_{i_\ell} > p_{i_{sat}} \qquad \text{(Reverse osmosis/hyperfiltration)}$$

At this point, the concentrations in the membrane at the interfaces are described by Eqs. (2.32) and (2.36).

It follows from the above that permeabilities measured for the different operating modes can be linked. For simplicity, consider permeation of a single component i. For gas permeation, the mass flux can be written as Eq. (2.64) as

$$J_i = \frac{P_i^G \left(p_{i_o} - p_{i_\ell} \right)}{\ell} \tag{2.91}$$

The identical expression for pervaporation was also derived earlier (Eq. (2.84)) as

$$J_i = \left(\frac{P_i^G}{\ell} \right) \left(p_{i_o} - p_{i_\ell} \right) \tag{2.92}$$

and the equivalent expression for reverse osmosis/hyperfiltration was derived (Eq. (2.38)) as

$$J_i = \frac{P_i^L}{\ell} \left\{ c_{i_o} - c_{i_\ell} \exp \left[\frac{-v_i(p_o - p_\ell)}{RT} \right] \right\} \tag{2.93}$$

For a one component feed, c_{i_o} is equal to c_{i_ℓ} and so Eq. (2.93) becomes

$$J_i = \frac{P_i^L}{\ell} \cdot c_{i_o} \left\{ 1 - \exp \left[\frac{-v_i(p_o - p_\ell)}{RT} \right] \right\} \tag{2.94}$$

The interconversion of gas permeability P_i^G and liquid permeability is given in Eq. (2.89), so Eq. (2.94) can be written as

$$J_i = \frac{P_i^G}{\ell} \cdot p_{i_{sat}} \left\{ 1 - \exp \left[\frac{-v_i(p_o - p_\ell)}{RT} \right] \right\} \tag{2.95}$$

Equations (2.91), (2.92), and (2.95) allow the membrane permeability P_i^G to be calculated from measurements in each of the different operating modes. This should allow the permeability obtained in one operating mode to be used to predict the expected results for another mode. This principle holds true, however, only if sorption of the permeant into the membrane does not affect the nature of the membrane itself. Sorption of high levels of the permeating component can lead to swelling and plasticization of the polymer of which the membrane is made, thereby affecting the permeation properties. In hyperfiltration and pervaporation measurements, the membrane is exposed to liquid permeant (unit activity) and swells to the maximum extent, and permeabilities are very high. In gas permeation, the degree of swelling is a function of component vapor pressure, and the permeability of most components increases sharply with their vapor pressure. In practice, the highest permeability measured in the gas permeation mode is usually less than the permeability for the same component measured in hyperfiltration or pervaporation experiments, because the membrane is usually exposed to vapor pressures lower than the saturation vapor pressure. One of the few exceptions is found with perfluoro polymer membranes, which generally have low levels of sorption even when exposed to liquid solvent. Gas permeation, pervaporation, and hyperfiltration permeability data are then similar.

An idealized representation of the connection between the three permeation processes, assuming that the permeability for all three modes is the same, is shown in Figure 2.19 [12]. This figure shows the transitions between the different operating regions as the feed and permeate pressures change. Three regions are present: the gas separation region, where the feed and permeate are both gases; the pervaporation region, where the feed is a liquid and the permeate a gas; and the RO/hyperfiltration region, where the feed and permeate are both liquids. The permeate flux through the membrane is plotted as a function of the normalized driving force,

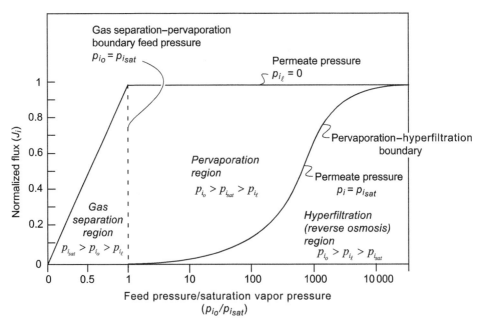

Figure 2.19 *Permeation through a membrane, expressed as normalized flux, as a function of normalized feed pressure* ($p_{i_o}/p_{i_{sat}}$). *The figure shows the smooth transition from gas separation, to pervaporation, to hyperfiltration. The curves shown are calculated using Eqs. (2.91), (2.92), and (2.95). Source: [12]/ John Wiley & Sons.*

measured by the ratio of the feed pressure to saturation vapor pressure ($p_{i_o}/p_{i_{sat}}$). A linear scale is used for feed pressure below the saturation vapor pressure, corresponding to the linear dependence of gas flux on feed pressure shown in Eq. (2.91). Above the saturation vapor pressure an exponential scale is used, because the hyperfiltration equation (Eq. (2.95)) shows the membrane flux to be an exponential function of feed pressure. The flux axis of Figure 2.19 has also been normalized by setting $J_{i_{max}}$ to 1 according to the equation:

$$J_{i_{max}} = \frac{P_i^G p_{i_{sat}}}{\ell} = \frac{P_i^L c_{i_o}}{\ell} = 1 \tag{2.96}$$

Two lines separate the regions of the figure. A vertical line identifies the gas separation/ pervaporation boundary, reached when the feed pressure (p_{i_o}) equals the saturation vapor pressure ($p_{i_{sat}}$) and the feed changes from a gas to a liquid. A second line identifies the pervaporation/hyperfiltration boundary, the point at which the permeate pressure $p_{i_\ell} = p_{i_{sat}}$, and the permeate changes from a gas to a liquid.

A third line shows the maximum achievable flux through the membrane. This corresponds to the point at which the permeate pressure p_{i_ℓ} is set to zero, that is, a hard vacuum on the permeate side. At low feed pressures, the membrane is operating in the gas separation region and as the normalized feed pressure ($p_{i_o}/p_{i_{sat}}$) increases, the gas flux also increases. The gas flux reaches its maximum value (arbitrarily set to 1) when the feed pressure reaches the saturation vapor pressure. The gas flux at this point is $P_i^G p_{i_{sat}}/\ell$. When the feed pressure is increased further, the feed pressure exceeds the saturation vapor pressure and the feed gas liquefies. The process then enters the pervaporation region. At this point, further increases in feed pressure do not increase the flux. This is consistent with the pervaporation flux Eq. (2.84), in which the feed pressure is set at the saturation vapor pressure, and the permeate pressure is set to zero. That is

$$J_i = \frac{P_i^G}{\ell}\left(p_{i_o} - p_{i_\ell}\right) = \frac{P_i^G}{\ell}\left(p_{i_{sat}} - 0\right) = \frac{P_i^G \cdot p_{i_{sat}}}{\ell} \tag{2.97}$$

The pervaporation/hyperfiltration boundary line in Figure 2.19 represents the membrane flux at a permeate pressure just above the saturation vapor pressure. Under these conditions, liquid forms on both sides of the membrane and Eq. (2.38) for reverse osmosis, repeated below as Eq. (2.98), can be used to calculate the membrane flux.

$$J_i = \frac{P_i^L}{\ell}\left\{ c_{i_o} - c_{i_\ell} \exp\left[\frac{-v_i(p_o - p_\ell)}{RT} \right] \right\} \tag{2.98}$$

Initially, the flux increases linearly with increasing feed pressure, but then asymptotically approaches a maximum value $J_{i_{max}}$ of $P_i^L c_{i_o}/\ell$ at very high feed pressures. This is consistent with Eq. (2.98), because as $(p_o - p_\ell) \to \infty$, then

$$J_i \to \frac{P_i^L}{\ell} \cdot c_{i_o} \tag{2.99}$$

and following Eq. (2.96), the maximum value of J_i is arbitrarily normalized to 1.

The ability of the solution-diffusion model to demonstrate the connection between the processes of gas separation, pervaporation, and RO is one of its great strengths. The performance of a membrane at any feed and permeate pressure can be represented as a point within this figure.

2.3   Structure–Permeability Relationships in Solution-Diffusion Membranes

In the preceding section, the effects of concentration and pressure gradient driving forces on permeation through membranes were described in terms of the solution-diffusion model and Fick's law. The resulting equations all contain a permeability term, P, which must be determined experimentally. This section describes how the nature of the membrane material affects permeant diffusion and sorption coefficients, which in turn determine membrane permeability. By analyzing the factors that determine membrane permeability, useful correlations and rules of thumb can be derived to guide the selection of membrane materials with the optimum flux and selectivity properties for a given separation. Most of the experimental data in this area have been obtained from membranes designed for gas separation applications. However, the same general principles apply to all polymeric solution-diffusion membranes.

The problem of predicting membrane permeability can be divided into two parts, because permeability is the product of the diffusion coefficient and the sorption coefficient:

$$P = D \cdot K \tag{2.100}$$

The sorption coefficient (K) in Eq. (2.100) is the term linking the concentration of a component in the fluid phase with its concentration in the membrane polymer phase. Because sorption is an equilibrium term, conventional thermodynamics can be used to calculate the sorption coefficients of many components in polymers to within a factor of 2 or 3 of the experimental value. However, diffusion coefficients (D) are kinetic terms that reflect the effect of the surrounding environment on the molecular motion of permeating components. Calculation of diffusion coefficients in liquids and gases is possible, but calculation of diffusion coefficients in polymers is much more difficult. In the long term, the best hope for accurate predictions of diffusion in polymers are the molecular dynamics calculations described in an earlier section. However, this technique is still under development and is currently limited to calculations of the diffusion of small gas molecules in amorphous polymers; the agreement between theory and experiment is modest. In the meantime, simple correlations based on polymer free volume must be used.

As a general rule, membrane material changes affect the diffusion coefficient of a permeant much more than the sorption coefficient. For example, Figure 2.20 shows some typical gas permeation data taken from a paper of Tanaka et al. [29]. The diffusion and sorption coefficients of four gases in a family of 18 related polyimides are plotted against each other. Both sorption and diffusion coefficients are fairly well grouped for each gas. However, for any one gas, the difference in diffusion coefficient from the highest to lowest value is approximately 100-fold, whereas the spread in sorption coefficients is only 2- to 4-fold. Changes in polymer chemistry affect both the sorption and diffusion coefficients, but the effect on the diffusion coefficient is much more profound. This same effect was apparent in the discussion of the RO data shown in Table 2.1. In that set of data, changing the membrane material chemistry changed the membrane sorption coefficients 10-fold, but the diffusion coefficients changed almost 1000-fold.

More detailed examination of the data shown in Figure 2.20 shows that the relative position of each polymer within the group of 18 is approximately the same for all gases. That is, the polymer

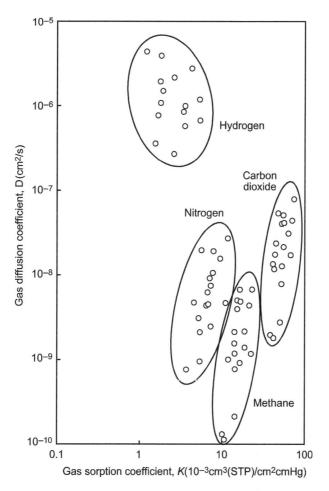

Figure 2.20 *Gas diffusion and sorption coefficients plotted for a family of 18 related polyimides. Source: Data of Tanaka et al. [29].*

with the highest diffusion coefficient for methane also has the highest diffusion coefficients for nitrogen, carbon dioxide, and hydrogen. The trend for the sorption or solubility coefficients is similar. As a general rule, changes in polymer chemistry and structure that change the diffusion coefficient or sorption coefficient for one gas change the properties for other gases in the same way. This is why permeability can be varied by several orders of magnitude by changing the membrane material, whereas changing selectivity by more than a factor of 10 is difficult.

As we saw in Eq. (2.51), selectivity is the product of mobility selectivity and solubility selectivity, that is

$$\alpha_{ij} = \left(\frac{D_i}{D_j}\right)\left(\frac{K_i}{K_j}\right)$$

and in general, most of the change in the value of membrane selectivity is due to the mobility selectivity term rather than the solubility term. Data for permeation of carbon dioxide and methane in very different membrane materials, shown in Table 2.2, illustrate this effect. For

Table 2.2   Diffusion and sorption selectivities for carbon dioxide/methane for a variety of polymers [29–32].

Polymer	P_{CO_2} (Barrer)	D_{CO_2} (10^{-8} cm^2/s)	K_{CO_2} (cm^3(STP)/cm$^3\cdot$atm)	K_{CH_4} (cm^3(STP)/cm$^3\cdot$atm)	Total selectivity α_{CO_2/CH_4}	Mobility D_{CO_2}/D_{CH_4}	Sorption K_{CO_2}/K_{CH_4}
Rubbery polymers							
Silicone rubber	3800	2200	1.29	0.42	3.2	1.1	3.1
Polyisoprene	150	1350	3.83	0.26	5.0	1.5	3.4
Glassy polymers							
Polyethylene terephthalate	17.2	4.46	2.9	0.83	27.3	7.8	3.5
Polystyrene	12.4	8.50	1.1	0.38	15.8	5.5	2.9
Polycarbonate	6.8	3.20	1.6	0.40	19	4.7	4.0
Polysulfone	5.6	2.00	2.1	0.59	22	5.9	3.7
PMDA-ODA polyimide	2.7	0.56	3.6	0.95	46	11.9	3.8

example, the permeability to carbon dioxide changes by more than 1000-fold between polyimide and silicone rubber. For these two materials, the diffusion coefficient varies by a factor of 4000, but the sorption coefficient only varies by a factor of 3. The sorption selectivity term is even more constant, varying only between 2.9 and 4.0 over the whole range of materials. Almost all of the 10-fold difference in CO_2/CH_4 selectivity is due to the mobility selectivity term, which varies from 1.1 for a soft, rubbery material (silicone rubber) to 11.9, for the most selective and lowest permeability polymer (PMDA-ODA polyimide). As with most membrane separation applications, there is a trade-off between high permeability, low selectivity materials and low permeability, high selectivity materials.

In the following sections, factors that determine the magnitude of diffusion and solubility coefficients in polymers are discussed.

2.3.1 Diffusion Coefficients

The Fick's law diffusion coefficient of a molecule permeating a medium is a measure of the frequency with which the molecule moves and the size of each movement. Therefore, the magnitude of the diffusion coefficient is governed by the restraining forces of the medium on the diffusing species. Isotopically labeled carbon in a diamond lattice has a very small diffusion coefficient. The carbon atoms of diamond move infrequently, and each movement is very small – only 1–2 Å. On the other hand, isotopically labeled carbon dioxide in a gas has an extremely large diffusion coefficient. The gas molecules are in constant motion and each jump is on the order of 1000 Å or more. Table 2.3 lists some representative values of diffusion coefficients in different media.

The main observation from Table 2.3 is the enormous range of values of diffusion coefficients – from 10^{-1} to 10^{-30} cm²/s. Diffusion in gases is well understood and is treated in standard textbooks dealing with the kinetic theory of gases [33, 34]. Diffusion in metals and crystals is a topic of considerable interest to the semiconductor industry, but not in membrane permeation. This book focuses principally on diffusion in liquids and polymers, in which the diffusion coefficients vary from about 10^{-5} to about 10^{-10} cm²/s.

2.3.1.1 Diffusion in Liquids

Liquids are simple, well-defined systems and provide the starting point for modern theories of diffusion. An early and still fundamentally sound equation was derived by Einstein who applied simple macroscopic hydrodynamics to diffusion at the molecular level. He assumed the diffusing solute to be a sphere moving in a continuous fluid of solvent, in which case it can be shown that

Table 2.3 *Typical diffusion coefficients in various media (25 °C).*

Permeant/material	Diffusion coefficient, D (cm²/s)
Oxygen in air (atmospheric pressure)	1×10^{-1}
Oxygen in water	3×10^{-5}
Oxygen in silicone rubber	3×10^{-5}
Oxygen in polysulfone	4×10^{-8}
Oxygen in polyester	5×10^{-9}
Sodium atoms in sodium chloride crystals	1×10^{-20}
Aluminum atoms in metallic copper	1×10^{-30}

$$D = \frac{kT}{6\pi a\eta} \tag{2.101}$$

where k is Boltzmann's constant, a is the radius of the solute, and η is the solution viscosity. This is known as the Stokes–Einstein equation. The equation is a good approximation for large solutes with radii greater than 5–10 Å, but, as the solute becomes smaller, the approximation of the solvent as a continuous fluid becomes less valid. In this case, there may be a slip of solvent at the solute molecule surface. A second limiting case assumes complete slip at the surface of the solute sphere; in this case

$$D = \frac{kT}{4\pi a\eta} \tag{2.102}$$

Thus, the Stokes–Einstein equation is perhaps best expressed as

$$D = \frac{kT}{n\pi a\eta} \qquad 4 \leq n \leq 6 \tag{2.103}$$

An important conclusion to be drawn from the Stokes–Einstein equation is that the diffusion coefficient of solutes in a liquid only changes slowly with molecular weight, because the diffusion coefficient is proportional to the reciprocal of the radius, which in turn is approximately proportional to the cube root of the molecular weight.

Application of the Stokes–Einstein equation requires a value for the solute radius. A simple approach is to assume the molecule is spherical, and calculate the solute radius from the molar volume of the chemical groups making up the molecule. Using values for the solute radius calculated this way, together with measured and known diffusion coefficients of solutes in water, Edward [35] constructed a graph of the coefficient n in Eq. (2.103) as a function of solute radius, as shown in Figure 2.21. With large solutes, n approaches 6; that is, Einstein's application of normal macroscopic fluid dynamics at the molecular level is a good approximation. However, when the solute radius falls below about 4 Å, water can no longer be regarded as a continuous fluid, and n falls below 6. Nonetheless, that an equation based on macroscopic hydrodynamic theory applies to molecules to the 4 Å level is an interesting result.

The Stokes–Einstein equation works well for diffusion of solutes in simple liquids, but fails in more complex fluids, such as a solution containing a high-molecular-weight polymer. Dissolving a polymer in a liquid increases the solvent viscosity, but the solute diffusion coefficient is not significantly affected. For example, as the concentration of poly(vinyl pyrrolidone) dissolved in water changes from 0 to 20 wt%, the viscosity of the solution increases by several orders of magnitude. However, the diffusion coefficient of sucrose only changes by a factor of 4 [36]. The long polymer chains of dissolved poly(vinyl pyrrolidone) molecules link distant parts of the aqueous solution and change the macroscopic viscosity of the fluid substantially, but, in the fluid immediately surrounding the diffusing sucrose molecule, the effect of polymer chain length is much less noticeable. This result illustrates the difference between the microscopic environment of the diffusing solute and the macroscopic environment measured by conventional viscometers. In simple liquids, the macroscopic and microscopic environments are the same, but in liquids containing dissolved macromolecules, or in gels and polymer films, the microscopic environment and the macroscopic environment can be very different.

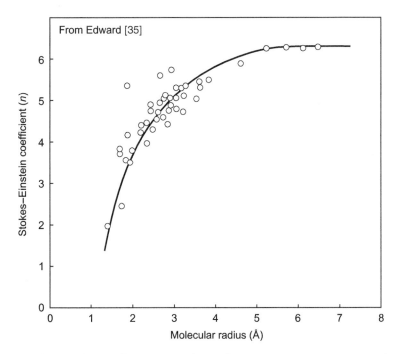

Figure 2.21 *Value of the coefficient* n *in the Stokes–Einstein equation required to achieve agreement between calculation and experimental solute diffusion coefficients in water. Source: Adapted from [35].*

2.3.1.2 *Diffusion in Polymers*

The concept that the local environment around the permeating molecule determines the diffusion coefficient of the permeant is key to understanding diffusion in polymer membranes. Polymers can be divided into two broad categories – rubbers and glasses. In a rubbery polymer, portions of the polymer chains are free to move because of thermal motion, and segments of the polymer backbone can also rotate around their axis; this makes the polymer soft and elastic. Thermal motion of these segments also leads to high permeant diffusion coefficients. In a glassy polymer, steric hindrance along the polymer backbone inhibits rotation and free motion of segments of the polymer. The result is a rigid, tough polymer. Thermal motion in this type of material is limited, so permeant diffusion coefficients are low. If the temperature of a glassy polymer is raised, the increase in thermal energy at some point becomes sufficient to overcome the steric hindrance restricting motion of the polymer backbone segments. At this temperature, called the *glass transition temperature* (T_g), the polymer changes from a glass to a rubber.

Figure 2.22 shows a plot of diffusion coefficient as a function of permeant molecular weight for permeants diffusing through a liquid (water), two soft rubbery polymers (natural rubber and silicone rubber), and a hard, stiff glassy polymer (polystyrene) [37]. For very small molecules, such as helium and hydrogen, the diffusion coefficients in all of the media are comparable, differing by no more than a factor of 2 or 3. These very small molecules only interact with one or two atoms in their immediate proximity. The local environment for these small solutes in the three polymers is not radically different to that in a liquid such as water. On the other hand, larger diffusing solutes with molecular weights of 200–300 and above have molecular diameters of 6–10 Å. Such solutes are in quite different local environments in the different media. In water,

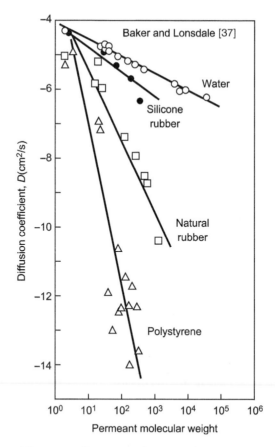

Figure 2.22 *Permeant diffusion coefficient as a function of permeant molecular weight in water, natural rubber, silicone rubber, and polystyrene. Diffusion coefficients of solutes in polymers usually lie between the value in silicone rubber, a very permeable polymer, and the value in polystyrene, a very impermeable material. Source: Adapted from [37].*

the Stokes–Einstein equation applies, and the resistance to movement of the solute is not much larger than that of a very small solute. In polymer membranes, however, several segments of the polymer chain are involved in each movement of the diffusing species. This type of cooperative movement is statistically unlikely; consequently, diffusion coefficients are much smaller than in liquid water. Moreover, the differences between the motion of polymer segments in the flexible rubbery membranes and in the stiff polystyrene membrane are large. The polymer chains in rubbers are considerably more flexible and rotate more easily than those in polystyrene. One manifestation of this difference in chain flexibility is the difference in elastic properties; another is the difference in diffusion coefficient.

An example of the change in diffusion coefficient as the matrix material changes is illustrated by Figure 2.23. In this example, the polymer matrix material is changed by plasticization of the polymer, ethyl cellulose, by the permeant, dichloroethane [38]. The resulting change in the diffusion coefficient is shown in the figure. The concentration of dichloroethane in the polymer matrix increases from very low levels (<1% dichloroethane) to very high levels (>90% dichloroethane). As the concentration of dichloroethane increases, the polymer changes from a glassy

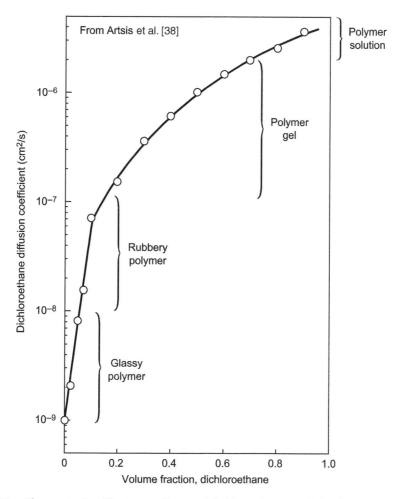

Figure 2.23 *Changes in the diffusion coefficient of dichloroethane in ethyl cellulose as a function of the volume fraction of dichloroethane dissolved in the polymer matrix. Source: Data of Artsis et al. [38].*

polymer to a rubbery polymer, to a solvent-swollen gel, and finally to a dilute polymer solution. Ethyl cellulose is a glassy polymer with a glass transition temperature of about 45–50 °C. At room temperature and low concentrations of dichloroethane (below about 5 vol%), the ethyl cellulose matrix is glassy, and the dichloroethane diffusion coefficient is in the range $1–5 \times 10^{-9}$ cm^2/s. As the dichloroethane concentration increases to above 5 vol%, enough solvent has dissolved in the polymer to reduce the glass transition temperature to below the temperature of the experiment. The polymer chains then have sufficient freedom to rotate, and the polymer becomes rubbery. As the dichloroethane concentration increases further, the polymer chain mobility also increases, as does the diffusion coefficient of dichloroethane. At 20 vol% dichloroethane, the diffusion coefficient is 1×10^{-7} cm^2/s, 100 times greater than the diffusion coefficient in the glassy polymer. Above 20 vol% dichloroethane, sufficient solvent is present to allow relatively large segments of the polymer chain to move. In the range 20–70 vol% dichloroethane, the matrix is best characterized as a solvent-swollen gel, and the diffusion coefficient of dichloroethane increases from 1×10^{-7} to 2×10^{-6} cm^2/s. Finally, at dichloroethane concentrations above 70 vol%, sufficient solvent is present for the matrix to be characterized as a polymer solution. In this final solvent concentration range, the increase in diffusion coefficient with further increases in dichloroethane concentration is relatively small.

Figures 2.22 and 2.23 show the significant difference between diffusion in liquids and in rubbery and glassy polymers. A great deal of work has been performed over the last two decades to achieve a quantitative link between the structure of polymers and their permeation properties. No such quantitative structure–property relationship is at hand or even in sight. What has been achieved is a set of semi-empirical rules that allow the permeation properties of related families of polymers to be correlated based on small changes in their chemical structures. The correlating tool most generally used is *fractional free volume* v_f (cm^3/cm^3). This free volume is the fraction of the space occupied by the polymer that is not occupied by the atoms that make up the polymer chains. The fractional free volume is usually defined as

$$v_f = \frac{v - v_o}{v} \tag{2.104}$$

where v is the specific volume of the polymer (cm^3/g), that is, the reciprocal of the polymer density, and v_o is the volume occupied by the molecules themselves (cm^3/g). The free volume of a polymer is the sum of the many small spaces between the polymer chains in these amorphous, non-crystalline materials.

The free volume of a polymer can be determined by measuring the specific volume, then calculating the occupied volume (v_o) of the groups that form the polymer. Tables of the molar volume of different chemical groups have been prepared by Bondi [39] and van Krevelen [40]. By summing the molar volume of all the groups in the polymer repeat unit, the occupied molar volume of the polymer can be calculated. The occupied volume obtained in this way is about 1.3 times larger than the Van der Waals volume of the groups. The factor of 1.3 occurs because some unoccupied space is inevitably present even in crystals at 0 K. The fractional free volumes of a number of important membrane materials are given in Table 2.4.

The concept of polymer free volume is illustrated in Figure 2.24, which shows polymer specific volume (cm^3/g) as a function of temperature. At high temperatures, the polymer is in the rubbery state. Even in the rubbery state, the polymer chains cannot pack perfectly, and some

Table 2.4 *Calculated fractional free volume for representative membrane materials at ambient temperatures (Bondi method).*

Polymer	Polymer type	Glass transition temperature, T_g (°C)	Fractional free volume (cm^3/cm^3)
Silicone rubber	Rubber	−129	0.16
Natural rubber	Rubber	−73	0.16
Polycarbonate	Glass	150	0.16
Poly(phenylene oxide)	Glass	167	0.20
Polysulfone	Glass	186	0.16
6FDA-ODA polyimide	Glass	300	0.16
Poly(4-methyl-2-pentyne) [PMP]	Glass	>250	0.28
Poly(1-trimethylsilyl-1-propyne) [PTMSP]	Glass	>250	0.34

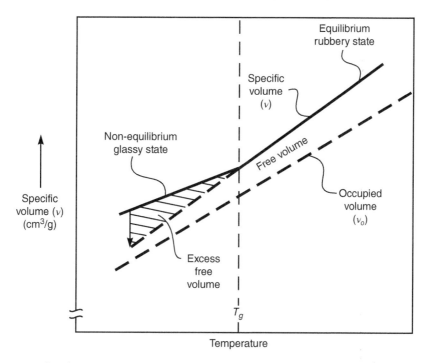

Figure 2.24   *The change in specific volume as a function of temperature for a typical polymer.*

unoccupied space – free volume – exists between the polymer chains. This free volume is over and above the space normally present between molecules in a crystal lattice. Although this free volume is only a few percent of the total volume, it is sufficient to allow motion of segments of the polymer backbone. In this sense a rubbery polymer, although solid at the macroscopic level, has some of the characteristics of a liquid. As the temperature of the polymer decreases, the free volume also decreases. At the glass transition temperature (T_g), the free volume is reduced to a point at which the polymer chains can no longer move freely. This transition is quite sharp, usually over a temperature range of 3–5 °C. Segmental motion is then reduced by several orders of magnitude. When the glass transition temperature is traversed, extra free volume elements between the polymer chains are frozen into the polymer matrix. As the polymer temperature is reduced further, its occupied volume will continue to decrease as the vibration energy of the groups forming the polymer decreases, but the free volume that is the difference between the occupied volume and the actual volume remains essentially constant. Therefore, a glassy polymer contains both the normal free volume elements caused by the incomplete packing of the groups making up the polymer chains and excess free volume elements frozen into the polymer matrix because motion of the polymer chains is very restricted. It is the existence of these tiny, frozen-in-place, excess free volume elements that contributes to the relatively high permeability of glassy polymers, and their ability to selectively permeate different-sized permeants at different rates.

The fractional free volume of most materials is small. For rubbers, the fractional free volume calculated by the Bondi method is generally about 10–15%. For glassy polymers, the fractional free volume is higher, generally in the range of 15–20% because of the excess free volume contribution. A number of polymers with extraordinarily rigid polymer backbones have been prepared, and their free volumes are correspondingly unusually high – as much as 25–35%

of the polymer volume is unoccupied space [41–43]. Permeation through these polymers is described in the next chapter.

Diffusivity, and hence permeability, of a polymer can be linked to the fractional free volume by the empirical equation

$$D = A \cdot \exp\left(\frac{B}{v_f}\right) \tag{2.105}$$

where D is the diffusivity and A and B are adjustable parameters. When applied within a single class of materials, the correlation between the free volume and gas diffusivity or permeability suggested by this equation is often good; an example is shown in Figure 2.25 [44]. When the correlation is broadened out to include diverse types of polymer, there is much more scatter. The relationship between the free volume and the sorption and diffusion coefficients of gases in polymers, particularly glassy polymers, has been an area of a great deal of experimental and theoretical work and is the subject of a number of reviews, but a predictive model has yet to emerge [45, 46]. A detailed discussion of permeation of gases in glassy polymers is given in a recent review by Merrick et al. [47].

A factor that complicates understanding the effect of free volume on permeation is that the excess free volume is not permanent. Even in a glassy polymer, some motion of the polymer chains occurs. This allows the polymer to slowly change to a lower excess free volume, higher density state. The loss of free volume is usually very slow in thick polymer films. This is because, to eliminate its space, a free volume element must move to the surface of the polymer, like a bubble leaving a liquid. In a thick film, the volume element must move a considerable distance to escape and so densification of the polymer is very slow. However, in membranes only a

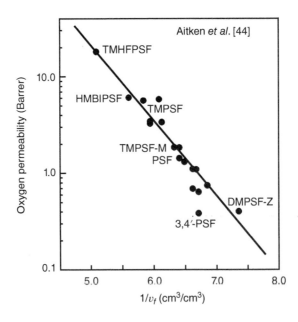

Figure 2.25 Correlation of the oxygen permeability coefficient for a family of related polysulfones with inverse fractional free volume (calculated using the Bondi method). Source: Reproduced with permission from [44]. Copyright (1992) American Chemical Society.

micron or so thick, the distance the free volume element must diffuse to escape is much less. In these thin membrane films, polymers can lose a large fraction of their excess free volume over a period of a few months to a few years. This loss of excess free volume causes the membrane permeability to decrease over time.

Huang and Paul were the first to measure the effects of loss of excess free volume on gas permeation [48, 49]. Some of their data for polyphenylene oxide (PPO) are shown in Figure 2.26. The oxygen permeability of thick PPO films is 20 Barrer and the oxygen/nitrogen selectivity is 4.4. The permeability of PPO films stored for long periods of time at 30 °C steadily decreases; the rate of loss of permeability is inversely proportional to the square of the film thickness. The permeability of a 25 μm thick film is reduced by only about 20% after 1 year of storage, but a 0.4 μm thick film loses almost two-thirds of its permeability in the same time. The oxygen/nitrogen selectivity increases from 4.4 to about 5.1 as the permeability falls. Refractive index measurements show that these changes in permeation are accompanied by an increase in polymer density.

Paul and coworkers and others performed similar experiments with several other polymers [50]. Some representative data measured with thin membranes are shown in Figure 2.27. Polysulfone (PSF), Matrimid® (a proprietary polyimide), and polyphenylene oxide are all widely used to make industrial gas separation membranes. As shown, for most materials the permeability declines to less than half of the original value over the first 10 000–25 000 hours of operation. However, much of this loss occurs during the first 1000 hours (about 40 days). Suppliers of membrane separation units and plants take this initial loss into account when designing equipment [47].

Figure 2.27 also shows data for PTMSP, the first member of the exceptionally high free volume class of materials now called polymers of intrinsic porosity (PIMs). The permeability of a thin PTMSP membrane falls to less than 10% of its initial value in less than a week. Other membranes of this class have the same problem. These materials have exceptional properties when made and tested as films more than 100 μm thick, but suffer catastrophic permeability

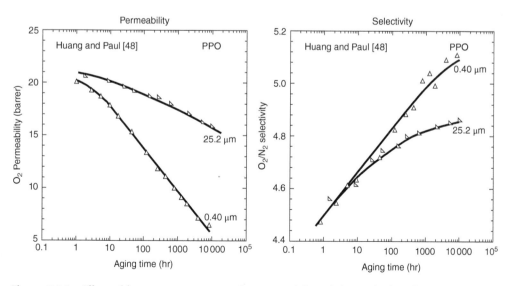

Figure 2.26 Effect of long-term storage on the permeability of thin polyphenylene oxide (PPO) films. Source: Some data of Huang and Paul. Reproduced with permission from [48]. Copyright (2004) Elsevier.

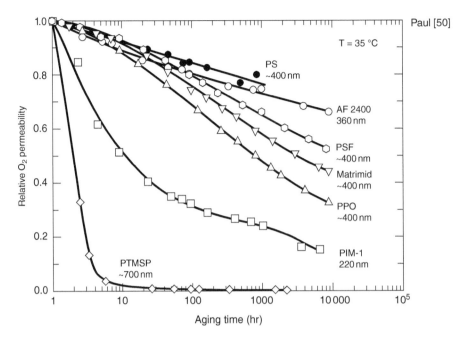

Figure 2.27   Measurements of the permeability loss from a range of polymers over time, as measured with very thin membranes over 10 000 hours (about 1 year) of operations. Source: Reproduced with permission from [50]. Copyright (2014) Elsevier.

declines when fabricated as thin membranes. Cross-linking and other techniques to reduce polymer chain mobility and permeability loss have been tried but with limited success.

2.3.2   Sorption Coefficients in Polymers

The second key factor determining permeability in polymers is the sorption coefficient. The data in Figure 2.20 and Table 2.2 show that sorption coefficients for a particular gas are relatively constant within a single family of related materials. In fact, sorption coefficients of gases in polymers are relatively constant for a wide range of chemically different polymers. Figure 2.28 plots methane sorption and diffusion coefficients in Tanaka's fluorinated polyimides [29], carboxylated polyvinyl trimethylsiloxane [51], and substituted polyacetylenes [52], all amorphous glassy polymers, and for a variety of substituted siloxanes, all rubbers [53]. The diffusion coefficients vary by more than 100 000, showing the extraordinary sensitivity to changes in the packing of the polymer chains and to their flexibility. In contrast, sorption coefficients vary by only a factor of 10 around a mean value of about $15 \times 10^{-3} \text{ cm}^3(\text{STP})/\text{cm}^3 \cdot \text{cmHg}$. The sorption selectivity term (K_i/K_j) for a gas pair is even more constant, even for very different polymers, as the data in Table 2.2 showed.

The sorption coefficients of gases in polymers remain relatively constant because, to a fair approximation, sorption in polymers behaves as though the polymers were ideal fluids. Gas sorption in a polymer is expressed from Eq. (2.60) as

$$c_{i_{(m)}} = K_i^G p_i \tag{2.106}$$

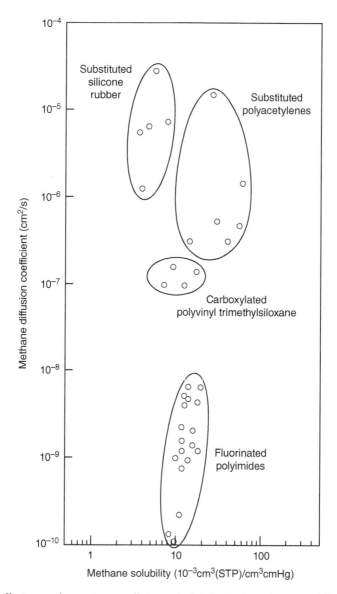

Figure 2.28 *Diffusion and sorption coefficients (solubility) of methane in different families of polymer materials. Diffusion coefficients change over a wide range but sorption coefficients are relatively constant. Source: Adapted from [29, 51–53].*

By substituting for the sorption coefficient K_i^G from Eq. (2.59), Eq. (2.106) can be written as

$$c_{i_{(m)}} = m_i \rho_m \frac{\gamma_i^G p_i}{\gamma_{i_{(m)}} p_{i_{sat}}} \tag{2.107}$$

Converting from concentration to mole fraction using

$$c_{i_{(m)}} = m_i \rho_m n_{i_{(m)}} \tag{2.108}$$

it follows that Eq. (2.108) can be written as

$$\frac{c_{i_{(m)}}}{\rho_m m_i} = n_{i_{(m)}} = \frac{\gamma_i^G p_i}{\gamma_{i_{(m)}} p_{i_{sat}}}$$

(2.109)

For an ideal gas dissolving in an ideal liquid, γ_i^G and $\gamma_{i_{(m)}}$ are both unity, so Eq. (2.106) can be written as

$$n_{i_{(m)}} = \frac{p_i}{p_{i_{sat}}}$$

(2.110)

where $n_{i_{(m)}}$ is the mole fraction of the gas sorbed in the liquid, p_i is the partial pressure of the gas, and $p_{i_{sat}}$ is the saturation vapor pressure at the pressure and temperature of the liquid. To apply Eq. (2.110), the gas saturation vapor pressure must be determined. This can be done by extrapolating from available vapor pressure data to the ambient range using the Clausius–Clapeyron equation. For some gases, the vapor pressure obtained does not correspond to a stable gas–liquid equilibrium because the gas is supercritical at ambient temperatures. However, the calculated value is adequate to calculate the fractional sorption using Eq. (2.110) [54]. At 25 °C, the saturation vapor pressure of methane extrapolated in this way is 289 atm. From Eq. (2.110), the mole fraction of methane dissolved in an ideal liquid is then 1/289 or 0.0035. The ideal solubility and measured solubilities of methane in a number of common liquids are given in Table 2.5. Although there is some spread in the data, particularly for polar solvents such as water or methanol, the overall agreement is remarkably good. A more detailed discussion of the solubility of gases in polymers is given by Petropoulos [55] and Doghieri et al. [56].

To apply the procedure outlined above to a polymer, it is necessary to use the Flory–Huggins theory of polymer solutions, which takes into account the entropy of mixing of solutes in polymers caused by the large difference in molecular size between the two components. The Flory–Huggins expression for the free energy of mixing of a gas in polymer solution can be written [57]

$$\Delta G = RT \ln \frac{p_i}{p_{i_{sat}}} = RT \left[\ln V_i + \left(1 - \frac{v_i}{v_j} \right)(1 - V_i) \right]$$

(2.111)

where ΔG is the change in Gibbs free energy brought about mixing, v_i and v_j are the molar volumes of the gas (i) and the polymer (j) respectively, and V_i the volume fraction of the

Table 2.5   Mole fraction of methane in various solvents at 25 °C and 1 atm.

Liquid	Methane solubility (mole fraction)
Ethyl ether	0.0045
Cyclohexane	0.0028
Carbon tetrachloride	0.0029
Acetone	0.0022
Benzene	0.0021
Methanol	0.0007
Water	0.00002

The solubility of methane in an ideal liquid under these conditions is 0.0035 [53].

polymer occupied by the sorbed gas. When $\nu_i = \nu_j$, that is, the gas and polymer molecules are approximately the same size, Eq. (2.111) reduces to Eq. (2.110), the ideal liquid case. When the molar volume of the gas is much smaller than the molar volume of the polymer, then $\nu_i/\nu_j \rightarrow 0$ and Eq. (2.111) becomes

$$\ln \frac{p_i}{p_{i_{sat}}} = \ln V_i + (1 - V_i) \tag{2.112}$$

Equation (2.112) can be rearranged to

$$V_i = \frac{p_i/p_{i_{sat}}}{\exp(1 - V_i)} \tag{2.113}$$

and since V_i is small, $\exp(1 - V_i)$ is approximately $\exp(1) \approx 2.72$. Eq. (2.113) then becomes

$$V_i = \frac{p_i/p_{i_{sat}}}{2.72} \tag{2.114}$$

Comparing Eqs. (2.110) and (2.114), we see that the volume fraction of gas sorbed by an ideal polymer is 1/2.72 of the mole fraction of the same gas sorbed in an ideal liquid.[4]

The results of such a calculation are shown in Table 2.6. In Figure 2.29, the calculated sorption coefficients in an ideal polymer from Table 2.6 are plotted against the average sorption coefficients of the same gases in Tanaka's polyimides [29]. The calculated values are within a factor of 2 of the experimental values, which is extremely good agreement considering the simplicity of Eq. (2.114).

As shown above, thermodynamics can qualitatively predict the sorption of simple gases in polymers to within a factor of 2 or 3. Moreover, Eq. (2.114) predicts that all polymers should have about the same sorption for the same gas and that sorption of different gases is inversely proportional to their saturation vapor pressures.

Another way of showing the same effect is to plot gas sorption against some convenient measure of saturation vapor pressure, such as the gas boiling point or critical temperature. Figure 2.30 shows a plot of this type for a typical glassy polymer (polysulfone), a typical rubber

Table 2.6   Solubility of gases in an ideal liquid and an ideal polymer (35 °C).

Gas	Calculated saturation vapor pressure, $p_{i_{sat}}$ (atm)	Ideal solubility in a liquid at 1 atm (mole fraction) (Eq. (2.110))	Ideal solubility in a polymer (volume fraction) (10^{-3} cm^3(STP)/cm$^3 \cdot$ cmHg) (Eq. (2.114))
N_2	1400	0.0007	2.8
O_2	700	0.0014	4.8
CH_4	366	0.0027	18.5
CO_2	79.5	0.0126	31.0

[4] V_i is the volume fraction of the gas sorbed in the polymer. To calculate the amount of gas sorbed in cm^3(STP)/cm^3, the molar density of the sorbed gas must be known. We assume this density is 1/MW (mol/cm^3).

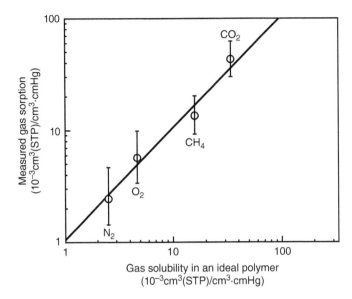

Figure 2.29 *Average sorption coefficients of simple gases in a family of 18 related polyimides plotted against the expected sorption in an ideal polymer calculated using Eq. (2.114). Source: Data from Tanaka et al. [29].*

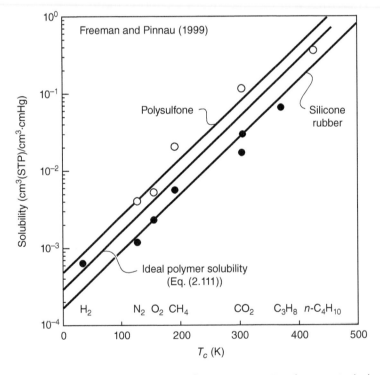

Figure 2.30 *Solubilities as a function of critical temperature (T_c) for a typical glassy polymer (polysulfone) and a typical rubbery polymer (silicone rubber) compared with values for the ideal solubility calculated from Eq. (2.114). Source: Reproduced with permission from [58]. Copyright (1999) American Chemical Society.*

(silicone rubber), and the values for the ideal solubility of a gas in a polymer calculated using Eq. (2.114) [58]. The figure shows that the difference in gas sorption values of polymers is relatively small and the values are grouped around the calculated value.

Although all of these predictions are qualitatively correct, the solubility axis in Figure 2.30 has a log scale, and the relatively small difference between the behavior of an ideal polymer and an actual polymer as shown in the figure is important in selecting the optimum material for a particular separation. The usual starting point for this fine-tuning is the dual-sorption model originally proposed by Barrer, Barrie and Slater [59]. This model has since been extended by Michaels et al. [60], Toi et al. [61], Koros et al. [62], and many others.

According to the dual-sorption model, gas sorption in a polymer (c_m) occurs in two types of sites. The first type of site is filled by gas molecules dissolved in the equilibrium free volume portion of a material at a concentration c_D. In rubbery polymers, this is the only population of dissolved gas molecules. In glassy polymers an additional type of site exists, populated by a concentration c_H of molecules dissolved in the excess free volume of the glassy polymer. The total sorption in a glassy polymer is then

$$c_m = c_D + c_H \tag{2.115}$$

The molecules dissolved in the equilibrium free volume will behave as in normal sorption in a liquid and their concentration can be related to the pressure in the surrounding gas by a linear Henry's law expression equivalent to Eq. (2.106)

$$c_D - K_D p \tag{2.116}$$

This fraction of the total sorption is equivalent to the value calculated in Eq. (2.114). The other fraction (c_H) is assumed to be sorbed into the excess free volume elements, which are limited, so sorption will cease when all the sites are filled. Sorption in these sites is best approximated by a Langmuir-type adsorption isotherm

$$c_H = \frac{c_H' b p}{1 + bp} \tag{2.117}$$

where c_H' is the saturation sorption concentration and b is a constant of proportionality. At high pressures $c_H \to c_H'$, and all excess free volume sites become filled.

From Eqs. (2.116) and (2.117), it follows that the total sorption can be written as

$$c_m = K_D p + \frac{c_H' b p}{1 + bp} \tag{2.118}$$

The form of the sorption isotherm predicted from the dual-sorption model is shown in Figure 2.31. Because the expressions for sorption contain three adjustable parameters, good agreement between theory and experiment is usually obtained.

Sometimes, much is made of the particular values of the constants b and K_D. However, these constants should be treated with caution because they depend totally on the starting point of the curve fitting exercise. That is, starting with an arbitrary value of c_H', the other constants b and K_D can usually be adjusted to obtain good agreement of Eq. (2.118) with experiment. If the starting

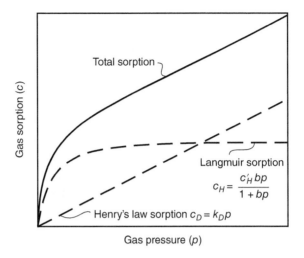

Figure 2.31   *An illustration of the two components that contribute to gas sorption in a glassy polymer according to the dual-sorption model. Henry's law sorption occurs in the equilibrium free volume portion of the polymer. Langmuir sorption occurs in the excess free volume between polymer chains that exists in glassy polymers.*

value for c'_H is changed, then equally good agreement between theory and experiment can still be obtained, but with different values of b and K_D [63].

Permeation of gases in glassy polymers can also be described in terms of the dual-sorption model. One diffusion coefficient (D_D) is used for the portion of the gas dissolved in the polymer according to the Henry's law expression and a second, somewhat larger, diffusion coefficient (D_H) for the portion of the gas contained in the excess free volume. The Fick's law expression for flux through the membrane has the form

$$J = -D_D \frac{dc_D}{dx} - D_H \frac{dc_H}{dx}$$
(2.119)

2.4   Conclusions

The solution-diffusion model provides a good description of permeant transport through membranes operating under dialysis, reverse osmosis, gas separation, and pervaporation conditions. The fundamental equations describing transport in all of these processes can be derived from simple, basic principles without resource to process-specific factors. These equations provide an accurate description of membrane behavior and the dependence of permeant transport on pressure, concentration, and the like. The general agreement of transport coefficients derived in all of these processes with each other, and the general universality of the approach are strong indications of the reliability of the model. Direct measurements of concentration gradients in membranes provide additional support for the model.

The creation of the unified theory able to rationalize transport in the dense membranes used in RO, pervaporation, and gas separation occurred over a 20-year period from about 1960 to 1980. Development of this theory is one of the successes of membrane science. The theory did not form overnight as the result of one single breakthrough, but rather as the result of a series of incremental steps. The paper of Lonsdale, Merten, and Riley [21] applying the solution-diffusion model to RO for the first time was important. Also important was the series of papers by Paul and coworkers showing the connection between hydraulic permeation (RO) and pervaporation [2–4, 19], and providing the experimental support for the solution-diffusion model as applied to these processes.

The solution-diffusion model uses Fick's law of diffusion (Eq. (2.1)) as its basis. However, Fick's law in the strictest sense is only valid for a two-component system comprised of a membrane and one diffusing component. Also, membranes highly swollen by the permeant require a 'frame-of-reference correction' because the difference in velocity between the stationary membrane material and the permeating components is not accounted for. Fortunately for all of us, the deviations from Fick's law are minor when the permeant concentrations in the membrane are small, and the equations presented in this chapter are applicable to the majority of membrane applications without significant error.

The application of Fick's law to the diffusion part of the solution-diffusion model has to be re-examined, however, when the membrane is highly swollen by the permeants. The frame-of-reference correction to Fick's law, described in papers by Paul [64, 65] and Kamaruddin and Koros [66], can be applied, but this does not extend the description to more than two components, and a membrane separation process has a minimum of three components: the membrane material and at least two permeants which are being separated. An alternative approach is to replace Fick's law in the solution-diffusion model by the Maxwell–Stefan diffusive transport equation. This equation is based on the relative velocities of the components of the system to one another. The frame-of-reference problem is then sidestepped. Those with a mathematical bent will find a readable introduction to the Maxwell–Stefan equation and its application to membrane processes is given in the book by Wesselingh and Krishna [14]. A recent paper by Paul [15] discusses the use of the Maxwell–Stefan equation for organic hyperfiltration processes.

Most readers are likely to find the Maxwell–Stefan approach heavy going. It is doubtful if the advantage of Maxwell–Stefan formalism will ever persuade the average membrane researcher to switch from the relative simplicity of Fick's law. Nonetheless, the Maxwell–Stefan approach has its supporters, particularly for use in fundamental investigations of membrane transport behavior.

The solution-diffusion model has been less successful at providing a link between the nature of membrane materials and their membrane permeation properties. This link requires an ability to calculate membrane diffusion and sorption coefficients. These calculations require knowledge of the molecular level of interactions of permeant molecules and their motion in the polymer matrix that is not yet available. Only semi-empirical correlations such as the dual-sorption model or free volume correlations are available. The best hope for future progress toward a priori methods of calculating permeant sorption and diffusion coefficients lies in computer-aided simulations of molecular dynamics, but accurate predictions using this technique are years – perhaps decades – away.

References

1. Meares, P. (1966). On the mechanism of desalination by reversed osmotic flow through cellulose acetate membrane. *Eur. Polym. J.* **2**: 241.
2. Paul, D.R. (1974). Diffusive transport in swollen polymer membranes. In: *Permeability of Plastic Films and Coatings* (ed. H.B. Hopfenberg), 35–48. New York: Plenum Press.

3. Paul, D.R. (1976). The solution-diffusion model for swollen membranes. *Sep. Purif. Meth.* **5**: 33.
4. Paul, D.R. and Ebra-Lima, O.M. (1970). Pressure-induced diffusion of organic liquids through highly swollen polymer membranes. *J. Appl. Polym. Sci.* **14**: 2201.
5. Yasuda, H. and Peterlin, A. (1973). Diffusive and bulk flow transport in polymers. *J. Appl. Polym. Sci.* **17**: 433.
6. Sourirajan, S. (1970). *Reverse Osmosis*. New York: Academic Press.
7. Fick, A. (1855). Über diffusion. *Poggendorff's Ann. Phys. Chem.* **94**: 59.
8. Smit, E., Mulder, M.H.V., Smolders, C.A. et al. (1992). Modeling of the diffusion of carbon dioxide in polyimide matrices by computer simulation. *J. Membr. Sci.* **73**: 247.
9. Charati, S.G. and Stern, S.A. (1998). Diffusion of gases in silicone polymers: molecular dynamic simulations. *Macromolecules* **31**: 5529.
10. Theodorou, D.N. (2006). Principles of molecular simulation of gas transport in polymers. In: *Materials Science of Membranes for Gas and Vapor Separations* (ed. Y. Yampolskii, I. Pinnau, and B.D. Freeman), 49–89. Chichester, England: Wiley.
11. Wijmans, J.G. and Baker, R.W. (1995). The solution-diffusion model: a review. *J. Membr. Sci.* **107**: 1.
12. Wijmans, J.G. and Baker, R.W. (2006). The solution-diffusion model: a unified approach to membrane permeation. In: *Materials Science of Membranes for Gas and Vapor Separation* (ed. Y. Yampolskii, I. Pinnau, and B.D. Freeman), 159–188. Chichester, England: Wiley.
13. Thorsen, T. and Holt, T. (2009). The potential for power production from salinity gradients by pressure retarded osmosis. *J. Membr. Sci.* **335**: 103–110.
14. Wesselingh, J.A. and Krishna, R. (2000). *Mass Transfer in Multi-Component Mixtures*. Delft, The Netherlands: Delft University Press.
15. Paul, D.R. (2004). Reformulation of the solution-diffusion theory of reverse osmosis. *J. Membr. Sci.* **241**: 371.
16. Theeuwes, F., Gale, R.M., and Baker, R.W. (1976). Transference: a comprehensive parameter governing permeation of solutes through membranes. *J. Membr. Sci.* **1**: 3.
17. Rosenbaum, S. and Cotton, O. (1969). Steady-state distribution of water in cellulose acetate membrane. *J. Polym. Sci.* **7**: 101.
18. Kim, S.N. and Kammermeyer, K. (1970). Actual concentration profiles in membrane permeation. *Sep. Sci.* **5**: 679.
19. Paul, D.R. and Paciotti, D.J. (1975). Driving force for hydraulic and pervaporation transport in homogeneous membranes. *J. Polym. Sci. Polym. Phys. Ed.* **13**: 1201.
20. Wijmans, J.G. (2004). The role of permeant molar volume in the solution-diffusion model transport equations. *J. Membr. Sci.* **237**: 39.
21. Lonsdale, H.K., Merten, U., and Riley, R.L. (1965). Transport properties of cellulose acetate osmotic membranes. *J. Appl. Polym. Sci.* **9**: 1341.
22. Merten, U. (1966). Transport properties of osmotic membranes. In: *Desalination by Reverse Osmosis* (ed. U. Merten), 15–54. Cambridge, MA: MIT Press.
23. Riley, R.L., Lonsdale, H.K., Lyons, C.R., and Merten, U. (1967). Preparation of ultrathin reverse osmosis membranes and the attainment of theoretical salt rejection. *J. Appl. Polym. Sci.* **11**: 2143.
24. Bhore, N., Gould, R.M., Jacob, S.M. et al. (1999). New membrane process debottlenecks solvent dewaxing unit. *Oil Gas J.* **97**: 67.
25. Kataoka, T., Tsuro, T., Nakao, S.-I., and Kimura, S. (1991). Permeation equations developed for prediction of membrane performance in pervaporation, vapor permeation and reverse osmosis based on the solution-diffusion model. *J. Chem. Eng. Jpn.* **24**: 326–333.

26. Wijmans, J.G. and Baker, R.W. (1993). A simple predictive treatment of the permeation process in pervaporation. *J. Membr. Sci.* **79**: 101.

27. Greenlaw, F.W., Prince, W.D., Shelden, R.A., and Thompson, E.V. (1977). Dependence of diffusive permeation rates by upstream and downstream pressures. *J. Membr. Sci.* **2**: 141.

28. Baker, R.W., Wijmans, J.G., and Huang, Y. (2010). Permeability, permeance and selectivity: a preferred way of reporting pervaporation performance data. *J. Membr. Sci.* **348** (1, 2): 356–352.

29. Tanaka, K., Kita, H., Okano, M., and Okamoto, K. (1992). Permeability and permselectivity of gases in fluorinated and non-fluorinated polyimides. *Polymer* **33**: 585.

30. Koros, W.J., Coleman, M.R., and Walker, D.R.B. (1992). Controlled permeability polymer membranes. *Annu. Rev. Mater. Sci.* **22**: 47.

31. Merkel, T.C., Bondar, V.I., Nagai, K. et al. (2000). Gas sorption, diffusion, and permeation in poly(dimethylsiloxane). *J. Polym. Sci. B Polym. Phys.* **38**: 415.

32. Teplyakov, V. and Meares, P. (1990). Correlation aspects of the selective gas permeabilities of polymeric materials and membranes. *Gas Sep. Purif.* **4**: 66.

33. Hirschfelder, J.O., Curtis, C.F., and Bird, B.B. (1954). *Molecular Theory of Gases and Liquids*. New York: Wiley.

34. Reid, R.C., Prausnitz, J.M., and Poling, B.E. (1987). *The Properties of Gases and Liquids*, 4e. New York: McGraw Hill.

35. Edward, J.T. (1970). Molecular volumes and the Stokes–Einstein equation. *J. Chem. Educ.* **47**: 261.

36. Nishijima, Y. and Oster, G. (1956). Diffusion in concentrated polymer solutions. *J. Polym. Sci.* **19**: 337.

37. Baker, R.W. and Lonsdale, H.K. (1974). Controlled release: mechanisms and rates. In: *Controlled Release of Biological Active Agents* (ed. A.C. Tanquary and R.E. Lacey), 15–72. New York: Plenum Press.

38. Artsis, M., Chalykh, A.E., Khalturinskii, N.A. et al. (1972). Diffusion of organic diluents into ethyl cellulose. *Eur. Polym. J.* **8**: 613.

39. Bondi, A. (1968). *Physical Properties of Molecular Crystals, Liquids, and Glasses*. New York: Wiley.

40. van Krevelen, D.W. (1990). *Properties of Polymers*. Amsterdam: Elsevier.

41. Budd, P.M., McKeown, N.B., Ghanem, B.S. et al. (2008). Gas permeation parameters and other physiochemical properties of a polymer of intrinsic microporosity: polybenzodioxane PIM-1. *J. Membr. Sci.* **325**: 851.

42. Pinnau, I. and Toy, L.G. (1996). Transport of organic vapors through poly[1-(trimethylsilyl)-1-propyne]. *J. Membr. Sci.* **116**: 199.

43. Srinivasan, R., Auvil, S.R., and Burban, P.M. (1994). Elucidating the mechanism(s) of gas transport in poly[1-(trimethylsilyl)-1-propyne] (PTMSP) membranes. *J. Membr. Sci.* **86**: 67.

44. Aitken, C.L., Koros, W.J., and Paul, D.R. (1992). Effect of structural symmetry on gas transport properties of polysulfones. *Macromol.* **25**: 3424.

45. Park, J.Y. and Paul, D.R. (1997). Correlation and prediction of gas permeability in glassy polymer membrane materials via a modified free volume based group contribution method. *J. Membr. Sci.* **125**: 29.

46. Matteucci, S., Yampolskii, Y., Freeman, B.D., and Pinnau, I. (2006). Transport of gases and vapors in glassy and rubbery polymers. In: *Materials Science of Membranes for Gas and Vapor Separation* (ed. Y. Yampolskii, I. Pinnau, and B.D. Freeman), 1–47. Chichester, England: Wiley.

47. Merrick, M.M., Sujanani, R., and Freeman, B.D. (2020). Glassy polymers: historical findings, membrane applications and unsolved questions. *Polymer* **211**: 123196.
48. Huang, Y. and Paul, D.R. (2004). Physical aging of thin glassy polymer films monitored by gas permeability. *Polymer* **45**: 8377.
49. Huang, Y. and Paul, D.R. (2007). Effect of film thickness on the gas-permeation characteristics of glassy polymer membranes. *Ind. Eng. Chem. Res.* **46**: 2342.
50. Tiwari, R.R., Smith, Z.P., Lin, H. et al. (2014). Gas permeation in thin films of high free-volume glassy perfluoropolymers, part 1: physical aging. *Polymer* **55**: 5788.
51. Platé, N.A. and Yampol'skii, Y.P. (1994). Relationship between structure and transport properties for high free volume polymeric materials. In: *Polymeric Gas Separation Membranes* (ed. D.R. Paul and Y.P. Yampol'skii), 155–208. Boca Raton, FL: CRC Press.
52. Masuda, T., Iguchi, Y., Tang, B.-Z., and Higashimura, T. (1988). Diffusion and solution of gases in substituted polyacetylene membranes. *Polymer* **29**: 2041.
53. Stern, S.A., Shah, V.M., and Hardy, B.J. (1987). Structure-permeability relationships in silicone polymers. *J. Polym. Sci. Polym. Phys. Ed.* **25**: 1263.
54. Denbigh, K. (1961). *The Principles of Chemical Equilibrium*. Cambridge: Cambridge University Press.
55. Petropoulos, J.H. (1994). Mechanisms and theories for sorption and diffusion of gases in polymers. In: *Polymeric Gas Separation Membranes* (ed. D.R. Paul and Y.P. Yampol'skii), 17–82. Boca Raton, FL: CRC Press.
56. Doghieri, F., Quinzi, M., Rethwisch, D.G., and Sarti, G.C. (2006). Predicting gas solubility in membranes through non-equilibrium thermodynamics for glassy polymers. In: *Materials Science of Membranes for Gas and Vapor Separation* (ed. B.D. Freeman, Y. Yampolskii, and I. Pinnau), 137–158. Chichester, England: Wiley.
57. Flory, P.J. (1953). *Principles of Polymer Chemistry*, 511. Ithaca, NY: Cornell University Press.
58. Freeman, B.D. and Pinnau, I. (1999). Polymer membranes for gas separation. *ACS Symp. Ser.* **733**: 6.
59. Barrer, R.M., Barrie, J.A., and Slater, J. (1958). Sorption and diffusion in ethyl cellulose. *J. Polym. Sci.* **27**: 177.
60. Michaels, A.S., Vieth, W.R., and Barrie, J.A. (1963). Solution of gases in polyethylene terephthalate. *J. Appl. Phys.* **34**: 1.
61. Toi, K., Morel, G., and Paul, D.R. (1982). Gas sorption in poly(phenylene oxide) and comparisons with other glassy polymers. *J. Appl. Sci.* **27**: 2997.
62. Koros, W.J., Chan, A.H., and Paul, D.R. (1977). Sorption and transport of various gases in polycarbonate. *J. Membr. Sci.* **2**: 165.
63. Morisato, A., Freeman, B.D., Pinnau, I., and Casillas, C.G. (1996). Pure hydrocarbon sorption properties of poly(1-trimethylsilyl-1-propyne) [PTMSP] and poly(1-phenyl-1-propyne) [PPP] and PTMSP/PPP blends. *J. Polym. Sci. Polym. Phys. Ed.* **34**: 1925.
64. Paul, D.R. (1973). Relation between hydraulic permeation and diffusion in homogenous swollen membranes. *J. Polym. Sci. B Polym. Phys.* **11**: 289.
65. Paul, D.R. (1974). Further comments on the relation between hydraulic permeation and diffusion. *J. Polym. Sci. B Polym. Phys.* **12**: 1221.
66. Kamaruddin, H.D. and Koros, W.J. (1997). Some observations about the application of Fick's first law for membrane separation of multicomponent mixtures. *J. Membr. Sci.* **135**: 147.

3

Microporous Membranes – Characteristics and Transport Mechanisms

Membrane Technology and Applications, Fourth Edition. Richard W. Baker.
© 2024 John Wiley & Sons Ltd. Published 2024 by John Wiley & Sons Ltd.

3.1   Introduction

In the preceding chapter, permeation through dense polymeric membranes was described. In these membranes, the permeating molecules dissolve in the membrane polymer as in a liquid. The molecules are contained in tiny spaces between the polymer chains that make up the membrane. A permeating molecule remains in one of these spaces until a chance thermal motion of the polymer opens a channel sufficiently large for the molecule to move to an adjoining space. This process of moving from space-to-space (diffusion) is controlled by the nature of the permeating molecule and the polymer matrix. Starting with Fick's law as the governing principle, we were able to develop a semi-rigorous treatment that offers an explanation of molecular permeation that is both (i) consistent with experimental results and (ii) enables the seemingly disparate processes of dialysis, reverse osmosis (RO), gas separation, and pervaporation to be understood in a unified way.

This chapter is focused on permeation of gases and liquids through microporous membranes, especially very finely microporous membranes. The theory of permeation through these membranes is far less developed, and it is difficult to see a clear path forward to that end. Transport through microporous membranes is affected by a variety of hard-to-compute effects. Whereas a solid polymer membrane can be considered essentially as the equivalent of a homogeneous, high-molecular-weight liquid, a porous membrane is made up of a matrix containing cavities that can vary in size, shape, quantity, uniformity of distribution, and degree of interconnection. The behavior of a molecule encountering such a structure is thus dependent on all of these variables. As a result, no adequate theories to explain their properties exist, and this chapter presents a much more qualitative approach than Chapter 2.

In microporous membranes, molecules travel through permanent connected pores in the membrane. In some cases, the pores may not be much bigger than the space between polymer chains, as in the solution-diffusion membranes just described, but in general the pores are larger and the matrix material is rigid enough that the pores are permanent on the time scale of membrane permeation. Most of the permeating component travels through the pores; very little passes through the matrix material. The membrane matrix can be a rigid polymer, a ceramic metal oxide, a glass, or an amorphous carbon.

Microporous membranes are usually characterized by their average pore diameter measured by various methods (positron annihilation measurements, BET nitrogen adsorption techniques, adsorption of molecular probes of varying size, and X-ray scattering measurements). The pore diameters obtained may not accurately represent reality however, because, as mentioned above, the pores in many membranes are non-uniform in shape and distribution, and constrictions between pores may be more important in regulating permeation than the dimensions of the pores themselves.

Microporous membranes are used in both gas and liquid separation processes, but because the processes are quite different, we will treat them separately. Gas separation processes generally use membranes with pore diameters below about 25 Å. Above this, the membranes have high permeances but separations between different gases are generally uninterestingly low. Liquid permeation processes use a wider range of membranes. Low molecular weight solvents and solutes can be separated by hyperfiltration using membranes with pore diameters between 3 and 30 Å. Ultrafiltration uses membranes with pores from 10 to 20 Å to about 200 Å. Membranes with pores greater than 200 Å are usually called microfiltration membranes. The pore ranges given above are used to guide the discussion of the different processes that follow, but there is no sudden transition of one process to another at a particular pore diameter.

3.2 Gas Separation in Microporous Membranes

3.2.1 Membrane Categories

Microporous membranes with a variety of compositions and structures have been investigated for their gas separation capabilities. Since composition and structure are key to understanding their properties, a review of the main types is in order before attempting to rationalize their transport properties.

In the discussion that follows, microporous membranes are divided into two broad groups, crystalline and amorphous, illustrated in Figure 3.1. The first group consists of (a) zeolites, (b) crystalline metal organic frameworks (MOFs) and (c) covalent organic frameworks (COFs). These materials have a structure that contains a regular repeating network of identical pores. The second group comprises amorphous materials, having a network of pores of variable size connected in a random way. Carbonized polymers and polymers of intrinsic porosity (PIMs) are members of this group.

The potential of these materials is illustrated by Figure 3.2, taken from Robeson's 2008 paper [1], which shows the permeability/selectivity trade-off for CO_2/CH_4 separations. A vast number of conventional polymers lead to the 2008 upper bound line. Also shown are a group

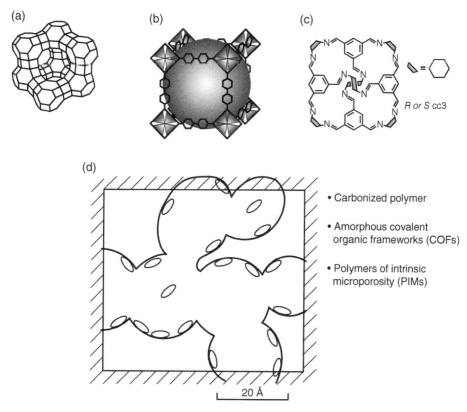

Figure 3.1 *Microporous membranes for gas separation fall into two groups, depending whether they are made from crystalline (a–c) or amorphous (d) materials.*

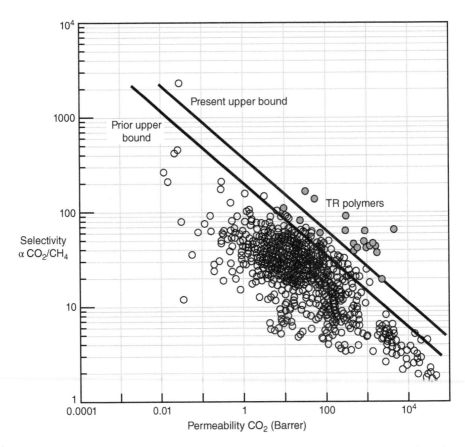

Figure 3.2 *Upper bound correlation for CO_2/CH_4 separation [1]. The TR polymer data shown is from Park et al. [2].*

of results obtained with carbonized polymer films (TR polymers [2]), which clearly transcend the upper bound. These membranes are finely microporous structures with pores in the 5–20 Å diameter range. It is this sort of result that has led to a surge of interest in microporous membrane materials.

3.2.2   Crystalline Finely Microporous Membranes

3.2.2.1   Zeolites

Zeolites are silicalite or aluminosilicate materials formed from a three-dimensional network of SiO_4 and AlO_4 tetrahedra. The tetrahedra are linked by shared oxygen atoms to form cages. In a zeolite structure, these individual cages are linked together in geometric forms that create pore openings with defined regular shapes and sizes. Figure 3.3 shows how one type of building block, the sodalite β-cage, can form several structures with pores ranging from 3 to 8 Å in diameter. There are over 250 known zeolite structures, of which 40 occur naturally [3].

Aluminosilicate structures carry a negative charge which is usually balanced by Na^+, K^+, or Ca^{2+} counter ions. Fine tuning of the pore openings of the zeolite structure is possible by exchanging the counter ions K^+ and Ca^{2+} for Na^+, for example.

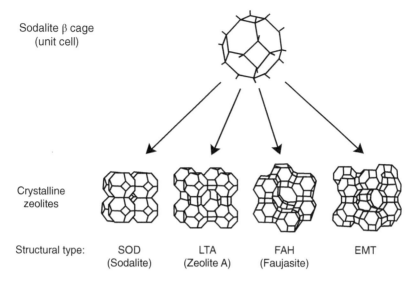

Figure 3.3 *Sodalite SiO$_4$/AlO$_4$ β-unit cells can link together to form a variety of zeolite structures, each with its own geometry and pore size. Source: Ramsay, J.D.F. and Kallus, S. (2000)/ Elsevier.*

The pioneer of artificial synthesis of zeolites was Barrer, and he was the first to investigate their diffusion properties and potential use for gas separations, starting in the 1940s. Zeolites are now used throughout industry as sorbents and catalysts. Zeolite membranes can be formed on microporous ceramic or stainless-steel supports by the *in situ* hydrothermal synthesis techniques illustrated in Figure 3.4. A support substrate is immersed in a precursor solution with the correct Si/Al stoichiometry, pH, and ionic strength. The solution is heated in a sealed autoclave for a predetermined time. The type of zeolite that will form is determined by many factors described elsewhere [4, 5]. Often, a structure-directing reagent is used to form a particular type of zeolite. For example, the tetrapropyl ammonium ion (TPA$^+$) leads exclusively to MFI or silicalite zeolites. After hydrothermal synthesis, the membrane is tested to see if it is gas tight. If not, the sample is reimmersed in the precursor solution and autoclaved a second time. When the sample is gas tight, it is calcined. The sample is then ready for evaluation.

An alternative technique of seeding the support structure with small crystals of the desired zeolite, developed by Tsapatsis, is shown in Figure 3.5 [6]. The seed crystals are prepared by separately synthesizing the required zeolite or grinding a sample to the required size (∼0.5 μm). The support must be thoroughly cleaned to promote uniform and good coverage of the seed crystal solution on the support. The support is then heated with an appropriate AlO$_4$/SiO$_4$ solution in an autoclave. In this process, somewhat simpler solutions can be used and structure-directing reagents may not be needed. The crystal nucleation technique generally leads to thinner and more defect-free zeolite layers. Nonetheless, zeolite membranes must normally be 10–50 μm thick to be defect-free. Even then, the membrane must be handled carefully to avoid damage.

Membranes of the type shown in Figure 3.5 are not easy to make at the industrial scale and are inherently expensive. Nevertheless, Mitsui and ECN, both in Japan, and Nine Heaven in China have commercialized zeolite membranes for solvent dehydration by pervaporation or vapor permeation [7, 8].

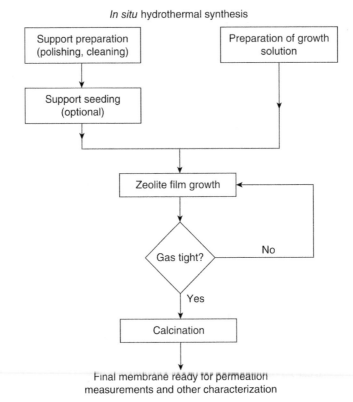

In situ hydrothermal synthesis

Figure 3.4   *Methods of preparing zeolite membranes. Source: Reproduced with permission from [5]. Copyright (2006) John Wiley & Sons. Ltd.*

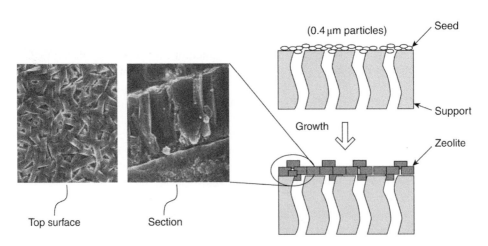

Figure 3.5   *Zeolite membrane formation by the seed nucleation and growth technique. Source: From [5], John Wiley & Sons.*

3.2.2.2   Metal Organic Frameworks (MOFs)

In the last decade, metal organic frameworks (MOFs) have become of considerable interest to membrane makers. MOFs are finely microporous crystalline structures, in the form of an inorganic metal cation framework connected by organic linkers. The pore structure and surface chemistry can be systematically tuned by changing the metal cations or the organic linkers, as illustrated in Figure 3.6.

In general, MOFs have higher fractional free volumes and higher BET surface areas than zeolites. The largest cavity in the MOF structure can be quite large, in the 10–20 Å range, but membrane permeation is controlled by the windows in the cavity walls, which are usually smaller. Because MOFs are crystalline, the diameter of these windows can be precisely measured. The MOF ZIF-8, for example, has a diameter of 11.6 Å with a nominal window pore diameter of 3.4 Å, yet the apparent pore diameter from membrane permeation measurements with gases of known diameter (C_2H_4, C_2H_6, C_3H_6, and C_3H_8) seems to be about 4.5 Å [9]. It appears that the organic linker molecules that make up the frame of the pore windows have some flexibility.

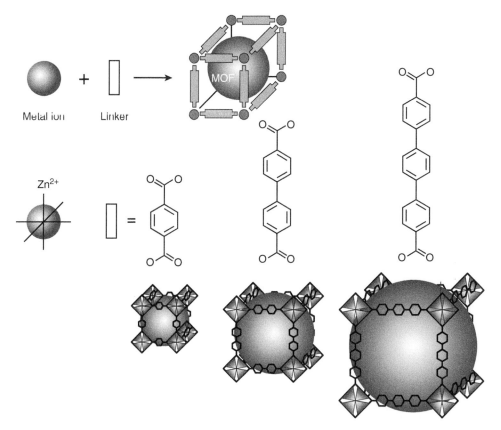

Figure 3.6 *A series of related MOFs formed from Zn^{2+} and benzenedicarboxylate oligomers. Increasing the size of the benzenedicarboxylate linker increases the pore diameter of the resulting MOF. The simple MOF cell shown forms the unit cells of a crystalline structure held together by the coordinating metal ions (Zn^{2+}) at the corners of the unit cell. Changing the metal ion to Ni^{2+}, Cu^{2+}, or Fe^{3+} is also possible, and using the same coordinating ligands will produce a different series of MOFs.*

Laboratory synthesis of MOFs is generally straightforward. By way of example, the synthesis of ZIF-8 is described below. ZIF-8 is a popular material for study because of its chemical stability and the availability of its precursor components. ZIF-8 is made by preparing a 3% solution of $Zn(NO_3)_2 \cdot 6H_2O$ in methanol, and a second solution of about 3% 2-methylimidozole in methanol. The solutions are mixed rapidly and stirred well. After about an hour, the reaction is complete and the precipitated solution is centrifuged and washed to remove excess methanol. The resulting ZIF-8 nanocrystals are dried by evaporating any remaining methanol at 70 °C, and are about 500 Å in diameter. Several thousand different MOFs have been synthesized in the last few years; the structures of some of the more important ones are illustrated in Table 3.1.

The most direct procedure for forming MOF membranes is similar to the seed nucleation procedure for zeolites. A dispersion of MOF seed crystals in a dilute solution of poly(ethyleneimine) (PEI) is coated onto an alumina porous support. The PEI chemically bonds the ZIF crystals to hydroxyl groups on the Al_2O_3 support. The seeded support is then suspended in a solution of ZIF precursor components in methanol and autoclaved at 100 °C for several hours. A ZIF layer grows on the seed crystals and eventually forms a continuous film 10–40 μm thick across the surface of the alumina support. A number of alternative methods of forming MOF membranes are being developed with the goal of producing membranes on lower cost polymeric supports [11]. The general approach is to form the MOF layer at the interface between a metal ion containing solution and an organic ligand containing solution. The process is similar to that used to make interfacial polymeric RO membranes. The resulting membranes can be as thin as 1–2 μm but are hard to make without defects.

Copious amounts of MOF research have been published. The data has been collected and discussed in several recent reviews [10–14]. By way of example, data from 17 different laboratories that made and tested membranes for the separation of propylene/propane mixtures using pure ZIF-8 membranes are listed in Table 3.2 [12]. There is a great deal of scatter, with propylene permeance ranging from 272 down to 0.8 gpu, and propylene/propane selectivity ranging from 152 to 0.8. That said, some experiments yielded exceptional permeances and selectivities that cannot be achieved by any polymer membrane. The data was created at different pressures, temperatures, and feed gas compositions, all of which can be expected to lead to some data spread. Nonetheless, the spread shown is far more than would be seen for a similar set of results obtained for a polymeric membrane, and reflects the difficulty of making MOF membranes. If the extremes (the four most selective and the four least selective membranes) are eliminated, the data is more closely grouped, with an average propylene permeance of 82 gpu and propylene/propane selectivity of 49. A stable membrane with properties in this range would be a strong candidate for scale up and industrial use.

3.2.2.3   *Crystalline Covalent Organic Frameworks (COFs)*

The results obtained with MOF membranes have prompted the development of similar, but purely organic, materials called covalent organic frameworks (COFs). COFs are large organic molecules with a molecular weight of between 1000 and 2000 Da. Their chemical structure is such that each molecule contains one or two tiny cavities. When a solution of these molecules is evaporated, the resulting solid sometimes forms as an amorphous precipitate, but can also form nanocrystals in which the molecular cavities link together to form an interconnected porous network.

COFs can be made by simple reactions of the type shown in Figure 3.7. For example, the COF called CC3 is produced by the reaction of four trialdehyde and six diamine molecules, the reaction shown in the figure [15, 16]. The reaction product is an 1112 Da molecule that contains two 3.5–4.5 Å cages. The BET internal surface area of the crystal form is about 700 m^2/g.

Table 3.1 Common name, alternative name and physical characteristics for commonly investigated MOFs [10].

Common name	Alternative name(s)	Metal cation(s)	Ligand	Limiting pore aperture (Å)	BET surface area (m²/g)
ZIF-8	$Zn(2\text{-methylimidazolate})_2$ or $Zn(Melm)_2$	Zn^{2+}	(2-methylimidazole)	3.4	1813
HKUST-1	$Cu_3(BTC)_2$	Cu^{2+}	benzen-1, 3, 5-tricarboxylate (BTC)	9	1500–2100
MIL-53	$M(OH)(BDC)$	Al^{3+}	1,4-benzenedicarboxylate (BDC)	8.5	1100–1500
MIL-101	$Cr_3O(H_2O)_2F(BDC)_3$	Cr^{3+}	1,4-benzenedicarboxylate (BDC)	12	2800–4230
MOF-74	$M_2(dobdc)$ or CPO-270-M	Mg^{2+}, Mn^{2+}, Fe^{2+}, Co^{2+}, Ni^{2+}, Zn^{2+}	2,5-diozide-1,4-benzenedicarboxylate (dobcd)	12	1277–1957
UiO-66	$Zr_6O_4(OH)_4(BDC)_6$	Zr^{4+}	1,4-benzenedicarboxylate (BDC)	6	1067

Table 3.2   *Permeance and selectivity data for propylene/propane separation by a pure ZIF-8 membrane.*

P/ℓ C_3H_6 (gpu)	P/ℓ C_3H_8 (gpu)	α C_3H_6/C_3H_8		
45.0	0.3	150.0		
109.9	0.7	152.0		
11.2	7.5×10^{-2}	150.0		
64.9	0.6	103.5		
1.3	1.5×10^{-2}	84.0		Average
170.2	2.4	70.8		P/ℓ C_3H_6 = 82 gpu
59.7	1.1	55.8		α C_3H_6/C_3H_8 = 49
55.0	1.2	46.0		
61.5	1.4	44.8		
24.2	0.6	40.9		
65.7	1.8	36.1		
25.8	0.8	31.6		
272.3	9.0	30.1		
44.8	2.8	16.0		
11.0	1.0	11.0		
0.8	0.2	4.0		
9.0	12.0	0.8		

Source: Data from 17 different laboratories (from a table collected by Qian et al. [12]).

Although the most studied application of these materials has been selective gas adsorption, some promising membrane applications have also been found. An attractive feature of low molecular weight COFs is that they are soluble in chloroform and other organic solvents. This opens up the possibility of making pure COF composite membranes by coating a dilute solution onto a polymeric or ceramic support. On solvent evaporation, the COF molecules phase separate and form a microcrystalline structure. Zhang et al. [17] have made membranes of this type with mixed gas propylene/propane selectivities of about 12 and permeabilities of 400 Barrer.

3.2.2.4   Mixed-Matrix Membranes

Although pure zeolite, MOF, and COF membranes have been made in the laboratory, it will be difficult to use them for large industrial separations. The problems are the high cost of the ceramic supports usually required and the difficulty of making defect-free membranes on a large scale. There may also be problems with stability in the environment of use, but this aspect has yet to be studied seriously.

One potential solution to the manufacturing issues mentioned above is to form what is known as a mixed-matrix membrane, which comprises a dispersion of individual fine crystalline particles in a polymer matrix. To make the membrane, small particles are stirred into a polymer solution. The solution can then be coated onto a porous support to form a composite structure, according to the standard procedures for making solution-coated membranes detailed in

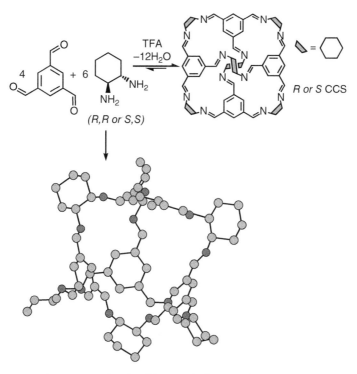

Figure 3.7 The simple cage molecule CC3 made by condensation of triformylbenzene and 1,2, diaminocyclohexane. The cage contains two cavities with a diameter of 3.5–4.5 Å Source: Adapted from [15] [16].

Chapter 4. The particles embedded in the polymer matrix provide the selective capabilities, and the mechanical strength is provided by the underlying support.

Figure 3.8 shows the selective layer of a mixed-matrix membrane loaded with zeolite, MOF or COF particles. At low particle loadings, the individual particles can be considered to be well separated. At higher loadings, some small islands of interconnected particles form; at even higher loadings, these islands grow and connect to form extended pathways. At loadings above a certain critical value, continuous channels form within the membrane, and almost all the microporous

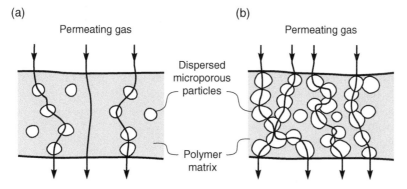

Figure 3.8 Gas permeation through mixed-matrix membranes containing different amounts of dispersed zeolites, MOFs, or COFs. (a) Diffusion below the percolation threshold and (b) Diffusion above the percolation threshold.

particles are connected. This particle loading is called the percolation threshold. A particle loading of 30 vol% or above is usually required to reach the percolation threshold. Once the percolation threshold is reached, most of the permeating gas permeates through the pores. The membrane permeance is generally high and the membrane selectivity is no longer affected by the properties of the membrane matrix material. Attempts have been made to model the improvement produced by increasing the concentration of particles in the polymer matrix; the equations that result are not simple. The topic has been reviewed by Vinh-Thang and Kaliaguine [18].

The first papers on polymer/zeolite mixed-matrix membranes for gas separation appeared in the 1990s. The potential of the idea was clear and so it attracted a good deal of interest. Koros, for example, focused on zeolites with small aperture sizes [19] such as zeolite 4 A, with an effective aperture size of 3.8–4.0 Å, the goal being to separate oxygen (Lennard–Jones diameter 3.47 Å) from nitrogen (Lennard–Jones diameter 3.8 Å). The theoretical oxygen/nitrogen selectivity for a zeolite was calculated to be 37 with an oxygen permeability of 0.8 Barrer. If it could be made, this would be an exceptional membrane.

Unfortunately, there were problems. First, obtaining zeolite particles of the appropriate size was not easy. To achieve useful permeances, the selective mixed-matrix layer of the membrane has to be less than 1–2 μm thick. This means the average diameter of the zeolite crystals dispersed in the polymer matrix has to be about 0.1 μm or less. In the laboratory, this problem can be circumvented by making the membranes 10–50 μm thick and using zeolite crystals 1–2 μm in diameter. Producing small diameter zeolite particles was difficult on the scale required for industrial use, and remains an unsolved problem. A second problem was poor adhesion between the zeolite particles and polymer matrix, which led to gas bypass around the edges of the zeolite. The bypass issue becomes more significant as the zeolite loading increases towards the percolation threshold. A partial solution is to chemically modify the outer surface of the zeolite by attaching organic groups more compatible with the polymer matrix. Experiments showed that this helped, but did not completely solve the problem.

By 2010, it was clear that achieving the hoped-for separations with mixed-matrix membranes based on zeolites was a difficult, and perhaps unsolvable problem. The best mixed-matrix membranes prepared in laboratory studies were only marginally better than the available polymeric membranes, and difficult to scale-up economically. The discovery of MOFs at this time gave the whole mixed-matrix approach a new life. MOFs have the porous structure of zeolites, but, because the network is mostly formed from organic groups, it proved much easier to disperse the MOF particles into the polymer matrix phase without creating bypass channels. It was also easier to make the nanocrystals required for thin membranes.

Within a few years of the first studies, reports describing membranes that far exceeded the permeability and selectivity of the best polymeric membranes began to appear. Multiple laboratories have reported propylene/propane selectivities in the 10–30 range. If these membranes can be scaled up and incorporated into modules, they could offer a long sought-after solution to this separation problem, and others as well. There remains the issue of membrane stability, not just the chemical and physical stability of the MOFs, but also the effect of contaminants in the feed gas. Industrial process streams are not the pristine mixtures of two components used in the laboratory; they typically contain a long list of contaminants in the ppm range. Zeolite membranes, when tested with industrial process streams, were poisoned by components that slowly went into the membrane and stayed there. The effect of these contaminants on MOF membranes is unknown, but is also likely to be bad.

3.2.3   Amorphous Microporous Membranes

One of the advantages of the crystalline microporous membranes described above is that the microporous structure is usually well characterized and can be manipulated in a systematic way to optimize the membrane permeability and selectivity. The amorphous materials of which the second category of finely microporous membranes is composed are harder to characterize and more difficult to adjust. On the other hand, they are easier to make into membranes that have the potential to be scaled up. The mixed-matrix approach is sometimes used, but seldom needed.

3.2.3.1   Polymers of Intrinsic Microporosity (PIMs)

PIMs are made from rigid backbone polymers in which chain mobility is extremely sterically hindered. The polymers are soluble in some organic solvents and can be solution-coated onto suitable substrates in the standard manner to form thin composite membranes. As the coating solvent evaporates, the polymer chains cannot rearrange to eliminate the spaces between them, and the result is a polymer film containing a random network of small, interconnected pores. In conventional glassy polymers, the free volume between the polymer chains is typically about 15% of the total polymer volume; in PIMs, it is 25% or more, enough to form a network of interconnected channels through the membrane.

The first PIM material was poly(1-trimethylsilyl-propyne), or PTMSP, the composition of which is shown in Figure 3.9. The double bonds in the main chain, together with severe steric hindrance from the bulky trimethylsilyl group, almost prohibit segmental rotation of the polymer backbone. As a consequence, the backbone is rigid, the glass transition temperature (T_g) is high and the fractional free volume is an exceptional 30% or so, meaning that one-third of the polymer is unoccupied space. When it was first made, PTMSP had the lowest density of any known polymer. Not unexpectedly, the gas permeation properties are anomalous.

Since PTMSP was first made, a large number of similar PIM materials, also known as superglassy polymers, have been synthesized [20–22]. PTMSP and the other two examples shown in Figure 3.9 have all been widely studied. Because of their exceptional free volume, PIMs as a class are all extremely permeable. Some representative data for PTMSP and PIM-1 are shown in Table 3.3. The permeation properties of polysulfone, a conventional glassy polymer, are included for comparison.

As can be seen, gas permeabilities in PIM membranes are orders of magnitude higher than those of conventional glassy polymers, and are also substantially higher than high permeability rubbery polymers such as poly(dimethyl siloxane), for many years the most permeable polymer

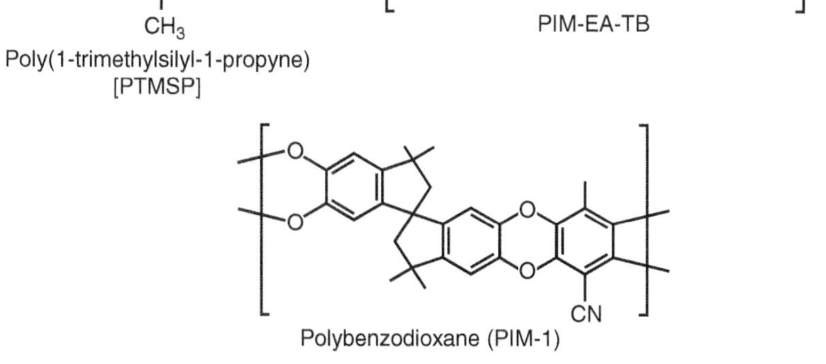

Figure 3.9   *The structures of three PIMs. Extremely stiff-backboned, rigid polymer chains pack poorly, leading to high fractional free volume.*

Table 3.3   *A comparison of the properties of two PIMs (PTMSP and PIM-1) with a high free volume conventional glassy polymer (polysulfone) and silicone rubber (polydimethyl siloxane), an exceptionally permeable rubbery polymer.*

Polymer	Free volume (%)	BET surface area (m^2/g)	N_2 permeability (Barrer)	Selectivity O_2/N_2
PTMSP	29	550	6300	1.5
PIM-1	25	830	340	3.8
Polysulfone	18	<1	0.25	5.6
Polydimethyl siloxane	16	<1	220	2.2

known. The extremely high free volume provides a sorption capacity into the pores as much as 10 times that of a conventional glassy polymer. Diffusion coefficients are in the range 10^3–10^6 cm^2/s, much greater than those observed in conventional dense materials. The combination of extremely high sorption and extremely high diffusion indicates that the gas transport mechanism is overwhelmingly pore flow, not solution-diffusion.

The finely nanoporous nature of PIMs is also demonstrated by the BET surface area measurements listed in Table 3.3. A conventional glassy polymer like polysulfone shows no evidence of any internal nanoporous structure; consequently, the BET surface area is very small. In contrast, PIMs have internal BET surface areas of 500–1000 m^2/g of polymer. This large internal surface area is comparable to that of carbon black, suggesting that the internal PIMs structure is similar to carbon black and like adsorbents. Further evidence of the microporous structure is that PIMs exhibit pore-blocking similar to that of microporous ceramic membranes when challenged with mixtures of condensable and non-condensable gases.

Although PIM membranes have been widely studied in the laboratory, they have found no industrial application. The main reason is physical aging, manifest by a rapid loss of most of their permeability over a matter of weeks or even days. Aging comes about because the polymer chains are not completely immobilized, enabling the material to gradually densify, eliminating most of the free volume through which gas transport takes place. Loss of free volume also occurs in conventional glassy polymers, but over a much longer time. A comparison of the permeability and stability of PTMSP and PIM-1 with polymers currently used in industrial processes is shown in Chapter 2, Figure 2.27. To date all attempts to stabilize these polymers by increasing their chain rigidity or cross-linking have failed.

A comparison of the fractional pore volume and internal surface area of some representative MOFs, zeolites, and PIMs is shown in Figure 3.10 [23]. In general, MOFs have higher free volumes and internal surface areas than zeolites or PIMs. The implication is that MOF membranes have higher permeabilities than zeolites or PIMs, and this is likely true. But all these materials have high enough permeabilities to be useful membranes. The key parameters hindering their use are ease of fabrication, physical and chemical stability, and the high cost of the finished membrane modules.

3.2.3.2   *Thermally Rearranged/Microporous Carbon Membranes*

The first microporous carbon membranes were produced by Barrer in the 1950s and 1960s by compressing high-surface-area carbon powders at very high pressures [24, 25]. The resulting porous plugs had pores of 10–30 Å diameter and were used to study diffusion of gases and vapors. Later, Koresh and Soffer produced microporous hollow fiber carbon membranes by pyrolyzing polyacrylonitrile or polyimide membranes in an inert atmosphere at 500–800 °C [26].

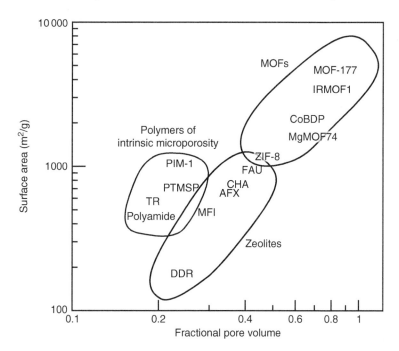

Figure 3.10   *Fractional pore volume and internal surface area of representative MOFs, zeolites and polymers of intrinsic microporosity (PIMs). Source: Adapted from [23].*

This technology was brought to the small module stage but failed commercially because the membranes were brittle and difficult to make defect-free on a large scale. However, because of their exceptional separation properties for some gas mixtures, microporous carbon membranes remain an active area of research for gas separation and some pervaporation applications, with groups at work in Japan [27–29], Korea [30], and the United States [31–33].

The general procedure used to convert a polymer to a microporous carbon membrane is shown in Figure 3.11. A wide variety of precursor polymers can be used. Polyacrylonitrile, poly(vinylidene chloride), and poly(furfuryl alcohol) are easily carbonized and were widely used in earlier work, but more recent work favors polyimides. As the precursor membrane is heated, there is a gradual loss of weight. Most polymers lose 10–20 wt% by the time the polymer has been heated to 300–500 °C. At this point, the polymer becomes yellow to brown. Heating at higher temperatures produces more weight loss; most polymers lose their hetero atoms at 600–1000 °C, and the membrane becomes black and denser. The permeance falls, but selectivity increases. Some results of Kita et al. [29] that illustrate this point are shown in Figure 3.12. Kita's base polymer membrane was a thin phenolic resin layer coated onto a microporous ceramic support. After heating to 500 °C, the membrane exhibits a hydrogen permeance of about 500 gpu, with H_2/CO_2 selectivity of only 3–4. After heating to 800 °C, the hydrogen permeance drops 100-fold, but the H_2/CO_2 selectivity rises to more than 20. The permeation properties of carbonized membranes are sometimes adjusted by a post-treatment step in which the membrane is heated in air or oxygen at 100–300 °C. A further weight loss occurs and the membrane appears to become a little more microporous. The permeance increases, but selectivity usually decreases.

Rao and Sirkar, at Air Products, made finely porous carbon membranes to separate light hydrocarbons from hydrogen [31]. The membranes had good C_1–C_4/H_2 selectivities, and could,

Preparation of carbon membranes

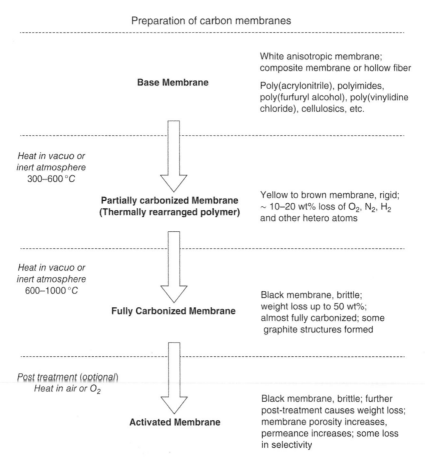

Base Membrane

White anisotropic membrane;
composite membrane or hollow fiber

Poly(acrylonitrile), polyimides,
poly(furfuryl alcohol), poly(vinylidine
chloride), cellulosics, etc.

*Heat in vacuo or
inert atmosphere
300–600 °C*

**Partially carbonized Membrane
(Thermally rearranged polymer)**

Yellow to brown membrane, rigid;
~ 10–20 wt% loss of O_2, N_2, H_2
and other hetero atoms

*Heat in vacuo or
inert atmosphere
600–1000 °C*

Fully Carbonized Membrane

Black membrane, brittle;
weight loss up to 50 wt%;
almost fully carbonized; some
graphite structures formed

*Post treatment (optional)
Heat in air or O_2*

Activated Membrane

Black membrane, brittle; further
post-treatment causes weight loss;
membrane porosity increases,
permeance increases; some loss
in selectivity

Figure 3.11 *Typical changes during the preparation of a microporous carbon membrane by pyrolysis.*

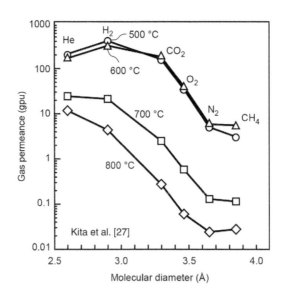

Figure 3.12 *Changes in the gas permeances of carbon membranes made from phenolic resin heated to different carbonizing temperatures. Source: Adapted from [29].*

in principle, be used to upgrade many refinery hydrogen, light hydrocarbon-containing gas streams otherwise usable only for fuel. Unfortunately, small amounts of C_{6+} hydrocarbons present in the feed gas were strongly, even permanently, adsorbed and accumulated in the membrane. The result was a steady loss of membrane permeance. Others have noticed similar problems. Koros, for example, reported that as little as a few ppm of C_{6+} hydrocarbons decreased microporous carbon membrane permeability by over 60% in less than a day. The same effect was noticed with low concentrations of water. The best target application for carbon membranes appears to be separation of olefin/paraffin mixtures, an expensive energy consuming separation to do by distillation, and where the high value of the products may justify the cost of pretreatment to remove trace contaminants.

3.3 Gas Separation: Transport Mechanisms

3.3.1 Surface Adsorption and Diffusion

An illustration of the type of porous structure used to model microporous membranes is shown in Figure 3.13. The membrane pores form a three-dimensional network of interconnected cavities. The average diameter varies; most membranes of interest have cavities with pore diameters in the 5–25 Å range. Typically, the pore volume represents 30–50% of the total membrane volume. An important characteristic of these materials is their very large internal surface area.

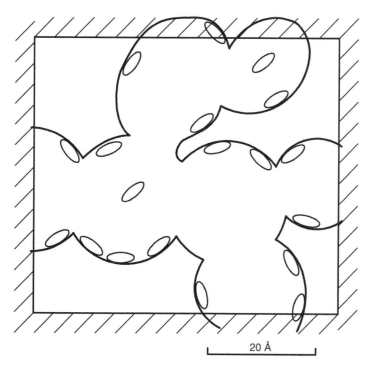

Figure 3.13   *An illustration of surface area gas adsorption in a microporous membrane formed from a network of approximately 20 Å diameter pores. A few molecules are in the free space of the pores, but most are adsorbed on the pore walls.*

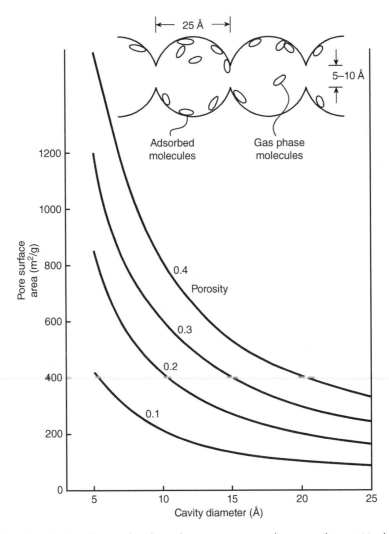

Figure 3.14   *The calculated internal surface of microporous membranes with porosities between 0.1 and 0.4 and average pore diameters between 5 and 25 Å.*

A set of simple calculations showing the internal surface area as a function of pore diameter and membrane porosity is given in Figure 3.14. The pores are assumed to be a series of linked spheres similar to those of Figure 3.13. The surface area in m^2/g of material is inversely proportional to the average pore radius. The figure shows that membranes with pores smaller than 25 Å in diameter have an internal surface area in the 100–1000 m^2/g range. These calculations are consistent with surface area measurements made by the BET method. Measurements of this type are shown in Table 3.4.

The BET surface areas of the microporous materials are in the 500–2000 m^2/g range. In contrast, the BET area of a typical glassy polymer, polysulfone, is less than 1 m^2/g. The high surface area of the membrane results in high gas adsorption values, which determine the membrane permeation properties. PIM membranes are made as cast polymer films and are visually essentially identical to solution-diffusion membranes, but a high BET surface area measurement is a clear distinguishing marker of their microporous nature.

Table 3.4 Comparison between surface area and pore volume for microporous and dense membranes.

Membrane type	Membrane material	Free volume (porosity) (cm^3/cm^3)	BET surface area (m^2/g)
PIM	PTMSP	0.30	550
PIM	PIM-1	0.25	830
MOF	ZIF-8	0.48	1800
TR polymer	Carbonized polyimide (TR-5)	0.19	710
Dense polymer (transport by solution-diffusion)	Polysulfone	0.18	<1

Figure 3.15 shows the adsorption of a series of light gases on a high surface area microporous activated carbon sample with a BET area of about 1200 m^2/g [34]. These adsorption isotherms are typical of other high surface area microporous materials. The adsorption of hydrogen is low, at atmospheric pressure only about 0.5 cm^3/g, which is approximately the same as the porosity of

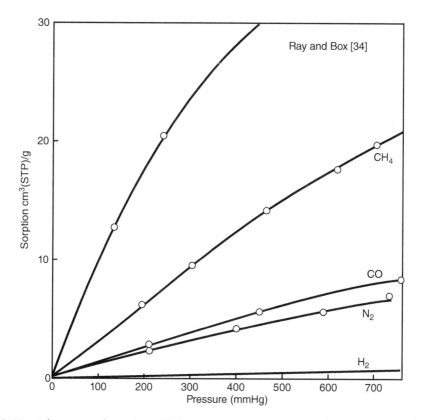

Figure 3.15 Adsorption of a series of light gases on microporous activated carbon with a BET surface area of about 1150 m^2/g. Source: Adapted from [34].

the material. In other words, essentially all of the hydrogen gas sorbed into the material remains as free gas in the open pores, with very little adsorbed onto the pore walls. This is common to different porous materials, enabling hydrogen (or helium, which has similar properties) to be used as a marker against which the relative sorption of other gases in the material can be measured. The difference between the hydrogen sorption and the gas sorption indicates the proportion of gas adsorbed on the pore surfaces. The figure shows that adsorption of the light gases N_2, CO, CH_4, and CO_2 is far higher than hydrogen. It follows that, with these gases, there are 5–50 molecules of adsorbed gas for every molecule in the gas-phase pore space. Permeation through microporous membranes thus takes place by a combination of rapid but numerically sparse gas-phase diffusion (J_g) and slower but numerically rich surface diffusion (J_s). The total flow (J_t) through microporous membranes can then be written as

$$J_t = J_g + J_s \tag{3.1}$$

3.3.2   Knudsen Diffusion

The gas phase contribution to permeation is linked to Knudsen diffusion. At atmospheric pressure, the mean free path, γ, of common gases is in the range 500–2000 Å. This means that within small pore membranes, the ratio of the pore radius, r, to the gas mean free path (r/γ) is much less than one. Diffusing gas molecules then have far more collisions with the pore walls then with other gas molecules. At every collision with the pore walls, the gas molecules are momentarily adsorbed and then reflected in a random direction. Molecule–molecule collisions are rare, so each gas molecule moves independently of all others. Hence, with gas mixtures in which the different species move at different average velocities, a separation is possible. For Knudsen diffusion, the gas flow in a membrane assumed to be made of cylindrical right capillaries is given by the equation

$$j_g = \frac{4r\varepsilon}{3} \cdot \left(\frac{2RT}{\pi m}\right)^{\frac{1}{2}} \cdot \frac{p_o - p_\ell}{\ell \cdot RT} \tag{3.2}$$

where m is the molecular weight of the gas, j_g is the flux in $gmol/cm^2 \cdot s$, ε is the porosity of the membrane, r is the pore radius, ℓ is the pore length, and p_o and p_ℓ are the absolute pressures of the gas species at the beginning of the pore ($x = 0$) and at the end ($x = \ell$).

It follows from Eq. (3.2) that the permeability of a membrane where a gas i flows by Knudsen diffusion is proportional to $1/\sqrt{m_i}$. The selectivity of this membrane $\alpha_{i/j}$, is given by the expression

$$a_{i/j} = \sqrt{\frac{m_j}{m_i}} \tag{3.3}$$

This result was first observed experimentally by Graham and is called Graham's law of diffusion. Knudsen diffusion membranes have been used to separate gas isotopes that are difficult to separate by other methods; for example, tritium from hydrogen, $C^{12}H_4$ from $C^{14}H_4$ and most importantly $U^{235}F_6$ from $U^{238}F_6$. The membrane selectivity for uranium mixtures is only 1.0043, so hundreds of separation stages are required to produce a complete separation. Nevertheless, at the height of the cold war, the US Atomic Energy Commission operated three plants fitted with microporous metal membranes that processed almost 20 000 tons per year of uranium.

In membranes with pores larger than 30–40 Å, the internal membrane surface area is low and the amount of adsorbed gas is small. Knudsen diffusion is then a significant component of the transport mechanism, especially for light gases. Some typical results illustrating the balance between surface diffusion and gas phase (Knudsen) diffusion, normalized relative to helium, are shown in Figure 3.16 [25, 35]. The light gases (H_2, He, N_2, O_2, CH_4) all have limited sorption on the pore surface, so the bulk of permeation is gas phase Knudsen diffusion. As the condensability of the gas increases (as measured by boiling point or critical temperature), the amount of surface adsorption increases and the contribution of surface diffusion to gas permeation increases. For butane, for example, 80% of the total gas permeation is due to surface diffusion.

The pores in the Vycor membrane are in the 50 Å diameter range; the carbon membrane has pores in the 40 Å range. This means the internal surface area of both these membranes is relatively small, at between 100 and 300 m^2/g. Because of the limited adsorption by the membrane, surface diffusion and Knudsen diffusion both contribute to permeation.

Smaller pore membranes, especially membranes with pores below 25 Å, are more commonly used for gas separation applications. In these membranes, permeation due to surface diffusion is much larger than the Knudsen diffusion term for all gases except hydrogen and helium. The gas phase contribution to permeation can then be ignored and the membrane flux can be written

$$J_s = D_s \cdot \frac{dc_s}{dx} \tag{3.4}$$

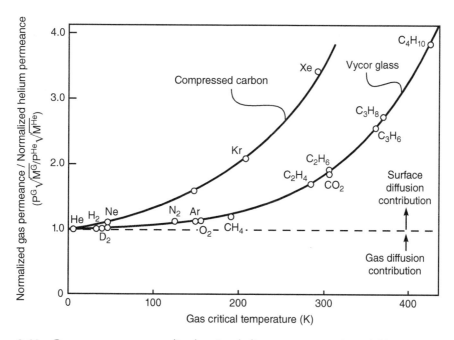

Figure 3.16   *Gas permeances, normalized against helium permeance, through Vycor microporous glass membranes [35] and compressed activated carbon membranes [36]. Since helium is essentially non-adsorbed onto the pore surfaces, the portion below the helium marker line represents gas phase Knudsen diffusion, and the portion above represents the extra contribution of surface diffusion.*

where c_s is the adsorbed gas concentration cm^3(STP)/cm^3 of membrane material and D_s is the surface diffusion coefficient. A concentration gradient of adsorbed gas is created along the pore surface, and adsorbed molecules move from one adjacent site to another. Surface diffusion coefficients are generally between 1×10^{-4} and 1×10^{-2} cm^2/s, intermediate between diffusion coefficients in a gas and a liquid. The nature of the adsorption surface seems important; energetically heterogeneous adsorption surfaces have lower diffusion coefficients than more homogeneous surfaces, but the process is not well understood. Much of the literature data is from the 1960s [24, 25, 36], and the subject has only been studied sporadically since then [37–40].

Although almost all gas molecules are adsorbed onto the membrane surface in membranes with pores less than 10 Å in diameter, surface diffusion is only one of the factors controlling membrane permeation. Two others are molecular sieving and pore blocking. These are important, particularly when such membranes are being used to separate gas mixtures.

3.3.3 Molecular Sieving

When the pore diameter of a microporous membrane is less than 10 Å, the pores begin to separate gases by molecular sieving. Molecular sieving occurs based on the relative sizes of molecules in the feed gas and their relative abilities to squeeze through pores that are not much larger than the permeating molecule. Data illustrating the effect of molecular sieving obtained with a microporous ZIF-8 membrane are shown in Figure 3.17 [9]. The permeances of pure gases through a 2.5 μm thick ZIF-8 membrane are shown plotted against the Lennard–Jones collision diameter of the gas molecule. Hydrogen, the smallest gas, has the highest permeance. Permeance

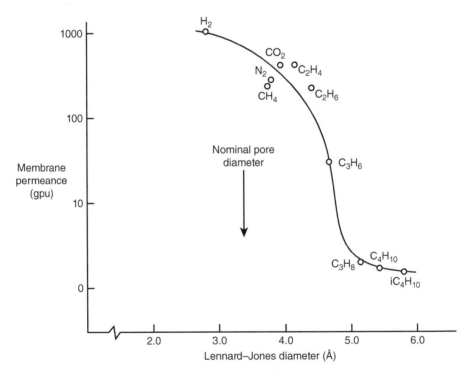

Figure 3.17   *Pure gas permeance through a microporous ZIF-8 metal organic framework membrane [9]. A sharp decline in permeance occurs when the molecular diameter approaches 4.5 Å.*

declines to about one third of hydrogen permeance as the gas molecules get larger: N_2, CH_4, CO_2, and ethane, for example, with molecular diameters of 3.8–4.4 Å. Further increases in permeant size produce a drastic reduction in permeance. Propylene is only 0.2 Å larger in diameter than ethane, but has 1/8 of its permeance, and propane, 0.4 Å larger than ethane, has 1/100 of its permeance.

The sudden reduction in permeance above a critical diameter of about 4.5 Å is thought to be a molecular sieving effect. The ZIF-8 crystal structure consists of 11.6 Å diameter cavities linked together by 3.4 Å diameter windows, which would suggest that any molecule of diameter greater than 3.4 Å would be blocked from permeating the lattice. Based on the data shown in Figure 3.17, the gas diameter cut off seems to be closer to 4.5 Å. This difference has been seen in other studies [41]. The discrepancy may be due in part to the difference between the effective gas kinetic diameter and the calculated Lennard–Jones diameter of the gas. But it also appears that the organic imidazole linkages that form the framework of the ZIF-8 crystals have some flexibility and can change their configuration enough to enable molecules larger than the nominal lattice diameter to pass.

Pure gas data of the type shown in Figure 3.17 have been measured with a number of PIM and TR polymers. The cutoff curves are seldom as clear cut as the ZIF-8 data, but there is good evidence for a significant molecular sieving contribution to permeation. Based on their pure gas properties, some of these membranes should be candidates for important separation problems – ethylene from ethane or propylene from propane, for example. Unfortunately, pure gas data are very poor predictors of mixed gas separation performance, often because of a phenomenon called pore blocking.

3.3.4 Pore Blocking

The ability of a microporous membrane to separate a gas mixture is difficult to estimate from pure gas data. The problem is pore blocking, illustrated in Figure 3.18. Pore blocking was first

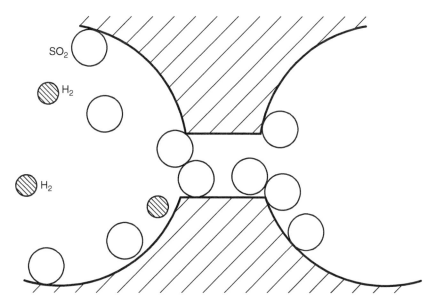

Figure 3.18   *An illustration of pore blocking. Pure gas measurements show that hydrogen is more permeable than SO_2. With gas mixtures, adsorbed SO_2 fills membrane constrictions and prevents hydrogen permeation, so that the membrane is selective for SO_2 over hydrogen.*

noticed in the 1960s by Ash et al., who measured permeation of sulfur dioxide/hydrogen mixtures through compressed carbon plugs [24]. The plugs had an average pore diameter in the 40 Å range, and were expected to be hydrogen selective, based on pure gas data. Even at low sulfur dioxide partial pressures, however, hydrogen permeated at about half the flux expected from pure gas measurements; at higher sulfur dioxide partial pressures (although still well below the sulfur dioxide saturation pressure), the hydrogen flux decreased sharply and a point was reached where no measurable permeation of hydrogen occurred. This decrease is shown in Figure 3.19. The cause is blocking of constrictions in the porous network by the adsorbed sulfur dioxide molecules. The sulfur dioxide could still permeate, but hydrogen was essentially excluded.

The effect of pore blocking is apparent in relatively large pore membranes such as those used by Ash et al. but is even more noticeable in smaller pore membranes. In zeolite or MOF membranes, pore blocking can completely transform the separation performance. For example, Table 3.5 shows the effect of pore blocking on the separation of hydrogen/butane mixtures in an MFI zeolite [42]. The smallest pore in the crystal structure is about 5.5 Å in diameter, enough to allow free permeance of hydrogen and partially hindered butane permeation. Flux data for both pure and mixed gases are shown. The pure gas data indicates that the membrane should be slightly hydrogen selective. However, in the presence of butane, the hydrogen flux decreases

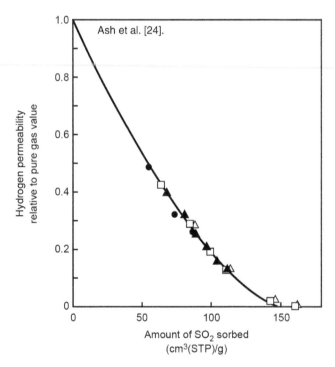

Figure 3.19 *Blocking of hydrogen in hydrogen/sulfur dioxide gas mixture permeation experiments with finely microporous carbon membranes as a function of the amount of sulfur dioxide adsorbed by the membrane. As sulfur dioxide sorption increases the hydrogen permeability is reduced until at SO_2 sorption of about 140 cm^3(STP)/g, the membrane is completely blocked and only sulfur dioxide permeates. Data obtained at several temperatures fall on the same master curve (● at 0 °C; ▲ at −10 °C; □ at −20.7 °C; △ at −33.6 °C). Source: Reproduced with permission from [24]. Copyright (1963) John Wiley & Sons.*

Table 3.5 *Gas permeation of hydrogen and n-butane at 25 °C in MFI zeolite membranes: pure gas data taken at 50 kPa, 50/50 mixed gas at 100 kPa, both from [42].*

Pure gas flux (m · mol/m² · s)		Mixed gas flux (m · mol/m² · s)	
Hydrogen	*n*-Butane	Hydrogen	*n*-Butane
5.0	2.5	0.1	1.9

50-fold. The butane flux also decreases, but only by a fraction, so the net result is to make the membrane strongly selective for butane over hydrogen. This type of result is best rationalized by a pore blocking mechanism.

Similar results are found in most other molecular sieving materials when treated with mixtures of condensable and less-condensable gases. Figure 3.20 shows another example of the phenomenon, in this case illustrated by data from hydrogen/propane mixed gas experiments using a membrane made from PTMSP, a PIM [20].

The pure gas permeabilities of propane and hydrogen are similar, so the PTMSP membrane would appear to have almost no selectivity for propane/hydrogen mixtures. The results when tested with gas mixtures are very different. As the fraction of propane in the propane/hydrogen feed gas is increased, the hydrogen permeability falls and the membrane becomes increasingly propane selective, reaching a propane/hydrogen selectivity of 25 at high propane partial pressures.

The condensable component causing pore blocking need not be present at a high concentration. Process developers using small-pore microporous membranes have shown they are susceptible to large flux declines from even small amounts of water vapor or heavy hydrocarbons. An extreme example of this effect is shown in Figure 3.21, which shows permeation of nitrogen through a finely microporous PTMSP membrane. The initial nitrogen flux is high and stable, but when just

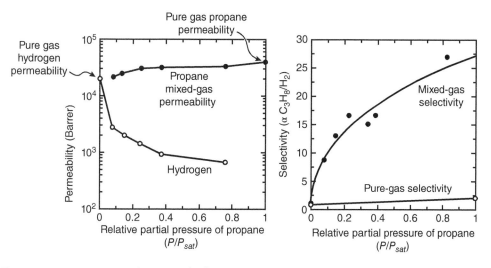

Figure 3.20 *Permeation of propane/hydrogen gas mixtures through PTMSP membranes [20]. The pure gas permeabilities of propane and hydrogen in this membrane are similar, but in gas mixtures, propane adsorption blocks hydrogen passage. As the content of propane in the mix increases, the membrane becomes increasingly propane selective.*

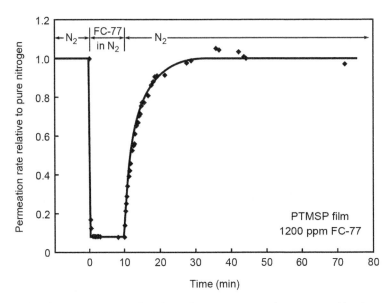

Figure 3.21   *The change in nitrogen flux through a PTMSP membrane caused by the presence of a condensable vapor in the feed gas. This behavior is characteristic of extremely finely microporous membranes. The condensable vapor adsorbs in the 5–15 Å diameter pores of the membrane, blocking the flow of the noncondensable nitrogen gas.*

1200 ppm of FC-77 (a condensable perfluoro octane/perfluoro decane mixture) is added, the nitrogen flux drops 20-fold. When the condensable vapor is removed from the feed gas, the nitrogen flux returns to its original value. The large change in permeation properties produced by minor feed gas components is hard to control and is one of the reasons that these membranes have yet to be used in industrial processes.

3.3.5   Summary

As described above, microporous membranes have a number of characteristic features that clearly distinguish their behavior from the solution-diffusion behavior described in Chapter 2. First, they have high free volumes, between 0.2 and 0.4. Some exceptional glassy polymer membranes have free volumes that overlap the lower end of this range, but the free volume elements are distributed as unconnected spaces between polymer chains, whereas in microporous membranes the elements form an interconnected porous network. From this arise the exceptionally high permeabilities that are the second characteristic feature. The third characteristic feature is the internal surface area, between 200 and 2000 m²/g, which is orders of magnitude higher than that of high-free-volume glassy polymers.

Finally, the phenomenon of blocking occurs only in materials having interconnected pores. If a mixture of a condensable gas and a non-condensable gas has a selectivity different than the pure gas data suggest and the mixed gas selectivity is primarily due to a reduction in the non-condensable gas permeance, the cause is pore blocking and the membrane is almost certainly microporous.

To date, work on the development of microporous membranes for use in gas separation applications has been simultaneously tantalizing and a disappointment. A significant part of the problem is that the multiple mechanisms of molecular transport are complex, overlapping, and hard to untangle. No satisfactory overall permeation theory that could guide microporous membrane development work exists. An additional problem is that much of the published data has been obtained with pure gases. Of course, this is necessary for a better understanding of these interesting materials. However, care needs to be exercised in drawing conclusions about their usability. When pure gas data are extrapolated to predict the expected separation of gas mixtures, the results have generally been overly optimistic and sometimes contrary to expectations. Whether a better understanding of the phenomena at play will lead to industrially useful gas permeation membranes remains unknown.

3.4   Liquid Permeation in Microporous Membranes

The use of finely porous membranes to separate suspended materials and dissolved solutes, such as proteins, colloids, viruses, and bacteria, from aqueous solutions has a long history, dating to the first use of membranes in biochemistry labs of the 1920s. Since then, membranes with smaller pores have been developed, and these are able to separate microsolutes, such as dyes or oligomers in the 200–1000 MW range, from aqueous and organic solvent solutions. The general category of microporous membranes used for liquid separations is broad, covering membranes with pore diameters from as small as 20–30 Å for the finest ultrafiltration membranes to the micron range for microfiltration. The mechanism of separation in all cases is filtration, with the larger suspended material or solute retained by the membrane and the smaller purified solvent being recovered on the permeate side. As discussed in the relevant chapters, the distinction between ultrafiltration and microfiltration is a loose one, and it is often (and certainly for the purpose of this chapter) more useful to characterize the membranes, their modes of operation and their applications in terms of whether the filter operates by screen filtration or depth filtration.

Methods of making microporous membranes for liquid separations are described in detail in Chapter 4. Membranes with pores smaller than about 100 Å are generally made by a variant of the Loeb–Sourirajan process. The membranes are anisotropic and the selective separating layer is only a few microns thick, formed at the surface facing the feed solution. These membranes are often called screen filters and separate by a molecular filtration mechanism.

Larger pore membranes with pores between 100 and 500 Å in diameter are usually isotropic; although some contaminants may be filtered out on the top surface, other components are caught at the pore constrictions in the interior of the membrane. These are called depth filters. The two types of membrane are illustrated in Figure 3.22.

3.4.1   Screen Filters

The mechanism of particle filtration by screen filters is relatively easily described mathematically and has been the subject of many studies; Wang and Lin have published a review of this work [43]. Ferry [44] was the first to model membrane retention by a screen filter. In his model, pores were assumed to be equal circular capillaries with a radius, r, which is large compared to the solvent molecule radius, such that the total area of the pore is available for transport of solvent. A solute molecule whose radius, a, is an appreciable fraction of the pore radius cannot approach nearer than one molecular radius to the pore wall, and thus a much smaller area is available for solute transport. The model is illustrated in Figure 3.23.

The area, A, of the pore available for solute transport is given by the equation

$$\frac{A}{A_o} = \frac{(r-a)^2}{r^2} \qquad (3.5)$$

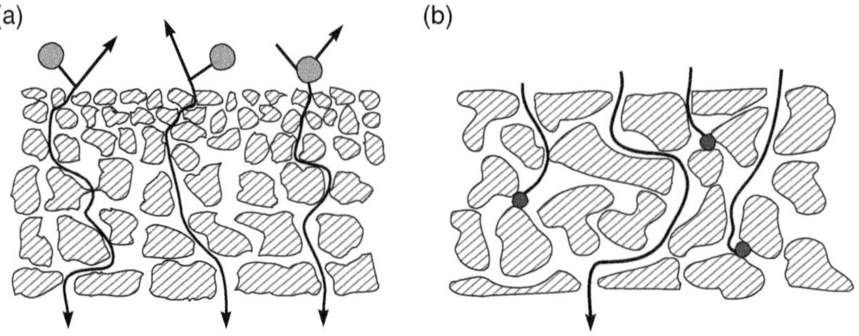

(a)   (b)

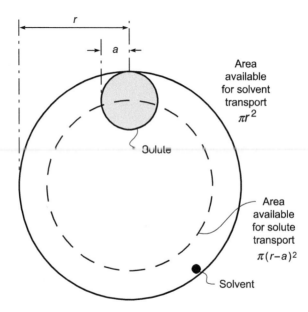

Figure 3.22 *Separation of particulates can take place (a) at the membrane surface by screen filtration or (b) by capture in the interior of the membrane by depth filtration.*

Figure 3.23 *The Ferry mechanical exclusion model of solute transport in small pores. Source: Adapted from [44].*

where A_o is the area of the pore available for solvent molecules. Later, Renkin [45] showed that Eq. (3.5) must be modified to account for the parabolic velocity profile of the fluid as it passes through the pore. The effective fractional pore area available for solutes in this case is

$$\left(\frac{A}{A_o}\right)' = 2\left(1 - \frac{a}{r}\right)^2 - \left(1 - \frac{a}{r}\right)^4 \tag{3.6}$$

where $(A/A_o)'$ is equal to the ratio of the solute concentration in the filtrate (c_ℓ) to the concentration in the feed (c_o), that is,

$$\left(\frac{A}{A_o}\right)' = \left(\frac{c_\ell}{c_o}\right) \tag{3.7}$$

It follows from Eq. (3.7) and the definition of solution rejection

$$\mathbf{R} = \left(\frac{c_o - c_\ell}{c_o}\right) \cdot 100\% \tag{3.8}$$

that the rejection of an ultrafiltration membrane is

$$\mathbf{R} = \left[1 - 2\left(1 - \frac{a}{r}\right)^2 + \left(1 - \frac{a}{r}\right)^4\right] \cdot 100\% \tag{3.9}$$

Equation (3.9) is known as the Ferry–Renkin equation, and can be used to estimate the pore size of ultrafiltration membranes from the rejection of a solute of known radius. Figure 3.24a shows curves calculated this way for membranes having nominal pore diameters of 50, 100

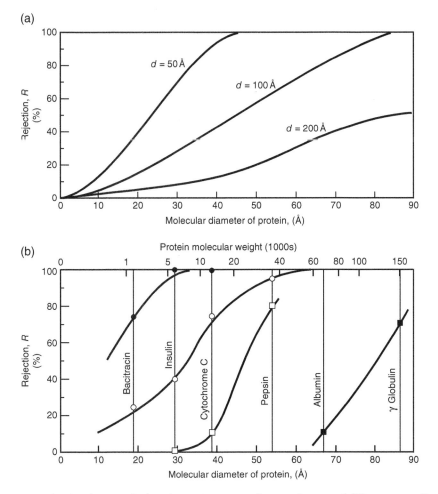

Figure 3.24 *Plot (a) shows calculated rejection curves for membranes of different pore diameters obtained by using the Ferry–Renkin Eq. (3.9). Plot (b) shows experimental rejection data for ultrafiltration obtained using solutions of proteins of known molecular diameter. The approximate pore diameter of the membranes can be obtained by comparing the two sets of curves.*

and 200 Å. Figure 3.24b shows experimental rejection results obtained with four membranes of increasing pore size. The rejection of a series of protein solutions is plotted against the protein molecular diameter determined by other characterization techniques. By comparing the experimental data with the theoretical curves using Eq. (3.9), an estimate of the pore diameter of the test membranes is possible. It appears that the most retentive membrane has a pore diameter of about 50 Å, while the least retentive membrane has a pore diameter of about 200 Å [46].

Globular proteins are usually the basis for this work because their molecular weights and molecular diameters are known. A list of some commonly used marker molecules is given in Table 3.6.

The cutoff of ultrafiltration membranes is usually characterized by solute molecular weight, but several other factors affect permeation. One important example is the shape of the molecule to be retained. When membrane retention measurements are performed with linear, water-soluble molecules such as polydextran, polyethylene glycol, or poly(vinyl pyrrolidone), the measured rejection is usually much lower than the rejection measured for proteins of the same molecular weight. It is believed that linear, water-soluble polymer molecules are able to snake through the membrane pores as illustrated in Figure 3.25. Protein molecules, however, exist in solutions as wound globular coils held together by hydrogen bonds. These globular molecules cannot deform to pass through the membrane pores and are therefore rejected. Some results showing the rejection of different molecules for a polysulfone ultrafiltration membrane are listed in the table accompanying Figure 3.25 [46]. The membrane shows significant rejection to globular protein molecules as small as pepsin (MW 35 000) and cytochrome c (MW 13 000), but is completely permeable to a flexible linear polydextran, with an average molecular weight of more than 100 000.

Another factor that affects permeation through ultrafiltration membranes, particularly with polyelectrolytes, is pH. For example, polyacrylic acid is usually very well rejected at pH 5 and above, but passes unhindered through the same membrane at pH 3 and below. The change in behavior is due to a change in configuration of the polyacid. At pH 5 and above, polyacrylic acid is ionized. In the ionized form, the negatively charged carboxyl groups along the polymer backbone repel each other; the polymer coil is then extended and relatively inflexible. In this form, the molecule cannot readily permeate the small pores of an ultrafiltration membrane. At pH 3 and below, the carboxyl groups along the polyacrylic acid polymer backbone are all

Table 3.6   *Marker molecules used to characterize ultrafiltration membranes.*

Species	Molecular weights (×1000)	Estimated molecular diameter (Å)
Sucrose	0.34	11
Raffinose	0.59	13
Vitamin B_{12}	1.36	17
Bacitracin	1.41	17
Insulin	5.7	27
Cytochrome C	13.4	38
Myoglobin	17	40
α-Chymotrysinogen	25	46
Pepsin	35	50
Ovalbumin	43	56
Bovine albumin	67	64
Aldolase	142	82
γ-Globulin	150	84

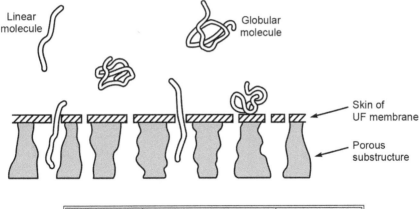

	Globular proteins		Linear polymer
Solute	Pepsin	Cytochrome *c*	Polydextran
MW (1000s)	35	13	100
Rejection (%)	90	70	0

Figure 3.25 *Ultrafiltration membranes are rated on the basis of nominal molecular weight cutoff, but the shape of the molecule to be retained has a major effect on retentivity. Linear molecules pass through a membrane, whereas globular molecules of the same molecular weight may be retained. The table shows typical results obtained with globular protein molecules and linear polydextran for the same polysulfone membrane. Source: Adapted from [46].*

protonated. The resulting neutral molecule is much more flexible and can pass through the membrane pores.

3.4.2 Depth Filters

The mechanism of particle capture by depth filtration is more complex than for screen filtration. Simple capture of particles by sieving at pore constrictions occurs, but adsorption of particles on the interior surface of the membrane is usually at least as important. Figure 3.26 shows four mechanisms that contribute to particle capture in depth membrane filters. The first and most obvious mechanism, simple sieving and capture of particles at constrictions in the membrane, is often a minor contributor to the total separation. The three other mechanisms are inertial capture, Brownian diffusion, and electrostatic adsorption [47, 48]. In all three cases, particles smaller than the diameter of the pore are captured by adsorption onto the internal surface of the membrane.

In inertial capture, particles in the flowing liquid cannot follow the fluid flow lines through tortuous pores. As a result, such particles are captured as they impact the pore wall. This capture mechanism does not affect tiny particles, which can be easily swept along with the solvent flow, but becomes more of an issue as the particle size increases, even though the particle may be much smaller than the pore diameter. In experiments with colloidal gold particles and depth filtration membranes with tortuous pores approximately 5 μm in diameter, Davis showed that 60% of 0.05 μm-diameter particles were captured [49]. In contrast, nucleation track membranes with 5 μm, almost straight-through pores (essentially no tortuosity) retained less than 1% of the particles.

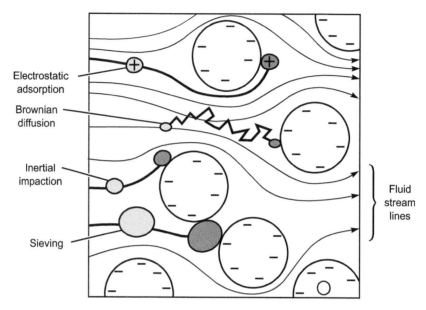

Figure 3.26 *Particle capture in liquid filtration by depth microfilters. Four capture mechanisms are shown: simple sieving; inertial impaction; Brownian diffusion and electrostatic adsorption.*

The third mechanism, capture by Brownian diffusion, mostly affects very small particles. Although easily carried along by the moving fluid, such particles are subject to Brownian motion, which from time to time brings them into direct collision contact with the pore walls, where they are captured by surface adsorption. A neat experimental observation that contrasts these two effects, albeit in the gas phase, is shown in Figure 3.27 [50]. The paradoxical result is that the most penetrating particles are those of intermediate diameter, which are relatively less affected than larger or smaller counterparts by either Brownian diffusion or inertial capture.

The final mechanism of capture in Figure 3.26 is electrostatic adsorption. This affects charged particles when the membranes also carry surface charges. Many common colloidal materials carry a slight negative charge, so membranes containing an excess of positive groups on the pore surfaces can provide enhanced removal. Several microfiltration membrane manufacturers produce this type of charged membrane. One problem is that the adsorption capacity of the charged groups is exhausted as filtration proceeds, and the retention then falls.

3.5   Conclusions and Future Directions

For many years, the principal applications of microporous membranes were ultrafiltration and microfiltration. The separation mechanism was clear and simple. In the last two decades, there has been a surge of interest in using such membranes, in their most finely porous embodiments, for gas separation. Pure gas permeation measurements show that these membranes are much more permeable and sometimes more selective than polymeric dense solution-diffusion membranes. Unfortunately, permeation with gas mixtures is usually very different and results from real world applications have been disappointing. Even small amounts of contaminant gases or other co-permeating components can produce anomalous results.

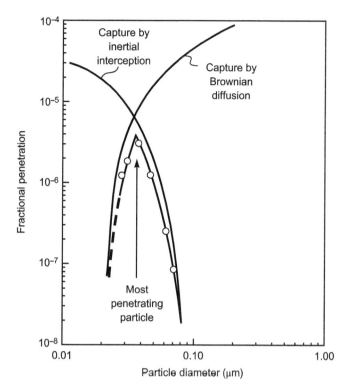

Figure 3.27   *Gas-borne particle penetration through an ultrathin PVDF (poly(vinylidene fluoride)) membrane. The most successful penetrants are those that are relatively unaffected by either Brownian diffusion or inertial capture. Source: Data from [50].*

None of the large numbers of new, finely microporous gas separation membranes reported in the last 15 years has made it to industrial use. This not a materials problem, and further studies of materials variants are unlikely to lead to a better result. Future investigations should focus on using the best-known materials formulated into thin membranes (less than 5 μm thick) and evaluated with gas mixtures representative of industrial gas streams and containing the kinds of contaminants typically found in such streams. Absent such studies these materials are likely to stay in the laboratory.

References

1. Robeson, L.M. (2008). The upper bound revisited. *J. Membr. Sci.* **320**: 390.
2. Park, H.B., Jung, C.H., Lee, Y.M. et al. (2007). Polymers with cavities tuned for fast selective transport of small molecules and ions. *Science* **318**: 254.
3. Ramsay, J.D.F. and Kallus, S. (2000). Zeolite membranes. In: *Membrane Science and Technology*, vol. **6** (ed. N.K. Kanellopoulos), 373–395. Elsevier.
4. Barrer, R.M. (1982). *Hydrothermal Chemistry of Zeolites*. London: Academic Press.
5. Gavalas, G.R. (2006). Zeolite membranes for gas and liquid separations. In: *Materials Science of Membranes for Gas and Vapor Separation* (ed. Y. Yampolskii, I. Pinnau, and B.D. Freeman), 307–336. Chichester, UK: Wiley.

6.   Lai, Z., Bonilla, G., Diaz, I. et al. (2003). Microstructural optimization of a zeolite membrane for organic vapor separation. *Science* **300**: 456.

7.   Sato, K., Sugimoto, K., and Nakane, T. (2008). Synthesis of industrial scale NaY zeolite membranes and ethanol permeating performance in pervaporation and vapor permeation up to 130 °C and 570 kPa. *J. Membr. Sci.* **310** (1, 2): 161–173.

8.   Sato, K., Sugimoto, K., and Nakane, T. (2008). Mass-production of tubular NaY zeolite membranes for industrial purpose and their application to ethanol dehydration by vapor permeation. *J. Membr. Sci.* **319** (1, 2): 244–255.

9.   Pan, Y. and Lai, Z. (2021). Sharp separation of C_2/C_3 hydrocarbon mixtures by zeolitic imidazolate framework (ZIF-8) membranes synthesized in aqueous solutions. *Chem. Commun.* **47**: 10275.

10. Galizia, M., Seok, W., Smith, Z.P. et al. (2017). Polymers and mixed matrix membranes for gas and vapor separation: a review and prospective opportunities. *Macromolecules* **50**: 7809.

11. Zhang, C., Wu, B.-H., Ma, M.-Q. et al. (2019). Ultrathin metal/covalent-organic framework membranes towards ultimate separation. *Chem. Soc. Rev.* **48**: 3811.

12. Qian, Q., Asinger, P.A., Asinger, P.A. et al. (2020). MOF-based membranes for gas separation. *Chem. Rev.* **120**: 8161.

13. Wu, T., Prasetya, N., and Li, K. (2020). Recent advances in aluminum-based metal-organic-frameworks (MOFs) and its membrane applications. *J. Membr. Sci.* **615**: 118493.

14. Kalaj, M., Bentz, K.C., Ayala, S. et al. (2020). MOF-polymer hybrid material: from simple composites to tailored architectures. *Chem. Rev.* **120**: 8267.

15. Chen, L., Reiss, P.S., Chong, S.Y. et al. (2009). Separation of rare gases and chiral molecules by selective binding in porous organic cages. *Nat. Mater.* **8**: 973.

16. Bushell, A.F., Budd, P.M., Attfield, M.P. et al. (2013). Nanoporous organic polymer/cage composite membranes. *Agnew. Chem. Int. Ed.* **52**: 1253.

17. Zhang, Q., Li, H., Chen, S. et al. (2020). Mixed-matrix membranes with solubile organic molecular cage for highly efficient C_3H_6/C_3H_8 separation. *J. Membr. Sci.* **611**: 118288.

18. Vinh-Thang, H. and Kaliaguine, S. (2013). Predictive models for mixed-matrix membrane performance: a review. *Chem. Rev.* **113**: 4980.

19. Zimmerman, C.M., Singh, A., and Koros, W.J. (1997). Tailoring mixed matrix composite membranes for gas separations. *J. Membr. Sci.* **137** (1, 2): 145–154.

20. Pinnau, I., Casillas, C.G., Morisato, A., and Freeman, B.D. (1996). Hydrocarbon/hydrogen mixed gas permeation in poly(1-trimethylsilyl-1-propyne) (PTMSP), poly(1-phenyl-1-propyne) (PPP), and PTMSP/PPP blends. *J. Polym. Sci., Part B Polym. Phys.* **34** (15): 2613–2621.

21. Budd, P.M., Ghanem, B.S., Makhseed, S. et al. (2004). Polymers of intrinsic microporosity (PIMs): robust, solution-processable, organicnanoporous materials. *Chem. Commun.* **2**: 230–231.

22. Carta, M., Croad, M., Malpass-Evans, R. et al. (2014). Triptycene induced enhancement of membrane gas selectivity for microporous Tröger's base polymers. *Adv. Mater.* **26**: 3526.

23. Krishna, R. and van Baten, J.M. (2011). In silico screening of metal-organic frameworks in separation applications. *Phys. Chem. Chem. Phys.* **13**: 10593.

24. Ash, R., Barrer, R.M., and Pope, C.G. (1963). Flow of adsorbable gases and vapours in microporous medium. *Proc. R. Soc. London, Ser. A* **271**: 1–18.

25. Ash, R., Baker, R.W., and Barrer, R.M. (1967). Sorption and surface flow in graphitized carbon membranes. *Proc. Roy. Soc.* **299**: 434.

26. Koresh, J.E. and Sofer, A. (1983). Molecular sieve carbon selective membrane. *Sep. Sci. Technol.* **18**: 723.

27. Morooka, S., Kusakabe, K., and Kusuki, Y. (2000). Microporous carbon membranes. In: *Recent Advances in Gas Separation by Microporous Ceramic Membranes, Membrane Science and Technology*, vol. **6** (ed. N.K. Kanellopoulos), 323–334. Amsterdam: Elsevier Science B.V.

28. Kita, H. (2006). Gas and vapor separation membranes based on carbon membranes. In: *Materials Science of Membranes for Gas and Vapor Separation* (ed. Y. Yampolskii, I. Pinnau, and B.D. Freeman), 337–354. Chichester: Wiley.

29. Kita, H., Nanbu, K., Maeda, H., and Okamoto, K.I. (2004). Gas separation and pervaporation through microporous carbon membranes derived from phenolic resin. In: *Advanced Materials for Membrane Separations, ACS Symposium Series*, vol. **876** (ed. I. Pinnau and B.D. Freeman), 203–217. Washington, DC: American Chemical Society.

30. Kim, Y.K., Park, H.B., and Lee, Y.M. (2004). Carbon molecular sieve membranes derived from thermally labile polymer containing blend polymers and their gas separation properties. *J. Membr. Sci.* **243**: 9–17.

31. Rao, M.B. and Sirkar, S. (1993). Nanoporous carbon membranes for separation of gas mixtures by selective surface flow. *J. Membr. Sci.* **85** (3): 253–264.

32. Williams, P.J. and Koros, W.J. (2008). Gas separation by carbon membranes. In: *Advanced Membrane Technology and Applications* (ed. N.N. Li, A.G. Fane, W.S.W. Ho, and T. Matsuura), 599–631. Hoboken, NJ: Wiley.

33. Jones, C.W. and Koros, W.J. (1995). Characterization of ultramicroporous carbon membranes with humidified feeds. *Ind. Eng. Chem. Res.* **34**: 158–163.

34. Ray, A.E. and Box, E.O. (1950). Adsorption of gases on activated charcoal. *Ind. Eng. Chem.* **42**: 1315.

35. Hwang, S.T. and Kammermeyer, K. (1975). *Membranes in Separations*, Techniques of Chemistry, vol. **7**. New York: Wiley.

36. Barrer, R.M. (1963). Diffusion in porous media. *Appl. Mater. Res.* **2**: 129.

37. Choi, J.-G., Do, D.D., and Do, H.D. (2001). Surface diffusion of absorbed molecules in porous media: monolayer, multilayer and capillary condensation regimes. *Ind. Eng. Chem. Res.* **40**: 4005.

38. Sircar, S. and Rao, M.B. (2000). Nanoporous carbon membranes for gas separation. In: *Recent Advances in Gas Separation by Microporous Ceramic Membranes* (ed. N.K. Kanellopoulos), 473. Elsevier.

39. Way, J.D. and Roberts, D.L. (1992). Hollow fiber inorganic membranes for gas separations. *Sep. Sci. Technol.* **27**: 29.

40. Fuertes, A.B. (2000). Adsorption-selective carbon membranes for gas separation. *J. Membr. Sci.* **177**: 9.

41. Fairen-Jimenez, D., Moggach, S.A., Wharmby, M.T. et al. (2011). Opening the gate: framework flexibility in ZIF-8 explored by experiments and simulations. *J. Am. Chem. Soc.* **133**: 8900.

42. Vroon, Z.A.E.P., Keizer, K., Gilde, M.J. et al. (1996). Transport properties of alkanes through thin zeolite MFI membranes. *J. Membr. Sci.* **113**: 293.

43. Wang, R. and Lin, S. (2021). Pore model for nanofiltration: history, theoretical framework, key predictions, limitations and prospects. *J. Membr. Sci.* **620**: 118809.

44. Ferry, J.D. (1936). Ultrafilter membranes and ultrafiltration. *Chem. Rev.* **18**: 373.

45. Renkin, E.M. (1955). Filtration, diffusion and molecular sieving through porous cellulose membranes. *J. Gen. Physiol.* **38**: 225.

46. Baker, R.W. and Strathmann, H. (1970). Ultrafiltration of macromolecular solutions with high-flux membranes. *J. Appl. Polym. Sci.* **14**: 1197.
47. Lukaszewicz, R.C., Tanny, G.B., and Meltzer, T.H. (1978). Membrane-filter characterizations and their implications for particle retention. *Pharm. Technol.* **2**: 77.
48. Meltzer, T.H. (1987). *Filtration in the Pharmaceutical Industry*. New York: Marcel Dekker.
49. Davis, R.H. and Grant, D.C. (1992). Theory for dead end microfiltration. In: *Membrane Handbook* (ed. W.S.W. Ho and K.K. Sirkar), 461–479. New York: Van Nostrand Reinhold.
50. Grant, D.C., Liu, B.Y.H., Fischer, W.G., and Bowling, R.A. (1989). Particle capture mechanisms in gases and liquids: an analysis of operative mechanisms. *J. Environ. Sci.* **42**: 43.

4

Membranes and Modules

Membrane Technology and Applications, Fourth Edition. Richard W. Baker.
© 2024 John Wiley & Sons Ltd. Published 2024 by John Wiley & Sons Ltd.

4.1   Introduction

Industrial interest in membrane separation processes was prompted by two developments: first, the ability to produce high-performance membranes on a large scale and second, the ability to package these membranes into compact modules. These breakthroughs took place in the 1960s to early 1970s, as part of the development of reverse osmosis (RO) and ultrafiltration. Adaptation of the technology to other membrane processes took place in the 1980s.

Several factors contribute to the successful fabrication of a high-performance membrane module. First, membrane materials with the appropriate chemical, mechanical, and permeation properties must be selected; this choice is very application specific. However, once the material has been selected, the technology required to fabricate and package the membrane is similar for many applications. This chapter focuses on general methods of forming membranes and membrane modules; the criteria used to select membrane materials are described in the chapters on the individual processes.

Preparation techniques are organized by membrane structure: isotropic membranes, anisotropic membranes, and ceramic membranes. Isotropic membranes have a uniform composition and structure throughout; such membranes can be porous or dense. Anisotropic (or asymmetric) membranes, on the other hand, consist of a number of layers, each with different structures and permeabilities. A typical anisotropic membrane has a relatively dense, thin surface layer supported on an open, much thicker microporous substrate. The surface layer performs the separation and is the principal barrier to flow through the membrane. The open support layer provides mechanical strength. Ceramic and other wholly or partly inorganic membranes can be either isotropic or anisotropic. However, these membranes are grouped separately from pure polymeric membranes because their preparation methods are so different.

The membrane classification scheme described above works fairly well. However, a major membrane preparation technique, phase separation, also known as phase inversion, is used to make both isotropic and anisotropic membranes. This technique is covered under anisotropic membranes.

4.2   Isotropic Membranes

4.2.1   Isotropic Nonporous Membranes

Nonporous (also referred to as dense) isotropic membranes are not commonly used in separation processes because they are relatively thick, making their fluxes too low for most practical purposes. However, they are used in laboratory work to characterize membrane properties and in a few processes where high permeance is not required. They are prepared by solution casting or thermal melt-pressing.

4.2.1.1   Solution Casting

Solution casting is often used to prepare small samples of membrane for laboratory characterization experiments. An even film of an appropriate polymer solution is spread across a flat plate with a casting knife. The casting knife consists of a steel blade, resting on two runners, arranged to form a precise gap between the blade and the plate onto which the film is cast. A typical handheld knife is shown in Figure 4.1. After casting, the solution is left to stand, and the solvent evaporates to leave a thin, uniform polymer film. A detailed description of many types of hand-casting knives and simple casting machines is given in the book of Gardner and Sward [1].

The polymer solution used for solution casting should be sufficiently viscous to prevent it from running over the casting plate, while still being spreadable; typical polymer concentrations are in the range 15–20 wt%. Preferred solvents are moderately volatile liquids, such as methyl ethyl ketone, ethyl acetate, and cyclohexane. Films cast from these solutions are dry within a few hours. When the solvent has completely evaporated, the dry film can be lifted from the glass plate. If the film adheres to the plate, soaking in a swelling non-solvent, such as water or alcohol, will usually loosen the film.

Solvents with high boiling points, such as dimethyl formamide or *N*-methyl pyrrolidone, are unsuitable for solution casting, because their low volatility requires long evaporation times, during which the cast film can absorb sufficient atmospheric water to precipitate the polymer, producing a mottled, hazy surface. Very volatile solvents, such as methylene chloride and tetrahydrofuran, can also cause problems. Rapid evaporation of the solvent cools the casting solution, causing gelation of the polymer. The result is a film with a mottled, orange-peel-like

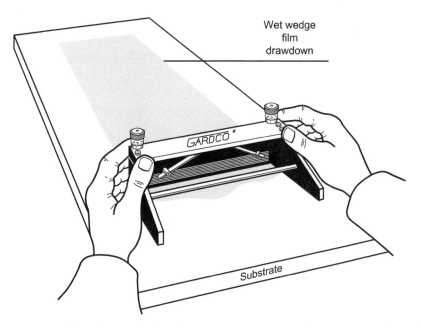

Figure 4.1   *A typical hand-casting knife. The micrometer screws on the runners on either side of the blade are used to adjust the gap between the blade and the casting plate. (Paul N. Gardner Company, Inc., Pompano Beach, FL).*

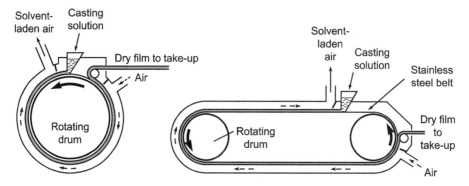

Figure 4.2   Machinery used to make solution-cast film on a commercial scale.

surface. Smooth films can be obtained with rapidly evaporating solvents by covering the cast film with a glass plate raised 1–2 cm above the film to slow evaporation.

Solution-cast film is produced on a larger scale for medical applications, battery separators, or other specialty uses with machinery of the type shown in Figure 4.2 [2]. Viscose film is made by this technique. The solution is cast onto the surface of a rotating drum or a continuous polished stainless-steel belt. These machines are generally enclosed to control water vapor pickup by the film as it dries and to minimize solvent vapor losses to the atmosphere.

4.2.1.2   Melt Extruded Film

Many polymers, including polyethylene, polypropylene, and nylons, do not dissolve in appropriate solvents at room temperature, so membranes cannot be made from them by solution casting. To prepare small pieces of film, a laboratory press, as shown in Figure 4.3, can be used. The polymer is compressed between two heated plates. Typically, a pressure of 2000–5000 psi is applied for 15 minutes, at a plate temperature just below the melting point of the polymer. Melt extrusion is also used on a very large scale to make dense films for packaging applications, either by extrusion as a sheet from a die or as blown film. Detailed descriptions of this equipment can be found in specialized monographs. A good overview is given in the article by Mackenzie [2].

4.2.2   Isotropic Microporous Membranes

Isotropic microporous membranes have much higher fluxes than isotropic dense membranes and are widely used as microfiltration membranes. Other significant uses are as inert spacers in battery and fuel cell applications and as the rate-controlling element in controlled drug delivery devices.

The most important type of microporous membrane is formed by one of the phase separation techniques discussed in the next section; most isotropic microporous membrane is made in this way. The remaining types are made by various proprietary techniques, the more important of which are described below.

4.2.2.1   Track-Etch Membranes

Track-etch membranes were developed by the General Electric Corporation Schenectady Laboratory [3]. The two-step preparation process is illustrated in Figure 4.4. First, a thin polymer film is irradiated with fission particles from a nuclear reactor or other radiation source.

Figure 4.3 *A typical laboratory press used to form melt-pressed membranes. Source: Carver, Inc., Wabash, IN.*

The massive particles pass through the film, breaking polymer chains and leaving behind a sensitized track of damaged polymer molecules. These tracks are much more susceptible to chemical attack than the base polymer material. When the film is passed through a solution that etches the polymer, the film is etched preferentially along the sensitized nucleation tracks, thereby forming pores. The exposure time of the film to radiation determines the number of membrane pores; the etch time determines the pore diameter [4]. A feature of the track-etch preparation technique is that the pores are uniform cylinders traversing the membrane at right angles. The membrane tortuosity is, therefore, close to one, and all pores have the same diameter. These membranes are almost perfect screen filters, and are widely used to measure the number and type of suspended particles in air or water. A known volume of fluid is filtered through the membrane,

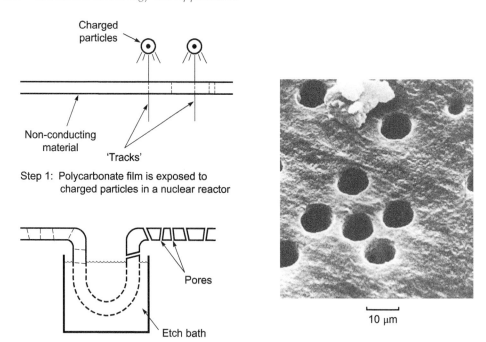

Charged
particles

Non-conducting
material

'Tracks'

Step 1: Polycarbonate film is exposed to
charged particles in a nuclear reactor

Pores

Etch bath

Step 2: The tracks left by the particles are
preferentially etched into uniform, cylindrical pores

10 µm

Figure 4.4 *Diagram of the two-step process to manufacture nucleation track membranes [6]/ American Association for the Advancement of Science and photograph of resulting structure. Source: Photograph from G.E. Healthcare.*

and all particles larger than the pore diameter are captured on the surface of the membrane, where they can be identified and counted. To minimize the formation of doublet holes produced when two nucleation tracks are close together, the membrane porosity is usually kept relatively low, about 5% or less. This low porosity results in low fluxes. General Electric assigned the technology to a spin-off company, Nuclepore Corporation, in 1972 [5]. Nuclepore® membranes remain the principal commercially available track-etch membranes. Polycarbonate or polyester films are usually used as the base membrane material and sodium hydroxide as the etching solution. Other materials can also be used; for example, etched mica has been used in research studies.

4.2.2.2   Expanded-Film Membranes

Expanded-film membranes are made from crystalline polymers by an orientation and annealing process. A number of manufacturers produce porous membranes by this technique. The original development began with a group at Celanese, who made microporous polypropylene membranes by this process under the trade name Celgard® [6]. In the first step of the process, a highly oriented film is produced by extruding polypropylene at close to its melting point, coupled with a very rapid drawdown. Polypropylene is semi-crystalline, and the crystallites formed on cooling the polymer are aligned in the direction of drawdown. After cooling and annealing, the film is stretched, at right angles to the drawdown direction, up to 300%. During this stretching, the amorphous regions between the crystallites are deformed, forming slit-like voids, 250–2500 Å wide. The pore size of the membrane is controlled by the rate and extent of the

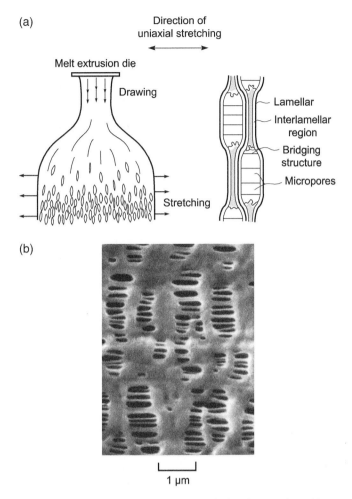

Figure 4.5 *(a) Preparation method of a typical expanded polypropylene film membrane, in this case Celgard®. (b) Scanning electron micrograph of the microdefects formed on uniaxial stretching of films. Source: Reproduced with permission from [6]. Copyright (1974) American Chemical Society.*

perpendicular stretching step. The formation process is illustrated in Figure 4.5. Celgard-type membranes are widely used as battery separators. Expanded-film membrane made from poly(tetrafluoroethylene) film and developed by W. L. Gore is familiar as Gore-Tex® [7], the breathable, waterproof fabric. The expanded-film process has also been adapted to the production of hollow fibers [8, 9]; Membrana produces this type of fiber for use in blood oxygenator equipment (Chapter 14) and membrane contactors (Chapter 13). The commercial success of these products has motivated other companies to produce similar materials [10].

4.2.2.3   Template Leaching

Template leaching is another method of producing isotropic microporous membranes from insoluble polymers. In this process, a homogeneous melt is prepared from a mixture of the polymer and a leachable component. To finely disperse the leachable component, the mixture

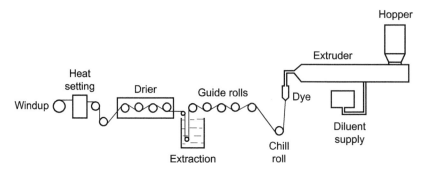

Figure 4.6   *Flow schematic of a melt extruder system used to make polypropylene membranes by template leaching. Source: Adapted from [12].*

is often homogenized, extruded, and pelletized several times before final extrusion as a thin film. After formation of the film, the leachable component is removed with a suitable solvent, thus forming a microporous membrane [11, 12]. The leachable component can be a soluble, low-molecular-weight solid, a liquid such as liquid paraffin, or even a polymeric material such as polystyrene. A drawing of a template leaching membrane production machine is shown in Figure 4.6.

4.3   Anisotropic Membranes

Anisotropic membranes are layered structures in which the porosity, pore size, or even membrane composition change from the top to the bottom surface of the membrane. Usually anisotropic membranes have a thin, selective surface layer supported on a thicker, highly permeable micro-porous layer. Because the selective layer is thin, membrane fluxes are high. The microporous support may itself be coated onto a nonwoven paper to provide the additional strength required for handling the membrane. The importance of anisotropic membranes was not recognized until Loeb and Sourirajan prepared the first high-flux, anisotropic reverse osmosis membranes around 1960, by what is now known as the Loeb–Sourirajan technique [13], although hindsight makes it clear that some of the membranes produced in the 1930s and 1940s were also anisotropic. Loeb and Sourirajan's discovery was a critical breakthrough in membrane technology. Their membranes were an order of magnitude more permeable than the isotropic membranes produced previously from the same materials. For a few years, the Loeb–Sourirajan technique was the only method of making anisotropic membranes, but the demonstrated benefits of the anisotropic structure encouraged the development of other methods. Improvements in preparation methods and properties were accelerated by the availability in the 1960s of the scanning electron microscope (SEM), which allowed the effects of changes in the formation process on membrane structure to be easily assessed.

The selective layer and the microporous support layer of Loeb–Sourirajan membranes are made from one polymer solution in a single step. The pore size and porosity change from the top to the bottom of the membrane, but the membrane material is everywhere the same. Anisotropic membranes made by other techniques often consist of layers of different materials which serve different functions. Often the support layer is itself asymmetric, with a finely porous surface that provides a smooth base onto which to coat the selective layer. Important examples of such composite membranes are reverse osmosis membranes made by the interfacial polymerization process developed by Cadotte [14] and gas separation membranes made by the solution-coating processes developed by Ward [15] and Riley [16].

The following sections cover three types of anisotropic membranes:

- **Phase separation membranes:** This category includes membranes made by the Loeb–Sourirajan technique. Also covered are a variety of related techniques, such as polymer precipitation by solvent evaporation, precipitation by absorption of water from the vapor phase, and precipitation by cooling.
- **Interfacial composite membranes:** This type of anisotropic membrane is made by polymerizing reactive monomers to form an extremely thin layer of polymer on the surface of a microporous support polymer.
- **Solution-coated composite membranes:** To prepare these membranes, one or more thin, dense polymer layers are solution coated onto the surface of a microporous support.

4.3.1   Phase Separation Membranes

The Loeb–Sourirajan technique is now recognized as a special case of a more general class of membrane preparation, best called the phase separation process, but sometimes called the phase inversion process or the polymer precipitation process. The term phase separation describes the process most clearly because it involves changing a liquid, one-phase, polymer/solvent casting solution into two separate phases: a solid, polymer-rich phase which forms the matrix of the membrane, and a liquid, polymer-poor phase that forms the membrane pores.

Precipitation of the cast polymer solution can be achieved in several ways, summarized in Table 4.1. Precipitation by immersion in a bath of water was the technique developed by Loeb and Sourirajan, but precipitation can also be caused by absorption of water from a humid atmosphere. A third method is to cast the film as a hot polymer solution. As the cast film cools, a point is reached at which precipitation (phase separation) occurs to form a microporous structure; this method is sometimes called thermally induced phase separation (TIPS). Finally, evaporation of one of the solvents in the casting solution can be used to cause precipitation. In this technique, the casting solution consists of a polymer dissolved in a mixture of a volatile good solvent and a less volatile non-solvent (typically water or alcohol). When a film of the solution is cast and allowed to evaporate, the volatile good solvent evaporates first, the film then becomes enriched in the non-volatile non-solvent and precipitates. Many combinations of these processes have also been developed. For example, a cast film placed in a humid

Table 4.1   *Phase separation membrane preparation procedures.*

Procedure	Process
Water precipitation (the Loeb–Sourirajan process)	The cast polymer solution is immersed in a non-solvent bath (typically water). Absorption of water and loss of solvent cause the film to rapidly precipitate from the top surface down
Water vapor absorption	The cast polymer solution is placed in a humid atmosphere. Water vapor absorption causes the film to precipitate
Thermal gelation	The polymeric solution is cast hot. Cooling causes precipitation
Solvent evaporation	A mixture of solvents is used to form the polymer casting solution. Evaporation of one of the solvents after casting changes the solution composition and causes precipitation

atmosphere can precipitate partly because of water vapor absorption but also because of evaporation of one of the more volatile components.

4.3.1.1   Polymer Precipitation by Water (the Loeb–Sourirajan Process)

The first phase separation membrane was developed at UCLA from 1958 to 1960 by Sidney Loeb, then working on his master's degree, and Srinivasa Sourirajan, then a post-doctoral researcher [13]. In their process, precipitation is induced by immersing the cast film of polymer solution in a water bath. In the original Loeb–Sourirajan process, a solution containing 20–25 wt % cellulose acetate dissolved in a mixture of water-miscible solvents was cast as a thin film on a glass plate. The film was left to stand briefly, to allow some of the solvent to evaporate, then immersed in a water bath to precipitate the film and form the membrane. The membrane was usually post-treated by annealing in a bath of hot water. The steps of the process are illustrated in Figure 4.7.

Variants of the Loeb–Sourirajan process, using different polymers, casting solutions, compositions, and precipitation step details, remain by far the most important membrane preparation technique [17]. The process is used to make reverse osmosis, ultrafiltration, and gas separation membranes, forming either the support structure or the entire membrane. The protocols used to make supports for reverse osmosis and gas separation membranes are chosen to produce a thin, porous skin, suitable for overcoating with a dense selective layer. The protocols for ultrafiltration membranes are chosen to produce a finely microporous top selective layer with pores in the 30–200 Å diameter range. The structure of the underlying porous support layer may also be different. Scanning electron micrographs of the typical sponge structure for an RO membrane and the finger structure for an ultrafiltration membrane are shown in Figure 4.8 [18]. These photographs show how small changes in the casting solution can produce major differences in membrane properties. Both membranes shown in the figure were prepared from an aromatic

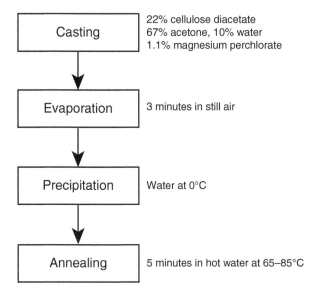

Figure 4.7   *Process scheme used by Loeb and Sourirajan to form the first water precipitation phase separation membranes. Source: Adapted from [13].*

(a)   Sponge structure cast from 22 wt% Nomex in dimethylacetamide

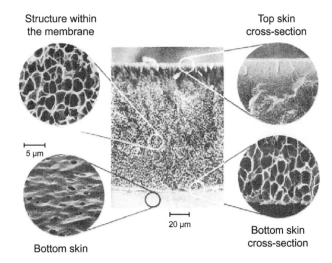

Structure within
the membrane

Top skin
cross-section

5 µm

20 µm

Bottom skin

Bottom skin
cross-section

(b)   Finger structure cast from 18 wt% Nomex in dimethylacetamide

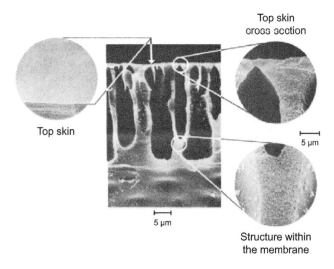

Top skin
cross section

Top skin

5 µm

5 µm

Structure within
the membrane

Figure 4.8 *Scanning electron micrographs of aromatic polyamide (Nomex®, Du Pont) Loeb-Sourirajan membranes cast from (a) 22 wt% and (b) 18 wt% polymer in dimethylacetamide. Source: Reproduced with permission from [18]. Copyright (1975) Elsevier.*

polyamide-dimethylacetamide casting solution (Nomex® from DuPont (Wilmington, DE)); only the polymer concentration in the solutions was different.

The membranes shown in Figure 4.8 were made by casting the polymer solution onto glass plates. This procedure is still used in the laboratory; for commercial production, large casting machines produce rolls of membrane up to 5000 m long and 1–2 m wide. A simplified diagram

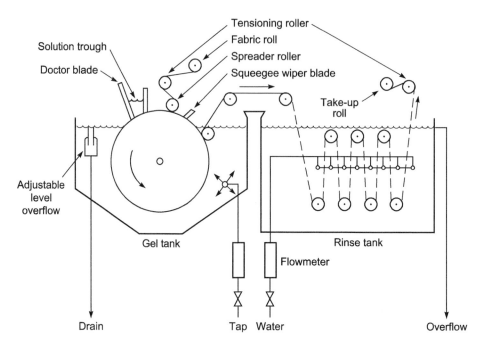

Figure 4.9 *Schematic of Loeb-Sourirajan membrane casting machine used to prepare reverse osmosis or ultrafiltration membranes. A knife and trough are used to coat the casting solution onto a nonwoven paper web. The coated web then enters the water filled gel tank, where the casting solution precipitates. After the membrane has formed, it is washed thoroughly to remove residual solvent before being wound up on the take-up roll.*

of a casting machine is shown in Figure 4.9. The polymer solution is cast onto a moving web of nonwoven paper, which provides mechanical strength for handling. The cast film is precipitated by immersion in a water bath. Water precipitates the top surface of the cast film almost instantaneously, forming the dense, selective skin. This skin slows entry of water into the underlying polymer solution, which precipitates more slowly, typically over 10–60 seconds, to form the more porous substructure. Depending on the polymer, the casting solution, and other parameters, the skin thickness varies from 0.5 to 2.0 µm. Casting machine speeds vary from as low as 1 to 2 m/min for slowly precipitating solutions, such as cellulose acetate, to 10 m/min for rapidly precipitating solutions, such as polysulfone. A listing of some casting solutions and precipitation conditions for historically important membranes made by the Loeb–Sourirajan technique is given in Table 4.2 [13, 17, 19–21].

Small casting machines often apply the casting solution with a doctor blade as shown in Figure 4.10a. Large industrial machines tend to use a slot die solution dispenser of the type shown in Figure 4.10b. In a slot die, a metering pump is used to push the casting solution through a precisely controlled gap between the flat plates. Membrane made by this procedure is less sensitive to defects in the paper support or variations in the support paper thickness. Also, the membrane thickness is easily controlled by the metering pump and is relatively independent of the viscosity of the coating solution. Development of the slot die technique has also made it possible to make multilayer castings as shown in Figure 4.10c. As the support paper moves under the dies, the first die lays down a film of casting solution. The second die then lays a second layer on top of the first. Normally, casting solutions will be made from the same polymer dissolved in different solvents, or at different concentrations in the same solvent. When precipitated in a

Table 4.2   *Historically important examples of conditions for preparation of phase separation (Loeb–Sourirajan) membranes.*

Casting solution composition	Precipitation conditions	Application and comments
22.2 wt% cellulose acetate (39.8 wt% acetyl polymer) 66.7 wt% acetone 10.0 wt% water 1.1 wt% magnesium perchlorate	3 min evaporation, precipitate into 0 °C water, anneal for 5 min at 65–85 °C	The first Loeb–Sourirajan RO membrane [13]
25 wt% cellulose acetate (39.8 wt% acetyl polymer) 45 wt% acetone 30 wt% formamide	0.5–2 min evaporation, precipitate in 0 °C water, anneal for 5 min at 65–85 °C	The Manjikian formulation widely used in early 1970s for RO membranes[19]
8.2 wt% cellulose acetate (39.8 wt% acetyl polymer) 8.2 wt% cellulose triacetate (43.2 wt% acetyl polymer) 45.1 wt% dioxane 28.7 wt% acetone 7.4 wt% methanol 2.5 wt% maleic acid	Up to 3 min evaporation at approximately10 °C, precipitate into 0 °C water, anneal at 85–90 °C for 3 min	A high-performance RO cellulose acetate blend membrane [20]. Similar formulations are still used to make gas separation membranes
15 wt% polysulfone (Udell P 1700) 85 wt% N-methyl-2-pyrrolidone	Cast into 25 °C water bath. No evaporation or annealing step necessary	An early ultrafiltration membrane formulation [17]. Similar polysulfone-based casting solutions are still widely used
20.9 wt% polysulfone 33.2 wt% dimethyl formamide 33.2 wt% tetrahydrofuran 12.6 wt% ethanol	Forced evaporation with humid air 10–15 s, precipitate into 20 °C water	A high-performance gas separation membrane with a completely dense nonporous skin ~1000 Å thick [21]

downstream step, a membrane with a gradation of pore sizes or compositions forms. This technique is used in the production of microfiltration membranes. The membrane surface with the largest pores faces the incoming feed and captures the largest contaminant particles. Smaller contaminant particles pass through this layer and are captured in the next more finely porous layer. By spreading the contaminant load throughout the whole membrane, the lifetime of the membrane is extended.

Since the work of Loeb and Sourirajan, development of the technology has proceeded on two fronts. Industrial makers have generally taken an empirical approach, making improvements based on trial-and-error experience. Concurrently, theories of membrane formation based on fundamental studies of the precipitation process have been developed. These theories originated with the early industrial developers at Amicon [17, 18, 22, 23] and were then taken up at a number of academic centers [24–26].

4.3.1.2   *Empirical Approach to Membrane Formation by Water Precipitation*

Over the years, several rules of thumb have developed to guide producers of phase separation membranes. These rules can be summarized as follows:

Choice of polymer. The ideal polymer is a tough, amorphous, non-brittle, thermoplastic with a glass transition temperature more than 50 °C above the expected use temperature. A high

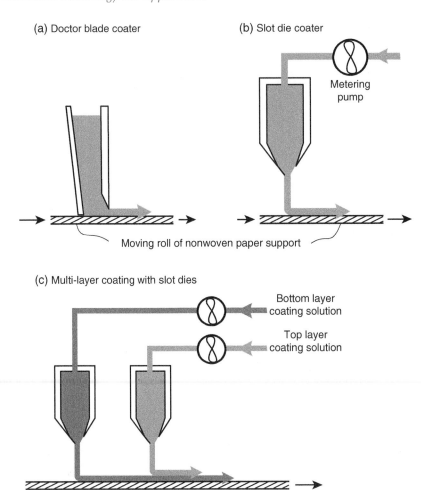

Figure 4.10 *Doctor blade coating solution onto a moving paper web (a) is simpler, but slot die coaters (b) give a better product and can be used to deposit multilayer coatings (c).*

molecular weight is important. Commercial polymers made for injection molding have molecular weights in the 30 000–40 000 Da range; for water precipitation, polymers with higher molecular weights are usually preferable. If the polymer is crystalline or a rigid glass, the resulting membrane may be too brittle and will break if bent during later handling. The polymer must also be soluble in a suitable water-miscible solvent. Polymers that meet these specifications include cellulose acetate, polysulfone, poly(vinylidine fluoride), polyetherimide, and aromatic polyamides.

Choice of casting solution solvent. Generally, the best solvents are aprotic solvents such as dimethyl formamide, *N*-methyl pyrrolidone, and dimethyl acetamide. These dissolve a wide variety of polymers, and form casting solutions that precipitate rapidly when immersed in water, to give porous, anisotropic membranes. Casting solutions using low-solubility-parameter solvents, such as tetrohydrofuran, acetone, dioxane, and ethyl formate, are generally not appropriate. Such casting solutions precipitate slowly and give relatively nonporous membranes. However, small amounts of these solvents may be added as casting solution modifiers (see below). Figure 4.11 illustrates the apparent correlation between solvent solubility parameter and membrane porosity

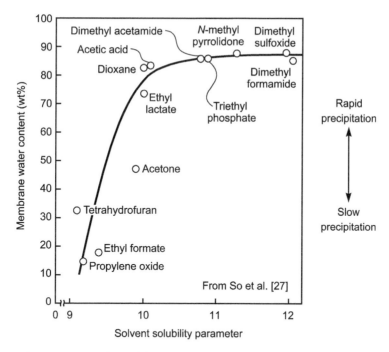

Figure 4.11 *The porosity (indicated by water content of the newly precipitated membrane) of cellulose acetate membranes cast from 15 wt% solutions with various solvents. The same trend of high porosity and rapid precipitation with high-solubility-parameter solvents was seen with a number of other membrane materials. Source: Reproduced with permission from [27]. Copyright (1971) John Wiley and Sons.*

as demonstrated by So et al. [27]. The percentage of water in the precipitated membrane on the *y*-axis is a measure of the membrane porosity, since the water-filled regions of the newly cast membrane form the pores.

Increasing the polymer casting solution concentration always reduces the porosity and flux of the membrane. Typical concentrations for porous ultrafiltration membranes are in the range of 15–20 wt%. Concentrations for reverse osmosis or gas separation membranes are higher, generally about 25 wt%, and casting solutions used to make hollow fiber membranes may contain as much as 35 wt% polymer.

Precipitation medium. Water is almost always the precipitation medium. Some work has been done with nonaqueous media, particularly to form hollow fiber membranes, where safe handling of the precipitation medium and control of atmospheric emissions are less problematic than in flat sheet casting. In general, the results obtained with nonaqueous precipitation have not justified the increased complexity of the process. Precipitation media such as methanol or isopropanol almost always precipitate the casting solution more slowly than water, and the resulting membranes are usually denser, less anisotropic, and exhibit lower fluxes than membranes precipitated with water.

The temperature of the water used to precipitate the casting solution is important; this temperature is controlled in commercial membrane plants. Generally low-temperature precipitation produces lower flux, more retentive membranes. For this reason, chilled water is sometimes used to prepare cellulose acetate RO membranes.

Casting solution modifiers. Membrane properties are often tailored by adding small amounts of modifiers to the casting solution. The casting solutions shown in Table 4.2 contain up to six components; modern commercial casting solutions may be yet more complex. Even though the solution may contain only 5–20 wt% modifiers, these can change the membrane performance significantly. This aspect of membrane preparation is a black art, and most practitioners have their preferred ingredients. Addition of low-solubility-parameter solvents normally produces denser, more retentive membranes, as does increasing the polymer concentration of the casting solution. Addition of salts such as zinc chloride and lithium chloride usually gives more open membranes. Polymeric additives – commonly poly(vinyl pyrrolidone) or poly(ethylene glycol) – may also be used; by making the casting solution more viscous, these polymers can eliminate the large finger pores shown in Figure 4.8. Also, although most of these water-soluble polymers and salts are removed during precipitation and washing of the membrane, a portion remains trapped, making the final membrane more hydrophilic and often less brittle.

The nature of the nonwoven support paper can be an important variable. Polyester paper is the most commonly used material, but nonwoven polypropylene and polyphenylene sulfide papers are also used. If the paper is too coarse, membrane pinholes can result; if the paper is too fine, adhesion between the paper and the microporous membrane layer will be poor.

The evaporation step. The Loeb–Sourirajan recipe for the first successful anisotropic membrane used a 3-minute evaporation step after the casting solution was coated and before the cast film was immersed into the water precipitation bath. This step provides time for some of the casting solution to penetrate into the porous support paper, to coat any surface defects in the paper and to smooth out irregularities and streaks formed in the casting solution. More importantly, it is also an opportunity to change the composition of the surface layer of the casting solution that is going to precipitate in the next step. How much change will occur depends on the evaporation time, the presence or absence of volatile components in the casting solution, and the nature of the ambient air directly above the cast film. Manipulating the evaporation step in a reproducible controlled way in laboratory glass-plate casting is difficult, but industrial equipment is often fitted with forced air evaporators and air knives, and the conditions under which evaporation occurs are more controlled and can be an important adjustable preparation parameter.

When developing membranes from a new polymer, practitioners of the empirical approach usually prepare a series of trial casting solutions based on past experience with similar polymers. Membrane films are made by casting onto glass plates, followed by precipitation in a water bath. The casting solutions most likely to yield good membranes are often immediately apparent. The rate of precipitation is important. Slow precipitation produces dense, more isotropic membranes; rapid precipitation produces porous, anisotropic membranes. The appearance and mechanical properties of the membrane surface – shine, brittleness, and membrane thickness – and the membrane thickness compared to the cast solution thickness also provide clues to the membrane structure. Based on these trials, one or more casting solutions will be selected for systematic parametric development. A useful discussion of casting solution selection and polymers commonly used to make phase separation membranes is given in Nunes and Peinemann [28]; Guillen et al. have also reviewed the recent literature [29].

Membranes prepared by precipitation into water will usually be collected as a damp roll of water-impregnated membrane. This wet membrane often requires drying before subsequent manufacturing operations, but simple air drying is usually not an option. The capillary forces produced by water evaporation cause the membrane pores to collapse and densify, and the membrane shrinks and curls.

A final rinse step, in which the water content of the membrane is exchanged with a solution of propylene glycol, can sometimes be used to solve the problem. The water content of the propylene glycol solution can evaporate during handling, but enough glycol remains to prevent densification. However, many membranes, such as those used in gas separation and pervaporation, must be completely dry before use. The usual drying procedure is to first exchange the water in the membrane with ethanol or isopropanol, then in a second step to exchange the alcohol mixture with a hydrocarbon, such as octane. The surface tension forces created by octane evaporation are much below those of water, so shrinkage and curl are avoided.

4.3.1.3   Theoretical Approach to Membrane Formation by Water Precipitation

Over the years, several approaches have been used to rationalize the formation of phase separation membranes. When the Loeb–Sourirajan process was first developed, the science behind it was shrouded in mystery, and many wild surmises as to what was happening were promulgated. The light went on with the adoption of explanations based on phase diagrams, as popularized by Michaels [17], Strathmann [18, 22, 23], and Smolders [24–26]. In this approach, composition changes as membrane formation takes place are tracked from a point representing the original casting solution to a point representing the composition of the final membrane. The casting solution composition moves to the final membrane composition by losing solvent and gaining water from the precipitation bath.

A simplified diagram for the components used to prepare Loeb–Sourirajan membranes is shown in Figure 4.12. The corners of the triangle represent the three pure components – polymer, solvent, and non-solvent (water); points within the triangle represent mixtures of the three components. The diagram represents the initial casting solution as having only two

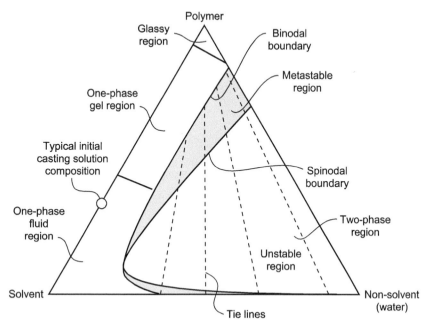

Figure 4.12   *A three-component phase diagram to rationalize the formation of Loeb–Sourirajan membranes. In the two-phase region, tie lines link the precipitated polymer-rich phase with its equilibrium polymer-poor phase.*

components: polymer and a single solvent. As was seen from Table 4.2, this is never true in reality, but serves as a conceptual illustration.

The diagram has two principal regions: a one-phase region, in which all components are miscible, and a two-phase region, in which the system separates into a solid (polymer-rich) phase and a liquid (polymer-poor) phase. The compositions of these phases are linked by tie lines. Between the one- and two-phase regions is a relatively narrow region of metastability. The casting solution starts as a point on the polymer/solvent axis, with zero water content. After immersion in the precipitation bath, the solution loses solvent to the bath and gains water from it. As a result, the casting solution composition moves across the one-phase region, through the metastable region, and into the two-phase region, where precipitation (phase separation) occurs. When all the solvent has diffused to the precipitation bath, the averaged composition of the two phases of the precipitated membrane reaches a point on the polymer/water axis.

The original approach of Michaels [17] and Strathmann et al. [22] was to present the process of membrane formation as a line through the phase diagram, as shown in Figure 4.13. During membrane formation, the composition changes from point A, the initial casting solution composition, to point D, which represents the averaged final membrane composition. Following the tie lines at this point gives the compositions, L and S, of the two phases. The figure shows that the polymer-poor phase, which makes the membrane pores, is just water (point L is at the water apex). The polymer-rich phase, which makes the membrane matrix, is water-swollen polymer of composition S. The position of D on the S–L tie line determines the overall porosity of

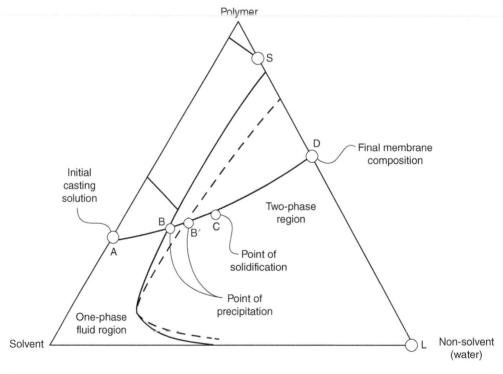

Figure 4.13 *Membrane formation in phase separation membranes was first rationalized as a path through the three-component phase diagram from the initial polymer casting solution (A) to the final membrane (D). Source: Adapted from [22].*

the membrane. Thus, the fraction of the final membrane that is pores is given by the ratio of the numerical lengths (S–D)/(S–L); the fraction that is membrane matrix is (L–D)/(S–L).

The point B is the point at which the composition crosses the binodal curve into the metastable region, but has yet to precipitate. The solution remains in the metastable region for the time required for the polymer chains to undergo the reorientation and aggregation needed for precipitation to occur. At point B′, the solution leaves the metastable region and precipitation begins. The time taken for solvent–water exchange before precipitation occurs can be measured, because the membrane turns opaque as precipitation begins. Depending on the casting solution composition, first precipitation may be almost instantaneous or may take as long as a few minutes. As solvent continues to be lost, the viscosity of the polymer-rich phase increases. At some point, the viscosity is high enough for the precipitated polymer to be regarded as a solid. This composition is at C in Figure 4.13. Once the precipitated polymer solidifies, further bulk movement of the polymer is hindered.

The precipitation path in Figure 4.13 is shown as a single line representing the average composition of the whole membrane. In fact, the rate of precipitation and the precipitation path differ at different points in the membrane. When the cast film of polymer solution is exposed to the precipitation medium, precipitation begins at the top surface. The surface layer precipitates rapidly, so the two phases formed there do not have time to agglomerate, and the resulting structure is finely microporous. The surface layer then becomes a barrier that slows further loss of solvent and imbibition of non-solvent by the cast film. The result is increasingly slow precipitation from the top to the bottom surface of the film. As precipitation slows, the average pore size increases, because the two phases formed on precipitation have more time to separate before the polymer phase gels. The differences between the precipitation rates and the pathway taken at different places in the casting solution mean that the precipitation process is best represented by the movement of a line, rather than a single point, through the phase diagram. This concept was developed in a series of papers by Smolders and coworkers at Twente University [24–26]. The movement of this line is illustrated in Figure 4.14 [24]. At time t_2, for example, a few seconds after the precipitation process has begun, the top surface of the polymer film has almost completely precipitated, and its composition is close to the polymer non-solvent axis. On the other hand, at the bottom surface of the film where precipitation has only just begun, the composition is still close to that of the original casting solution.

The precipitation pathways taken by the solutions at the top and bottom surfaces of the membrane are shown in Figure 4.14. The solution at the top surface of the membrane enters the gel region of the phase diagram before the precipitation boundary is reached. On precipitation, the polymer gel densifies, but no micropores are formed. The solution at the bottom surface remains in the one-phase fluid region of the phase diagram until the precipitation boundary is reached. When precipitation occurs, the solution is fluid and separate into polymer-rich and polymer-poor phases. The polymer-poor phase eventually forms the pores in the final microporous membrane.

In Figure 4.14, the precipitation pathway enters the two-phase region of the phase diagram above the critical point at which the binodal and spinodal lines intersect. This is important, because it means that precipitation will occur as a liquid droplet in a continuous polymer-rich phase. If overly dilute casting solutions are used, the precipitation pathway enters the two-phase region below the critical point, and precipitation produces polymer gel particles in a continuous liquid phase, forming a membrane that is weak and powdery.

The simplified treatment of membrane formation using the three-component phase diagrams above is about as far as this approach can be usefully taken. Much effort has been made to calculate the pathways through the phase diagrams and to use the results to predict the effect of membrane formation variables on the fine membrane structure. As a quantitative predictor

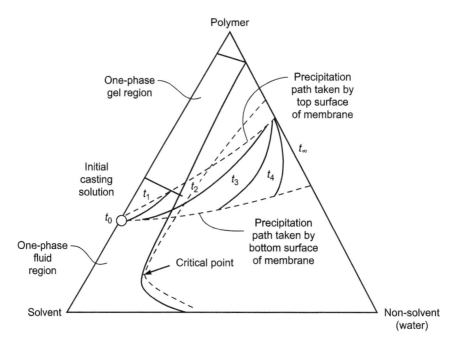

Figure 4.14 *The surface layer of water-precipitation membranes precipitates faster than the underlying substrate. The precipitation pathway is best represented by the movement of a line through the three-component phase diagram. Source: Adapted from [24].*

of membrane performance, this approach has failed. However, as a tool to rationalize the complex interplay of factors determining membrane performance in a qualitative way, the phase diagram approach, albeit making for a heavy read, has proven useful. For those up to the challenge, the literature is reviewed in detail elsewhere [29, 30].

4.3.1.4   Polymer Precipitation by Cooling

Perhaps the simplest preparation technique for phase separation membranes is thermal gelation, often called TIPS, in which a film is cast from a hot, one-phase polymer/solvent solution. As the cast film cools, the polymer precipitates, and the solution separates into a polymer matrix phase containing dispersed pores filled with solvent. Because cooling is usually uniform throughout the cast film, the resulting membranes are relatively isotropic microporous structures, with pores that can be controlled to within 0.1–10 μm. Macrovoids that often occur in Loeb–Sourirajan membranes are usually avoided. A number of microfiltration membranes are made using TIPS.

The precipitation process is described in an early Akzo patent to Castro [31] and can be represented by the phase diagram shown in Figure 4.15. This is a simplified drawing of the actual phase diagram, described later in papers by Lloyd et al. [32, 33] and others [34–36]. The phase diagram shows the metastable region between the bimodal and spinodal phase boundaries discussed in reference to Figure 4.12, with additional complications caused by the crystalline nature of many of the polymers used to form thermal phase separation membranes. The pore volume in the final membrane is determined mainly by the initial composition of the solution, because this determines the ratio of the polymer to liquid phase in the cooled film. However, the spatial distribution and size of the pores are determined largely by the rate of cooling. In general,

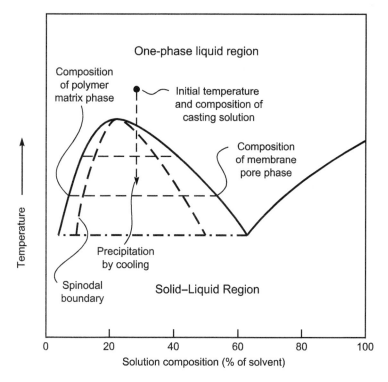

One-phase liquid region

Composition of polymer matrix phase

Initial temperature and composition of casting solution

Composition of membrane pore phase

Precipitation by cooling

Spinodal boundary

Solid–Liquid Region

Temperature

Solution composition (% of solvent)

0   20   40   60   80   100

Figure 4.15 *Phase diagram showing the composition pathway traveled by the casting solution during precipitation by cooling (thermal gelation).*

more rapid cooling produces smaller membrane pores and greater membrane anisotropy [33, 36]. A review of the more recent literature is given by Tang et al. [37].

Membrane preparation by thermal gelation is possible with many polymers, but is mainly used to make membranes from polyethylene, polypropylene, and poly(vinylidene fluoride). The technique was first developed and commercialized by Akzo [36], which produced micro-filtration polypropylene and poly(vinylidene fluoride) membranes marketed under the trade name Accurel®. Flat sheet and hollow fiber membranes were made. The original polypropylene membranes were prepared from a solution of polypropylene in *N,N*-bis(2-hydroxyethyl) tallow amine. The amine and polypropylene formed a clear solution at temperatures above 100–150 °C. If the solution was cooled slowly, an open cell structure of the type shown in Figure 4.16a resulted, with interconnecting passageways between cells in the micron size range. If the solution was cooled and precipitated rapidly, a much finer structure was formed, as shown in Figure 4.16b. The rate of cooling was, therefore, a key parameter determining the final structure of the membrane. The anisotropy of the membranes can be increased by cooling the top and bottom surface of the cast film at different rates.

A schematic diagram of a commercial-scale TIPS process is shown in Figure 4.17. The hot polymer solution is cast onto a water-cooled chill roll, causing the polymer to precipitate. The precipitated film is passed through an extraction tank containing methanol, ethanol, or isopropanol to remove the solvent. Finally, the membrane is dried, sent to a laser inspection station, trimmed, and rolled up.

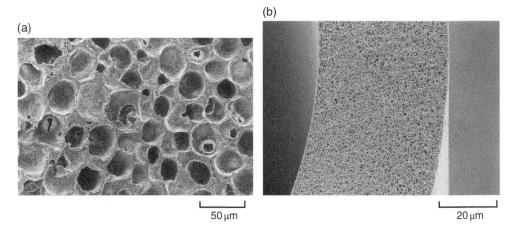

Figure 4.16 *Polypropylene structures (a) Type I: open cell structure formed at low cooling rates, (b) Type II: fine structure formed at high cooling rates. Source: Reproduced with permission from [36]. Copyright (1985) American Chemical Society.*

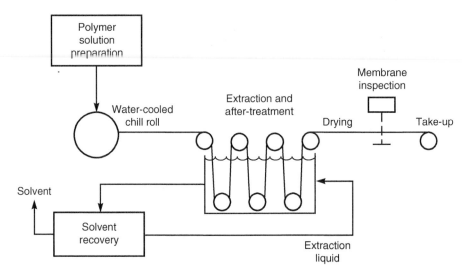

Figure 4.17 *Equipment to prepare microporous membranes using the polymer precipitation by cooling technique. Source: Reproduced with permission from [36]. Copyright (1985) American Chemical Society.*

4.3.1.5   Polymer Precipitation by Solvent Evaporation

This technique was used by Bechhold, Elford, Pierce, Ferry, and others almost a century ago. In its simplest form, a polymer is dissolved in a two-component solvent mixture consisting of a volatile solvent, such as methylene chloride or acetone, in which the polymer is readily soluble, and a less volatile non-solvent, typically water or an alcohol. The polymer solution is cast onto a glass plate and left to evaporate in air. As the volatile solvent evaporates, the casting solution is

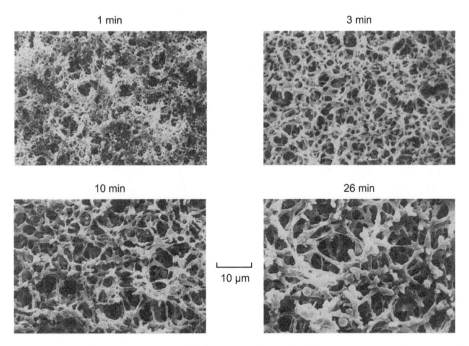

Figure 4.18 *SEM photomicrographs of the bottom surface of cellulose acetate membranes cast from a solution of acetone (volatile solvent) and 2-methyl-2,4-pentanediol (nonvolatile non-solvent). The evaporation time before the structure is fixed by immersion in water is shown. Source: Reproduced with permission from [38]. Copyright (1994) Elsevier.*

enriched in the nonvolatile non-solvent, so the polymer precipitates, forming the membrane structure. The process can be continued until the membrane has completely formed, or it can be stopped, and the membrane structure fixed, by immersing the cast film in a precipitation bath of water or other non-solvent. Precipitation by evaporation is much slower than precipitation by water bath immersion. As a result, the membranes are only modestly anisotropic and have large pores. Scanning electron micrographs of membranes made by this process are shown in Figure 4.18 [38].

Many factors determine the porosity and pore size of membranes formed by the solvent evaporation method. As Figure 4.18 shows, if the membrane is fixed after a short evaporation time, the resulting structure will be finely microporous. If the evaporation step is prolonged, the average pore size will be larger. Increasing the non-solvent content of the casting solution or decreasing the polymer concentration tends to increase porosity. It is important for the non-solvent to be completely incompatible with the polymer. If partially compatible non-solvents are used, the precipitating polymer phase contains sufficient residual solvent to allow it to flow, and the pores will collapse as the solvent evaporates. The result is a dense rather than a microporous film.

4.3.1.6 *Polymer Precipitation by Absorption of Water Vapor*

Preparation of microporous membranes by solvent evaporation alone is not widely practiced. However, a combination of solvent evaporation and absorption of water vapor from a humid atmosphere is an important method of making microfiltration membranes. The processes involve proprietary casting formulations not normally disclosed by membrane developers. However,

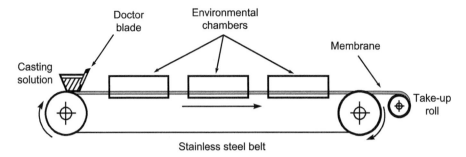

Figure 4.19   *Schematic of casting machine used to make microporous membranes by water vapor absorption. A casting solution is deposited as a thin film on a moving stainless-steel belt moving at about 1–ft/min. The film passes through a series of humid and dry chambers, where the solvent evaporates from the solution, and water vapor is absorbed. This precipitates the polymer, forming a microporous membrane that is taken up on a collection roll.*

during the development of composite membranes at Gulf General Atomic, Riley et al. prepared this type of membrane and described the technology in some detail in a series of Office of Saline Water Reports [39]. These reports remain some of the best descriptions of the technique. The casting solutions are complex and often contain 5–10 components. A typical composition taken from Riley's report comprises 8.1% wt% cellulose nitrate, 1.3 wt% cellulose acetate, 49.5 wt% acetone (a volatile good solvent), 22.3 wt% ethanol, and 14.7 wt% *n*-butanol (nonvolatile poor solvents), 2.6 wt% water (a non-solvent), 0.3 wt% Triton X-100 (a surfactant solution modifier), and 1.2 wt% glycerin (a polymer plasticizer).

The type of equipment used by Riley et al. is shown in Figure 4.19. The casting solution is cast onto a moving stainless-steel belt. The cast film then passes through a series of environmental chambers. Warm, humid air is usually circulated through the first chamber, where the film loses the volatile solvent by evaporation and simultaneously absorbs water. A key issue when making microfiltration membranes is to avoid formation of a dense surface skin on the air side of the membrane. Dense skin formation is generally prevented by incorporating sufficient polymer non-solvent in the casting solution. Polymer precipitation and formation of two phases then occur after a small portion of the volatile solvent component in the mixture has evaporated. The total precipitation process is slow, taking about 10–30 minutes to complete. Typical casting speeds are on the order of 1–5 ft/min. To allow higher casting speeds the casting machine must be very long – commercial machines can be up to 100 ft. After precipitation in the humid chamber(s), the membrane passes to drying chamber(s), through which hot, dry air is circulated to evaporate the remaining solvent and dry the film. The formed membrane is then wound onto a take-up roll. The membrane structure is more isotropic and more microporous than membranes precipitated by immersion in water, making it suitable for use in microfiltration. Membranes made by the water vapor absorption/solvent evaporation precipitation process often have the characteristic nodular form shown in Figure 4.20. A discussion of some of the practical considerations involved in making this type of membrane is given in a book by Zeman and Zydney [40].

4.3.2   Interfacial Polymerization Membranes

Loeb and Sourirajan's work stimulated the development of other novel techniques for making asymmetric structures. One of the most important was interfacial polymerization, an entirely

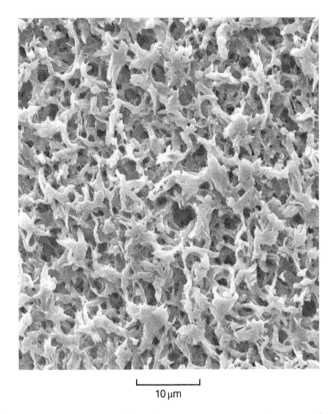

10 μm

Figure 4.20 *Characteristic structure of a phase-separation membrane made by water vapor absorption and solvent evaporation. Source: Reproduced with permission of Millipore Corporation, Billerica, MA.*

new method developed by John Cadotte, then at North Star Research. The general technique is illustrated in Figure 4.21. First, an aqueous solution of a reactive prepolymer, such as a polyamine, is deposited in the pores of a microporous support membrane. The amine-loaded support is then immersed in a solution of a reactant in a water-immiscible solvent, such as a diacid or triacid chloride in hexane. The amine and acid chloride react at the interface of the two immiscible solutions to form a densely crosslinked, extremely thin (0.1 μm or less) membrane layer. A thicker, less crosslinked, more permeable hydrogel layer forms under this surface layer and fills the pores of the support membrane to a depth of a micron or so, as shown in the figure. The nature of the microporous support affects the membrane properties significantly. It should be very finely porous, to withstand the high pressures that will be applied when the membrane is in use for RO, but must also have a high surface porosity, so that it is not a barrier to water flow. Typically, a polysulfone ultrafiltration membrane is used. Because the layer formed at the interface is so thin and highly crosslinked, the finished composite membranes exhibit both high permeances and high selectivities.

Early SEM micrographs of interfacial polymerization membranes showed a relatively smooth featureless membrane surface. Modern microscopes show the corrugated surface illustrated in Figure 4.22 [41]; though the structure is typical, its cause is not well understood. One possibility is that the heat of the interfacial reaction vaporizes a portion of the organic solvent [41]. Osmotic

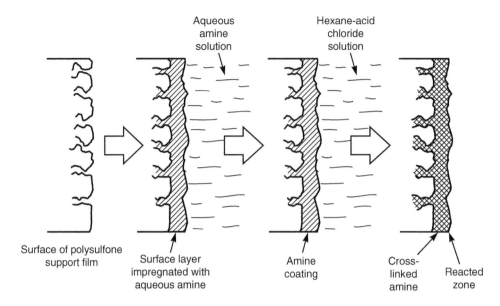

Figure 4.21   *The interfacial polymerization process. The microporous film is impregnated with an aqueous amine solution, then treated with a multivalent crosslinking agent dissolved in a water-immiscible organic fluid, such as hexane or Freon-113. An extremely thin crosslinked polymer film forms at the interface of the two solutions.*

flow across the interfacial layer has also been cited as a mechanism. The surface structure of these membranes has been described in a number of papers [42].

The first membranes made by Cadotte were based on the reaction of a poly(ethyleneimine) (in water) and toluene-2,4-diisocyanate or isophthaloyl chloride (in hexane). The structure of the membrane, known as NS100, is shown in Figure 4.23. The NS100 membranes were much more salt selective than the best cellulose acetate Loeb–Sourirajan membranes and also had better water fluxes. However, they were sensitive to even trace amounts (ppb levels) of chlorine, commonly used as an antibacterial agent in water. The chlorine caused chain cleavage of the polymer at the amide bonds, resulting in loss of salt rejection properties.

The interfacial polymerization process was later refined by Cadotte et al. at FilmTec Corp. [14, 44], Riley et al. at UOP [45], and Kamiyama et al. [46] at Nitto in Japan, and different chemistries were developed. The FT-30 membrane produced by reaction of phenylenediamine with trimesoyl chloride, developed by Cadotte when at FilmTec (now a division of Dow Chemical), is particularly important. This membrane, which has a high water flux and consistent salt rejections of greater than 99.5% with seawater [44, 47], made single-pass seawater desalination possible. A more detailed description of the chemistry of interfacial composite membranes is given in the discussion of RO membranes in Chapter 6 and in a review by Petersen [48] and Lau et al. [49].

Production of interfacial composite membranes in the laboratory is relatively easy, but development of equipment for their large-scale manufacture required some ingenuity. The problem is the fragility of the interfacial surface film. One early solution to this problem is illustrated in Figure 4.24. The support membrane passes from the aqueous amine bath to the organic acid chloride bath and then through a drying/curing oven. The transfer rollers are arranged so that the surface on which the interfacial reaction occurs never contacts a roller until leaving the oven, when the interfacial selective layer is completely formed. The finished

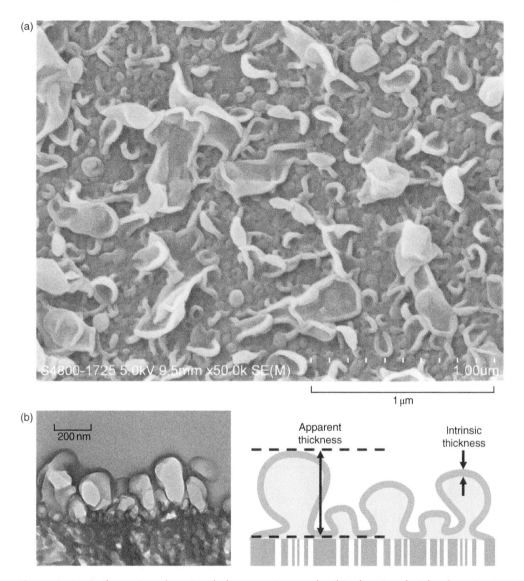

(a)

(b)

Apparent thickness

Intrinsic thickness

Figure 4.22 *Surface (a) and sectional electron micrographs (b) of an interfacial polymerization membrane from a paper by Peng et al. The drawing shows the generally expected congealed structure of the interfacial polyamide on the microporous support membranes. Source: Reproduced with permission from [41]. Copyright (2021) Elsevier.*

composite membrane is coated with a protective solution of a water-soluble polymer, such as poly(vinyl alcohol). When this solution is dried, the membrane is wound onto a take-up roll. The poly(vinyl alcohol) layer protects the membrane from damage during subsequent cutting and rolling into spiral-wound modules. When the module is used for the first time, the feed water washes off the protective layer to expose the membrane, and the module is ready for use. At the present time, very large machines – sometimes 50–100 m long – are used to make 5000-m rolls of membrane at speeds of 20 m/min.

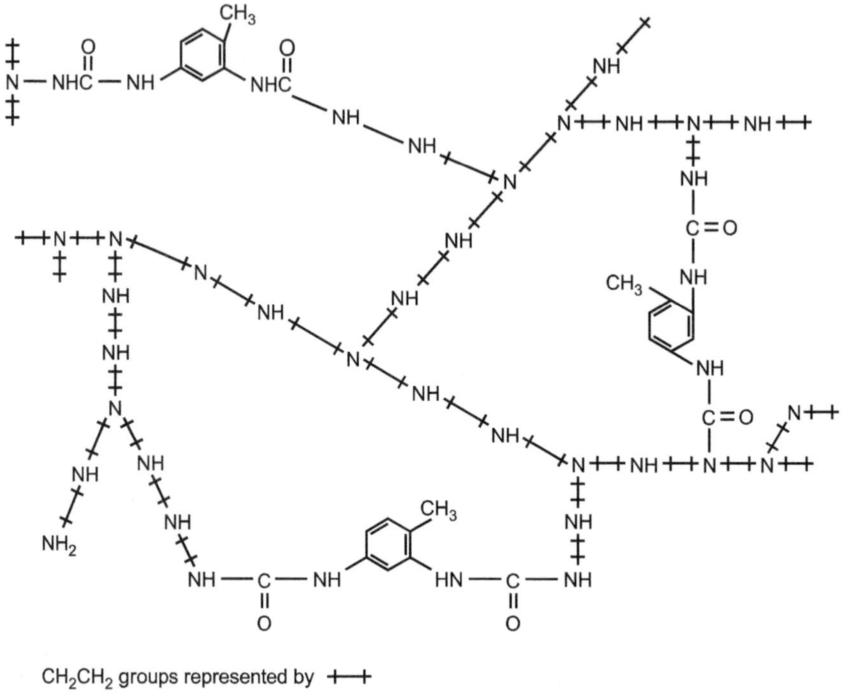

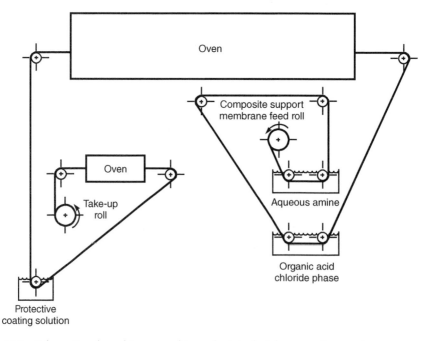

CH₂CH₂ groups represented by +—+

Figure 4.23 Idealized structure of the NS100 membrane, made from poly(ethylenimine) crosslinked with toluene 2,4-diisocyanate. The membrane had almost twice the water flux and one-fifth the salt leakage of the best reverse osmosis membranes then available. *Source: Reproduced with permission from [43] John Wiley and Sons. Copyright (1977) NRC Research Press.*

Figure 4.24 *Schematic of machinery used to make interfacial composite membranes.*

Interfacial polymerization membranes are widely used in reverse osmosis and nanofiltration, but not in gas separation, because of the water-swollen hydrogel filling the pores of the support membrane. In RO, this layer is hydrated and offers little resistance to water flow, but when the membrane is dried for use in gas separation, the gel becomes a rigid glass with a low gas permeability [50]. As a result, defect-free interfacial composite membranes usually have low gas fluxes, although their selectivities can be good.

4.3.3 Solution-Coated Composite Membranes

Another important type of anisotropic membrane is made by solution coating a thin selective layer onto a preformed microporous support membrane. Composite membranes of this type are widely used in gas separation and pervaporation. One of their main advantages is that the microporous support layer can be made in one operation and the selective layer added in a second step, enabling each layer to be tailored to the function it must perform. The support is usually made from a tough amorphous thermoplastic with good mechanical properties. Polysulfone, poly(vinylidene fluoride), poly(acrylonitrile), and poly(etherimide) are often used. The selective layer must have the required permeability and selectivity, but its mechanical strength is relatively unimportant; composite membranes with a selective layer of soft silicone rubber are used successfully in gas separation processes that operate at 100 bar transmembrane pressure. Another advantage of composite membranes is that, because the selective layer is so thin (typically 0.2–2.0 µm), only one or two grams of selective polymer are needed per m^2 of membrane. This allows tailormade polymers costing US$20–50 per gram or even more to be used.

In the laboratory, composite membranes can be made by hand wicking or, in a more reproducible way, with a small spin-coating machine. Spin coating was developed in the electronics industry to coat photoresists and photolithographic films onto silicon wafers. An excess of dilute polymer solution is placed on the substrate, which is then rotated at high speed. Fluid spins off the edge of the rotating substrate until the desired film thickness is achieved. The coating layer thickness can be decreased by increasing the rotation speed or decreasing the polymer concentration in the applied solution. A schematic of a typical laboratory spin coater is shown in Figure 4.25.

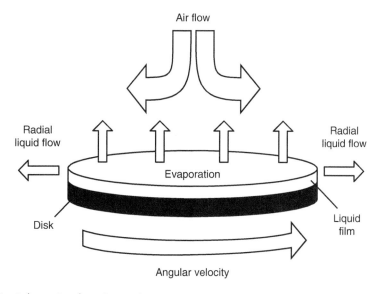

Figure 4.25 *Schematic of a spin-coating apparatus.*

Industrially, most membranes are prepared by the meniscus-coating method developed by Riley [16, 51] and others [52] and shown in Figure 4.26. The support must be clean, defect-free and very finely microporous at the surface, to prevent penetration of the coating solution into the pores. Ultrafiltration membranes meet these conditions, and are often used as supports. The suitable support is coated directly with a dilute polymer solution layer 30–60 μm thick, which after evaporation leaves a thin selective polymer layer 0.2–2 μm thick. Obtaining defect-free films by this technique requires considerable attention to the coating solution parameters and the operating procedures, as well as to the properties of the support. More recent developments have been reviewed by Peng et al. [53] and Xie et al. [54].

Because the selective layer in modern composite membranes is so thin, resistance to transport in the support layer can contribute to the total resistance to transport through the membrane. Not

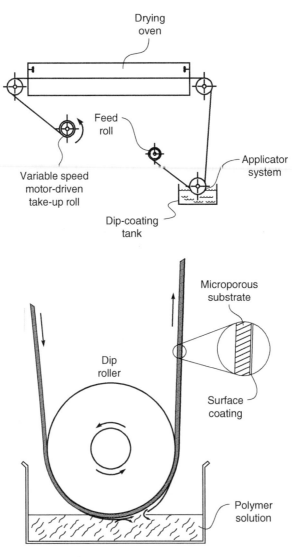

Figure 4.26   *Schematic diagram of a meniscus-coating apparatus.*

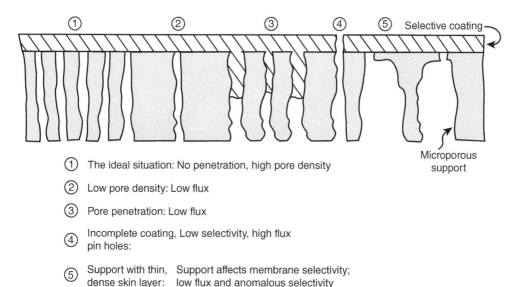

① The ideal situation: No penetration, high pore density

② Low pore density: Low flux

③ Pore penetration: Low flux

④ Incomplete coating, Low selectivity, high flux
    pin holes:

⑤ Support with thin,   Support affects membrane selectivity;
    dense skin layer: low flux and anomalous selectivity

Figure 4.27 *Support membrane properties that affect composite membrane performance.
Source: After Heinzelmann [56].*

only does the resistance of the support decrease the flux through the membrane, but it also degrades the selectivity of the overall membrane [55]. To approach the intrinsic selectivity of the polymer chosen for the selective layer, at least 90% of the resistance to flow should lie within that layer. As well as the support having a high flux, the surface pores must be small enough to support the thin selective layer under high pressure, and must also be close together so that the permeating components do not need to take a long tortuous path to reach the pore. These requirements can be difficult to meet simultaneously.

Figure 4.27 shows some of the problems of solution coating onto microporous supports: (2) low pore density, (3) pore penetration, (4) incomplete coating caused by dust particles or irregularities in the support surface, and (5) regions of the support membrane where the membrane pores come close to the membrane surface but are separated from the selective layer by a dense patch of support polymer. In this last case, some of the permeating gas will permeate through open pores, but some will permeate through the thin support polymer layer before reaching the surrounding pores. The support material selectivity can then affect the overall membrane selectivity. In short, creating the ideal support membrane is often as difficult as finding and depositing the ideal selective layer.

One of the most intractable problems is low support surface porosity (pore density). To minimize pore penetration and to provide the required mechanical strength, the average pore diameter should be less than 500 Å, and preferably less than 200 Å. This type of membrane can be made by the Loeb–Sourirajan technique. However, there is a trade-off between surface porosity and pore diameter; membranes with pore diameters of 500 Å or less typically have a surface porosity of less than 5%. Figure 4.28 illustrates the effect of low pore density in thin composite gas separation membranes in which the selective layer thickness is ℓ and the pore diameter is d. When the selective layer is comparatively thick and the ratio ℓ/d is greater than about 3, gas molecules permeating the membrane not directly above the pore have to travel a greater and less direct distance to the pore than those that approach 'head on', but the difference

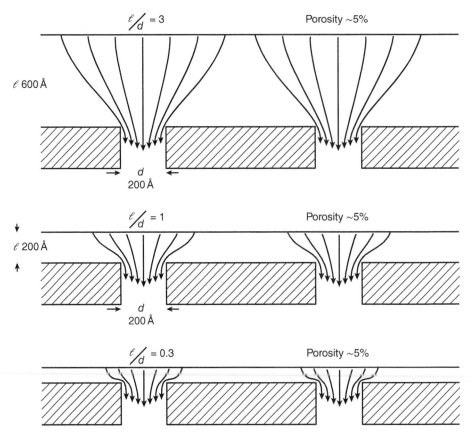

Figure 4.28 *The support porosity and the ratio of the composite membrane thickness (ℓ) to pore diameter (d) affect membrane processes. At a porosity of 5%, membrane permeance is significantly affected by the membrane porosity at ℓ/d ratios of less than 3. At ℓ/d ratios of 1 or less, the membrane permeance is a fraction of the ideal permeance estimated from the permeance in the absence of support effects.*

is small and the effect of low porosity on gas permeation is minor. As the ℓ/d ratio decreases, the effect becomes more pronounced, and when the ratio is significantly less than about 1, the travel path for molecules not in direct line with the pore becomes so constricted that only the membrane area directly above the pore contributes to gas permeation.

Hao and Wijmans [57, 58] have used computational fluid dynamics (CFD) to determine the effect of the support layer porosity and pore size on the overall permeance and selectivity. Their calculated results and some experimental data are shown in Figure 4.29. The selective layer is silicone rubber, which has a CO_2 permeability of 3400 Barrer. Membranes with selective layers between 0.08 and 10 μm were prepared on the same support; the average pore diameter of the support was about 780 Å (0.078 μm). The figure shows that membranes with a selective layer thicker than about 3 μm, and hence ℓ/d ratio of about 40, track the expected intrinsic permeance line, and the support layer, with a surface porosity of about 4.2%, has very little effect on the membranes permeance. At selective layer thickness below 3 μm, however, the measured permeance falls below the expected value. A 0.25 μm thick membrane, with an ℓ/d of about 3, has a permeance of just over 5000 gpu, one-third of the expected intrinsic value of about 15 000 gpu.

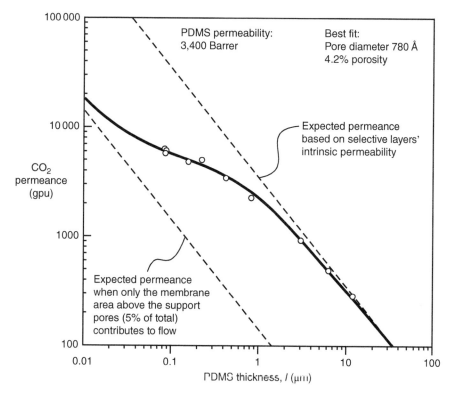

Figure 4.29 *Effect of the selective layer thickness on the overall membrane permeance. Source: Adapted from [58].*

The permeance of very thin membranes (thickness 0.08 μm, ℓ/d ratio of 1) is about 7000 gpu, whereas the expected intrinsic value should be about 45 000 gpu, suggesting that only the membrane directly above a membrane pore (about 5% of the total area) is contributing to flow.

One solution to the above problems is to use an intermediate gutter layer of a highly permeable polymer between the microporous support and the selective layer. The gutter layer is made of a material more permeable than the thin selective layer. As its name implies, this layer acts as a conduit to transport material to the support membrane pores. Finally, because the selective layer of the composite membrane is often very thin and correspondingly delicate, such membranes are often protected by a sealing layer, also formed from a highly permeable material, to protect the membrane from damage during handling. A schematic and a scanning electron micrograph of a multilayer composite membrane are shown in Figure 4.30. A discussion of the issues involved in preparing this type of membrane on a large scale is found in some recent papers by Dong and Sheng [59, 60]

4.3.4   Repairing Membrane Defects

In preparing anisotropic membranes, the goal is to make the selective layer as thin as possible, but still defect free. In good quality membranes, whether integral asymmetric or composite, a thickness as low as 1000 Å can be achieved. With layers as thin as this, defects caused by gas bubbles, dust particles, and support imperfections can be difficult to eliminate. Such defects may not significantly affect the performance of liquid separation membranes used in

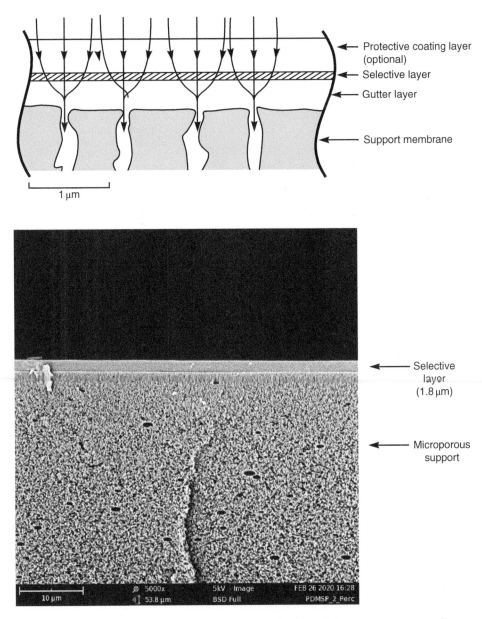

Figure 4.30   *Schematic and scanning electron micrograph of a multilayer composite membrane on a microporous support. Source: Courtesy of Membrane Technology and Research, Inc.*

ultrafiltration and reverse osmosis, but can be disastrous for gas separation. Browall [61] solved this problem for a composite membrane by overcoating the membrane with a second thin layer of a highly permeable polymer to seal defects, as shown in Figure 4.31.

Later, Henis and Tripodi [62] showed that defects in anisotropic Loeb–Sourirajan membranes could be overcome in a similar way. A sufficiently thin coating does not change permeation through the underlying selective layer but does plug defects, through which simple convective

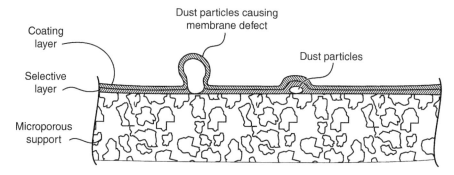

Figure 4.31   *Method developed by Ward, Browall and others at General Electric to seal membrane defects in composite membranes. Source: Adapted from [61].*

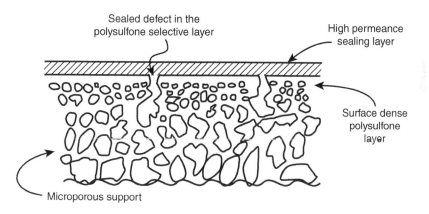

Figure 4.32   *Schematic of Henis and Tripodi sealing technique. Source: Adapted from [62].*

gas flow can occur. Henis and Tripodi's membrane is illustrated in Figure 4.32. The base membrane was a polysulfone Loeb–Sourirajan membrane. The silicone rubber layer is many times more permeable than the polysulfone selective layer and does not function as a selective barrier but rather plugs defects, thereby reducing non-diffusive gas flow. The flow of gas through the portion of the silicone rubber layer over the defect is high compared to the flow through the defect-free portion of the membrane. However, because the area of defects is very small, the total gas flow through the defects is negligible. Because the membrane no longer has to be defect-free, the skin can be made much thinner, and the resultant increase in flux more than compensates for the slight reduction thereof due to the sealing layer.

4.4 Ceramic and Glass Membranes

4.4.1 Ceramic Membranes

Several companies have developed microporous inorganic ceramic membranes for ultrafiltration and microfiltration applications. These membranes are made from aluminum, titanium, or silicon oxides, generally by a slip coating-sintering procedure, and have pore diameters in the

0.01–10 μm range. They have the advantages of being chemically inert and stable at high temperatures, conditions under which polymer membranes fail. This stability makes them particularly suitable for food, biotechnology, and pharmaceutical applications, in which membranes are repeatedly steam sterilized and cleaned with aggressive solutions [28, 63].

In the slip coating-sintering process, a porous ceramic support tube is made by pouring a dispersion of a fine-grain ceramic material and a binder into a mold and sintering at high temperature. The pores between the particles that make up the support tube are large. One surface of the tube is then coated with a suspension of finer particles in a solution of a cellulosic polymer or poly(vinyl alcohol), (known as a slip), which acts as a binder and viscosity enhancer to hold the particles in suspension. When dried and sintered at high temperatures, the coating forms a finely microporous surface layer. Usually several slips are applied sequentially, each layer being formed from a suspension of progressively finer particles. The result is a finely porous anisotropic structure. Most commercial ceramic ultrafiltration membranes are made this way, generally in the form of tubes or perforated blocks. A scanning electron micrograph of the surface of this type of multilayer membrane is shown in Figure 4.33.

The slip coating-sintering procedure can be used to make membranes with pore diameters as small as 100–200 Å, suitable for use in ultrafiltration or as supports for denser membranes used in gas separation. Other techniques, particularly sol–gel methods, are used to produce more finely porous membranes with pores from 10 to 100 Å. In the sol–gel process, the slip coating procedure described above is taken to the colloidal level. Generally, the substrate to be coated with the sol–gel is a microporous ceramic tube formed by the slip coating-sintering technique described above. This support is then solution coated with a colloidal or polymeric gel of an

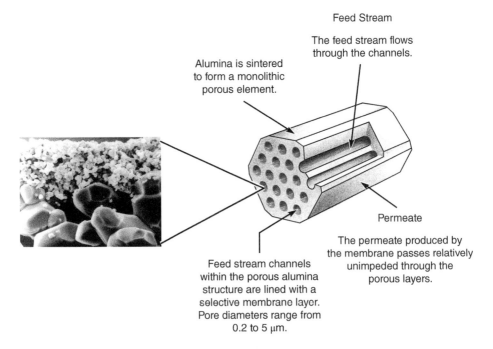

Feed Stream

The feed stream flows through the channels.

Alumina is sintered to form a monolithic porous element.

Feed stream channels within the porous alumina structure are lined with a selective membrane layer. Pore diameters range from 0.2 to 5 μm.

Permeate

The permeate produced by the membrane passes relatively unimpeded through the porous layers.

Figure 4.33 *Cross-sectional scanning electron micrograph of a three-layered alumina membrane/ support (pore sizes of 0.2, 0.8, and 5 μm, respectively). Source: Reproduced with permission from [64]. Copyright (1988) Elsevier.*

inorganic hydroxide. These solutions are prepared by controlled hydrolysis of metal salts or metal alkoxides to hydroxides.

Sol–gel methods fall into two categories, depending on how the colloidal coating solution is formed. The processes are shown schematically in Figure 4.34 [64–66]. In the particulate–sol method, a metal alkoxide dissolved in alcohol is hydrolyzed by addition of excess water or acid. The precipitate of $Al(OH)_3$ that results is maintained as a hot solution for an extended period, during which the precipitate forms a stable colloidal solution. This process is called peptization. The colloidal solution is then cooled and coated onto the microporous support membrane. The layer formed must be dried carefully to avoid cracking the coating. In the final step the film is sintered at 500–800 °C. The overall process can be represented as:

Precipitation: $\quad Al(OR)_3 + H_2O \rightarrow Al(OH)_3$

Peptization: $\quad Al(OH)_3 \rightarrow \gamma\text{-}Al_2O_3 \times H_2O \text{ (Bohmite) or } \delta\text{-}Al_2O_3 \times 3H_2O \text{ (Bayerite)}$

Sintering: $\quad \gamma\text{-}Al_2O_3 \times H_2O \rightarrow \gamma\text{-}Al_2O_3 + H_2O$

In the polymeric sol–gel process, partial hydrolysis of a metal alkoxide dissolved in alcohol is accomplished by adding the minimum of water to the solution. The active hydroxyl groups on

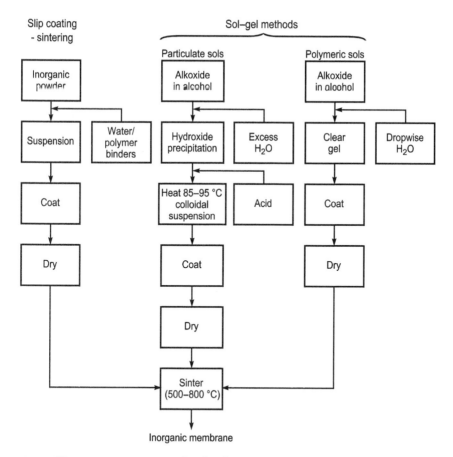

Figure 4.34 *Slip coating-sintering and sol–gel processes used to make ceramic membranes. Source: Reprinted with permission from [65]. Copyright (2000) Elsevier.*

the alkoxides react to form an inorganic polymer precursor molecule that can be coated onto the ceramic support. On drying and sintering, the metal oxide film forms. Chemically the polymeric sol–gel process can be represented as:

$$\text{Hydrolysis:} \quad \text{Ti}(\text{OR})_4 + 2\text{H}_2\text{O} \rightarrow \text{Ti}(\text{OR})_2(\text{OH})_2 + 2\text{ROH}$$

$$\text{Polymerization:} \quad n\text{Ti}(\text{OR})_2(\text{OH})_2 \rightarrow \left[\text{Ti}(\text{OR})_2 - \text{O}\right]_n + \text{H}_2\text{O}$$

$$\begin{array}{c} | \\ \text{O} \\ | \end{array}$$

$$\text{Crosslinking:} \quad \left[\text{Ti}(\text{OR})_2 - \text{O}\right]_n \rightarrow \left[\text{Ti}(\text{OH})_2 - \text{O}\right] \left[\text{Ti}(\text{OH}) - \text{O}\right]$$

Depending on the starting material and the coating procedure, a wide range of membranes can be made by sol–gel processes. The problem of cracking the films on drying and sintering can be alleviated by adding small amounts of a polymeric binder to the coating solution. The coating process may be repeated several times to give a defect-free film. Sol–gel membranes are the subject of research interest, particularly for gas separation applications, but so far have found limited commercial use. They are useful in a number of small industrial applications such as nanofiltration/ultrafiltration of aggressive aqueous solutions or organic solvent mixtures. The membranes are also used as microporous supports for the zeolite, MOF and TR carbon membranes described in Chapter 3. A number of reviews covering the general area of ceramic membrane preparation and use have appeared [28, 63, 67].

4.4.2 Microporous Glass Membranes

Microporous glass membranes in the form of tubes and fibers were made for many years by Corning, PPG, and Schott. The membranes were used in various laboratory applications until 2018 when the last supplier (Corning) discontinued production. The leaching process used to make this type of membrane has been described by Beaver [68]. The starting material is a glass containing 30–70% silica, as well as oxides of zirconium, hafnium, or titanium, and extractable materials. The extractable materials comprise one or more boron-containing compounds and alkali metal oxides and/or alkaline earth metal oxides. Glass hollow fibers produced by melt extrusion are treated with dilute hydrochloric acid at 90 °C for 2–4 hours to leach out the extractable materials, washed to remove residual acid, and then dried.

4.5 Other Membranes

There are a number of specialized membrane types whose use is currently restricted to one particular separation process. Rather than divide these small topic areas into two chapters, I have combined both membrane preparation and membrane use into a single discussion in the specific process chapter. This includes:

Finely microporous membranes (zeolites, MOFs, COFs, PIMs and carbon membranes)	Chapter 3: Microporous Membranes
Mixed matrix membranes	Chapter 3: Microporous Membranes
Liquid membranes	Chapter 12: Carrier Transport
Ion conducting and metal membranes	Chapter 15: Other Membrane Processes

4.6   Hollow Fiber Membranes

The membrane preparation techniques described so far were mostly developed to produce flat-sheet membranes. However, these techniques can be adapted to produce membranes in the form of thin tubes or fibers. An important advantage of hollow fiber membranes is that compact modules with high membrane surface areas can be formed. However, this advantage is offset by the generally lower fluxes compared to flat-sheet membranes made from the same materials. Nonetheless, the development of hollow fiber membranes by Mahon and the group at Dow Chemical in 1966 [69] and their later commercialization by Dow, Monsanto, DuPont, and others represents one of the major events in membrane technology. A good review of the early development of hollow fiber membranes is given by Baum et al. [70]. Other reviews are given by Moch [71], McKelvey et al. [72], and Chung [73].

Like their flat-sheet counterparts, hollow fiber membranes can be isotropic or anisotropic, but most are anisotropic, with a dense selective layer on the outside or inside surface of a microporous support. The dense surface layer can be either integral with the fiber or a separate coating. The diameters of hollow fiber membranes vary over a wide range, from 50 to 3000 μm. Several thousand fibers must be packed into bundles and potted into tubes to form a membrane module; modules with a surface area of even a few square meters require kilometers of fibers. Because a module must contain no broken or defective fibers, hollow fiber production requires high reproducibility and good quality control.

The types of hollow fiber membranes in production are illustrated in Figure 4.35. Fibers less than 200 μm diameter are usually called hollow fine fibers. Such fibers can withstand high hydrostatic pressures applied from the outside without collapsing, so they are used in reverse osmosis or high-pressure gas separation, in which the applied pressure can be 60 bar or more.

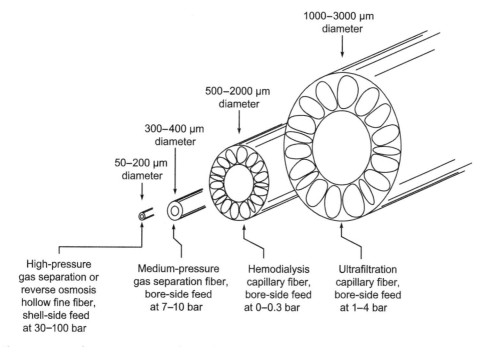

Figure 4.35   *Schematic (not to scale) of the principal types of hollow fiber membranes.*

The feed fluid is applied to the outside (shell side) of the fibers, and the permeate is removed down the fiber bore. When the fiber diameter is greater than 200–500 µm, the feed fluid is commonly applied to the inside bore of the fiber, and the permeate is removed from the outer shell. This technique is used for low-pressure gas separation, nitrogen from air for example, and for hemodialysis and ultrafiltration. Fibers with a diameter of greater than 500 µm are called capillary fibers and are limited to low-pressure applications.

Two methods are used to prepare hollow fibers: solution spinning and melt spinning. The most common process is solution spinning or wet spinning, in which a 20–30 wt% polymer solution is extruded and precipitated into a non-solvent, generally water. Fibers made by solution spinning have the anisotropic structure of Loeb–Sourirajan membranes. This technique is used to make relatively large, porous fibers for hemodialysis and ultrafiltration. In the alternative technique of melt spinning, a hot polymer melt is extruded from a die, cooled, and solidified in air, then immersed in a quench tank. Melt-spun fibers are usually denser and have lower fluxes than solution-spun fibers, but, because the fiber can be stretched after it leaves the die, very fine fibers can be made. Melt-spun fibers can also be produced at high speeds. The technique is used to make hollow fine fibers for high-pressure reverse osmosis and gas separation, and for spinning polymers insoluble in convenient solvents, such as poly(trimethylpentene). The distinction between solution spinning and melt spinning has gradually faded over the years. To improve fluxes, solvents, and other modifiers are generally added to melt spinning dopes; as a result, spinning temperatures have fallen considerably. Many melt spun fibers are now produced from spinning dopes containing as much as 30–60 wt% solvent, and require the spinner to be heated to only 70–100 °C to make the dope flow. These fibers are often cooled and precipitated by spinning into a water bath, which helps to form an anisotropic structure.

The first hollow fiber spinneret system was devised by Mahon at Dow [69]. Mahon's spinneret consisted of two concentric capillaries, the outer capillary having a diameter of approximately 400 µm, and the central capillary having an outer diameter of approximately 200 µm and an inner diameter of 100 µm. Polymer solution is forced through the outer capillary while air or liquid is forced through the inner one. The rate at which the core fluid is injected into the fibers relative to the flow of polymer solution governs the ultimate wall thickness of the fiber. Figure 4.36 shows a cross-section of this type of spinneret, which is still widely used to produce the large-diameter

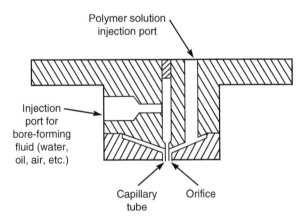

Figure 4.36 *Twin-orifice spinneret design used in solution-spinning of hollow fiber membranes. Polymer solution is forced through the outer orifice; bore-forming fluid is forced through the inner capillary.*

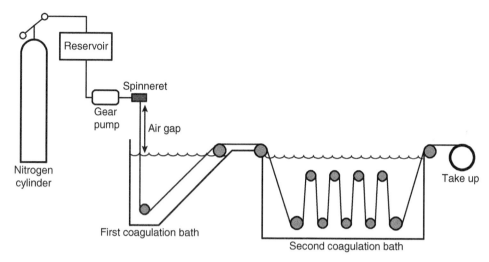

Figure 4.37 *A complete hollow fiber solution-spinning system. The fiber is spun into a coagulation bath, where the polymer spinning solution precipitates forming the fiber. The fiber is then washed, dried, and taken up on a roll.*

fibers used in ultrafiltration. Experimental details of this type of spinneret can be found elsewhere [72, 73]. A complete hollow fiber spinning system is shown in Figure 4.37.

In the laboratory, a fiber spinning system will usually only have a single spinneret, but industrial systems can have as many as 50 spinnerets or more manifolded together using the same spinning solution. These systems operate continuously for weeks or months at a time. Individual spinnerets may be taken offline if malfunctions occur, while the remaining units continue to operate.

The evaporation time between the solution exiting the spinneret and entering the coagulation bath is a critical variable, as are the compositions of the bore fluid and the coagulation bath. The position of the dense anisotropic skin can be adjusted by varying the bath and bore solutions. For example, if water is used as the bore fluid and the coagulation bath contains some solvent, precipitation will occur first and most rapidly on the inside surface of the fiber. If the solutions are reversed so that the bore solution contains some solvent and the coagulation bath is water, the skin will tend to be formed on the outside surface of the fiber. In many cases precipitation will begin on both surfaces of the fiber, and a dense layer will form on both inside and outside surfaces, as shown in Figure 4.38. This ability to manipulate

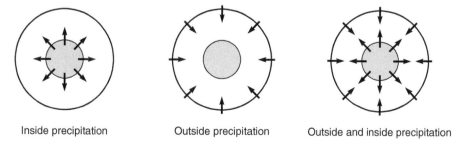

Figure 4.38 *Depending on the bore fluid and the composition of the coagulation bath, the selective skin layer can be formed on the inside, the outside or both sides of the hollow fiber membrane.*

the position of the dense skin is important, because in use the skin should normally face the feed fluid to be separated.

Hollow fiber spinning dope usually has a higher polymer concentration and is more viscous than the casting solutions used to form equivalent flat-sheet membranes. Because hollow fiber membranes must withstand the applied pressure of the process without collapsing, the mechanical demands are higher than for flat sheets (which can be cast on a support paper), and a finer, stronger, higher density support structure is required. The more concentrated casting solutions yield membranes with thicker skins and hence lower fluxes, but the low cost of producing a large membrane area in hollow fiber form can compensate for the lower flux.

Hollow fiber spinning dopes and preparation procedures vary more widely than flat-sheet recipes; some representative dopes and spinning conditions taken from the patent literature are given in Table 4.3.

The two-orifice spinneret shown in Figure 4.36 is used to form fibers from a single polymer solution. Multilayer composite fibers, in which the microporous shell provides mechanical strength, but the selective layer is a coating of a different material, require more complicated methods, various of which are described in the patent literature [77–79]. A common method is to use a double spinneret of the type shown in Figure 4.39 [80, 81]. Although considerable optimization is needed, and delamination at the interface between the two polymer layers can be a problem, this type of procedure can yield a high-quality product with a precisely controlled structure [82, 83]. Scanning electron micrographs of a two-layer pervaporation hollow fiber made this way are shown in Figure 4.40 [84]. Like their flat-sheet equivalents, composite hollow fibers bring the benefits of tailoring of each layer to its function and ability to use expensive materials for the selective layer. Ube, Praxair, Air Products and Medal all produce this type of fiber for gas separation.

Another type of composite fiber is illustrated in Figure 4.41 [85]. These fibers consist of one or two coating layers deposited on a knitted tubular braid, of the type once commonly used to cover electrical wires. The braid increases the strength of the hollow fibers substantially.

The first braid-supported membranes were reported by Hayano et al. of Asahi [86] in 1977, but the importance of the innovation was not widely recognized until Zenon introduced their Zeeweed hollow fibers in the 1990s [87]. The spinneret shown in Figure 4.41 applies two coating solutions to the braid. The first (inner) solution has a lower concentration of polymer so it penetrates into the braid, achieving good adhesion thereto. The second (outer) solution has a higher polymer concentration and forms the selective layer after precipitation in the water bath. Production of these robust fibers was a key step in the development of submerged membrane reactors (described in more detail in Chapter 7). Xia et al. have published a useful review [88].

4.7   Membrane Modules

A single industrial membrane plant may require thousands, or even hundreds of thousands, of square meters of membrane; the ability to package membrane into compact modules at reasonable cost is, therefore, essential. The earliest module designs were based on simple filtration technology and consisted of flat sheets of membrane held in a type of filter press; these are called plate-and-frame modules. Membranes in the form of 1–3 cm diameter tubes were developed at about the same time. Both designs are still used, but because of their high cost, they have been largely displaced in most applications by two other designs – the spiral-wound module and the hollow fiber module.

Table 4.3 *Preparation parameters for various anisotropic hollow fiber membranes.*

Intended use	Gas separation	Ultrafiltration	Reverse osmosis	Pervaporation (water/ethanol)
Membrane	Outside-skinned, fine microporous substrate 50 μm diameter [74]	Inside skin, capillary [75]	80 μm diameter fine fiber [69]	Capillary fiber 2 mm dia post imidized at 150–250°C ~ 6 h. [76]
Dope composition	• 37 wt% polysulfone (Udel® P3500) • 36 wt% *N*-methyl pyrrolidone • 27 wt% propionic acid	25cwt% polyacrylonitrile-vinyl acetate copolymer 68 wt% dimethyl formamide 7 wt% formamide	• 69 wt% cellulose triacetate • 17.2 wt% sulfolane • 13.8 wt% poly(ethylene glycol) (MW 400)	• 13 wt% polyamic acid • 1 wt% polyvinyl pyrrolidone • 17 wt% glycerol • 69 wt% *N*-methyl pyrrolidone
Spinning temp (°C)	15–100	65	200	20
Bore fluid	Water	• 10 wt% dimethyl formamide in water	Air	Water
Precipitation bath	Water	• 40 wt% dimethyl formamide in water	No bath; fiber forms on cooling. Solvents removed in later extraction step	Water
Precipitation temp (°C)	25–50	4	30	30

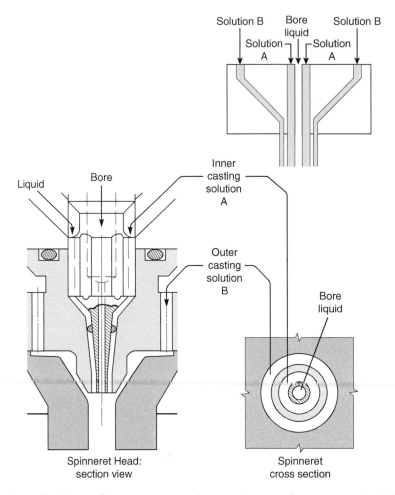

Figure 4.39 *A double capillary spinneret used to produce two-layer composite hollow fibers. Source: After Kopp et al. [80].*

Despite the importance of module technology, many researchers are astonishingly uninformed about module design issues. In part this is because module technology has been developed within companies, and developments are only found in patents, which are often difficult to read. The following sections give an overview of the principal module types, followed by a section summarizing the factors governing selection of particular types for different membrane processes.

4.7.1   Plate-and-Frame Modules

An early plate-and-frame design proposed by Stern et al. [89] for Union Carbide plants to recover helium from natural gas is shown in Figure 4.42. Membrane, feed spacers, and product spacers are layered together between two end plates. The feed mixture is forced across the surface of the membrane. A portion passes through the membrane, enters the permeate channel, and makes its way to a central permeate collection manifold.

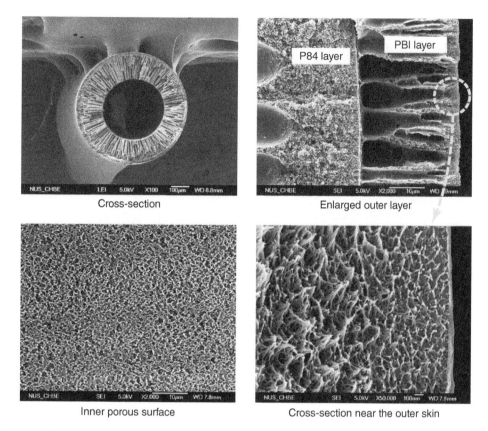

Figure 4.40 *Scanning electron micrographs of dual-layer hollow fiber membranes made from an inner core of P84 polyimide and an outer shell of PBI. Source: Reproduced with permission from [84]. Copyright (2007) Elsevier.*

Plate-and-frame units have been developed for electrodialysis and pervaporation systems, as wells as for some RO and ultrafiltration applications with highly fouling feeds. An example of an RO unit is shown in Figure 4.43 [90].

4.7.2 Tubular Modules

The use of tubular modules is now generally limited to ultrafiltration, for which resistance to membrane fouling (due to good fluid hydrodynamics) outweighs their high cost. Typically, the tubes consist of a porous nonwoven paper or fiberglass support with the membrane formed on the inside of the tubes. The first tubular membranes were between 2 and 3 cm in diameter, but now as many as 19 smaller tubes, each 0.5–1 cm in diameter, are nested inside a single, larger tube that serves as the pressure vessel, as shown in Figure 4.44. This provides a larger membrane area in the same size module housing.

In a typical system, many modules are manifolded in series. The permeate is removed from each tube and sent to a permeate collection header. A drawing of a 30-tube system is shown in Figure 4.45. The feed solution is pumped through all 30 tubes connected in series. This maintains a high fluid velocity in the tubes, which helps to control membrane fouling.

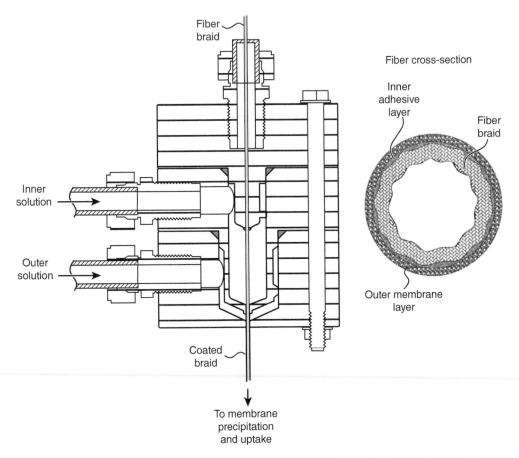

Figure 4.41 *Dual-layer spinneret used to make braid-reinforced hollow fiber membranes. Source:*
[86] / MDPI / CC BY 4.0.

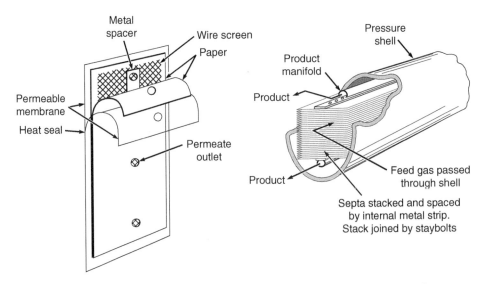

Figure 4.42 *Early plate-and-frame module for the separation of helium from natural gas. Source:*
Reproduced with permission from [89]. Copyright (1965) American Chemical Society.

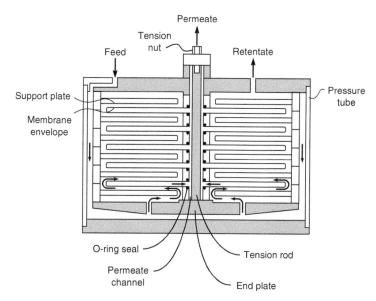

Figure 4.43 *Schematic of a plate-and-frame module. Plate-and-frame modules provide good flow control on both the permeate and feed side of the membrane, but the large number of spacer plates and seals lead to high module costs. The feed solution is directed across each plate in series. Permeate enters the membrane envelope and is collected through the central permeate collection channel. Source: Reproduced from [90] with permission from Elsevier, copyright (1996).*

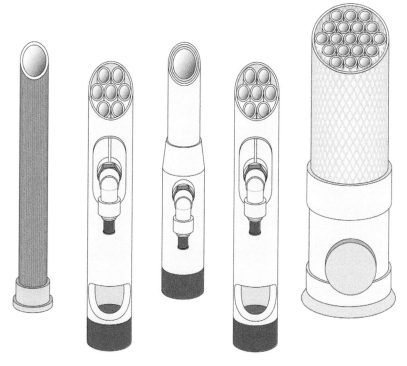

Figure 4.44 *Typical tubular ultrafiltration module design. The membrane is usually cast on a porous fiberglass or paper support, which is then nested inside a plastic or steel support tube.*

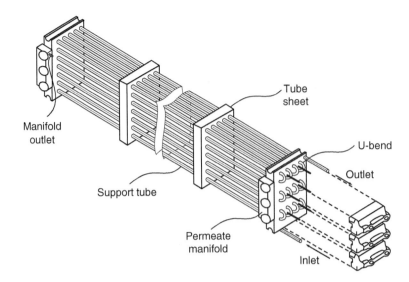

Figure 4.45 *Exploded view of a tubular ultrafiltration system in which 30 tubes are connected in series. Permeate from each tube is collected in the permeate manifold.*

4.7.3 Spiral-Wound Modules

Spiral-wound modules were used in a number of early artificial kidney designs, but were fully developed for industrial use at Gulf General Atomic. This work, directed at reverse osmosis, was carried out under the sponsorship of the Office of Saline Water [91–93]. The design shown in Figure 4.46 is the simplest form, consisting of an envelope of spacers and membrane wound around a perforated central collection tube; the module is placed inside a tubular pressure vessel. Feed passes axially down the module across the membrane envelope. A portion of the feed permeates into the membrane envelope, where it spirals towards the center and exits through the collection tube.

Small laboratory spiral-wound modules typically consist of a single membrane envelope wrapped around the collection tube, as shown in Figure 4.46, and contain $0.2–1.0 \, \text{m}^2$ of membrane. This type of module is widely used in 'under-the-sink' nanofiltration modules to remove calcium and other divalent ions from hard drinking water. Industrial-scale modules contain many membrane envelopes, each with an area of $1–2 \, \text{m}^2$, wrapped around the central collection pipe. The multi-envelope design developed at Gulf General Atomic by Bray [92] and others is illustrated in Figure 4.47. Multi-envelope designs minimize the pressure drop encountered by the permeate fluid traveling towards the central pipe. If a single membrane envelope was used for a large-membrane-area module, the path taken by the permeate to reach the central collection pipe would be several meters long, resulting in a large pressure drop along the permeate collection channel. By using multiple short envelopes, the pressure drop in any one envelope is kept at a manageable level. For many years, the standard industrial reverse osmosis/gas separation spiral-wound module was 8-in. in diameter and 40-in. long. However, there is a trend towards increasing the module diameter, and large reverse osmosis plants now sometimes use 16-in. diameter modules. The approximate membrane area and number of membrane envelopes used in industrial 40-in. long spiral-wound modules are given in Table 4.4.

Spiral-wound module

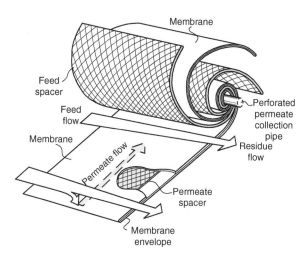

Spiral-wound module cross section

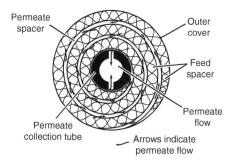

Figure 4.46 *Exploded view and cross-section drawings of a spiral-wound module. Feed fluid (liquid or gas) passes across the membrane surface. A portion passes through the membrane and enters the membrane envelope where it spirals inward to the central perforated collection pipe. A mesh spacer holds the feed and permeate channels open. One fluid enters the module (the feed) and two fluids leave (the residue and the permeate) Spiral-wound modules are commonly used for reverse osmosis, ultrafiltration and high-pressure gas separation.*

Four to six spiral-wound modules are normally connected in series inside a single pressure vessel (tube). A typical 8-in. diameter tube containing six modules has 150–$250\,\mathrm{m}^2$ of membrane area. An exploded view of a vessel containing two modules is shown in Figure 4.48 [93]. The end of each module is fitted with an anti-telescoping device (ATD) to prevent the module leaves shifting under the feed-to-residue pressure difference required to force feed fluid through the module. The ATD is also fitted with a tight rubber seal between the module and the pressure vessel to prevent fluid bypassing the module in the gap between the module and the vessel wall. In some food industry applications, it may be desirable to allow a small portion of the feed solution to bypass the module to prevent bacteria growing in the otherwise stagnant fluid. One way to achieve this bypass is by perforating the ATD [94].

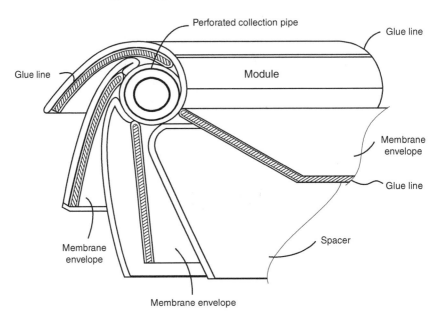

Figure 4.47   *Multi-envelope spiral-wound module [92], used to avoid excessive pressure drops on the permeate side of the membrane. 16-in. diameter modules may have as many as 50 membrane envelopes, each with a membrane area of 1–2 m².*

Table 4.4   *Typical membrane area and number of membrane envelopes for 40-in. long spiral-wound modules of different diameter.*

Module diameter (in.)	4	6	8	16
Number of membrane envelopes	4–6	6–10	15–30	50–100
Membrane area (m²)	3–6	6–12	20–40	80–150

The thickness of the membrane spacers used for different applications causes the variation in membrane area.

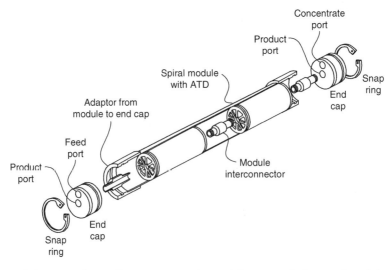

Figure 4.48   *Schematic of a small pressure vessel containing two spiral-wound modules. Typically, four to six modules are installed in a single pressure vessel.*

4.7.4   Hollow Fiber Modules

Hollow fiber modules have two basic geometries. The first is shell-side feed, illustrated in Figure 4.49a. This design is preferred for high-pressure applications and usually operates in the 20–60 bar range. In such a module, a loop or a closed bundle of fibers is contained in a

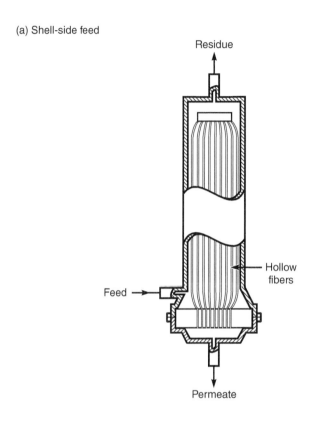

(a) Shell-side feed

Residue

Feed →

Hollow
fibers

Permeate

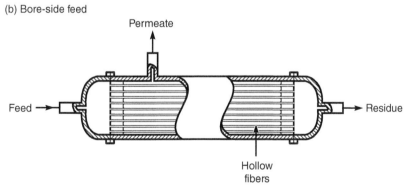

(b) Bore-side feed

Permeate

Feed →

→ Residue

Hollow
fibers

Figure 4.49   *Two types of hollow fiber modules used for gas separation, reverse osmosis, and ultrafiltration. (a) Shell-side feed modules are generally used for high-pressure applications up to 60–70 bar. Fouling on the feed side of the membrane can be a problem; feed pretreatment to remove particulates is required. (b) Bore-side feed modules are generally used for feed streams up to 10 bar, for which good flow control to minimize feed-side fouling and concentration polarization is desired*

pressure vessel. The system is pressurized from the shell side; permeate passes through the fiber wall and exits through the open fiber ends. Because the fiber wall must support a considerable hydrostatic pressure, the fibers usually have small diameters and thick walls, typically 50 μm internal diameter and 100–200 μm outer diameter.

The second type is bore-side feed, illustrated in Figure 4.49b. This design is preferred for lower pressure applications and typically operates at pressures below 10 bar. The fibers are open at both ends, and the feed fluid is circulated through the fiber bores. Capillary fibers are used to minimize pressure drop along the feed channels.

In bore-side-feed modules, it is important to ensure that all of the fibers have identical fiber diameters and permeances. The flow of fluid through the fiber bore is proportional to the fiber diameter to the fourth power, whereas the membrane area only changes by the second power. This means that small differences (±10% or less) in fiber diameter can produce large differences in the fraction of the feed permeating the membrane, and hence in overall module performance [95, 96]. Even a few overly large or small fibers can have a significantly deleterious impact. This is particularly important in the production of nitrogen from air and in hemodialysis, in both of which high levels of removal of the permeable component in a single pass are desired.

Concentration polarization is well controlled in bore-side feed modules. The feed solution passes directly across the active surface of the membrane, and no stagnant dead spaces are produced. This is far from the case in shell-side feed modules, in which flow channeling and stagnant areas between fibers cause significant concentration polarization problems [97]. Also, any suspended particulate matter in the feed solution is easily trapped in these stagnant areas, leading to irreversible fouling of the membrane. Baffles to direct the feed flow have been tried but are not widely used. A more common method of minimizing concentration polarization is to direct the feed flow normal to the direction of the hollow fibers, as shown in Figure 4.50. This produces a cross-flow configuration with relatively good flow distribution across the fiber surface. Several membrane modules may be connected in series, so high feed solution velocities can be used. A number of variants on this basic design have been patented [98, 99] and are reviewed by Koros and Fleming [100].

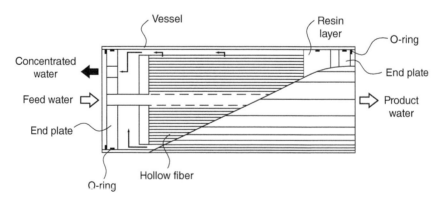

Figure 4.50 *A cross-flow hollow fiber module used to obtain better flow distribution and reduce concentration polarization (the Toyobo Hollosep RO module). Feed enters through the perforated central pipe and flows towards the module shell. Many hollow fiber gas separation modules also have this geometry.*

Table 4.5   *Effect of fiber diameter on membrane area and the number of fibers in a module 8 in. in diameter and 40 in. long.*

Module use	High-pressure RO and gas separation	Low-pressure gas separation		Ultrafiltration	
Fiber diameter (μm)	100	250	500	1000	2000
Number of fibers/module (thousands)	1000	250	40	10	2.5
Membrane area (m^2)	315	155	65	32	16
Packing density (cm^2/cm^3)	100	50	20	10	5

Twenty-five percent of the module volume is filled with fiber. A spiral-wound module of this size contains approximately 20–40 m^2 of membrane area and has a packing density of 6–13 cm^2/cm^3.

A second problem in shell-side-feed hollow fine fibers is permeate side parasitic pressure drops. The bore channel that carries the permeate is so narrow, and presents such a resistance to fluid passage, that a significant pressure drop develops along its length, reducing the pressure difference across the membrane that provides the driving force for permeation. In separation of mixtures of relatively impermeable components, such as oxygen and nitrogen, the pressure drop is small and unimportant. But in separations involving highly permeable gases, such as hydrogen or carbon dioxide, the pressure drop can be a significant fraction of the total applied pressure. Permeate-side pressure drops also tend to develop in spiral-wound modules. However, because the permeate channels are wider, pressure drops are usually smaller and less significant.

The greatest single advantage of hollow fiber modules is the packing density. The magnitude of this advantage is illustrated in Table 4.5, which shows the calculated membrane area contained in an 8-in. diameter, 40-in. long module. A spiral-wound module of this size contains about 20–40 m^2 of membrane area; the equivalent hollow fiber module, filled with fibers with a diameter of 100 μm, contains approximately 300 m^2 of membrane area, 10 times the area in a spiral-wound module. As the diameter of the fibers in the module increases, the module membrane area decreases; capillary modules have almost the same area as equivalent-sized spiral-wound modules.

Table 4.5 also shows the huge numbers of hollow fibers required for high-surface-area modules. A hollow fine fiber module with an area of 300 m^2 will contain 1000 km of fiber, packaged in the module as 1 million fibers each 1 m long. Sophisticated, high-speed automated equipment is required to produce the fibers and modules. A hollow fiber spinning operation may have 50–100 spinnerets. In general, the capital investment for a plant can only be considered when large numbers of modules are to be produced on a round-the-clock basis. The technology is maintained as a trade secret by the companies involved.

4.7.5   Other Module Types

4.7.5.1   *Vibrating and Rotating Modules*

In the module designs described thus far, the feed stream is pumped across the surface of the membrane at high velocity to control concentration polarization. A few vibrating or rotating modules, in which the membrane moves as the fluid flows across its surface, have been developed. One such design is shown in Figure 4.51 [102–104]. Vibration of the membrane at high speed creates agitation of the feed solution directly at the membrane surface. These modules can ultrafilter extremely concentrated, viscous solutions that cannot be treated by conventional

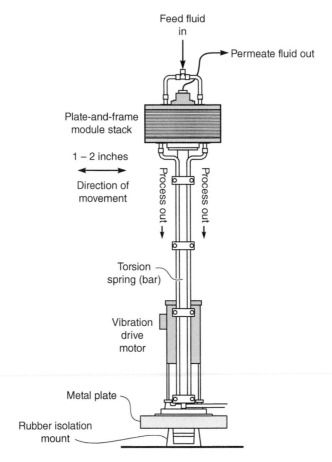

Figure 4.51 *New Logic International vibrating plate-and-frame module [101]. A motor taps a metal plate (the seismic mass) at 60 times/s. A bar connects the mass to a plate-and-frame membrane module, which vibrates by 1–2 in. at the same frequency. By shaking the module, high turbulence is induced in the feed flowing through the module. Turbulence occurs directly at the membrane surface, providing good control of membrane fouling.*

arrangements. The modules are expensive compared to alternative designs and this limits their application to high-value separations that cannot be performed by other processes. Currently available equipment is reviewed by Jaffrin [101].

4.7.5.2 *Submerged Modules*

The development of modules designed to enable frequent backflushing and air scrubbing, and able to operate at low pressures while submerged in heavily contaminated water, has transformed water purification at sewage treatment plants. The design and operation of these modules is described in Chapter 7, Ultrafiltration.

4.8 Module Selection

The suitability of a module for a particular separation must balance a number of considerations. The principal factors are summarized in Table 4.6.

Table 4.6   *Factors affecting module suitability for a given separation.*

Factor	Hollow fine fibers	Capillary fibers	Spiral-wound	Plate-and- frame	Tubular
Manufacturing cost (US$/m^2)	5–20	10–50	10–100	50–200	50–200
Concentration polarization and fouling control	Poor	Good	Moderate	Good	Very good
Permeate-side pressure drop	High	Moderate	Moderate	Low	Low
Suitability for high-pressure operation	Yes	No	Yes	Yes	Marginal
Limited to specific types of membrane material	Yes	Yes	No	No	No

Cost – always important – is difficult to quantify. For example, spiral-wound modules for reverse osmosis are produced, in large volumes using automated machinery, by three or four manufacturers, resulting in strong competition. Modules costs can then be in the US$10/m^2 range. Similar modules used in gas separation or ultrafiltration are produced in much lower numbers, often with more expensive materials, and so are more expensive. Hollow fiber modules are significantly cheaper, per square meter of membrane, than spiral-wound or plate-and-frame modules, but are best produced for high volume applications that justify the expense of developing and building the spinning and module fabrication equipment. This cost advantage is also often offset by the lower fluxes of the membranes compared with their flat-sheet equivalents. An estimate of module manufacturing cost is given in Table 4.6; the selling price can be two to three times higher.

The importance of the other factors listed in the table depends on the application – concentration polarization control is particularly important for liquid separations, less so for gas; hollow fibers cannot be made from some materials; the pressure conditions under which the module will operate may be the paramount consideration – and so on. The types of modules generally used in some of the major membrane processes are listed in Table 4.7.

Table 4.7   *Module designs most commonly used in the major membrane separation processes.*

Application	Module type
RO: seawater	Spiral-wound modules dominate. Only one significant hollow fine fiber producer remains
RO: industrial and brackish water	Spiral-wound modules widely used; bore-side capillary fibers used in low-pressure nanofiltration applications
Ultrafiltration	Tubular, capillary and spiral-wound modules all used. Tubular generally limited to highly fouling feeds (automotive paint), spiral-wound to clean feeds (ultrapure water)
Gas separation	Hollow fibers for high volume applications with low flux, low selectivity membranes in which concentration polarization is easily controlled (nitrogen from air)
	Spiral-wound when fluxes are higher, feed gases more contaminated and concentration polarization a problem (natural gas separations, vapor permeation)
Pervaporation	Plate-and-frame in the first systems and still used. Spiral-wound and capillary modules being introduced with polymeric membranes. Ceramics use tubular

In RO, spiral-wound modules dominate. Plate-and-frame and tubular modules are limited to a few applications in which membrane fouling is particularly severe, for example, in food applications or processing heavily contaminated industrial wastewater. The hollow fine fiber modules used in the past have now been almost completely displaced by spiral-wound modules, which are inherently more fouling resistant, and require less feed pretreatment.

For ultrafiltration, hollow fine fibers have never been seriously considered because of their susceptibility to fouling. If the feed solution is extremely fouling, tubular systems are still used. Recently, however, spiral-wound modules with improved resistance to fouling have been developed; these modules are increasingly displacing the more expensive tubular systems. This is particularly the case with cleaner feed solutions, for example, in the ultrafiltration of boiler feed water or municipal water to make ultrapure water for the electronics industry. Capillary systems are also used in some ultrafiltration applications.

For high-pressure gas separations, hollow fine fibers have a major segment of the market. They are the lowest cost option per unit membrane area, and their poor resistance to fouling is not a problem because gaseous feed streams can easily be filtered to remove oil mist and particulates. Also, gas separation membranes are often made from rigid glassy polymers, such as polysulfones, polycarbonates, and polyimides, which are easily formed into hollow fine fibers. Spiral-wound modules are used to process natural gas streams, which often contain condensable components that would foul hollow fine fiber modules.

Spiral-wound modules are more common in low-pressure gas separation applications, such as the production of oxygen-enriched air. In these applications, the feed is at low ambient pressure, and the permeate is at atmospheric pressure or a modest vacuum. Parasitic pressure drops on the permeate side of the membrane work against hollow fine fiber modules for such applications.

Pervaporation operates under constraints similar to those for low-pressure gas separation. Pressure drops on the permeate side of the membrane must be small, and many pervaporation membrane materials are rubbery, so both spiral-wound modules and plate-and-frame systems are in use. Plate-and-frame systems are competitive in this application despite their high cost, primarily because they can be operated at high temperatures with relatively aggressive feed solutions, conditions under which spiral-wound modules might fail.

4.9   Conclusions and Future Directions

The technology to fabricate ultrathin high-performance membranes and to package them into high-surface-area modules has steadily improved during the modern membrane era. As a result, the inflation-adjusted cost of membrane separation processes has decreased very significantly over the years. The first reverse osmosis membranes made by Loeb–Sourirajan processes had an effective thickness of 0.2–0.4 μm. Currently, several techniques are used to produce commercial membranes with a thickness of 0.1 μm or less. The permeability and selectivity of membrane materials have also increased two- to three-fold during the same period. As a result, membranes today have 5–10 times the flux and better selectivity than membranes available 30 years ago. In gas separation too, membranes with thicknesses of 0.1 μm are used in industrial applications. These trends are continuing. Composite membranes with an effective thickness of less than 0.05 μm have been made in the laboratory using advanced preparation techniques.

As a result of the improvements in membrane permeance, concentration polarization and membrane fouling have increased in importance over time. All membrane processes are affected by these problems, so the development of membrane modules with improved fluid flow (to minimize concentration polarization) and modules formed from membranes that can be easily cleaned (to control fouling) is likely to become an increasingly important area of research for manufacturers.

References

1. Gardner, H.A. and Sward, G.G. (1950). *Physical and Chemical Examination of Paints, Varnishes, Lacquers, and Colors*, 11e. Maryland: H.A. Gardner Laboratory.
2. Mackenzie, K.J. (2015). Film and sheeting material. In: *Kirk-Othmer Encyclopedia of Chemical Technology*, 5e (ed. R.E. Kirk and D.F. Othmer). John Wiley -Interscience.
3. Fleischer, R.L., Alter, H.W., Furman, S.C. et al. (1972). Particle track etching. *Science* **172**: 225.
4. Spohr, R. *Ions tracks and Microtechnology. Principles and Applications*. Braunschweig: Vieweg & Sohn Verlagsgessellschaft mbH.
5. Gingrich, J.E. (1988). The Nuclepore Story. *The 1988 Sixth Annual Membrane Technology/ Planning Conference*, Cambridge, MA.
6. Bierenbaum, H.S., Isaacson, R.B., Druin, M.L., and Plovan, S.G. (1974). Microporous polymeric films. *Ind. Eng. Chem. Proc. Res. Dev.* **13**: 2.
7. Gore, R.W. (1977). Porous products and process therefor. US Patent 4,187,390, issued 5 February 1980.
8. Kim, J., Jang, T.S., Kwon, Y.D. et al. (1994). Structural study of microporous polypropylene hollow fiber membranes made by the melt-spinning and cold-stretching method. *J. Membr. Sci.* **93**: 209.
9. Kim, J., Kim, S.S., Park, M., and Jang, M. (2008). Effects of precursor properties on the preparation of polyethylene hollow fiber membranes by stretching. *J. Membr. Sci.* **318** (1, 2): 201–209.
10. Mizutani, H., Nakamura, S., Kaneko, S., and Okayama, K. (1993). Microporous polypropylene sheets. *Ind. Eng. Chem. Res.* **32**: 221.
11. Chau, C.C. and Im, J.-H. (1988). Process of making a porous membrane. US Patent 4,874,568, issued 17 October 1989.
12. Lopatin, G., Yen, L.Y. and Rogers, R.R. (1987). Microporous membranes from polypropylene. US Patent 4,874,567, issued 17 October 1989.
13. Loeb, S. and Sourirajan, S. (1963). Sea water demineralization by means of an osmotic membrane. In: *Saline Water Conversion II*, Advances in Chemistry Series, vol. **38** (ed. R.F. Gould), 117–132. Washington, DC: American Chemical Society.
14. Cadotte, J.E. (1985). Evolution of composite reverse osmosis membranes. In: *Materials Science of Synthetic Membranes*, ACS Symposium Series, vol. **269** (ed. D.R. Lloyd), 273–294. Washington, DC: American Chemical Society.
15. Ward, W.J. III, Browall, W.R., and Salemme, R.M. (1976). Ultrathin silicone rubber membranes for gas separations. *J. Membr. Sci.* **1**: 99.
16. Riley, R.L., Lonsdale, H.K., Lyons, C.R., and Merten, U. (1967). Preparation of ultrathin reverse osmosis membranes and the attainment of theoretical salt rejection. *J. Appl. Polym. Sci.* **11**: 2143.
17. Michaels, A.S. (1968). High flow membranes. US Patent 3,615,024, issued 26 October, 1971.
18. Strathmann, H., Kock, K., Amar, P., and Baker, R.W. (1975). The formation mechanism of anisotropic membranes. *Desalination* **16**: 179.

19. Manjikian, S. (1967). Desalination membranes from organic casting solutions. *Ind. Eng. Chem. Prod. Res. Dev.* **6**: 23.

20. Saltonstall, Jr. C.W. (1969). Development and testing of high-retention reverse-osmosis membranes. International Conference PURAQUA, Rome, Italy.

21. Pinnau, I. and Koros, W.J. (1988). Defect free ultra high flux asymmetric membranes. US Patent 4,902,422, issued 20 February 1990.

22. Strathmann, H., Scheible, P., and Baker, R.W. (1971). A rationale for the preparation of Loeb–Sourirajan-type cellulose acetate membranes. *J. Appl. Polym. Sci.* **15**: 811.

23. Strathmann, H. and Kock, K. (1977). The formation mechanism of phase inversion membranes. *Desalination* **21**: 241.

24. Wijmans, J.G. and Smolders, C.A. (1986). Preparation of anisotropic membranes by the phase inversion process. In: *Synthetic Membranes: Science, Engineering, and Applications* (ed. P.M. Bungay, H.K. Lonsdale, and M.N. de Pinho), 39–56. Dordrecht: D. Reidel.

25. Altena, F.W. and Smolders, C.A. (1982). Calculation of liquid-liquid phase separation in a ternary system of a polymer in a mixture of a solvent and nonsolvent. *Macromolecules* **15**: 1491.

26. Reuvers, A.J., Van den Berg, J.W.A., and Smolders, C.A. (1987). Formation of membranes by means of immersion precipitation. *J. Membr. Sci.* **34**: 45.

27. So, M.T., Eirich, F.R., Strathmann, H., and Baker, R.W. (1973). Preparation of anisotropic Loeb–Sourirajan membranes. *Polym. Lett.* **11**: 201.

28. Nunes, S.P. and Peinemann, K.-V. (2006). Membrane materials and membrane preparation. In: *Membrane Technology in the Chemical Industry*, 2e (ed. K.-V. Peinemann and S.P. Nunes), 1–90. Weinheim, Germany: Wiley-VCH.

29. Guillen, G.R., Pan, Y., Li, M., and Hoek, E.M.V. (2011). Preparation and characterization of membrane formed by nonsolvent phase separation: a review. *Ind. Eng. Chem. Res.* **50**: 3798.

30. van de Witte, P., Dijkstra, P.J., van den Berg, J.W.A., and Feijen, J. (1996). Phase separation processes in polymer solutions in relation to membrane formation. *J. Membr. Sci.* **117** (1, 2): 1–31.

31. Castro, A.J. (1978). Methods for making microporous products. US Patent 4,247,498, issued 27 January 1981.

32. Lloyd, D.R., Barlow, J.W., and Kinzer, K.E. (1988). Microporous membrane formation via thermally-induced phase separation. In: *New Membrane Materials and Processes for Separation*, AIChE Symposium Series, vol. **261** (ed. K.K. Sirkar and D.R. Lloyd), 84. New York: AIChE.

33. Lloyd, D.R., Kim, S.S., and Kinzer, K.E. (1991). Microporous membrane formation in thermally induced phase separation. *J. Membr. Sci.* **64**: 1.

34. Vadalia, H.J.C., Lee, H.K., Meyerson, A.S., and Levon, K. (1994). Thermally induced phase separations in ternary crystallizable polymer solutions. *J. Membr. Sci.* **89**: 37.

35. Caneba, G.T. and Soong, D.S. (1985). Polymer membrane formation through the thermal-inversion process. *Macromolecules* **18**: 2538.

36. Hiatt, W.C., Vitzthum, G.H., Wagener, K.B. et al. (1985). Microporous membranes via upper critical temperature phase separation. In: *Materials Science of Synthetic Membranes*, ACS Symposium Series, vol. **269** (ed. D.R. Lloyd), 229. Washington, DC: American Chemical Society.

37. Tang, Y., Lin, Y., Ma, X., and Wang, X. (2021). A review on microporous polyvinylidene membrane fabricated via thermally induced phase separation for MF/UF applications. *J. Membr. Sci.* **639**: 119759.

38. Zeman, L. and Fraser, T. (1994). Formation of air-cast cellulose acetate membranes. *J. Membr. Sci.* **87**: 267.

39. Riley, R.L., Lonsdale, H.K., LaGrange, L.D. and Lyons, C.R. (1969). Development of ultra-thin membranes, Office of Saline Water Research and Development Progress Report No. 386, PB# 207036.

40. Zeman, L.J. and Zydney, A.L. (1996). *Microfiltration and Ultrafiltration: Principles and Applications*. New York: Marcel Dekker.

41. Peng, L., Jiang, Y., Wen, L. et al. (2021). Does interfacial vaporization of organic solvent affect the structure and separation properties of polyamide membranes? *J. Membr. Sci.* **625**: 119173.

42. Lim, Y.S., Goh, K., Lai, G.S. et al. (2021). Unraveling the role of support membrane chemistry and pore properties on the formation of thin-film composite polyamide membrane. *J. Membr. Sci.* **640**: 119805.

43. Rozelle, L.T., Cadotte, J.E., Cobian, K.E., and Kopp, C.V. Jr. (1977). Nonpolysaccharide membranes for reverse osmosis: NS-100 membranes. In: *Reverse Osmosis and Synthetic Membranes* (ed. S. Sourirajan), 249–262. Ottawa, Canada: National Research Council.

44. Cadotte, J.E. (1979). Interfacially synthesized reverse osmosis membrane. US Patent 4,277,344, issued 7 July 1981.

45. Riley, R.L., Fox, R.L., Lyons, C.R. et al. (1976). Spiral-wound poly(ether/amide) thin-film composite membrane system. *Desalination* **19**: 113.

46. Kamiyama, Y., Yoshioka, N., Matsui, K., and Nakagome, E. (1984). New thin-film composite reverse osmosis membranes and spiral-wound modules. *Desalination* **51**: 79.

47. Larson, R.E., Cadotte, J.E., and Petersen, R.J. (1981). The FT-30 seawater reverse osmosis membrane-element test results. *Desalination* **38**: 473.

48. Petersen, R.J. (1993). Composite reverse osmosis and nanofiltration membranes. *J. Membr. Sci.* **83**: 81.

49. Lau, W.J., Ismail, A.F., Misdan, N., and Kassim, M.A. (2012). A recent Progress in thin film composite membrane: a review. *Desalination* **287**: 190.

50. Ali, Z., Wang, Y., Ogieglo, W. et al. (2021). Gas separation and water desalination performance of defect-free interfacially polymerized para-linked polyamide thin-film composite membranes. *J. Membr. Sci.* **618**: 118572.

51. Riley, R.L., Lonsdale, H.K., and Lyons, C.R. (1971). Composite membranes for seawater desalination by reverse osmosis. *J. Appl. Polym. Sci.* **15**: 1267.

52. Forester, R.H. and Francis, P.S. (1968). Method of producing an ultrathin polymer film laminate. US Patent 3,551,244, issued 29 December 1970.

53. Peng, L.E., Yang, Z., Long, L. et al. (2022). A critical review of porous substrates of TFC polyamide membranes: mechanisms, membrane performance and future perspectives. *J. Membr. Sci.* **641**: 119871.

54. Xie, K., Fu, Q., Qiao, G.G., and Webley, P.A. (2019). Recent Progress on fabrication of polymeric thin film gas separation membranes for CO_2 capture. *J. Membr. Sci.* **572**: 38.

55. Pinnau, I., Wijmans, J.G., Blume, I. et al. (1988). Gas separation through composite membranes. *J. Membr. Sci.* **37**: 81.

56. Heinzelmann, W. (1991). Fabrication methods for pervaporation membranes. In: *Proceedings of the Fifth International Conference on Pervaporation Processes in the Chemical Industry*. Englewood, NJ: Bakish Material Corporation.

57. Wijmans, J.G. and Hao, P. (2015). Influence of the porous support on diffusion in composite membranes. *J. Membr. Sci.* **494**: 78.

58. Hao, P., Wijmans, J.G., He, Z., and White, L.S. (2020). Effect of pore location and pore size of the support membrane on the performance of composite membranes. *J. Membr. Sci.* **594**: 117465.

59. Dong, S., Wang, Z., Sheng, M. et al. (2020). Scaling up of defect-free flat membrane with ultra-high permeance used for intermediate layer of multilayer composite membrane and oxygen enrichment. *Sep. Purif. Tech.* **239**: 116580.

60. Sheng, M., Dong, S., Qiao, Z. et al. (2021). Large-scale preparation of multilayer composite membranes for post-combustion CO_2 capture. *J. Membr. Sci.* **636**: 119595.

61. Browall, W.R. (1975). Method for sealing breaches in multilayer ultrathin membrane composites. US Patent 3,980,456, issued 14 September 1976.

62. Henis, J.M.S. and Tripodi, M.K. (1980). A novel approach to gas separation using composite hollow fiber membranes. *Sep. Sci. Technol.* **15**: 1059.

63. Li, K. (2007). *Ceramic Membranes for Separation and Reaction*. Chichester, UK: Wiley.

64. Larbot, A., Fabre, J.P., and Guizard and Cot, L. (1988). Inorganic membranes obtained by sol-gel techniques. *J. Membr. Sci.* **39**: 203.

65. Ramsay, J.D.F. and Kallus, S. (2000). Zeolite membranes. In: *Membrane Science and Technology*, vol. **6** (ed. N.K. Kanellopoulos), 373–395. Elsevier.

66. Anderson, M.A., Gieselmann, M.J., and Xu, Q. (1988). Titania and alumina ceramic membranes. *J. Membr. Sci.* **39**: 243.

67. Merlet, R.B., Pizzoccaro-Zilamy, M.-A., Nijmeijer, A., and Winnurst, L. (2020). Hybrid ceramic membranes for organic solvent nanofiltration: state-of-the-art and challenges. *J. Membr. Sci.* **599**: 117839.

68. Beaver, R.P. (1986) Method of production porous hollow silica-rich fibers. US Patent 4,778,499, issued 18 October 1988.

69. Mahon, H.I. (1960). Permeability separatory apparatus, permeability separatory membrane element, method of making the same and process utilizing the same. US Patent 3,228,876, issued 11 January 1966.

70. Baum, B., Holley, W. Jr., and White, R.A. (1976). Hollow fibers in reverse osmosis, dialysis, and ultrafiltration. In: *Membrane Separation Processes* (ed. P. Meares), 187–228. Amsterdam: Elsevier.

71. Moch, I. Jr. (1995). Hollow fiber membranes. In: *Encyclopedia of Chemical Technology*, 4e, vol. **13** (ed. R.E. Kirk and D.F. Othmer), 312. New York: Wiley-Interscience Publishing.

72. McKelvey, S.A., Clausi, D.T., and Koros, W.J. (1997). A guide to establishing fiber macroscopic properties for membrane applications. *J. Membr. Sci.* **124**: 223.

73. Chung, T.S.N. (2008). Fabrication of hollow fiber membranes by phase inversion. In: *Advanced Membrane Technology and Applications* (ed. N.N. Li, A.G. Fane, W.S.W. Ho, and T. Matsuura), 821–839. Hoboken, NJ: Wiley.

74. Malon, R.F. and Cruse, C.A. (1989). Anisotropic gas separation membranes having improved strength. US Patent 5,013,767, issued 7 May 1991.

75. Takao, S. (1980). Process for producing acrylonitrile separation membranes in fibrous form. US Patent 4,409,162, issued 11 October 1983.

76. Cranford, R. and Roy, C. (2004) Solvent resistant asymmetric integrally skinned membranes. Canadian patent application CA2004/001047, published 27 January 2005.

77. Haubs, M. and Hassinger, W. (1990). Method and apparatus for applying polymeric coating. US Patent 5,156,888, issued 20 October 1992.

78. Puri, P.S. (1988). Continuous process for making coated composite hollow fiber membranes. US Patent 4,863,761, issued 5 September 1989.

79. Sluma, H-D., Weizenhofer, R., Leeb, A. and Bauer, K. (1991) Method of making a multilayer capillary membrane. US Patent 5,242,636, issued 7 September 1993.

80. Kopp, C.V., Streeton, R.J.W. and Khoo, P.S. (1990). Extrusion head for forming polymeric hollow fiber. US Patent 5,318,417, issued 7 June 1994.

81. Kusuki, Y., Yoshinaga, T. and Shimazaki, H., (1989). Aromatic polyimide double layered hollow filamentary membrane and process for producing the same. US Patent 5,141,642, issued 25 August 1992.

82. Widjojo, N., Chung, T.S., and Krantz, W.B. (2007). A morphological and structural study of Ultem/P84 copolyimide dual-layer hollow fiber membranes with delamination-free morphology. *J. Membr. Sci.* **294**: 132–146.

83. Pereira, C.C., Nobrega, R., Peinemann, K.-V., and Borges, C.P. (2003). Hollow fiber membranes obtained by simultaneous spinning of two polymer solutions: a morphological study. *J. Membr. Sci.* **226**: 35–50.

84. Wang, K.Y., Chung, T.-S., and Rajagopalan, R. (2007). Dehydration of tetrafluoropropanol (TFP) by pervaporation via novel PBI/BTDA-TDI/MDI co-polyimide (P84) dual-layer hollow fiber membranes. *J. Membr. Sci.* **287**: 60–66.

85. Ji, J. (2005). Method for producing defect free composite membranes. US Patent 7,081,273 Issued 25 July, 2006.

86. F. Hayano, Y. Hashino and K. Ichikawa, Semipermeable composite membranes. US Patent 4,061,821, issued 14 December, 1976.

87. Mailvaganam, M., Fabbricino, L., Rodrigues, C.F. and Donnelly, A.R. (1993). Hollow fiber semipermeable membrane of tubular braid, US Patent 5,472,607, issued 5 December 1995,

88. Xia, L., Ren, J., and McCutcheon, J.R. (2019). Braid reinforced thin film composite hollow fiber nanofiltration membranes. *J. Membr. Sci.* **585**: 109.

89. Stern, S.A., Sinclair, T.F., Gareis, P.J. et al. (1965). Helium recovery by permeation. *Ind. Eng. Chem.* **57**: 49.

90. Günther, R., Perschall, B., Reese, D., and Hapke, J. (1996). Engineering for high pressure reverse osmosis. *J. Membr. Sci.* **121**: 95.

91. Westmoreland, J.C. (1964). Spirally wrapped reverse osmosis membrane cell. US Patent 3,367,504, issued 6 February 1968.

92. Bray, D.T. (1965). Reverse osmosis purification apparatus. US Patent 3,417,870, issued 24 December 1968.

93. Kremen, S.S. (1977). Technology and engineering of ROGA spiral-wound reverse osmosis membrane modules. In: *Reverse Osmosis and Synthetic Membranes* (ed. S. Sourirajan), 371–386. Ottawa, Canada: National Research Council Canada.

94. Parekh, B.S. (ed.) (1988). *Reverse Osmosis Technology*. New York: Marcel Dekker.

95. Crowder, R.O. and Cussler, E.L. (1997). Mass transfer in hollow fiber modules with non-uniform fibers. *J. Membr. Sci.* **134**: 235.

96. Lemanski, J. and Lipscomb, G.G. (2000). Effect of fiber variation on the performance of counter-current hollow fiber gas separation modules. *J. Membr. Sci.* **167**: 241.

97. Lemanski, J. and Lipscomb, G.G. (2002). Effect of Shell-side flows on the performance of hollow fiber gas separation modules. *J. Membr. Sci.* **195**: 215.

98. Eckman, T.J. (1994). Hollow fiber cartridge. US Patent 5,470,469 issued 28 November 1995.

99. de Filippi, R.P. and Pierce, R.W. (1967). Membrane device and method. *US Patent 3,536,611 issued* 27 October 1970.

100. Koros, W.J. and Fleming, G.K. (1993). Membrane based gas separation. *J. Membr. Sci.* **83**: 1.
101. Jaffrin, M.Y. (2008). Dynamic shear-enhanced membrane filtration: a review of rotating discs, rotating membranes and vibrating systems. *J. Membr. Sci.* **324**: 7.
102. Culkin, B., Plotkin, A., and Monroe, M. (1998). Solve membrane fouling with high-shear filtration. *Chem. Eng. Prog.* **94**: 29.
103. Al Akoum, O., Jaffrin, M.Y., Ding, L. et al. (2002). An hydrodynamic investigation of microfiltration and ultrafiltration in a vibrating membrane module. *J. Membr. Sci.* **197**: 37.
104. Akoum, O., Jaffrin, M.Y., Ding, L.H., and Frappart, M. (2004). Treatment of dairy process waters using a vibrating filtration system and NF and RO membranes. *J. Membr. Sci.* **235**: 111–122.

5

Concentration Polarization

5.1 Introduction

In membrane separation processes, a gas or liquid mixture contacts the feed side of the membrane, and a permeate enriched in one of the components of the mixture is withdrawn from the downstream side of the membrane. Because feed mixture components permeate at different rates, concentration gradients can form in the feed and permeate fluids such that the composition at the membrane surfaces is not the same as in the bulk fluid. This changes permeation through the membrane. The phenomenon is called concentration polarization. Figure 5.1 illustrates a reverse osmosis (RO) experiment in which a pressurized mixture of salt and water contacts the membrane surface. The membrane has a higher permeability for water than salt. There is a bulk flow of salt and water to the membrane surface, but the water permeates preferentially and most of the salt is retained. As a consequence, the solution immediately adjacent to the membrane surface becomes enriched in salt on the feed side of the membrane. This concentration polarization increases the concentration difference for salt across the membrane, increasing its flux. The effect is to lower the membrane selectivity. A similar phenomenon occurs in other processes that involve transport of heat or mass across an interface. Mathematical descriptions of

Membrane Technology and Applications, Fourth Edition. Richard W. Baker.
© 2024 John Wiley & Sons Ltd. Published 2024 by John Wiley & Sons Ltd.

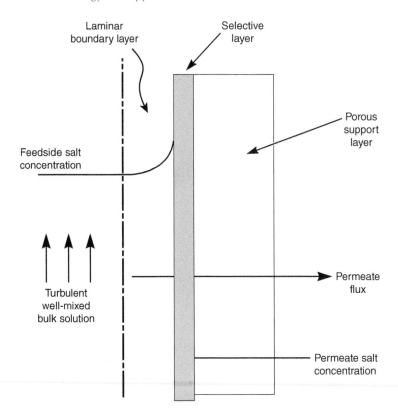

Figure 5.1 *Concentration gradients formed when an RO membrane is used to separate salt from water.*

these processes can be found in monographs on heat and mass transfer; for example, in the books of Carslaw and Jaeger [1]; Bird, Stewart, and Lightfoot [2]; and Crank [3].

The importance of concentration polarization depends on the process. Concentration polarization can significantly affect membrane performance in reverse osmosis, but it is usually well controlled in industrial systems. On the other hand, membrane performance in ultrafiltration, electrodialysis, and some pervaporation processes is seriously affected by concentration polarization.

In most membrane processes, there is volume fluid flow in one direction through the membrane, and the permeate-side composition depends only on the ratio of the components that cross the membrane. When this is the case, concentration gradients only form on the feed side of the membrane. Most of this chapter is devoted to discussion of the phenomenon in those terms. Membrane contactor processes (in which the permeate fluid composition is a combination of flow through the membrane and a permeate sweep flow) can have concentration polarization on both sides of the membrane. These processes are discussed towards the end of the chapter.

Two approaches have been used to describe the effect of concentration polarization. One has its origins in the dimensional analysis used to solve heat transfer problems. In this approach, the resistance to permeation across the membrane and the resistance in the fluid layers adjacent to the membrane are treated as resistances in series. Nothing is assumed about the thickness of the various layers or the transport mechanisms taking place.

Using this model and the assumption that concentration polarization occurs only on the feed side of the membrane, the flux J_i across the combined resistances of the feed side boundary layer and the membrane can be written as

$$J_i = k_{ov}\left(c_{i_b} - c_{i_p}\right) \tag{5.1}$$

where k_{ov} is the overall mass transfer coefficient, c_{i_b} is the concentration of component i in the bulk feed solution, and c_{i_p} is the concentration of component i in the bulk permeate solution. Likewise, the flux across the boundary layer is also J_i and can be written as

$$J_i = k_{b\ell}\left(c_{i_b} - c_{i_o}\right) \tag{5.2}$$

where $k_{b\ell}$ is the fluid boundary layer mass transfer coefficient, and c_{i_o} is the concentration of component i in the fluid at the feed/membrane interface, and the flux across the membrane can be written as

$$J_i = k_m\left(c_{i_o} - c_{i_p}\right) \tag{5.3}$$

where k_m is the mass transfer coefficient of the membrane.

Since the overall concentration drop $(c_{i_b} - c_{i_p})$ is the sum of the concentration drops across the boundary layer and the membrane, a simple restatement of the resistances-in-series model using the terms of Eqs. (5.1) through (5.3) is

$$\frac{1}{k_{ov}} = \frac{1}{k_m} + \frac{1}{k_{b\ell}} \tag{5.4}$$

When the fluid layer mass transfer coefficient ($k_{b\ell}$) is large, the resistance $1/k_{b\ell}$ of this layer is small, and the overall resistance is determined only by the membrane. When the fluid layer mass transfer coefficient is small, the resistance term $1/k_{b\ell}$ is large, and becomes a significant fraction of the total resistance to permeation. The overall mass transfer coefficient (k_{ov}) then becomes smaller, and the flux decreases. The boundary layer mass transfer coefficient is thus an arithmetical fix used to correct the membrane permeation rate for the effect of concentration polarization. Nothing is revealed about the causes of concentration polarization.

The boundary layer mass transfer coefficient is known from experiments to depend on many system properties; this dependence can be expressed as an empirical relationship of the type

$$k_{b\ell} = \text{constant } Q^\alpha h^\beta D^\gamma T^\delta..... \tag{5.5}$$

where, for example, Q is the fluid velocity through the membrane module, h is the feed channel height, D is the solute diffusion coefficient, T is the feed solution temperature, and so on. Empirical mass transfer correlations obtained this way can be used to estimate the performance of a new membrane unit by extrapolation from an existing body of experimental data [4–7]. However, these correlations have a limited range of applicability and must be reformulated with different coefficients for each new process and module design. The correlations cannot be used to obtain an a priori estimate of the magnitude of concentration polarization for a new process. This approach also does not provide insight into the dependence of concentration polarization on membrane properties. More detailed and more sympathetic descriptions of the mass transfer approach have been given by Cussler [8] and by Zeman and Zydney [9].

The second approach to concentration polarization, and the one used in this chapter, is to model the phenomenon by assuming that a thin layer of unmixed fluid, thickness δ, exists between the membrane surface and the well-mixed bulk solution. The concentration gradients that control concentration polarization form in this layer. This boundary layer film model oversimplifies the fluid hydrodynamics occurring in membrane modules and still contains one adjustable parameter, the boundary layer thickness. Nonetheless, this simple approach gives a physical description of the process that can explain most of the experimental data.

5.2   Boundary Layer Film Model

The usual starting point for the boundary layer film model is illustrated in Figure 5.2, which shows the velocity profile in a fluid flowing through the feed channel of an ultrafiltration or reverse osmosis membrane module. The average velocity of the fluid flowing down the channel is normally on the order of 10–20 cm/s. This velocity is far higher than the average velocity of the fluid flowing at right angles through the membrane, which is typically 10–20 µm/s. However, the velocity in the channel is not uniform. Friction at the fluid-membrane surface reduces the fluid velocity next to the membrane to essentially zero; the velocity increases as the distance from the membrane surface increases. Thus, the fluid flow velocity in the middle of the channel is high, the flow there is turbulent, and the fluid is well mixed. The velocity in the boundary layer next to the membrane is much lower, flow is laminar, and mixing occurs by diffusion. Concentration gradients due to concentration polarization are assumed to be confined to the boundary layer.

In any process, if one component is enriched at the membrane surface, then mass balance dictates that a second component is depleted at the surface. By convention, concentration polarization effects are described by considering the concentration gradient of the minor component, whether this is the more permeable or the less permeable component. Figure 5.3 shows the two possible general cases. In Figure 5.3a, the minor component is rejected by the membrane (as, for example, salt in reverse osmosis). In Figure 5.3b, the minor component permeates preferentially (as, for example, water in ethanol dehydration by pervaporation).

In the case of desalination of water by reverse osmosis, as in Figure 5.3a, the salt concentration c_{i_o} adjacent to the membrane surface is higher than the bulk solution concentration c_{i_b} because RO membranes preferentially permeate water and retain salt. Water and salt are brought toward

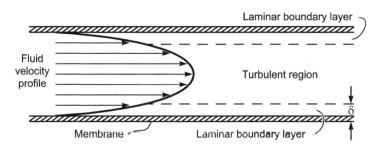

Figure 5.2   *Fluid flow velocity through an ultrafiltration/reverse osmosis membrane module channel is non-uniform, being fastest in the middle and essentially zero adjacent to the membrane. In the film model of concentration polarization, concentration gradients formed due to transport through the membrane are assumed to be confined to the laminar boundary layer.*

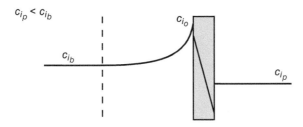

(a) Component enriched at membrane surface
(for example, salt in desalination of water by reverse osmosis)

(b) Component depleted at membrane surface
(for example, water in dehydration of ethanol by pervaporation)

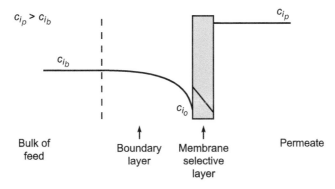

Figure 5.3 *Concentration gradients formed as a result of permeation through a selective membrane. By convention, concentration polarization is usually represented by the gradient of the minor component – salt in the reverse osmosis example (a) and water in the pervaporation example (dehydration of an ethanol solution) (b).*

the membrane surface by the flow of solution through the membrane, designated J_v.[1] Water and a little salt permeate the membrane, but most of the salt is retained and accumulates at the membrane surface, until a sufficient concentration gradient has formed to allow it to diffuse back to the bulk solution. Steady state is then reached.

In the case of dehydration of ethanol by pervaporation, as in Figure 5.3b, the water concentration c_{i_o} adjacent to the membrane surface is lower than the bulk solution concentration c_{i_b} because the pervaporation membrane preferentially permeates water and retains ethanol. Water and ethanol are brought towards the membrane surface by the flow of solution through the membrane. Water and a little ethanol permeate the membrane, but most of the ethanol is retained

---

[1] In this chapter, the term J_v is the volume flux ($cm^3/cm^2 \cdot s$) through the membrane measured at the feed-side conditions of the process.

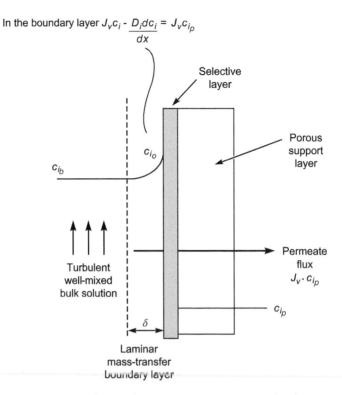

In the boundary layer $J_v c_i - \dfrac{D_i dc_i}{dx} = J_v c_{i_p}$

Figure 5.4 *Salt concentration gradients adjacent to a reverse osmosis desalination membrane. The mass balance equation for solute flux across the boundary layer is the basis of the film model description of concentration polarization.*

and accumulates at the membrane surface, until a sufficient concentration gradient has formed to allow it to diffuse back to the bulk solution. An equal and opposite water gradient must form; thus, water becomes depleted at the membrane surface.

The formation of these concentration gradients can be expressed in mathematical form. Figure 5.4 shows the steady-state salt gradient that forms across a reverse osmosis membrane.

The salt flux through the membrane is given by the product of the permeate volume flux J_v and the permeate salt concentration c_{i_p}. For dilute liquids, the permeate volume flux is within one or two percent of the volume flux on the feed side of the membrane, because the densities of the two solutions are almost equal. This means that, at steady state, the net salt flux at any point within the boundary layer must be equal to the permeate salt flux, $J_v c_{i_p}$. In the boundary layer, this net salt flux is equal to the convective salt flux towards the membrane, $J_v c_i$, minus the diffusive salt flux away from the membrane expressed by Fick's law ($D_i dc_i/dx$). So, from simple mass balance, transport of salt at any point within the boundary layer can be described by the equation

$$J_v c_i - D_i dc_i/dx = J_v c_{i_p} \qquad (5.6)$$

where D_i is the diffusion coefficient of the salt, x is the coordinate perpendicular to the membrane surface, and J_v is the volume flux in the boundary layer generated by permeate flow through the membrane. The mass balance equation (Eq. (5.6)) can be integrated over the thickness of the boundary layer to give the well-known polarization equation derived by Brian [10] for RO:

$$\frac{c_{i_o} - c_{i_p}}{c_{i_b} - c_{i_p}} = \exp(J_v\delta/D_i) \tag{5.7}$$

In this equation, c_{i_o} is the concentration of solute in the feed solution at the membrane surface, and δ is the thickness of the boundary layer. An alternative form of Eq. (5.7) replaces the concentration terms by an enrichment factor E, defined as c_{i_p}/c_{i_b}. The enrichment obtained in the absence of a boundary layer, that is, the intrinsic enrichment of the membrane, E_o, is then defined as c_{i_p}/c_{i_o}, and Eq. (5.7) can be written as

$$\frac{1/E_o - 1}{1/E - 1} = \exp(J_v\delta/D_i) \tag{5.8}$$

In the case of RO, the actual and intrinsic enrichment factors of salt (E and E_o) are much less than 1.0, typically about 0.01, because the membrane rejects salt almost completely and permeates water. For other processes, such as dehydration of aqueous ethanol by pervaporation, the enrichment factor for water will be greater than 1.0, and perhaps as high as 10–20, because the membrane selectively permeates the water.

The increase or decrease in the concentration of the permeating component at the membrane surface c_{i_o}, compared to the bulk solution concentration c_{i_b}, determines the extent of concentration polarization. The ratio of the two concentrations, c_{i_o}/c_{i_b}, is called the concentration polarization modulus and is a useful measure of the extent of concentration polarization. When the modulus is 1.0, the concentration at the membrane surface (c_{i_o}) is equal to the bulk concentration (c_{i_b}), and no concentration polarization occurs. As the modulus deviates farther from 1.0, the effect of concentration polarization on membrane selectivity and flux becomes increasingly important. From the definitions of E and E_o, the concentration polarization modulus is equal to E/E_o, that is, the actual enrichment factor divided by the intrinsic enrichment factor, and from Eqs. (5.7) and (5.8), the modulus can be written as

$$\frac{c_{i_o}}{c_{i_b}} = \frac{\exp(J_v\delta/D_i)}{1 + E_o[\exp(J_v\delta/D_i) - 1]} \tag{5.9}$$

Depending on the intrinsic enrichment (E_o) of the membrane, the modulus can be larger or smaller than 1.0. For RO, E_o is less than 1.0 and the concentration polarization modulus is normally between 1.1 and 1.5; that is, the concentration of salt at the membrane surface is 1.1–1.5 times larger than it would be in the absence of concentration polarization. The salt leakage through the membrane and the osmotic pressure that must be overcome to produce a flow of water are increased proportionately. Fortunately, modern RO membranes are extremely selective and permeable, and can still produce useful desalted water under these conditions. In other processes, such as pervaporation or ultrafiltration, the concentration polarization modulus may be as large as 5–10 or as small as 0.2–0.1, and may seriously affect the membrane performance.

Equation (5.9) shows the factors that determine the magnitude of concentration polarization, namely the boundary layer thickness δ, the intrinsic membrane enrichment E_o, the volume flux through the membrane J_v, and the diffusion coefficient of the solute in the boundary layer fluid D_i. The effect of changes in each of these parameters on the concentration gradients formed in the

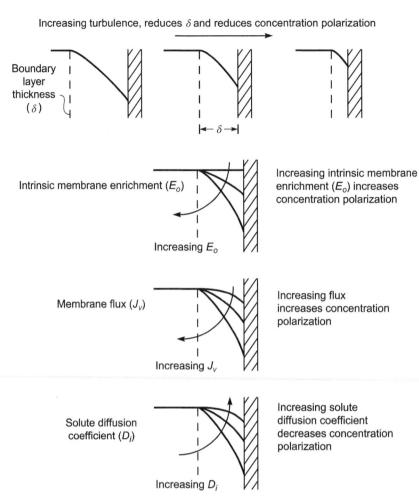

Figure 5.5   *The effect of changes in boundary layer thickness δ, intrinsic membrane enrichment E_o, membrane flux J_v, and solute diffusion D_i on concentration gradients in the stagnant boundary layer ($E_o > 1$).*

membrane boundary layer is illustrated graphically in Figure 5.5 for a process in which the intrinsic enrichment (E_o) is greater than 1.0, for example dehydration of ethanol by pervaporation.

Of the four factors that affect concentration polarization, the one most easily changed is the boundary layer thickness δ. As δ decreases, Eq. (5.9) shows that the concentration polarization modulus becomes exponentially smaller. Thus, the most straightforward way of minimizing concentration polarization is to reduce the boundary layer thickness by increasing turbulent mixing at the membrane surface. Factors affecting turbulence in membrane modules are described in detail in the review of Belfort et al. [11]. The most direct technique to promote mixing is to increase the feed flow velocity past the membrane surface. Therefore, many membrane modules operate at relatively high feed velocities. However, the energy consumption of the feed pumps places a practical upper limit on this approach. Membrane spacers made of open mesh plastic netting are also widely used to promote turbulence by disrupting fluid flow in the module channels, as shown in Figure 5.6. The selection of appropriate feed channel spacers is

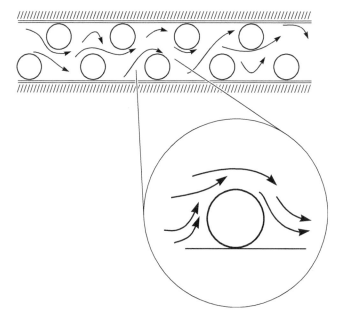

Figure 5.6 *Flow dynamics around the spacer netting often used to promote turbulence in a membrane module and reduce concentration polarization.*

an important issue for membrane module producers. Most producers select their spacers based on module performance results. In recent years, attempts have been made to put spacer selection on a more scientific basis using laboratory measurements and computer-aided design [12–14]. Pulsing the feed fluid flow through the membrane module is another technique used to control concentration polarization [15].

As has been explained and discussed above, the intrinsic enrichment E_o also affects concentration polarization. In pervaporation of organics from water, for example, concentration polarization is much more important when the solute is toluene (with an intrinsic enrichment E_o of 5000 over water) than when the solute is methanol (with an intrinsic enrichment E_o less than 5).

Another important characteristic of Eq. (5.9) is that it is the intrinsic enrichment E_o produced under actual operating conditions, not the selectivity, that determines the concentration polarization modulus. Enrichment and selectivity are linked but are not identical. This distinction is illustrated by the separation of hydrogen from inert gases in ammonia plant purge gas streams, which typically contain 30% hydrogen. The selectivity of the membrane for hydrogen over the inert gases (nitrogen, methane and argon) in this case is often between 100 and 200. The high selectivity means that the membrane permeate is 97% hydrogen; even so, because the feed gas contains 30% hydrogen, the enrichment E_o is only 97/30, or 3.3, so the concentration polarization modulus is negligible. On the other hand, as hydrogen is removed, its concentration in the feed gas falls. When the feed gas contains 5% hydrogen, the permeate will be 90% hydrogen and the intrinsic enrichment 90/5 or 18. Under these conditions, concentration polarization may affect the membrane performance.

Equation (5.9) and Figure 5.5 show that concentration polarization increases exponentially as the total volume flow J_v through the membrane increases. This is one of the reasons why modern spiral-wound reverse osmosis modules are operated at low pressures. Modern reverse osmosis membranes have two to five times the water permeability, at equivalent salt selectivities, of the first-generation cellulose acetate membranes. If membrane modules containing these new membranes were operated at the same pressures as the early cellulose acetate hollow fiber modules, two to five times the desalted water throughput could be achieved with the same number of modules. However, at such high fluxes, spiral-wound modules suffer from excessive concentration polarization, which leads to increased salt leakage and scale formation. This is one of the reasons modern modules are operated at about the same volume flux as the early modules, but at lower applied pressures. This reduces energy costs and controls concentration polarization.

The final parameter in Eq. (5.9) that determines the value of the concentration polarization modulus is the diffusion coefficient D_i of the solute away from the membrane surface. The size of the solute diffusion coefficient explains why concentration polarization is a greater factor in ultrafiltration than in reverse osmosis. Ultrafiltration membrane fluxes are usually higher than reverse osmosis fluxes, but the difference between the values of the diffusion coefficients of the retained solutes is more important. In reverse osmosis the solutes are dissolved salts, whereas in ultrafiltration the solutes are colloids and macromolecules. The diffusion coefficients of these high-molecular-weight components are about 100 times smaller than those of salts.

In Eq. (5.9), the balance between convective transport J_v and diffusive transport D_i/δ in the membrane boundary layer is characterized by the dimensionless ratio $J_v\delta/D_i$, commonly called the Peclet number. When the Peclet number is large ($J_v \gg D_i/\delta$), convective transport overwhelms diffusive transport, and the concentration polarization modulus is large. When the Peclet number is small ($J_v \ll D_i/\delta$), convection is easily balanced by diffusion in the boundary layer, and the concentration polarization modulus is close to unity.

Wijmans et al. [16] calculated the concentration polarization modulus using Eq. (5.9) as a function of the Peclet number. The resulting informative plot is shown in Figure 5.7. The figure is divided into two regions, depending on whether the concentration polarization modulus, c_{i_o}/c_{i_b}, is smaller or larger than 1.

- The polarization modulus is smaller than 1 when the permeating minor component is enriched in the permeate. In this case, the component becomes depleted in the boundary layer, for example, in the dehydration of ethanol by pervaporation shown in Figure 5.3b.

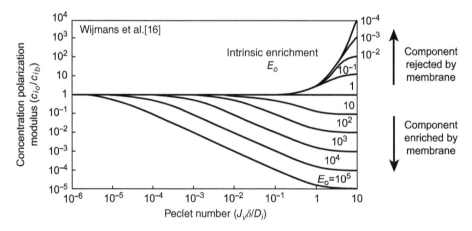

Figure 5.7 *Concentration polarization modulus c_{i_o}/c_{i_b} as a function of the Peclet number $J_v\delta/D_i$ for a range of values of the intrinsic enrichment factor E_o. Lines calculated through Eq. (5.9). This figure shows that components that are enriched by the membrane ($E_o > 1$) are affected more by concentration polarization than components that are rejected by the membrane ($E_o < 1$). Source: Adapted from [16].*

- The polarization modulus is larger than 1 when the permeating minor component is depleted in the permeate. In this case, the component is enriched in the boundary layer, for example, in the reverse osmosis of salt solutions shown in Figure 5.3a.

As might be expected, the concentration polarization modulus increases as the Peclet number increases; in other words, as the convective volume flux term J_v becomes increasingly large relative to the diffusion term D_i/δ, concentration polarization becomes increasingly significant. At high values of the ratio $J_v \delta/D_i$, the exponential term in Eq. (5.9) increases toward infinity, and the concentration polarization modulus c_{i_o}/c_{i_b} approaches a limiting value of $1/E_o$.

A striking feature of Figure 5.7 is its asymmetry with respect to enrichment and rejection of the minor component by the membrane. This means that, under comparable conditions, concentration polarization is much larger when the minor component of the feed is preferentially permeated by the membrane than when it is rejected. This follows from the form of Eq. (5.9). Consider the case when the Peclet number $J_v\delta/D_i$ is 1. The concentration polarization modulus expressed by Eq. (5.9) then becomes

$$\frac{c_{i_o}}{c_{i_b}} = \frac{\exp(1)}{1 + E_o[\exp(1) - 1]} = \frac{2.72}{1 + E_o(1.72)} \tag{5.10}$$

For components rejected by the membrane ($E_o \leq 1$), the enrichment E_o produced by the membrane lies between 1 and 0. The concentration polarization modulus c_{i_o}/c_{i_b} then lies between 1 (no concentration polarization) and a maximum value of 2.72. That is, the flux of the less permeable component cannot be more than 2.72 times higher than that in the absence of concentration polarization. In contrast, for a component enriched by the membrane in the permeate ($E_o \geq 1$), no limit to the magnitude of concentration polarization exists. For dilute solutions (c_{i_b} small) and selective membranes, the intrinsic enrichment can be 100–1000 or more. The concentration polarization modulus can then change from 1 (no concentration polarization) to close to 0 (complete concentration polarization). These two cases are illustrated in Figure 5.8.

5.2.1 Determination of the Peclet Number

Equation (5.9) and Figure 5.7 are powerful tools to analyze the importance of concentration polarization in membrane separation processes. However, before these tools can be used, the appropriate value to be assigned to the Peclet number $J_v\delta/D_i$ must be determined. The volume flux J_v through the membrane is easily measured, so determining the Peclet number becomes a problem of measuring the coefficient D_i/δ.

One approach to the boundary layer problem is to determine the ratio D_i/δ experimentally. This can be done using a procedure first proposed by Wilson [17]. The starting point for Wilson's approach is Eq. (5.8), which can be written as

$$\ln\left(1 - \frac{1}{E}\right) = \ln\left(1 - \frac{1}{E_o}\right) - J_v\delta/D_i \tag{5.11}$$

The boundary layer thickness δ in Eq. (5.11) is a function of the feed fluid velocity u in the module feed flow channel; thus, the term D_i/δ can be expressed as

$$\frac{D_i}{\delta} = k_o u^n \tag{5.12}$$

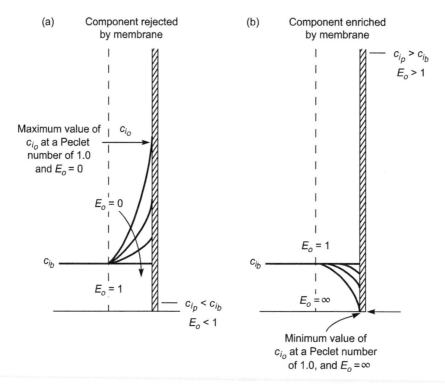

Figure 5.8 *Concentration gradients that form adjacent to the membrane surface for components (a) rejected or (b) enriched by the membrane. The Peclet number, characterizing the balance between convection and diffusion in the boundary layer, is the same $J_v\delta/D_i = 1$. When the component is rejected, the concentration at the membrane surface c_{i_o} cannot be greater than 2.72 c_{i_b}, irrespective of the membrane selectivity. When the minor component permeates the membrane, the concentration at the membrane surface can decrease to close to zero, so the concentration polarization modulus can become very large.*

where u is the superficial velocity in the feed flow channel and k_o and n are adjustable coefficients. Equation (5.11) can then be rewritten as

$$\ln\left(1 - \frac{1}{E}\right) = \ln\left(1 - \frac{1}{E_o}\right) - \frac{J_v}{k_o u^n} \tag{5.13}$$

Equation (5.13) can be used to calculate the dependence of system performance on concentration polarization. One method is to use data obtained with a single module operated at various feed solution velocities. A linear regression analysis is used to fit data obtained at different feed velocities to obtain an estimate for k_o and E_o; the exponent n is adjusted to minimize the residual error. Figure 5.9 shows some data obtained in pervaporation experiments with dilute aqueous toluene solutions and silicone rubber membranes [18]. Toluene is considerably more permeable than water through these membranes. In Figure 5.9, when the data were regressed, the best value for n was 0.96. The values of E_o, the intrinsic enrichment of the membrane, and k_o obtained by regression analysis are 3600 and 7.1×10^{-4}, respectively. The boundary layer coefficient D_i/δ, is then given by

$$\frac{D_i}{\delta} = 7.1 \times 10^{-4} u^{0.96} \tag{5.14}$$

where u is the superficial velocity in the module.

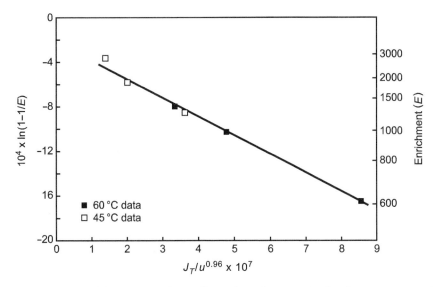

Figure 5.9 *Derivation of the mass transfer coefficient by Wilson's method. Toluene/water enrichments are plotted as a function of feed solution superficial velocity in pervaporation experiments. Enrichments were measured at different feed solution superficial velocities with spiral-wound membrane modules. Source: Reproduced with permission from [18]. Copyright (1976) Elsevier.*

A second method of determining the coefficient (D_i/δ) and the intrinsic enrichment of the membrane E_o is to use Eq. (5.11). The term $\ln(1 - 1/E)$ is plotted against the permeate flux measured at constant feed solution flow rates, but different permeate pressures or feed solution temperatures. This type of plot is shown in Figure 5.10 for data obtained with aqueous trichloroethane solutions in pervaporation experiments with silicone rubber membranes.

The coefficients D_i/δ obtained from the slopes of the lines at each velocity in Figure 5.10 can then be plotted as a function of the feed superficial velocity. The data show that the ratio D_i/δ varies with the superficial velocity according to the equation

$$\frac{D_i}{\delta} = 9 \times 10^{-4} u^{0.8} \tag{5.15}$$

From Eqs. (5.14) and (5.15), the value of the term D_i/δ at a fluid velocity of 30 cm/s is $1.6-1.8 \times 10^{-2}$ cm^2/s. Based on a trichloroethane diffusion coefficient in the boundary aqueous layer of 2×10^{-5} cm^2/s, this yields a boundary layer thickness of 10–15 μm. This boundary layer thickness is in the same range as values calculated for RO with similar modules.

5.3 Concentration Polarization in Liquid Separation Processes

The effect of concentration polarization on specific membrane processes is discussed in the individual application chapters. A brief comparison of the magnitude of concentration polarization for processes involving liquid feed solutions is given in Table 5.1. The key simplifying assumption is that the calculated boundary layer thickness for all processes is 20 μm.

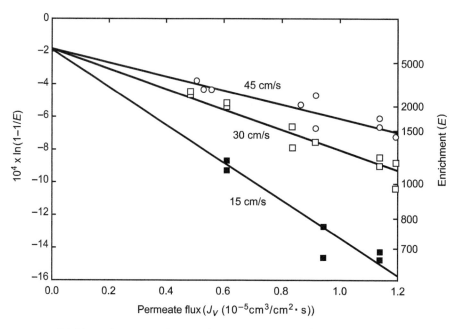

Figure 5.10 *Trichloroethane enrichment [ln(1 − 1/E)] as a function of permeate flux J_v in pervaporation experiments with silicone rubber membranes in spiral-wound modules using solutions of 100 ppm trichloromethane in water at different feed solution flow rates. The ratio D_i/δ can be calculated from the slope of the lines using Eq. (5.13). Source. Reproduced with permission from [18]. Copyright (1976) Elsevier.*

Table 5.1 *Representative values of the concentration polarization modulus calculated for a variety of liquid separation processes.*

Process	Typical intrinsic enrichment, E_o	Typical flux in engineering units and as J_v (10^{-3} cm/s)	Diffusion coefficient (10^{-6} cm^2/s)	Peclet number, $J_v\delta/D_i$	Concentration polarization modulus (Eq. (5.9))
RO					
Seawater desalination	0.01	50 l/m^2·h (1.4)	10	0.28	1.3
Brackish water desalination	0.01	90 l/m^2·h (2.3)	10	0.46	1.5
Ultrafiltration					
Protein separation	0.01	50 l/m^2·h (1.4)	0.5	5.6	70
Pervaporation					
Ethanol dehydration	20	0.1 kg/m^2·h (0.003)	20	0.0003	1.0
VOC from water	2000	1.0 kg/m^2·h (0.03)	20	0.003	0.14
Coupled transport					
Copper from water	1000	60 mg/cm^2·min (0.001)	10	0.0002	0.8

For these calculations a boundary layer thickness of 20 μm, typical of that in most spiral-wound membrane modules, is assumed.

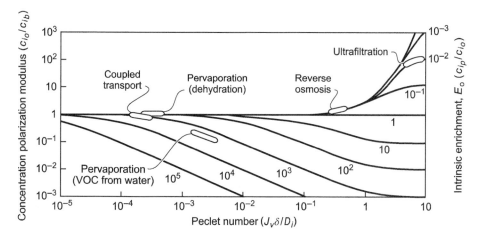

Figure 5.11 *Peclet numbers and intrinsic enrichments for the membrane separation processes shown in Table 5.1 superimposed on the concentration polarization plot of Wijmans et al. Source: Reproduced with permission from [16]. Copyright (1976) Elsevier.*

This value is typical for separation of solutions with spiral-wound modules in reverse osmosis, pervaporation, and ultrafiltration. Tubular, plate-and-frame, and bore-side feed hollow fiber modules, because of their better flow velocities, generally have lower calculated boundary layer thicknesses. Hollow fiber modules with shell-side feed generally have larger calculated boundary layer thicknesses because of their poor fluid flow patterns.

Table 5.1 shows typical enrichments and calculated Peclet numbers. In this table, as, elsewhere, it is important to recognize the difference between enrichment and separation factor. The enrichments shown are calculated for the minor component. For example, in the dehydration of ethanol, a typical feed solution of 96% ethanol and 4% water yields a permeate containing about 80% water; the enrichment, that is, the ratio of the permeate to feed concentration, is about 20. In Figure 5.11, the calculated Peclet numbers and enrichments shown in Table 5.1 are plotted on the Wijmans graph to show the relative importance of concentration polarization for the processes listed.

In coupled transport and solvent dehydration by pervaporation, concentration polarization effects are generally modest and controllable, with a concentration polarization modulus of 1.5 or less. In reverse osmosis, the Peclet number of 0.3–0.5 was calculated on the basis of typical fluxes of current RO membrane modules, which are 50–90 l/m²h. Concentration polarization modulus values in this range are between 1.0 and 1.5.

Figure 5.11 shows that ultrafiltration and pervaporation for the removal of organic solutes from water are both seriously affected by concentration polarization. In ultrafiltration, the low diffusion coefficient of macromolecules produces a concentration of retained solutes 70 times the bulk solution volume at the membrane surface. At these high concentrations, macromolecules precipitate, forming a gel layer at the membrane surface and reducing flux. The effect of this gel layer on ultrafiltration membrane performance is discussed in Chapter 6.

In the case of pervaporation of dissolved volatile organic compounds (VOCs) from water, the magnitude of the concentration polarization effect is a function of the enrichment factor. The selectivity of pervaporation membranes to different VOCs varies widely, so the intrinsic enrichment and the magnitude of concentration polarization effects depend strongly on the

Table 5.2　Enrichment factors measured for the pervaporation of VOCs from dilute solutions with silicone rubber spiral-wound modules.

Solute	Enrichment (E_o)
Trichloroethylene	5700
Toluene	3600
Ethyl acetate	270
Isopropanol	18

solute. Table 5.2 shows experimentally measured enrichment values for a series of dilute VOC solutions treated with silicone rubber membranes in spiral-wound modules [18]. When these values are superimposed on the Wijmans plot as shown in Figure 5.12, the concentration polarization modulus varies from 1.0, that is, no concentration polarization, for isopropanol, to 0.1 for trichloroethane, which has an enrichment of 5700.

As previously described, concentration polarization in liquid separations is often managed by maintaining a high feed flow rate across the surface of the membrane, reducing the boundary layer thickness and increasing the Peclet number. For this reason, membrane systems that contain several membrane modules often have the modules arranged in series (high feed velocity) rather than in parallel (low feed velocity). In some processes, particularly ultrafiltration, inter-stage feed pumps may be used to maintain high feed velocities through the modules.

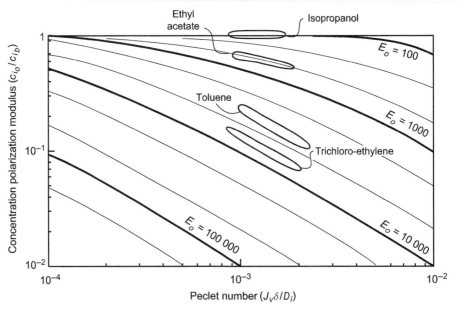

Figure 5.12　*A portion of the Wijmans plot shown in Figure 5.7 expanded to illustrate concentration polarization in pervaporation of dilute aqueous organic solutions using silicone rubber membranes. With solutes such as toluene and trichloroethylene, high intrinsic enrichments produce severe concentration polarization. Concentration polarization is much less with solutes such as ethyl acetate (enrichment 270), and is essentially eliminated with isopropanol (enrichment 18). Source: Reproduced with permission from [18]. Copyright (1976) Elsevier.*

5.4   Concentration Polarization in Gas Separation Processes

Concentration polarization in gas separation processes has not been widely studied; the effect can usually be assumed to be small because of the high diffusion coefficients of gases. In calculating the expression for the concentration polarization modulus of gases, the simplifying assumption that the volume fluxes on each side of the membrane are equal cannot be made. The starting point for the calculation is the mass-balance equation (Eq. (5.6)), which for gas permeation is written

$$J_{v_f} c_i - \frac{D_i dc_i}{dx} = J_{v_p} c_{i_p} \tag{5.16}$$

where J_{v_f} is the volume flux of gas on the feed side of the membrane and J_{v_p} is the volume flux on the permeate side. These volume fluxes ($cm^3/cm^2 \cdot s$) can be linked by correcting for the pressure on each side of the membrane using the expression

$$J_{v_f} p_o = J_{v_p} p_\ell \tag{5.17}$$

where p_o and p_ℓ are the gas pressures on the feed and permeate sides of the membrane. Hence,

$$J_{v_f} \frac{p_o}{p_\ell} = J_{v_f} \varphi = J_{v_p} \tag{5.18}$$

where φ is the pressure ratio p_o/p_ℓ across the membrane. Substituting Eq. 5.18 into Eq. 5.16 and rearranging gives

$$-D_i \frac{dc_i}{dx} = J_{v_f} \left(\varphi c_{i_p} - c_i \right) \tag{5.19}$$

Integrating across the boundary layer thickness, as before, gives

$$\frac{c_{i_o}/\varphi - c_{i_p}}{c_{i_b}/\varphi - c_{i_p}} = \exp \left(\frac{J_{v_f} \delta}{D} \right) \tag{5.20}$$

For gases, the enrichment terms, E and E_o, are most conveniently expressed in volume fractions, so that

$$E_o = \frac{c_{i_p}}{p_\ell} \frac{p_o}{c_{i_o}} = \frac{c_{i_p}}{c_{i_o}} \varphi \tag{5.21}$$

and

$$E = \frac{c_{i_p}}{p_\ell} \cdot \frac{p_o}{c_{i_b}} = \frac{c_{i_p}}{c_{i_b}} \cdot \varphi \tag{5.22}$$

Equation (5.20) can then be written as

$$\exp \left(\frac{J_{v_f} \delta}{D_i} \right) = \frac{1 - 1/E_o}{1 - 1/E} \tag{5.23}$$

which on rearranging gives

$$E/E_o = c_{i_o}/c_{i_b} = \frac{\exp\left(J_{v_f}\delta/D_i\right)}{1 + E_o\left[\exp\left(J_{v_f}\delta/D_i\right) - 1\right]} \tag{5.24}$$

Equation (5.24) has the same form as the expression for the concentration polarization modulus of liquids, Eq. (5.9).

When Eq. (5.24) is used to calculate the concentration polarization modulus, the Peclet number is very small, so little polarization is expected for most gas separation applications. Only in a few applications, such as the separation of VOCs from air, or vapor-phase separation of water from ethanol, where very high permeance membranes are used, can measurable concentration polarization be observed or expected. Channeling, in which a portion of the feed gas completely bypasses contact with the membrane through poor flow distribution in the module, can also reduce module efficiency in a way that is difficult to separate from concentration polarization. Channeling is more noticeable in gas permeation modules than in liquid permeation modules.

5.5 Concentration Polarization in Membrane Contactors and Related Processes

The description of concentration polarization so far has been limited to situations where net volume flow occurs through the membrane, and nothing is present on the permeate side except material that has permeated. However, processes exist in which fluids are introduced and circulate on both sides of the membrane, raising the possibility for concentration polarization to occur on both sides. Examples of this type of process are described in the chapter on membrane contactors; other examples are found in forward osmosis, pressure-retarded osmosis (PRO), membrane distillation, and gas dehydration.

A familiar process of this type is dialysis, illustrated in Figure 5.13 as it applies to treatment of blood in an artificial kidney, covered in Chapter 14. Urea, creatinine, and other nitrogen-containing metabolites from the blood permeate into an isotonic saline solution. The pressure and concentration of the isotonic saline are controlled so that there is small or no volume fluid flow through the membrane. Under these conditions, significant concentration gradients of the metabolites occur on the feed (blood) and permeate (saline) sides. On the blood side, metabolite concentrations decrease toward the membrane surface to below the bulk values; on the saline side they are higher at the surface than in the bulk saline solution. The effects are not equal on both sides of the membrane. Blood must be pumped slowly to avoid hemolysis. No such limitation occurs on the saline side of the membrane, so concentration polarization on this side is more easily controlled. The effects of polarization on both sides of the membrane are additive and reduce the metabolite removal rate.

A good example of a process in which there is a particularly large and asymmetric effect on performance brought about by concentration polarization is provided by electrodialysis. In electrodialysis, a voltage is applied across a stack of charged membranes as a way to desalt water. The concentration of salts on the brine side of the membrane is much higher than that on the dilute side, from which ions are removed by the voltage driving force. The resulting concentration gradients are shown in Figure 5.14.

The fluids on either side of the membrane are circulated at about the same rate so the stagnant boundary layers on either side of the membrane have about the same thickness. As the sodium

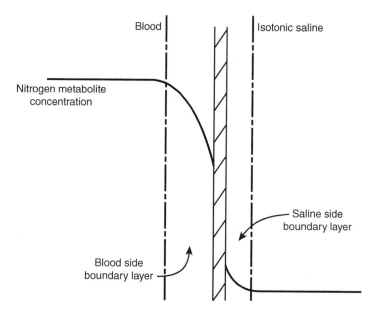

Figure 5.13 *Treatment of blood in an artificial kidney, showing the decrease in metabolite concentration in the boundary layer on the blood side, and the elevated concentration compared with the bulk saline solution on the permeate side. The feed and permeate side boundary layer thicknesses differ because the circulated saline is more turbulent than the circulated blood.*

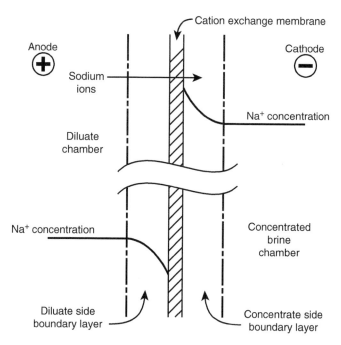

Figure 5.14 *Concentration polarization in electrodialysis. The absolute difference in sodium ion concentration across the boundary layer is the same on both sides of the membrane, but the proportionate effect in the diluate chamber is greater.*

ions migrate under the influence of the applied voltage, approximately equal and opposite concentration gradients build up on both sides of the membrane. However, the effect of these gradients is very different. On the diluate side, the ion concentration at the membrane surface is about half that in the bulk solution. (For example, the bulk solution concentration might be 2000 ppm, and the surface concentration 1000 ppm.) This 50% drop reduces the ion transport through the membrane by an equivalent amount. On the brine concentrate side, the gradient from surface to bulk is an equal and opposite 1000 ppm. However, the bulk concentration is typically enriched about 20-fold compared with the diluate, and is in the few percent range (say 4% – the drawing is not to scale). Thus the same 1000 ppm gradient represents a much smaller proportional change in the bulk concentration (only 2.5% change using the example numbers) and has a correspondingly small effect. Attempts to reduce concentration polarization on the concentrate side would provide a barely noticeable benefit and are generally not worthwhile. In contrast, control of concentration polarization on the diluate side is critical to the operation, and is discussed in Chapter 11.

Another factor determining the magnitude of concentration polarization effect on flux is the anisotropy of the membrane. Many membranes are asymmetric, with a dense surface layer on a more open support. Concentration polarization in the boundary layer adjacent to the dense selective layer is much easier to control than the polarization that may occur in the stagnant fluid layer on the microporous support side. The orientation of the membrane can then have a large effect on performance. If there is transmembrane volume fluid flow, the dense surface should face the incoming flow. The dense surface will also normally face the more dilute solution. The effect of membrane asymmetry on concentration polarization is important in PRO processes and is discussed in Chapter 13.

5.6   Conclusions and Future Directions

Most membrane processes are affected by concentration polarization, even if only to a slight extent, and for some applications the problem dominates process performance. Concentration polarization can be controlled by increasing the turbulence of the feed fluid, but this approach has practical limits. In ultrafiltration and electrodialysis, liquid recirculation pumps already represent a major capital cost of a new plant, and consume 20–40% of the operating power. In recent years, there has been a trend toward operating microfiltration and ultrafiltration membrane modules at low feed pressures. This lowers the flux through the membrane, but makes concentration polarization much easier to control. Air sparging and pulsed feed flow are also now widely used to promote increased turbulence in the feed solution.

As materials and fabrication techniques continue to improve, membrane fluxes and selectivities are likely to increase, leading to increased concentration polarization. Thus, the deleterious effects of concentration polarization on performance are likely to remain with us and to become more important, and innovative methods to deal with the issue will no doubt continue to be sought and developed.

References

1.   Carslaw, H.S. and Jaeger, J.C. (1947). *Conduction of Heat in Solids*. London: Oxford University Press.
2.   Bird, R.B., Stewart, W.E., and Lightfoot, E.N. (1960). *Transport Phenomena*. New York: Wiley.

3. Crank, J. (1956). *The Mathematics of Diffusion*. London: Oxford University Press.

4. Porter, M.C. (1972). Concentration polarization with membrane ultrafiltration. *Ind. Eng. Chem. Prod. Res. Dev.* **11**: 234.

5. Lepore, J.V. and Ahlert, R.C. (1988). Fouling in membrane processes. In: *Reverse Osmosis Technology* (ed. B.S. Parekh), 141–184. New York: Marcel Dekker.

6. Wickramasinghe, S.R., Semmens, M.J., and Cussler, E.L. (1992). Mass transfer in various hollow fiber geometries. *J. Membr. Sci.* **69**: 235.

7. Mi, L. and Hwang, S.T. (1999). Correlation of concentration polarization and hydrodynamic parameters in hollow fiber modules. *J. Membr. Sci.* **159**: 143.

8. Cussler, E.L. (1997). *Diffusion Mass Transfer in Fluid Systems*, 2e. New York, NY/Cambridge, UK: Cambridge University Press.

9. Zeman, L.J. and Zydney, A.L. (1996). *Microfiltration and Ultrafiltration: Principles and Applications*. New York: Marcel Dekker.

10. Brian, P.L.T. (1966). Mass transport in reverse osmosis. In: *Desalination by Reverse Osmosis* (ed. U. Merten), 161–292. Cambridge, MA: MIT Press.

11. Belfort, G., Davis, R.H., and Zydney, A.L. (1994). The behavior of suspensions and macromolecular solutions in cross-flow microfiltration. *J. Membr. Sci.* **96**: 1.

12. Da Costa, A.R., Fane, A.G., and Wiley, D.E. (1994). Spacer characterization and pressure drop modeling in spacer-filled channels for ultrafiltration. *J. Membr. Sci.* **87**: 79.

13. Schwinge, J., Neal, P.R., Wiley, D.E. et al. (2004). Spiral wound modules and spacers: review and analysis. *J. Membr. Sci.* **242** (1, 2): 129.

14. Li, F., Meindersma, W., de Haan, A.B., and Reith, T. (2005). Novel spacers for mass transfer enhancement in membrane separations. *J. Membr. Sci.* **253** (1, 2): 1.

15. Jaffrin, M.Y., Gupta, B.B., and Paullier, P. (1994). Energy savings pulsatile mode cross-flow filtration. *J. Membr. Sci.* **86**: 281.

16. Wijmans, J.G., Athayde, A.L., Daniels, R. et al. (1996). The role of boundary layers in the removal of volatile organic compounds from water by pervaporation. *J. Membr. Sci.* **109**: 135.

17. Wilson, E.E. (1915). A basis for rational design of heat transfer apparatus. *Trans. ASME* **37**: 47.

18. Baker, R.W., Wijmans, J.G., Athayde, A.L. et al. (1997). The effect of concentration polarization on the separation of volatile organic compounds from water by pervaporation. *J. Membr. Sci.* **137**: 159.

6

Reverse Osmosis (Hyperfiltration)

Membrane Technology and Applications, Fourth Edition. Richard W. Baker.
© 2024 John Wiley & Sons Ltd. Published 2024 by John Wiley & Sons Ltd.

6.1   Introduction and History

Reverse osmosis (RO) is a process for desalting water using membranes that are permeable to water but essentially impermeable to salt. Pressurized water containing dissolved salts contacts the feed side of the membrane; water depleted of salt is withdrawn as a low-pressure permeate. The ability of membranes to separate small solutes from water has been known for a very long time. Pfeffer, Traube, and others studied osmotic phenomena with ceramic membranes as early as the 1850s. In 1931, a process using membranes to desalt water was patented, and the term reverse osmosis was coined for this process [1]. Reverse osmosis is the dominant application of the technology described in this chapter, but the same equations and similar process flow diagrams can also be used to describe the separation of dissolved solutes from organic solutions, so the process is more broadly described as hyperfiltration. The bulk of this chapter will describe water separations (reverse osmosis) but we will close with a section on hyperfiltration of organic liquid mixtures. These applications are growing and are the subject of current research activity.

Modern interest in reverse osmosis dates from the work of Reid and Breton, who in 1959 showed that cellulose acetate films could perform this type of separation [2]. Their films were 5–20 μm thick, so fluxes were very low, but by pressurizing the feed salt solution to 65 bar, they obtained permeate water from which 98% of salt had been removed. The breakthrough discovery that made reverse osmosis a practical process was the development of the Loeb–Sourirajan anisotropic cellulose acetate membrane [3]. This membrane had 20 times the flux of the best membrane of Reid and Breton and equivalent rejections. With these membranes, water desalination by reverse osmosis became a potentially practical process. In 1965, the first commercial-scale plant was installed to desalinate brackish water at Coalinga, California. The first membrane modules were of tubular or plate-and-frame design, but within a few years Westmoreland, Bray, and others at the San Diego Laboratories of Gulf General Atomics had developed spiral-wound modules [4, 5]. Later, DuPont, building on earlier work of Dow, introduced polyaramide hollow fine fiber reverse osmosis modules under the name Permasep®.

Anisotropic cellulose acetate membranes were the industry standard through the 1960s to the mid-1970s, until Cadotte, then at North Star Research, developed the interfacial polymerization method of producing composite membranes [6]. Interfacial composite membranes had far higher salt rejections, combined with good water fluxes. Fluid Systems introduced the first commercial interfacial composite membrane in 1975. The construction of a commercial-scale seawater desalination plant at Jeddah, Saudi Arabia using these membranes was a milestone in reverse osmosis development. Later, at FilmTec, Cadotte developed a fully aromatic interfacial composite membrane based on the reaction of phenylenediamine and trimesoyl chloride [7]. This membrane has become the current industry standard. Over the past 40 years, the performance of membranes and membrane modules has steadily improved. These improvements and better process designs have cut the cost of sea water desalination to below $0.50/m^3 of water. The energy used by the process has also been reduced from 6.1 kWh/m^3, for the Jeddah plant, to around 2–3 kWh/m^3 for a new plant fitted with state-of-the-art membrane modules and equipped with energy recovery devices on the high-pressure side [8–11].

The next development, beginning in the mid-1980s, was the introduction of low-pressure nanofiltration membranes by all of the major RO companies [12]. These membranes are used to separate divalent ions from monovalent ions in water-softening applications, or to produce ultrapure water for the pharmaceutical and electronics manufacturing industries.

Currently, approximately 90 million m^3/day of water are desalted by reverse osmosis. Half of this capacity is installed in the Middle East and other desert regions to produce municipal water from wastewater, brackish groundwater or the sea. The remainder is installed in the United States, Europe, and Japan, principally to produce ultrapure industrial water or to bring contaminated water to drinking water standards. More than 20 seawater desalination plants with capacities of more than 100 000 m^3/day are now in operation. These plants contain up to 1 million m^2 of membrane. The world's current largest reverse osmosis plant, in Sorek, Israel, produces 700 000 m^3/day of desalted water, enough for a city of 1–2 million people.

The interfacial composite membrane has displaced the Loeb–Sourirajan anisotropic cellulose acetate membrane in almost all seawater and brackish water desalination applications. Cellulose acetate membranes, though they retain only a small market share, are still used in small industrial applications where their robust nature is valued.

An important advance expanding the technology beyond water treatment has been the development of a similar membrane process to separate solutes from organic solvents. This technology, normally referred to by the more general term of hyperfiltration, is currently used primarily for small-scale, high-value separations, although a few industrial plants have been installed. The first of these was developed by Grace Davison in conjunction with Mobil Oil for Mobil's Beaumont, Texas, refinery. The 3 million gal/day plant was installed in 1998 to separate a solution of methyl ethyl ketone and lube oil in an oil dewaxing process.

Some of the milestones in the development of the reverse osmosis industry are summarized in Figure 6.1.

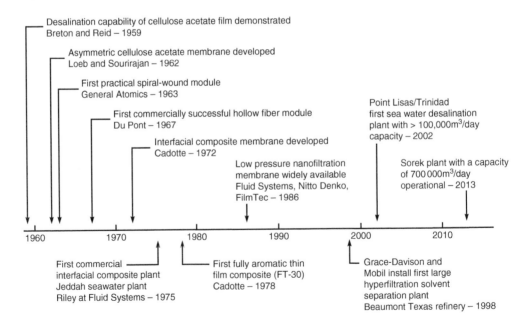

Figure 6.1   Milestones in the development of reverse osmosis.

6.2 Theoretical Background

Salt and water permeate reverse osmosis membranes according to the solution–diffusion transport mechanism described in Chapter 2. The water flux, J_i, is linked to the pressure and concentration gradients across the membrane by the equation

$$J_i = A(\Delta p - \Delta \pi) \tag{6.1}$$

where Δp is the pressure difference across the membrane, $\Delta \pi$ is the osmotic pressure differential across the membrane, and A is a constant. As this equation shows, at a low applied pressure, when $\Delta p < \Delta \pi$, water flows from the dilute to the concentrated-salt-solution side of the membrane by normal osmosis. When $\Delta p = \Delta \pi$, no flow occurs, and when the applied pressure is higher than the osmotic pressure, $\Delta p > \Delta \pi$, water flows from the concentrated to the dilute-salt-solution side of the membrane.

The salt flux, J_j, across a reverse osmosis membrane is described by the equation

$$J_j = B\left(c_{j_o} - c_{j_\ell}\right) \tag{6.2}$$

where B is the salt permeability constant and c_{j_o} c_{j_ℓ}, respectively, are the salt concentrations on the feed and permeate sides of the membrane. The concentration of salt in the permeate solution (c_{j_ℓ}) is usually much smaller than the concentration in the feed (c_{j_o}), so Eq. (6.2) can be simplified to

$$J_j = Bc_{j_o} \tag{6.3}$$

It follows from the above equations that the water flux is proportional to the applied pressure, but the salt flux is independent of pressure. This means that the membrane becomes more selective as the pressure increases. Selectivity can be measured in a number of ways, but most commonly, it is measured as the salt rejection coefficient $\mathbb{R}$, defined as

$$\mathbb{R} = \left[1 - \frac{c_{j_\ell}}{c_{j_o}}\right] \times 100\% \tag{6.4}$$

The salt concentration on the permeate side of the membrane can be related to the membrane fluxes by the expression

$$c_{j_\ell} = \frac{J_j}{J_i} \times \rho_i \tag{6.5}$$

where ρ_i is the density of water (g/cm^3). By combining Eqs. (6.1)–(6.3) the membrane rejection can be expressed as

$$\mathbb{R} = \left[1 - \frac{\rho_i \cdot B}{A(\Delta p - \Delta \pi)}\right] \times 100\% \tag{6.6}$$

The effects of the most important operating parameters on membrane water flux and salt rejection are shown schematically in Figure 6.2 [13]. The effect of feed pressure on membrane performance is

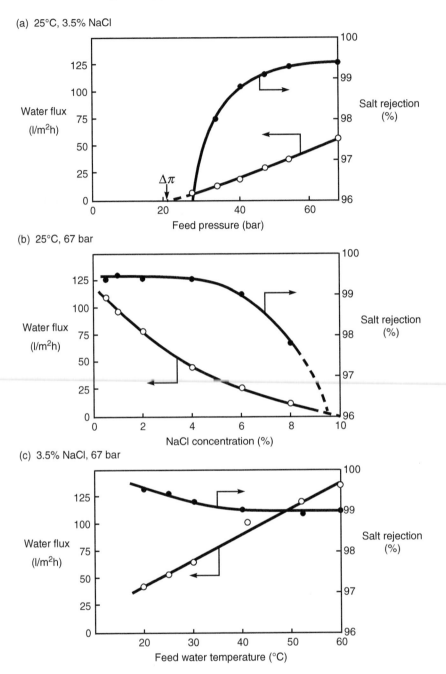

Figure 6.2 *Effect of feed pressure (a), feed salt concentration (b) and feed temperature (c) on the properties of a seawater desalination membrane (SW-30). Source: Adapted from [13].*

shown in Figure 6.2a. As predicted by Eq. (6.1), at a pressure equal to the osmotic pressure of the feed (23 bar), the water flux is zero; thereafter, it increases linearly as the pressure is increased. The salt rejection also extrapolates to zero at a feed pressure of 23 bar as predicted by Eq. (6.6), but increases very rapidly with increased pressure to reach salt rejections of more than 99% at an applied pressure of about 45 bar (twice the feed solution osmotic pressure).

The effect of increasing the concentration of salt in the feed solution on membrane performance is illustrated in Figure 6.2b. Increasing the salt concentration effectively increases the osmotic pressure term in Eq. (6.1); consequently, at a constant feed pressure of 67 bar, the water flux falls with increasing salt concentration. The water flux approaches zero when the salt concentration is about 10 wt%, at which point the osmotic pressure equals the applied hydrostatic pressure. The salt rejection also extrapolates to zero rejection at this point, but increases rapidly with decreasing salt concentration. Salt rejections of more than 99% are achieved at salt concentrations below 6%, corresponding to a net applied pressure ($\Delta\rho - \Delta\pi$) of about 27 bar.

The effect of temperature on salt rejection and water flux illustrated in Figure 6.2c is more complex. Transport of salt and water, represented by Eqs. (6.1) and (6.3) is an activated process, and both water and salt fluxes increase with increasing temperature. As Figure 6.2c shows, the effect of temperature on the water flux of membranes is significant; the water flux doubles as the temperature is increased from 30 to 60 °C. However, the effect of temperature on the salt flux is even more marked. This means that the salt rejection coefficient, proportional to the ratio B/A in Eq. (6.6), actually declines slightly as the temperature increases.

Measurements of the type shown in Figure 6.2 are typically obtained with small laboratory test cells. A laboratory test system is illustrated in Figure 6.3. Such systems are often used in general membrane quality control tests with a number of cells, arranged in series, through which fluid is pumped. The system is usually operated with a test solution of 0.2–1.0% sodium chloride at pressures ranging from 10 to 40 bar. The storage tank and flow recirculation rate are large enough that changes in concentration of the test solution due to loss of permeate can be ignored.

Some confusion can occur over the rejection coefficients quoted by membrane module manufacturers. The intrinsic rejection of good quality seawater membranes measured in a laboratory test system is often in the range 99.7–99.8%, whereas the same membrane in module form may have a salt rejection of 99.6–99.7%. This difference is due to small membrane defects introduced during module production and to concentration polarization, which has a small but measurable effect on rejection by the module. Manufacturers call the module value the nominal rejection.

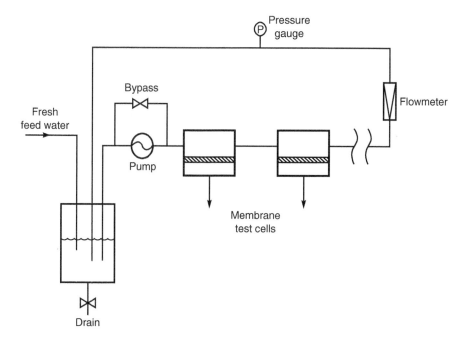

Figure 6.3   *Flow schematic of a high-pressure laboratory RO test system.*

However, manufacturers will generally only guarantee a lower figure, for example 99.5%, for the initial module salt rejection, to take into account variations between modules. To complicate matters further, module performance generally deteriorates slowly during the module lifetime due to membrane compaction, membrane fouling, and membrane degradation from hydrolysis, chlorine attack, or membrane cleaning. A decrease in the membrane flux by 20% over the module lifetime is not unusual, and the rejection can fall by 0.1–0.2%. Reverse osmosis system manufacturers allow for this decline in performance when designing systems. The lifetime of a membrane module in a modern well-maintained seawater desalination plant is about 7–10 years.

6.3   Membrane Materials

A number of materials and techniques have been used to make reverse osmosis membranes. The target of much of the early work was desalination of seawater (approximately 3.5 wt% salt), which required membranes with salt rejections of greater than 99.3% to produce an acceptable permeate containing less than 500 ppm salt. Early membranes could only meet this target performance when operated at very high pressures, up to 100 bar. As membrane performance has improved, the operating pressure has dropped to 50–60 bar. Membranes to desalt brackish water and wastewater feeds with salt concentrations of 0.1–0.5 wt% have also been developed. For these applications, membranes are typically operated at pressures in the 10–20 bar range, with a target salt rejection of about 99%. Another application is production of ultrapure water for industrial use. The feed in this case is often municipal drinking water, containing only 100–200 ppm of dissolved salts, mostly as divalent ions. The target membrane performance may be 98–99% sodium chloride rejection, but more than 99.5% divalent ion rejection. These membranes are operated at low pressures, typically in the 8–12 bar range. Many manufacturers tailor the properties of the same membrane material to meet the requirements of different applications. Inevitably, a trade-off between flux and rejection is involved.

A brief description of the commercially important membranes in current use follows. More detailed descriptions can be found in specialized reviews [11, 12]. Petersen's review on interfacial composite membranes, although now a little dated, is particularly worth noting [14].

6.3.1   Cellulosic Membranes

The first high-performance reverse osmosis membranes were made from cellulose acetate. The flux and rejection of cellulose acetate membranes have now been surpassed by interfacial composite membranes. Nevertheless, cellulose acetate membranes still maintain a small market in industrial applications because they are easy to make, mechanically tough, and resistant to degradation by chlorine and other oxidants, which is a significant problem with interfacial composite membranes. Cellulose acetate membranes can tolerate continuous exposure of up to 1 ppm chlorine, so chlorination can be used to sterilize the feed water, a major advantage with feed streams having significant bacterial loading.

The water and salt permeabilities of cellulose acetate membranes are extremely sensitive to the degree of acetylation of the polymer used to make the membrane [15, 16]. Fully substituted cellulose triacetate (44.2 wt% acetate) has an extremely high water-to-salt permeability ratio, but a low water permeability, so these membranes have very good salt rejection but low water fluxes. Nonetheless, they are still used in some seawater desalination plants because salt rejections of about 99.6% are attainable with a seawater feed. However, most commercial cellulose acetate membranes use a polymer containing 39.8 wt% acetate, equal to a degree of acetylation of 2.45. These membranes generally achieve 97–98% sodium chloride rejection and have good fluxes.

Permeability data measured with thick films of cellulose acetate show that membranes should be able to achieve salt rejections of 99.6% or more. In practice, this theoretical rejection is very difficult to obtain [17]. Figure 6.4 shows the salt rejection properties of 39.8 wt% acetate membranes made by the Loeb–Sourirajan process [18]. The freshly formed membranes have very high water fluxes of almost 300 l/ m^2 · h, but almost no rejection of sodium chloride. The membranes appear to have a finely microporous structure and are permeable to quite large solutes such as sucrose. The salt rejection can be greatly improved by heating the membrane in a bath of hot water for a few minutes. This annealing procedure is used with all cellulose acetate membranes and modifies the membrane by eliminating the micropores and producing a denser, more salt-rejecting skin. The water flux decreases, and the sodium chloride rejection increases. The temperature of this annealing step determines the final properties of the membrane. A typical rejection/flux curve for various annealed membranes is shown in Figure 6.4. Because their properties change on heating, annealed cellulose acetate membranes are not used above about 35 °C. The membranes hydrolyze slowly over time, so the feed water is usually adjusted to pH 4–6, the range in which the membranes are most stable.

In preparing membranes by the Loeb–Sourirajan technique, the casting solution composition is critically important. Other important process parameters are the time of evaporation before precipitation, the temperature of the precipitation bath, and the temperature of the annealing step. Most early membranes were made of 39.8 wt% acetate polymer, because this material was readily available and had the most convenient solubility properties. Over the years, Saltonstall and others developed better membranes by blending the 39.8 wt% acetate polymer with small

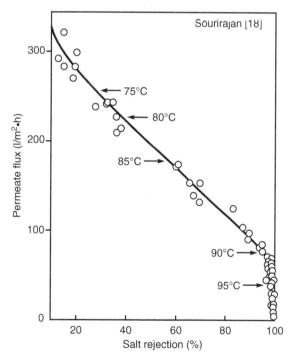

Figure 6.4 *The effect of annealing temperatures on the flux and rejection of cellulose acetate membranes. The annealing temperature is shown on the figure (Cellulose diacetate membranes tested at 100 bar with 0.5 M NaCl). Source: Reproduced with permission from [18]. Copyright (1970) Elsevier.*

amounts of triacetate polymer (44.2 wt% acetate) or other cellulose esters such as cellulose acetate butyrate [19]. These blends are generally used to form current cellulose acetate membranes. Blend membranes with seawater salt rejections of 99.0–99.5%, close to the theoretical maximum, can be made for specialized applications, but the flux of these membranes is modest. However, most applications of cellulose acetate membranes do not require such high salt rejections, so the typical commercial cellulose acetate membranes used to process industrial water streams have good fluxes and sodium chloride rejection of about 96%.

6.3.2 Noncellulosic Loeb–Sourirajan Membranes

During the 1960s and 1970s, the Office of Saline Water sponsored development of noncellulosic RO membranes made by the Loeb–Sourirajan process, but few matched the properties of cellulose acetate. Following the development of interfacial composite membranes by Cadotte, this line of research was largely abandoned. Nevertheless, some commercially successful noncellulosic membranes were developed. In particular, aromatic polyamide membranes were developed by several groups, including those at Toray [20], Chemstrand (Monsanto) [21], and Permasep (DuPont) [22], all in hollow fiber form. These membranes have good seawater salt rejections (up to 99.5%), but low water fluxes, in the 2–6 l/m$^2 \cdot$h range. The Permasep membrane, made in hollow fine fiber form to overcome the low water permeability, was produced under the names B-10 and B-15 for seawater desalination until the year 2000. The structure of the Permasep B-15 polymer is shown in Figure 6.5. The polymer contains a few sulfonic acid groups to make it more hydrophilic to raise its water flux. Polyamide membranes, like the interfacial composite membranes discussed below, are susceptible to degradation by chlorine because of their amide bonds.

Loeb–Sourirajan membranes based on sulfonated polysulfone and substituted poly(vinyl alcohol), produced by Hydranautics (Nitto Denko Corporation), found a commercial market as high-flux, low-rejection membranes in water softening applications, because their divalent ion rejection is high. These membranes were chlorine-resistant and withstood up to 40 000 ppm $\cdot$ h of chlorine exposure without degradation.[1] The structures of the polymers used by Hydranautics are shown in Figure 6.6.

Another group of membranes with good permeation properties was produced by condensation of furfuryl alcohol with sulfuric acid. The first membrane of this type was made by Cadotte at North Star Research and was known as the NS200 membrane [23]. These membranes were made by coating a microporous polysulfone support membrane with an aqueous solution of furfuryl alcohol followed by concentrated sulfuric acid. The coated support was then heated to 140 °C. The furfuryl alcohol formed a polymerized, crosslinked layer on the polysulfone support; the

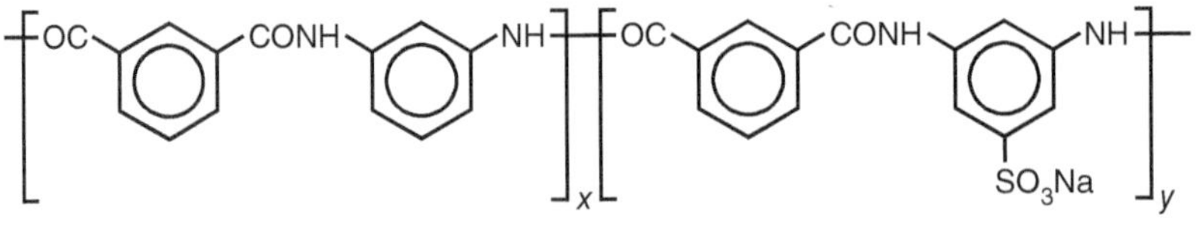

Figure 6.5 *Aromatic polyamide used by DuPont in its Permasep B-15 hollow fine fibers. Source: Adapted from [22].*

---

[1] The ability of a reverse osmosis membrane to withstand chlorine attack without showing significant loss in rejection is measured in ppm $\cdot$ h. This is the product of chlorine exposure expressed in ppm and the length of exposure expressed in hours. Thus, 1000 ppm $\cdot$ h is 1 ppm chlorine for 1000 h or 10 ppm chlorine for 100 h or 1000 ppm chlorine for 1 h, and so on.

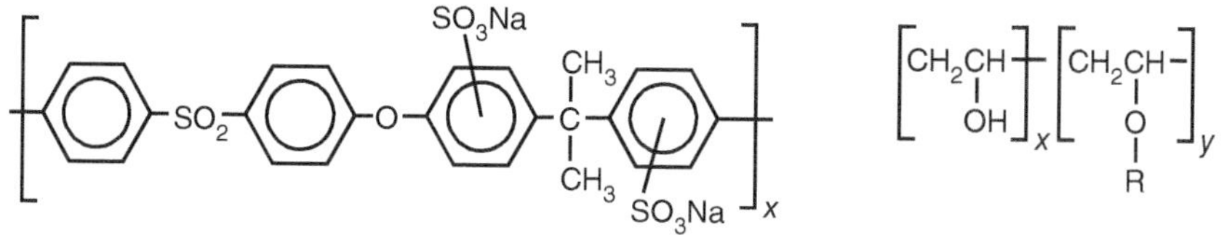

Figure 6.6 *Membranes based on sulfonated polysulfone and substituted poly(vinyl alcohol) are produced by Hydranautics (Nitto Denko Corporation) for nanofiltration applications.*

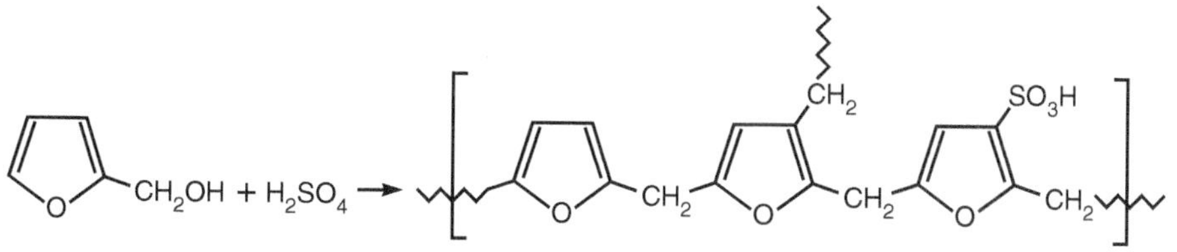

Figure 6.7 *A possible polymerization scheme for the NS200 condensation membrane.*

membrane was completely black. The chemistry of condensation and reaction is complex, but a possible polymerization scheme is shown in Figure 6.7.

The NS200 membranes had exceptional properties, including seawater salt rejections of up to 99.6% and fluxes of 40 $l/m^2 \cdot h$ at 30 bar. Unfortunately, the membranes were extremely sensitive to oxidants, and lost their excellent properties after only a few hundred hours of operation unless the feed water was completely free of dissolved chlorine and oxygen. A great deal of work was devoted to stabilizing the membrane, with little success.

Later, Kurihara and coworkers [24] at Toray produced a related membrane, using 1,3,5-tris (hydroxyethyl) isocyanuric acid as a comonomer. A possible reaction scheme is shown in Figure 6.8. This membrane, commercialized by Toray under the name PEC-1000, has the highest rejection of any membrane developed, with salt rejections of 99.9% and fluxes of 20 $l/m^2 \cdot h$ at 65 bar. The membrane also has the highest known rejections to low-molecular-weight organic solutes, typically more than 95% from relatively concentrated feed solutions. Unfortunately, these exceptional selectivities are accompanied by the same sensitivity to dissolved oxidants as the NS200 membrane. This problem was never completely solved, so the PEC-1000 membrane, despite its unsurpassed rejection properties, is no longer commercially available.

6.3.3 Interfacial Composite Membranes

The discovery by Cadotte and his coworkers that high-flux, high-rejection reverse osmosis membranes can be made by interfacial polymerization [6, 7], was a seminal development that led to the modern industry. Some steps along the way are illustrated in Table 6.1.

The first membranes made by Cadotte had salt rejections in tests with 3.5 wt% sodium chloride solutions (synthetic seawater) of greater than 99% and fluxes of 30 $l/m^2 \cdot h$ at a pressure of 100 bar. The membranes could also be operated at temperatures above 35 °C, the temperature ceiling for the Loeb–Sourirajan cellulose acetate membranes then in use. Modern interfacial composite membranes are significantly better. Typical membranes, tested with 3.5% sodium chloride solutions, have a salt rejection of 99.7% and a water flux of 50 $l/m^2 \cdot h$ at 35 bar; this is less than half the salt passage of the previous cellulose acetate membranes and twice the water flux. The rejection of low-molecular-weight dissolved organic solutes by interfacial membranes is also far better than by cellulose acetate. The only drawback of interfacial composite

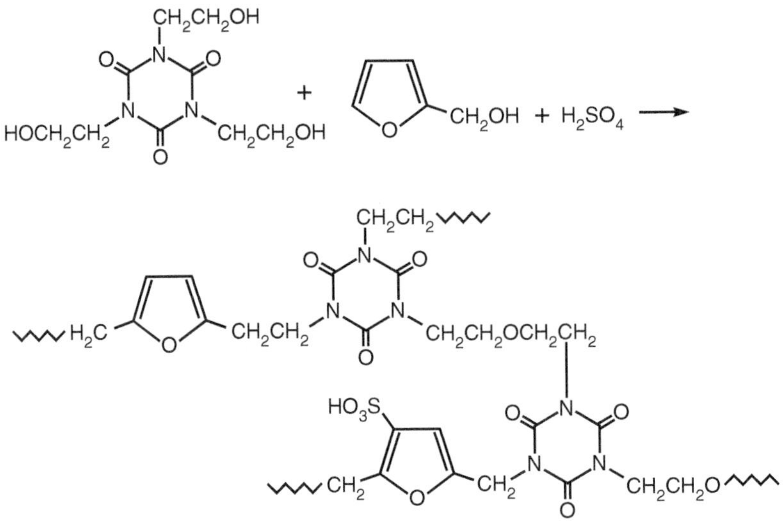

Figure 6.8 *Reaction sequence for Toray's PEC-1000 membrane.*

Table 6.1 *Characteristics of major interfacial polymerization RO membranes.*

Membrane	Developer	Properties
NS100 Polyethylenimine crosslinked with toluene 2,4-diisocyanate	Cadotte et al. [25] North Star Research	The first interfacial composite membrane achieved sea water desalination characteristics of >99% rejection, water flux 30 $l/m^2 \cdot h$ at 100 bar with seawater
PA 300/RC-100 Epamine (epichlorohydrin-ethylenediamine adduct) crosslinked with isophthaloyl chloride or toluene 2,4-diisocyanate	Riley et al. [26] Fluid Systems, San Diego	PA 300, based on isophthaloyl chloride (IPC), was introduced first, but RC-100, based on toluene 2,4-diisocyanate (TDI), proved more stable. This membrane was used at the first large RO sea water desalination plant (Jeddah, Saudi Arabia)
NF40 and NTR7250 Piperazine crosslinked with trimesoyl chloride	Cadotte FilmTec [7] and Kamiyama Nitto Denko [27]	The first all-monomeric interfacial membrane. Only modest seawater desalination properties, but a good brackish water membrane. More chlorine-tolerant than earlier membranes because of the absence of secondary amine bonds
FT-30/SW-30 *m*-Phenylenediamine crosslinked with trimesoyl chloride	Cadotte FilmTec [7]	An all-aromatic, highly crosslinked structure giving exceptional salt rejection and very high fluxes. By tailoring the preparation techniques, brackish water or sea water membranes can be made. Seawater version has a salt rejection of 99.5–99.7% at 55 bar. Brackish water version has >99% salt rejection at 40 $l/m^2 \cdot h$ and 15 bar. All the major RO companies produce variations of this membrane

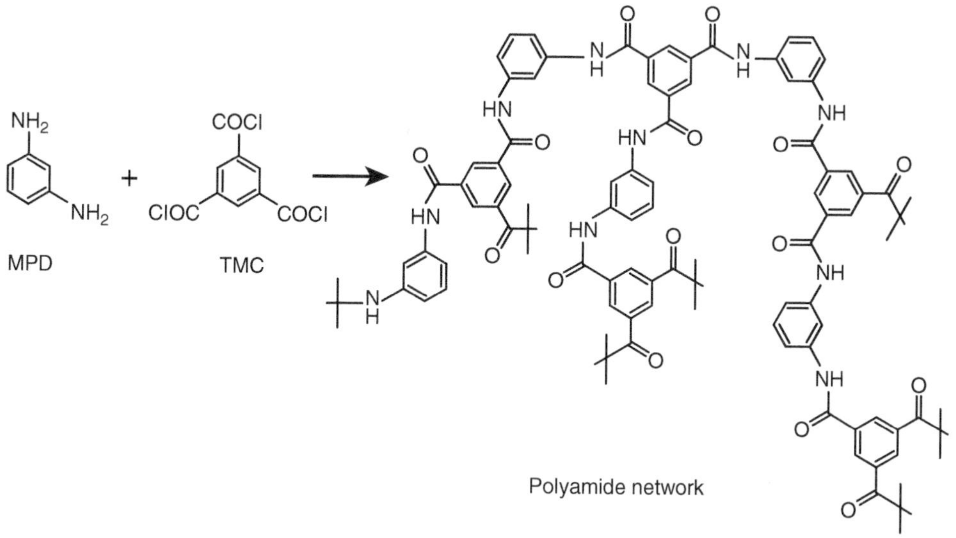

Figure 6.9 *Chemical structure of the FT-30 membrane developed by Cadotte using the interfacial reaction of phenylenediamine with trimesoyl chloride.*

membranes, and a significant one, is the rapid, permanent loss in selectivity that results from exposure to even ppb levels of chlorine or hypochlorite disinfectants. Although their chlorine resistance has improved over the years, these membranes still cannot be used with feed water containing more than a few ppb of chlorine.

The chemistry of the FT-30 membrane, which has an all-aromatic structure based on the reaction of phenylenediamine and trimesoyl chloride, is widely used. This chemistry, first developed by Cadotte [7] and shown in Figure 6.9, is now the industry standard and is used in modified form by all the major reverse osmosis membrane producers.

For a few years after the development of the first interfacial composite membranes, it was believed that the amine portion of the reaction chemistry had to be polymeric to obtain good membranes. This is not the case, and the monomeric amines, piperazine, and phenylenediamine, are used to form membranes with very good properties. Interfacial composite membranes contain urea or amide bonds, which are subject to degradation by chlorine attack. Chlorine appears to first replace the hydrogen atoms of secondary amide groups in the polymer. This mode of attack is slowed significantly if tertiary aromatic amines are used and the membranes are highly crosslinked. A slower, but ultimately more destructive, mode of attack is direct attack of the aromatic rings in the polymer [28]. Chemistries based on all-aromatic or piperazine structures are moderately chlorine tolerant and can withstand exposure to ppb levels of chlorine for prolonged periods or exposure to ppm levels for a few days. Early interfacial composite membranes, such as the NS100 or PA300, showed significant degradation at a few hundred ppm · h. Current membranes, such as the fully aromatic FilmTec™ FT-30 or the Hydranautics ESPA® membrane, can withstand up to several thousand ppm · h chlorine exposure. A number of chlorine tolerance studies have been made over the years; a discussion of the literature, and the ways the problem can be minimized, is given in some recent reviews [29, 30]. Heavy metal ions, such as iron, appear to strongly catalyze chlorine degradation. For example, the FT-30 fully aromatic membrane is somewhat chlorine resistant in heavy-metal-free water, but in natural waters, which normally contain heavy metal ions, chlorine resistance is low. The rate of chlorine attack is also pH sensitive.

6.4 Membrane Performance

General guidelines for membrane selectivity can be summarized as follows:

1. Multivalent ions are retained better than monovalent ions. Although the absolute values of the salt rejection vary over a wide range, the ranking for the different salts is the same for all membranes. In general, the order of rejection of ions by reverse osmosis membranes is as shown below.

   For cations:

   $$Fe^{3+} > Ni^{2+} \approx Cu^{2+} > Mg^{2+} > Ca^{2+} > Na^+ > K^+$$

   For anions:

   $$PO_4^{3-} > SO_4^{2-} > HCO_3^- > Br^- > Cl^- > NO_3^- \approx F^-$$

2. Dissolved gases such as ammonia, carbon dioxide, sulfur dioxide, oxygen, chlorine, and hydrogen sulfide always permeate well.
3. Rejection of weak acids and bases is highly pH dependent. When the acid or base is in the ionized form, the rejection will be high, but in the nonionized form, rejection will be low [31]. Data for a few weak acids are shown in Figure 6.10. At pHs above the acid pK_a, the solute rejection rises significantly, but at pHs below the pK_a, when the acid is in the neutral form, the rejection falls.
4. Rejection of neutral organic solutes generally increases with the molecular weight (or diameter) of the solute. Components with molecular weights above 100 Da are well rejected by all reverse osmosis membranes. Although differences between the rejection of organic solutes by different membranes are substantial, as the data in Figure 6.11 show, the rank order is generally consistent between membranes.
5. Negative rejection coefficients, that is, a higher concentration of solute in the permeate than in the feed, are occasionally observed, for example, for phenol and benzene with cellulose acetate membranes [33].

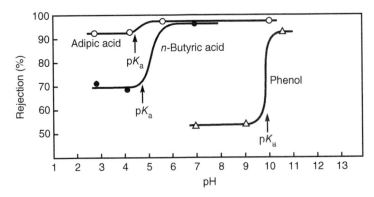

Figure 6.10 *Effect of pH on rejection of organic acids. Solute rejection increases at the pK_a as the acid converts to the ionized form. data Source: from T. Matsuura and S. Sourirajan [31].*

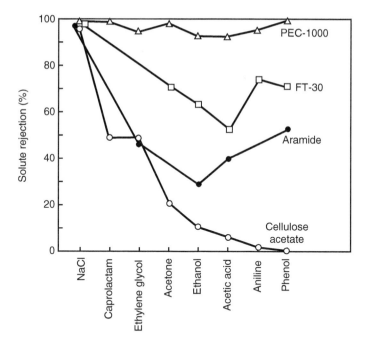

Figure 6.11 *Organic rejection data for the PEC-1000 membrane compared to FT-30, anisotropic aramid, and anisotropic cellulose membranes. Source: Reproduced with permission from [32]. Copyright (1990) Elsevier.*

6.5 Reverse Osmosis Membrane Categories

Reverse osmosis membranes can be grouped into four main categories:

- Seawater desalination membranes, used for 2–5 wt% salt solutions at pressures of 40–60 bar.
- Brackish water desalination membranes, used for 0.1–0.5 wt% salt solutions at pressures of 10–20 bar.
- Low-pressure nanofiltration membranes, used to produce ultrapure water from 200 to 5000 ppm salt solutions at pressures of 5–10 bar.
- Hyperfiltration membranes, used to separate solutes from organic solvent solutions.

6.5.1 Seawater Desalination Membranes

The relative performances of membranes produced for the seawater desalination market are shown in Figure 6.12, a plot of sodium chloride rejection as a function of membrane flux. The figure is divided into two sections by a dotted line at a rejection of 99.3%, generally considered to be the minimum rejection that can produce potable water from seawater in a single-stage plant. Membranes with lower sodium chloride rejections can be used to desalinate seawater, but at least a portion of the product water must be treated in a second-stage operation to achieve the target average permeate salt concentration of less than 500 ppm.

As Figure 6.12 shows, Toray's PEC-1000 crosslinked furfuryl alcohol membrane has by far the best sodium chloride rejection combined with good fluxes. This explains the sustained

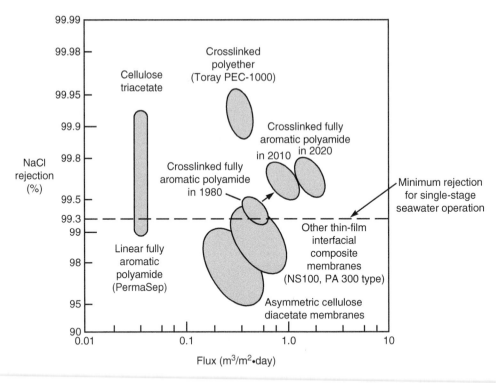

Figure 6.12 *Performance characteristics of membranes operating on seawater at 60 bar and 25 °C. The improvement in fully aromatic polyamide membranes from the initial measurements of Cadotte (1980) to today's (2020) industry standard is shown. Source: Modified from Riley [34].*

interest in this membrane despite its extreme sensitivity to dissolved chlorine and oxygen in the feed water. Hollow fine fiber membranes made from cellulose triacetate by Toyobo and aromatic polyamides by DuPont are also in the one-stage seawater desalination performance range, but the water fluxes of these membranes are low. However, because large-surface-area, hollow fine fiber, reverse osmosis modules can be produced economically, these membranes remained competitive until 2000, when DuPont ceased production. Toyobo still produces cellulose triacetate hollow fibers, particularly for use in the Red Sea-Persian Gulf region, where evaporation can increase the salt concentration in seawater to as much as 5% and the feed seawater is warm. The high salt rejection of cellulose triacetate membranes allows good water recovery to be achieved even from these high-osmotic-pressure feeds. With this exception, all new seawater desalination plants are based on interfacial composite membranes of the fully aromatic type, produced by Dow, Hydranautics (Nitto Denko) and Toray. Figure 6.12 shows the properties of the first-generation interfacial composite membrane (the NS100 and PA 300 type) and the first fully aromatic membrane when it first became available in about 1980. The improvement in the performance of this last membrane over the following 40 years is also shown. Membrane rejections have increased to 99.5–99.7% and fluxes have increased to more than 1.0 m^3/m^{2}day (40 l/m$^2 \cdot$ h). Even the best Loeb–Sourirajan cellulose diacetate membranes are not suitable for one-stage seawater desalination because their maximum salt rejection is less than 99%.

In the last 10–15 years, the presence of boron in desalinated water has become a problem for reverse osmosis plant operators. Seawater contains about 5 ppm boron. In the past, boron levels of 1–2 ppm were acceptable in the desalinated product, but most countries now require less than 1 ppm, and the WHO guidelines suggest a target of less than 0.5 ppm.

In seawater, boron can exist as boric acid, $B(OH)_3$, or as borate ion $B(OH)_4^-$.

$$B(OH)_3 + H_2O \leftrightharpoons B(OH)_4^- + H^+$$

The pK_a of boric acid is around 8.6 in seawater, so boric acid is the dominant form in neutral or acid solutions; in alkaline solutions above pH 9, the borate ion dominates. Since the hydrated ion is larger than the uncharged acid, boron rejection is pH dependent. At pH of 9 or above, modern seawater membranes can have a boron rejection of greater than 95%, enough to meet the 0.5 ppm target; at lower pHs rejection can fall to 80–90%. Raising the alkalinity of the feed water is not possible because of scaling issues, so nanofiltration pre- or posttreatment to remove boron is sometimes required, which increases cost. Membranes able to achieve adequate boron removal under neutral to slightly acid conditions have been reported and are likely to be installed in the future [35].

6.5.2 Brackish Water Desalination Membranes

Brackish water, such as arises from groundwater aquifers or estuaries, generally has a salt concentration in the 1000–5000 ppm range. The objective of desalination in this case is to convert 80–90% of the feed water to a desalted permeate, containing 200–500 ppm salt, and a concentrated brine that can be reinjected into the ground, sent to an evaporation pond, or discharged to the sea. In this application, membranes with 95–98% sodium chloride rejection are often adequate. For this reason, some brackish water plants still use cellulose acetate membranes, although interfacial composite membranes, which have better fluxes and rejections under comparable operating pressures, are now the norm. Composite membranes are always preferred for large operations, such as municipal drinking water plants, which can be built to handle the chlorine sensitivity of the membrane.

6.5.3 Nanofiltration Membranes

The goal of most of the early work on reverse osmosis was to produce desalination membranes with sodium chloride rejections greater than 99%. More recently membranes with lower sodium chloride rejections but much higher water permeabilities have been produced. These membranes, with properties that fall into a transition region between pure reverse osmosis membranes and pure ultrafiltration membranes, are called loose RO, low-pressure RO, or more commonly, nanofiltration membranes [12]. Typically, nanofiltration membranes have sodium chloride rejections between 20% and 80%; divalent ion rejections are much higher, in the 95–98% range. The membranes also reject 90% of organic solutes above 200 Da. Although some nanofiltration membranes are based on cellulose acetate, most are interfacial composites. The preparation procedure used to form these membranes can result in acid groups attached to the polymeric backbone. Neutral solutes such as lactose, sucrose, and raffinose are not affected by the presence of charged groups and their rejection increases in proportion to solute size. Nanofiltration membranes with molecular weight cut-offs to neutral solutes between 150 and 1500 Da are produced.

The rejection of salts by nanofiltration membranes is complicated and depends on both molecular size and Donnan exclusion effects caused by the acid groups attached to the polymer backbone [36, 37]. The phenomenon of Donnan exclusion is described in more detail in Chapter 11. In brief, fixed charged groups on the polymer backbone tend to exclude ions of the same charge, particularly multivalent ions, while being freely permeable to ions of the opposite charge, particularly

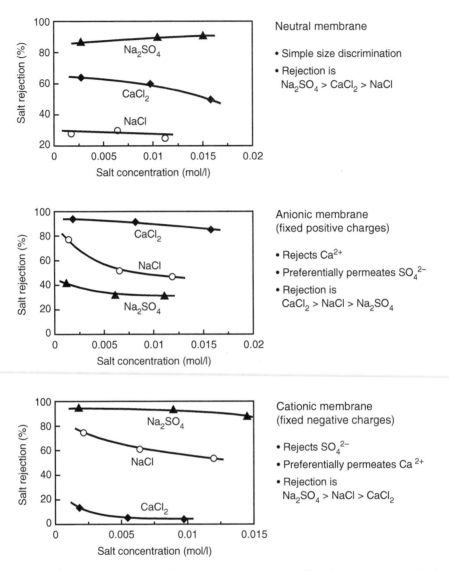

Figure 6.13 *Salt rejection with neutral, anionic and cationic nanofiltration membranes showing the effect of Donnan exclusion and solute size on relative rejections. Source: Data of Peters et al. [37].*

multivalent ions. Some results obtained by Peters et al. that illustrate the type of results that can be produced are shown in Figure 6.13 [37], in which the permeation properties of neutral, positively charged, and negatively charged membranes are compared.

The neutral nanofiltration membrane rejects the various salts in proportion to molecular size, so the order of rejection is simply

$$Na_2SO_4 > CaCl_2 > NaCl$$

The anionic nanofiltration membrane has positive groups attached to the polymer backbone. These positive charges repel positive cations, particularly divalent cations such as Ca^{2+}, while attracting negative anions, particularly divalent anions such as SO_4^{2-}. The result is an order of salt rejection

$$CaCl_2 > NaCl > Na_2SO_4$$

The cationic nanofiltration membrane has negative groups attached to the polymer backbone. These negative charges repel negative anions, such as SO_4^{2-}, while attracting positive cations, particularly divalent cations such as Ca^{2+}. The result is an order of salt rejection

$$Na_2SO_4 > NaCl > CaCl_2$$

Many nanofiltration membranes follow these rules, but often the behavior is more complex. Nanofiltration membranes frequently combine both size and Donnan exclusion effects to maximize the rejection of all salts and solutes. These membranes have high rejections and high water permeances at low salt concentrations, but lose their selectivity at salt concentrations above 1000 or 2000 ppm salt in the feed water. The membranes are therefore used to remove low levels of salt from already relatively clean water. The membranes are usually operated at low pressures of 8–12 bar.

The comparative performance of high-pressure, high-rejection reverse osmosis membranes, medium-pressure brackish water desalting membranes, and low-pressure nanofiltration membranes is shown in Table 6.2. Generally, the performance of a membrane with a particular salt can be estimated reliably once the performance of the membrane with one or two marker salts, such as sodium chloride and magnesium sulfate, is known. The rejection of dissolved neutral organic solutes is less predictable. For example, the PEC-1000 membrane had rejections of greater than 95% for almost all dissolved organics, whereas the organic solute rejections of even the best cellulose acetate membrane are usually no greater than 50–60%.

6.5.4   Organic Solvent Separating Membranes

A newer, and still developing, application of hyperfiltration membranes is in the processing of nonaqueous (organic solvent) solutions. Separating mixtures of organic solvents is difficult because of the high osmotic pressures involved. The separation of a 50/50 mixture of toluene/methylcyclohexane, for example, requires overcoming an osmotic pressure of more than

Table 6.2   *Properties of current good-quality commercial membranes.*

Parameter	Seawater membrane (SW-30)	Brackish water membrane (CA)	Nanofiltration membrane (NTR-7250)
Pressure (bar)	60	30	10
Solution concentration (%)	2–5	0.1–0.5	0.05
Rejection (%)			
NaCl	99.7	97	60
$MgCl_2$	99.9	99	89
$MgSO_4$	99.9	99.9	99
Na_2SO_4	99.8	99.1	99
$NaNO_3$	90	90	45
Ethylene Glycol	70	—	—
Glycerol	96	—	—
Ethanol	—	20	20
Sucrose	100	99.9	99.0

130 bar, implying operating pressures in the 200 bar range. This issue was discussed in Chapter 2. Osmotic pressures are less of a problem if the retained component has a relatively high molecular weight, above 200 Da, and is also present at low concentrations. The osmotic pressure is then in the 1–5 bar range and typical operating pressures are between 10 and 20 bar.

Most of the applications developed to date have used membranes that are freely permeable to organic solvents in the 50–150 Da range, but have significant rejections to solutes with molecular weights greater than 250–300 Da. Because the membranes must retain their mechanical and permeation properties in organic solvent mixtures, only a limited number of polymers are suitable as membrane materials. Some of the most widely used are listed in Table 6.3 [38]. These polymers do not swell or dissolve in most common solvents, but do dissolve in aprotic solvents such as NMP or DMAc, enabling anisotropic membranes to be prepared from them by the Loeb–Sourirajan phase-inversion technique. The solvent resistance of the membrane is sometimes improved by a post-formation crosslinking step.

An alternative membrane formation process is to form a microporous support membrane from one of the materials listed in Table 6.3 and then coat the membrane surface with a thin selective and crosslinked surface layer to form a composite membrane. In this case, the selective layer can be made from a wide variety of materials, depending on the separation to be performed. The membrane permeance and rejection change substantially depending on the permeating solvent. Solvents that swell the membrane the most have the highest fluxes and lowest selectivities. Some results that illustrate this effect for a silicone rubber composite membrane are shown in Figure 6.14. The rejected solute is a large polynuclear aromatic (MW = 330). The rejection at 8 bar is in the range 20–50%, depending on the solvent in which the solute is dissolved [39].

Table 6.3   Polymers used to prepare solvent stable, asymmetric membranes.

Polymer	Abbreviation	Molecular structure
Polyacrilonitrile	PAN	
Poly(vinylidene fluoride)	PVDF	
Polyimide (Matrimid®)	PI	
Polyimide (Lenzing P84)	PI	
Poly (etherimide)	PEI	
Poly(ether ether ketone)	PEEK	

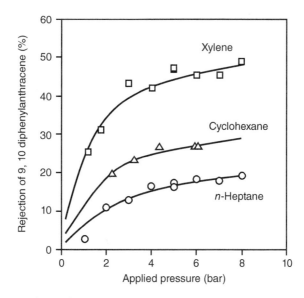

Figure 6.14 *Rejection of 9,10-diphenylanthracene dissolved in different hydrocarbon solvents by a silicone rubber composite membrane. The solvent that swells the membrane the most (n-heptane) gives rise to the lowest rejection. Source: Reproduced with permission from [39]. Copyright (2005) Elsevier.*

This type of silicone rubber composite membrane has found some commercial application in removing asphaltenes and other color bodies from refinery naphtha streams.

Another category of nanofiltration membrane uses rigid solvent-resistant materials to form finely porous membranes. In the 1980s, Nitto Denko developed polyimide-based ultrafiltration membranes that found a small use in separating polymers and pigments from toluene, ethyl acetate, hexane, and other solvents in waste paint and polymer solutions [40]. (More recently, ceramic membranes have been developed for the same type of separation.) The first commercial membranes were developed by W.R. Grace and produced under the trade name Starmem® [41, 42]. These membranes were made by a modified Loeb–Sourirajan process from Matrimid®-type polyimides. Later, Evonik-MET introduced DuraMem® and Puramem® membranes. These membranes are based on P84 polyimide, an extremely rigid, high-T_g polymer that remains glassy and relatively unswollen even in aggressive solvents. This type of membrane typically has a molecular weight cutoff of 150–900.

The permeation properties of the membrane can be measured with the simple test system shown in Figure 6.3. An easy and fast test to determine permeance and membrane integrity is to use a solvent containing a few hundred ppm of a dye such as Cresol Red (MW = 382) or Bromothymol Blue (MW = 624). At such low concentrations, the dye solution has the same flux as the pure solvent and dye rejection is usually in the 99% range. The test shows membrane defects as bright dye spots on the membrane. After membrane integrity has been established, the molecular weight cut-off of the membrane can be measured using a dilute solution incorporating oligomers, such as oligostyrene, of different molecular weights. By comparing the chromatogram peaks for the feed solution and the membrane permeate, a rejection molecular weight cut-off curve can be obtained. Figure 6.15 shows results obtained by Livingston [43], who popularized this test method. Test solutions using a series of low molecular weight n-alkanes or mixtures of quaternary ammonium salts can also be used with similar results.

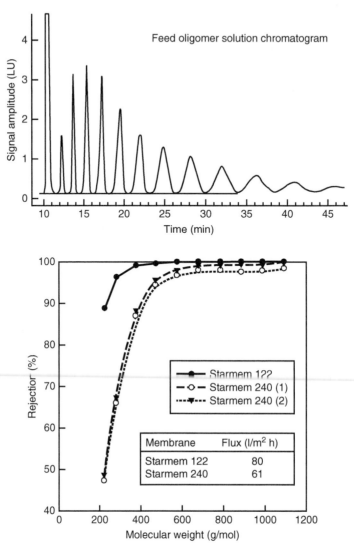

Figure 6.15 *Molecular weight cutoff curves obtained with a mixture of styrene oligomers. Source: Reproduced with permission from [43]. Copyright (2007) Elsevier.*

6.6 Membrane Modules

Currently, the type of spiral-wound module most commonly used for reverse osmosis is 8 in. in diameter, 40 in. in length, and contains about 40 m^2 of membrane. The industry is in the process of switching to 16-in. diameter, 40-in. long modules, containing about 150 m^2 of membrane. These larger modules achieve economies of scale and are used in new plants. The module production process used by the major manufacturers is completely automated and so production costs are low. Costs in 2020 were around $400 for an 8-in. diameter module, or approximately $10/m^2 of membrane. Five to seven modules are housed inside a filament-wound, fiber-glass-reinforced

Figure 6.16   *Skid-mounted RO plant able to produce 700 000 gal/day of desalted water. Source: Reproduced with the permission of Christ Water Technology.*

plastic tube. Longer modules, up to 60 in. in length, are produced by some manufacturers but have not been widely adopted. The module elements can be removed from the pressure vessels and exchanged as needed. A photograph of a typical skid-mounted system is shown in Figure 6.16. A typical spiral-wound 8-in. diameter membrane module will produce 8000–10 000 gal/day of permeate, so each of the 75-module skids shown in Figure 6.16 has a capacity of about 700 000 gal/day (2700 m^3/day).

Hollow fine fiber modules made from cellulose triacetate or aromatic polyamides were commonly used in the past for seawater desalination. These modules incorporated the membrane around a central tube, and feed solution flowed outward to the shell. Because the fibers were extremely tightly packed inside the pressure vessel, flow of the feed solution was slow. As much as 40–50% of the feed could be removed as permeate in a single pass through the module. However, the low flow and many constrictions meant that extremely good pretreatment of the feed solution was required to prevent membrane fouling from scale or particulates. A schematic illustration of such a hollow fiber module was shown in Chapter 4, Figure 4.50. The use of hollow fiber modules for this application has declined in the last 20 years and spiral-wound modules now have the bulk of the market.

6.7   Membrane Fouling and Control

Membrane fouling is the main cause of permeant flux decline and loss of product quality in reverse osmosis systems, so fouling control dominates system design and operation. The cause and prevention of fouling depend greatly on the feed water being treated, and appropriate control procedures must be devised for each plant. In general, sources of fouling can be divided into four principal categories: silt, scale, bacteria, and organic. More than one category may occur in the same plant.

Fouling control involves pretreatment and regular cleaning. Fouling by particulates (silt), bacteria, and organics such as oil is generally controlled by a suitable pretreatment procedure; this type of fouling affects the first modules in the plant the most. Fouling by scale is worse with more concentrated feed solutions; the last modules in the plant are most affected because they are exposed to the most concentrated feed water.

6.7.1   Silt

Early desalination plants used a combination of multilayer sand filtration, flocculation, antiscalants and bactericides to control fouling. These techniques are still used, but the fouling problem in some new plants has been much reduced by the use of an ultrafiltration membrane to treat the feed water before it enters the desalination unit. The ultrafiltration unit removes almost all of the bacteria and suspended solids. Savings can then be made by reducing the other pretreatment operations needed.

Silt is formed by suspended particulates of all types that accumulate on the membrane surface. Typical sources of silt are organic colloids, iron corrosion products, precipitated iron hydroxide, algae, and fine particulate matter. A good predictor of the likelihood of a particular feed water to produce fouling by silt is the silt density index (SDI) of the feed water. The SDI, an empirical measurement (ASTM Standard D-4189-82, 1987), is the time required to filter a fixed volume of water through a standard 0.45 μm pore size microfiltration membrane. Suspended material in the feed water that plugs the microfilter increases the sample filtration time, giving a higher SDI. The test procedure is illustrated in Figure 6.17.

An SDI of less than 1 means the reverse osmosis system can run for several years without colloidal fouling. An SDI of less than 3 means the system can run several months between cleanings. An SDI of 3–5 means particulate fouling is likely to be a problem and frequent, regular cleaning will be needed. An SDI of more than 5 is unacceptable and indicates that additional pretreatment is required. The maximum tolerable SDI also varies with membrane module design. Spiral-wound modules generally require an SDI of less than 5, whereas hollow fine fiber modules are more susceptible to fouling and require an SDI of less than 3.

To avoid fouling by suspended solids, some form of feed water filtration is required. All reverse osmosis units are fitted with a 0.45 μm cartridge filter in front of the high-pressure pump, but a sand filter, sometimes supplemented by addition of a flocculating chemical, such as alum or a cationic polymer, may be required to meet the target SDI value. Groundwaters usually have very low SDI values, and cartridge filtration is often sufficient. Surface or seawater may have an SDI of up to 200, requiring flocculation, coagulation, and deep-bed multimedia filtration before treatment.

6.7.2   Scale

Scale is caused by precipitation of dissolved metal salts in the feed water on the membrane surface. As salt-free water is removed in the permeate, the concentration of ions in the feed increases until at some point the solubility limit of some components is exceeded. Salt then precipitates on the membrane surface as scale. The proclivity of a particular type of feed water to produce scale can be determined by performing an analysis of the feed water and calculating the expected concentration factor in the brine. The ratio of the product water flow rate to feed water flow rate is called the recovery rate, equivalent to the term stage-cut used in gas separation.

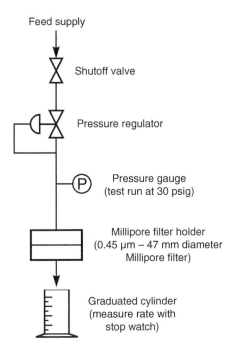

Feed supply

Shutoff valve

Pressure regulator

Pressure gauge
(test run at 30 psig)

Millipore filter holder
(0.45 μm – 47 mm diameter
Millipore filter)

Graduated cylinder
(measure rate with
stop watch)

(1) Measure the amount of time required for 500 ml of feed water to flow through a 0.45 micrometer Millipore filter (47 mm in diameter) at a pressure of 30 psig.

(2) Allow the feed water to continue flowing at 30 psig applied pressure and measure the time required for 500 ml to flow through the filter after 5, 10 and 15 minutes.

(3) After completion of the test, calculate the SDI by using the equation below.

$$SDI = \frac{100\ (1 - T_i / T_f)}{T_t}$$

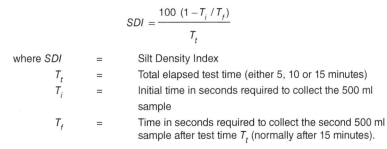

where SDI = Silt Density Index

T_t = Total elapsed test time (either 5, 10 or 15 minutes)

T_i = Initial time in seconds required to collect the 500 ml sample

T_f = Time in seconds required to collect the second 500 ml sample after test time T_t (normally after 15 minutes).

Figure 6.17 *The silt density index (SDI) test (ASTM Standard D-4189-82, 1987).*

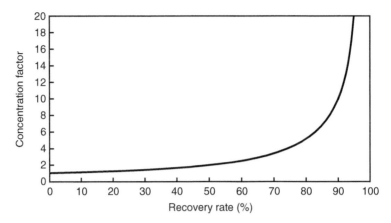

Figure 6.18 *The effect of water recovery rate on the brine solution concentration factor.*

$$\text{Recovery rate} = \frac{\text{Product flow rate}}{\text{Feed flow rate}} \qquad (6.7)$$

Assuming all the ions remain in the brine solution, the concentration factor is given by

$$\text{Concentration factor} = \frac{1}{1 - \text{Recovery rate}} \qquad (6.8)$$

The relationship between brine solution concentration factor and water recovery rate is shown in Figure 6.18. With plants that operate below a concentration factor of 2, that is, 50% recovery rate, scaling is not normally a problem. However, many brackish water plants operate at recovery rates of 80–90%, implying salt concentration factors of 5–10. Salt concentrations on the brine side of the membrane may then be above the solubility limit. In order of importance, the salts that most commonly form scale are

- calcium carbonate;
- calcium sulfate;
- silica complexes;
- barium sulfate;
- strontium sulfate;
- calcium fluoride.

Scale control is complex; the particular procedure depends on the composition of the feed water. Fortunately, calcium carbonate scale, by far the most common problem, is easily controlled by acidifying the feed or by using an ion exchange water softener to exchange calcium for sodium. Alternatively, an antiscalant chemical such as a polycarboxylate, polyacrylate, polyphosphonate, or polyphosphate can be added. Antiscalants interfere with the precipitation of the insoluble salt and maintain the salt in solution even when the solubility limit is exceeded [44]. Polymeric antiscalants may also be used, sometimes in combination with a dispersant to break up any flocs that occur.

Silica can be a particularly troublesome scalant because no effective antiscalant or dispersant is available. The solubility of silica is a strong function of pH and temperature, but in general the brine should not exceed 120 ppm silica. Once formed, silica scale is difficult to remove.

6.7.3 Biofouling

Biological fouling is the growth of bacteria on the membrane surface. The susceptibility of membranes to biological fouling is a strong function of the membrane composition. Cellulose acetate is an ideal nutrient for bacteria and such membranes can be completely destroyed by a few weeks of uncontrolled bacterial attack; feed water to cellulose acetate membranes must always be sterilized. Polyamide hollow fibers are also somewhat susceptible to bacterial attack; thin-film composite membranes are generally more resistant. Periodic treatment of these membranes with a bactericide usually controls biological fouling. Thus, control of bacteria is essential for cellulose acetate membranes and desirable for polyamides and composite membranes [29, 30]. Because cellulose acetate can tolerate up to 1 ppm chlorine, sufficient chlorination is used to maintain 0.2 ppm free chlorine. Chlorination can also be used to sterilize the feed water to polyamide and interfacial composite membranes, but residual chlorine must then be removed because the membranes are chlorine-sensitive. Dechlorination is generally achieved by adding sodium metabisulfate. In ultrapure water systems, water sterility is often maintained by UV sterilizers. The development of low-cost ultrafiltration/microfiltration membrane processes that remove particulates and all bacteria has encouraged the use of these membranes as a pretreatment step for new reverse osmosis plants.

6.7.4 Organic Fouling

Organic fouling is the attachment of materials such as oil or grease to the membrane surface. Such fouling may occur accidentally in municipal drinking water systems, but is more common in industrial applications in which reverse osmosis is used to treat a process or effluent stream. Removal of the organic material from the feed water by filtration or carbon adsorption is required.

6.7.5 Pretreatment

An example of a complete pretreatment flow scheme for a seawater reverse osmosis plant is shown in Figure 6.19. The water is controlled for pH, scale, particulates, and biological fouling. First, the feed water is treated with chlorine to sterilize it and the pH is adjusted to a value of 5–6. A polyelectrolyte is added to flocculate suspended matter, and two multilayer depth filters then remove suspended materials. The water is dechlorinated by dosing with sodium bisulfite,

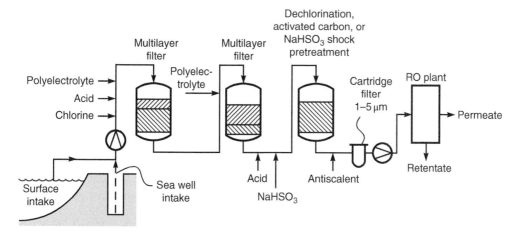

Figure 6.19   *Flow scheme showing the pretreatment steps in a typical seawater RO system.*

then passed through an activated carbon bed. As a final check, the pH is adjusted a second time, and the water is filtered through a 1–5-μm cartridge filter. The water is then ready to be fed to the reverse osmosis modules. Obviously, such pretreatment is expensive and may represent as much as one-third of the operating and capital cost of the plant. Nonetheless, this type of pretreatment was used by essentially all seawater desalination plants until about 2010. In recent years, the development of lower cost and more reliable ultrafiltration systems has resulted in some newer plants replacing much of the treatment train shown with a single ultrafiltration unit. This unit removes all suspended solids and bacteria, so chlorination of the feed water is not needed.

6.7.6   Membrane Cleaning

Good pretreatment is essential to achieve long membrane life, but an appropriate cleaning schedule is also necessary. Membrane cleaning is usually done when the membrane flux has declined by about 10–15% from its initial value. In plants with good pretreatment, cleaning may be needed only two to four times a year; problem feeds require more frequent cleaning. As with pretreatment, the specific cleaning procedure is a function of the feed water chemistry, the type of membrane, and the type of fouling. A typical regimen consists of flushing the membrane modules by recirculating a cleaning solution at high speed through the module, followed by a soaking period, followed by a second flush, and so on. The chemical cleaning agents commonly used are acids, alkalis, chelating agents, detergents, formulated products, and sterilizers.

Acid cleaning agents, such as hydrochloric, phosphoric, or citric acids, remove common scaling compounds effectively. With cellulose acetate membranes, the pH of the cleaning solution should not go below 2.0, or hydrolysis of the membrane will occur. Oxalic acid is particularly effective for removing iron deposits. Acid cleaners are not very effective against sulfate scales, for which a chelating agent such as ethylenediaminetetraacetic acid (EDTA) may be used.

To remove bacteria, silt or precipitates from the membrane, alkalis combined with surfactant cleaners are often used. Biz® and other laundry detergents containing enzyme additives are useful for removing biofoulants and some organic foulants. Most large membrane module producers now distribute their own pre-formulated products, which are mixtures of cleaning compounds, to users. These products are designed for various common feed waters and often provide a better solution to membrane cleaning than devising a cleaning solution for a specific feed.

Sterilization of the membrane system is also required to control bacterial growth. For cellulose acetate membranes, chlorination of the feed water is sufficient to control bacteria. Feed water to polyamide or interfacial composite membranes need not be sterile, because these membranes are usually fairly resistant to biological attack. Periodic shock disinfection using formaldehyde, peroxide, or peracetic acid solutions as part of a regular cleaning schedule is often enough to prevent biofouling.

Repeated cleaning gradually degrades reverse osmosis membranes. Most manufacturers now supply membrane modules with a multiyear limited warranty, depending on the application. Well-designed and maintained plants with good feed water pretreatment can usually expect membrane lifetimes of 5 years, and lifetimes of more than 7–10 years are not unusual. As membranes approach the end of their useful life, the water flux will normally have dropped by 20% from its initial value, and the salt rejection will have begun to fall. At this point, operators may try to 'rejuvenate' the membrane by treatment with a dilute polymer solution. This surface treatment plugs microdefects and restores salt rejection [45]. Typical treatment polymers are poly(vinyl alcohol)/vinyl acetate copolymers or poly(vinyl methyl ether). In this procedure, the membrane modules are carefully cleaned and then flushed with dilute solutions of the rejuvenation polymer. The exact mechanism of rejuvenation is unclear.

6.8   Applications

Applications fall into two categories: reverse osmosis, in which the solvent is water, and hyperfiltration, in which the solvent is an organic compound. To date, water treatment is overwhelmingly the larger application. Approximately 90 million m³/day of desalted water are produced by reverse osmosis plants around the world. Of that water, about half is produced by seawater desalination plants. These are large installations; there are more than 20 plants in operation around the world that each produce more than 100 000 m³/day of desalted water. Another 20–30% is water treated and used in diverse industrial areas, including electronics manufacturing, food processing, pharmaceutical production, electroplating, and others. The third major application, and one that has grown significantly in the last two decades, is in water reclamation. The goal in this case is removal of trace contaminants, salts, endocrine-disrupting chemicals, pesticide residues and the like from otherwise unusable waters to bring them up to municipal drinking water standards [8].

In contrast to reverse osmosis, the second category, hyperfiltration, remains a small, specialized market, and one in which the users themselves may have developed the relevant membrane technology. Though currently limited by a host of technical issues, hyperfiltration has considerable scope, and applications illustrating its potential are discussed at the end of the chapter.

The relative cost of reverse osmosis compared with other desalting technologies (ion exchange, electrodialysis, and multi-effect evaporation) is shown in Figure 6.20. The operating costs of electrodialysis and ion exchange scale almost linearly in proportion to the salt concentration of the feed. Therefore, these technologies are best suited to low-salt-concentration feed streams.

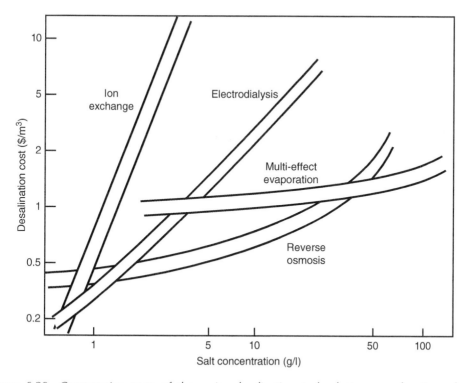

Figure 6.20 *Comparative costs of the major desalination technologies as a function of salt concentration. These costs should be taken as a guide only; site-specific factors can affect costs significantly. Source: Adapted from [46].*

On the other hand, the cost of multi-effect evaporation is relatively independent of the salt concentration and is mainly proportional to the mass of water to be evaporated. Thus, desalination by evaporation is best performed with concentrated salt solution feeds. Reverse osmosis costs increase significantly with salt concentration but at a lower rate than electrodialysis costs. The result is that reverse osmosis is the lowest-cost process for streams containing between 0.2% and 5.0% salt. However, site-specific factors or plant size may often make the technology the best approach, even for more dilute feed water streams.

6.8.1   Seawater Desalination

Seawater has a salt concentration of 3.2–5.0%, depending on the region of the world. Because of this high salinity, only membranes with salt rejections of 99.3% or more can produce potable water (water containing less than 500 ppm dissolved salt) from seawater in a single pass. Application of the first-generation cellulose acetate membranes, with rejections of only 97–99%, to seawater treatment was, therefore, very limited. With the development of interfacial composites, suitable seawater membranes became available, and many plants have been installed. These membranes can produce permeate water that meets the WHO standard of <500 ppm salt in a single pass. Most municipalities in the developed world require drinking water of a higher standard, containing less than 100–200 ppm salt. This quality of water usually requires a two-pass or partial two-pass system, typically having a single stage seawater system and a single stage brackish water system connected in series. The high-pressure seawater system removes almost all of the salt. The low-pressure brackish water system then removes enough residual salt to achieve the 100–200 ppm salt target. The concentrate from the brackish water system is recycled to the feed of the seawater system [9–11].

Early membranes were not competitive for very large seawater projects and multi-effect evaporation was usually used for plants of larger than $100\,000$ m^3/day capacity. These plants were often powered by steam from an adjacent electric power cogeneration unit. Cost reductions, improvements in process design, and the inherent flexibility of membrane systems, which provide easy startup/shutdown and turndown capabilities, have significantly improved the competitive position of reverse osmosis, which now has a majority of the large plant desalination market. Currently, about half a dozen new large seawater plants producing more than $100\,000$ m^3/day are built each year. The flexibility of membrane systems that allow easy startup/shutdown and turndown capability are additional advantages.

Early seawater plants operated at very high pressures, up to 100 bar; as membranes improved, operating pressures dropped to 60 bar. The osmotic pressure of seawater is about 23 bar, and the osmotic pressure of the rejected brine can be as much as 40 bar, so osmotic pressure markedly affects the net operating pressure in a plant. This effect is illustrated in Figure 6.21.

To avoid having to operate at excessively high pressures to overcome the high brine osmotic pressure, typical seawater plants operate at recovery rates of only 35–45%. At these low recovery rates, more than half of the feed water leaves the plant as pressurized brine. Because of the high pressures involved in seawater desalination, recovery of compression energy from the high-pressure brine stream is almost always worthwhile. In older plants, Pelton wheel units, able to recover 60–85% of the brine energy, were used. More recently, turbines or isobaric energy recovery systems have been used. These more efficient options can recover 90–95% of the energy contained in the high-pressure brine. The energy consumption of the plant is then in the range of 2.0–2.5 kWh/m^3 of water, far below the energy required for evaporation technology.

Improvements in membrane technology, both for filtration pretreatments and for the reverse osmosis unit itself, have more than kept pace with inflation, so that inflation adjusted water

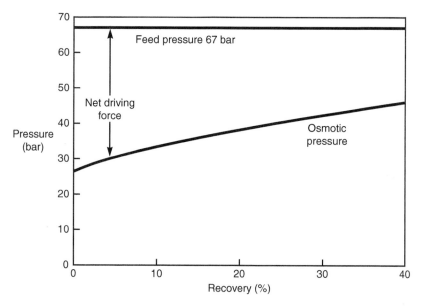

Figure 6.21   *Effect of water recovery on the seawater feed osmotic pressure and net driving pressure of a plant operating at 67 bar.*

production costs have actually fallen over the last 20 years. An analysis of the incremental improvements that have driven and continue to drive cost reduction is given by B. Penate and L. Garcia-Rodriguez [47]. The approximate operating costs for a seawater reverse osmosis plant in 2005 are given in Table 6.4. The initial capital cost of a plant, expressed as a function of operating capacity, is about US\$800–1000/m$^3 \cdot$ day (US\$4/gal $\cdot$ day), which translates over the plant life to about US\$0.25/m^3 of operating cost, and represents the biggest single cost component. The next largest component is power consumption; all other costs, including module replacement, labor, and chemicals amount together to only 20% of the cost. As the table indicates, a large new seawater plant can now produce water at a cost of around US\$0.50/m^3. Water from brackish water plants costs even less, in the region of US\$0.20/m^3. Completely current cost data are hard to obtain, but it is likely that costs in 2023 are even lower.

Seawater usually contains 4–6 ppm boron. Until 2000, reverse osmosis membranes had boron rejections of about 70–80%, which meant that water produced from seawater could contain 1–2 ppm boron. Drinking water standards for boron now require no more than 0.5–1.0 ppm, so

Table 6.4   *Operating costs for large sea water RO plants in 2005 [11].*

Product water cost component	(US\$/m^3)
Capital cost	0.25
Electric power (0.60/kWh)	0.17
RO membrane replacement (5 yr membrane life)	0.03
Chemicals	0.01
Maintenance and spare parts	0.02
Labor	0.04
Total product water cost	0.52

Capital costs are approximately US\$800–1000/m$^3 \cdot$ day capacity.

improved treatment to increase boron removal is needed. All of the major membrane manufacturers have improved the boron rejection of their membranes and boron rejections of 90% at neutral pH can now be achieved. Nonetheless, the permeate desalted water may still require further treatment. One solution is to use a nanofiltration system to filter boron from a portion of the desalinated water product.

6.8.2   Brackish Water Desalination

The salinity of brackish ground water is usually between 2000 and 10 000 ppm. The World Health Organization (WHO) recommendation for potable water is a salinity of no greater than 500 ppm, so in many cases only 90% of the salt must be removed from the feed. Early cellulose acetate membranes could achieve this removal easily, and treatment of brackish water was one of the first successful applications of reverse osmosis. The osmotic pressure of brackish water is approximately 0.8 bar per 1000 ppm salt. Hence the osmotic pressure of a raw brackish water feed is typically between about 1.6 and 8 bar, and the osmotic pressure remains moderate, even at the end of a train of modules when most of the feed water has permeated the membrane. In consequence, high levels of water recovery, typically 85–90%, are possible without having to resort to the high operating pressures used in seawater treatment.

Operational limitations in brackish water plants are generally due to scaling; calcium, sulfate, and silica-containing ions present in the feed become highly concentrated in the brine stream and can precipitate on the membrane surface. Disposal of the concentrated brine, which represents 10–15% by volume of feed water, is a significant problem. This has motivated research into increasing the water recovery of brackish water plants to 95–98%. Cohen and coworkers [48], for example, have proposed a two-stage membrane process. In the first stage of this scheme, 85–90% of the water permeates in the normal way. Calcium, silica, and other potential scalants in the retained brine are then precipitated and removed by addition of sodium hydroxide. The treated brine is reacidified, antiscalants added, and a further fraction of the water removed, thus achieving an overall water recovery of up to 98%. The final concentrate stream, representing only 2% of the volume of the feed, is small enough to be sent to an evaporation pond.

A simplified flow scheme for a brackish water reverse osmosis plant is shown in Figure 6.22. In this example, the brackish water is contaminated with suspended solids, so flocculation, followed by sand and cartridge filtration, is used to remove particulates. The pH of the feed solution is adjusted, the water is sterilized by chlorination to prevent bacterial growth on the membranes, and an antiscalant is added to inhibit precipitation of multivalent salts on the membrane. Finally, if chlorine-sensitive interfacial composite membranes are used, sodium bisulfite is added to remove excess chlorine before the water contacts the membrane.

A feature of the system design shown in Figure 6.22 is the staggered arrangement of the module pressure vessels. The volume of the feed water is reduced as water is removed in the permeate, and the number of modules arranged in parallel is reduced accordingly. In the example shown, the feed water passes initially through four module tubes in parallel, then through two, and finally through a single module tube in series. This is called a 'Christmas tree' or 'tapered module' design and maintains a high average feed solution velocity through the modules.

The operating pressure of brackish water reverse osmosis systems has gradually fallen over the past 20 years as the permeability and rejection rates of membranes have steadily improved. The first plants operated at pressures of 50 bar, but typical brackish water plants now operate at

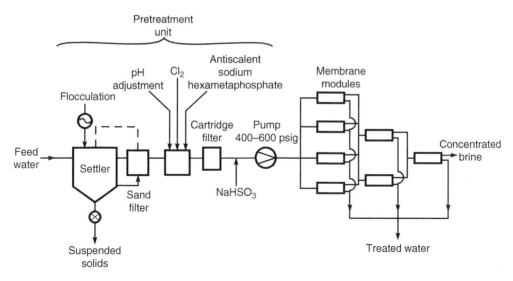

Figure 6.22   *Flow schematic of a typical brackish water RO plant. The plant contains seven pressure vessels, each containing six membrane modules. The pressure vessels are in a 'Christmas tree' array to maintain a high feed velocity through the modules.*

pressures in the range 10–20 bar. Capital costs of brackish water plants have stayed remarkably constant for almost 20 years; the rule of thumb of US\$250–500/m^3 · day (US\$1–2/gal · day) capacity is still true. Accounting for inflation, this reflects a very large reduction in real costs resulting from the better performance of modern membranes.

6.8.3   Industrial Applications

6.8.3.1   Ultrapure Water

Production of ultrapure water for the electronics industry is an established and growing application of reverse osmosis. The usual feed is municipal drinking water, which usually contains less than 100 ppm dissolved solids. However, the electronics industry requires water of extraordinarily high purity for wafer production and the like, so extensive treatment of municipal water is required. Table 6.5 shows the target water quality for a modern wafer manufacturing plant compared to that of typical municipal drinking water.

The first ultrapure water reverse osmosis system was installed at a Texas Instruments plant in 1970, as a pretreatment unit to an ion exchange process. Systems have increased in complexity as the needs of the industry for ever better quality water have increased. The flow scheme for a typical modern ultrapure water treatment system is shown in Figure 6.23. The plant comprises a complex array of operations, each requiring careful maintenance. As the key part of the process, the reverse osmosis plant typically removes more than 98% of all the salts and dissolved particulates in the feed water. Because the feed water is dilute, the reverse osmosis unit can operate at very high recovery rates – often 95% or more. Carbon adsorption then removes dissolved organics, followed by ion exchange to remove final trace amounts of ionic impurities. Bacterial growth is a major problem in ultrapure water systems; sterility is maintained by continuously recirculating the water through UV sterilizers and cartridge microfilters.

Table 6.5 *Ultrapure water specifications for typical wafer manufacturing process and levels normally found in drinking water (ASTM D-5127-99).*

	Ultrapure water (ε-1.1 type)	Typical drinking water
Resistivity at 25 °C	18.2	—
TOC (ppb)	<5	5000
Particles/L by laser > 0.1 μm	<100	—
Bacteria/100 ml by culture	<0.1	<30
Silica, dissolved (ppb)	<0.1	3000
Boron (ppb)	<0.02	40
Ions (ppb)		
Na^+	<0.02	3000
K^+	<0.02	2000
Cl^-	<0.05	10 000
F^-	<0.05	—
NO_3^-	<0.05	—
SO_4^{2-}	<0.02	15 000
Total ions	<0.1	<100 000

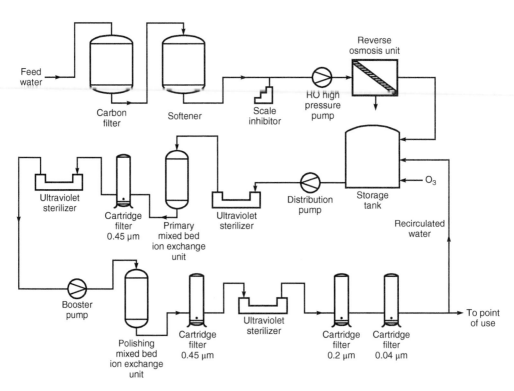

Figure 6.23 *Flow schematic of an ultrapure water treatment system. The reverse osmosis unit operates at a high water recovery rate, so the small concentrate stream is usually discharged to the sewer [49].*

6.8.3.2 Industrial Wastewater Treatment

In principle, treatment of industrial wastewaters to control environmental pollution should be a major application for reverse osmosis; in practice, reliability problems arising from membrane fouling have inhibited its widespread use. The most common applications are those in which chemicals of high value can be recovered. A good example is the recovery of nickel from nickel-plating rinse tanks, shown schematically in Figure 6.24. Watts nickel-plating baths contain high concentrations of nickel and other plating chemicals. After plating, a conveyor belt moves the parts through a series of connected rinse tanks. Water circulates through these tanks to rinse the parts free of nickel for the next plating operation. A typical countercurrent rinse tank produces a wastewater stream containing 2000–3000 ppm nickel; besides being a pollution problem, the wastewater contains valuable metal that would be lost if the water were discharged without treatment. This is an ideal application for reverse osmosis, because the rinse water is at nearly neutral pH, in contrast to many plating rinse waters, which are very acidic. The reverse osmosis unit produces permeate water, containing only 20–50 ppm nickel, which can be reused, and a small nickel concentrate stream that can be sent to the plating tank. Although the concentrate is more dilute than the plating tank drag-out, evaporation from the hot plating bath tank compensates for the extra water.

6.8.3.3 Water Reclamation and Reuse

In the early days of membrane development, membranes were expected to be widely used in the tertiary treatment of water (the third step to produce drinking water from sewage). At that time, the high cost of membrane technology kept this application from developing. At current production costs of only US$0.30–0.50/m^3 of water, the idea now makes economic sense in many water-limited regions of the world [8]. Despite remaining psychological barriers, plants

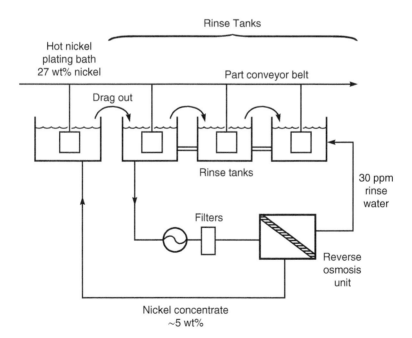

Figure 6.24 *Flow scheme showing the use of a reverse osmosis system to control nickel loss from rinse water produced in a countercurrent electroplating rinse tank.*

have been installed in Japan, Singapore, the Middle East, and the United States. The city of Los Angeles has operated Water Factory 21, a plant that has been a test bed for water reclamation trials, for many years [50]. The plant is in Orange County, California, an arid region where the principal local surface water source, the Colorado River, has a total salinity of 750 ppm. The plant includes a reverse osmosis system and treats secondary sewage to produce good-quality water, which is reinjected into the aquifer below the county. The water becomes mixed with natural groundwater, and thereby supplies drinking water elsewhere in the county. The development of membrane bioreactor technology for wastewater treatment (see Chapter 7) has led to the use of low-pressure reverse osmosis/nanofiltration units as a final treatment step in an increasing number of municipal wastewater reclamation projects. The bioreactor removes the bulk of the contaminants, producing water containing about 100 ppm total dissolved solids, mostly as chloride and sulfate salts, and 10–20 ppm total organic carbon. The reverse osmosis unit has 98–99% rejection for the organic carbon and reduces the dissolved salts to less than 50 ppm [8].

Nanofiltration membranes usually have high rejections to most dissolved organic solutes with molecular weights above 100–200 Da and good salt rejection at salt concentrations below 1000–2000 ppm salt. The membranes are also two- to five-fold more permeable than brackish and seawater RO membranes, so they can be operated at pressures as low as 3–5 bar and still produce useful fluxes. For these reasons, nanofiltration membranes are widely used as point-of-use drinking water treatment units in southern California and elsewhere in the southwestern United States. Some municipal and well water in this region contains on the order of 700 ppm dissolved salt and trace amounts of agricultural run-off contaminants. Many households use small 0.5-m^2 spiral-wound nanofiltration modules (under-the-sink modules) to filter this water, using the 2–3 bar pressure of tap water to provide the driving force.

6.8.4   Organic Solvent Separations

The use of hyperfiltration/nanofiltration membranes to process organic solvent mixtures has been the subject of a good deal of research interest in the last 15 years, but the technology has been slow to take off as an industrial process. The economic driver to develop industrial applications is clear. RO can produce essentially pure water from seawater at a cost of around US$0.50/m^3 of water. In stark contrast, the separation of organic solvents by distillation can cost US$10–50/m^3 solvent. In principle, huge cost savings are possible by replacing distillation with membrane filtration. However, there are problems.

In water treatment, a small group of membrane types can be used in diverse applications; little membrane or module customization is needed for a new application. This is not the case when treating organic solvent mixtures; change the solvent and the membrane, as well as glues, spacers and other module components, will often need to change too. The process equipment needed for an industrial solvent separating system is intrinsically more expensive than that for water treatment. Water treatment systems often use plastic components for process vessels, pipes, and valves; organic solvents are flammable and a good deal more toxic, so stainless steel is usually required. The rotating equipment, instrumentation, and controls must also be of a higher standard and able to operate in a potentially explosive environment.

To date, most industrial organic nanofiltration applications have fallen into two broad categories [38, 51, 52]:

- large bulk separations in refineries and petrochemical plants, where a partial separation is good enough
- small, demanding, high-value separations in the pharmaceutical and fine chemicals industries

Some of the first applications are described below. The total market is still small, probably in the US\$20–40 million/year range, but there is room for growth. The leading membrane and process suppliers are Evonik-MET, SolSep BV, and Koch Membrane Systems [38, 53, 54].

6.8.4.1 Bulk Separations

The first large solvent nanofiltration system was installed by Mobil Oil using membranes supplied by Grace Membrane Systems. The system, installed in 1998, was used in a refinery lube oil plant [53, 54]. Vacuum residual oil is a high boiling oil fraction produced by the main refinery vacuum fractionation column. The oil consists of C_{10} to C_{12} hydrocarbons together with high molecular weight naphthalene and other polycyclic aromatics. In the conventional process, shown in Figure 6.25a, the oil is diluted with 3–10 times its volume of methyl ethyl ketone or toluene. The solvent/oil solution is cooled, and the high molecular weight aromatic contaminants precipitate out as a waxy solid and are removed by a drum filter. The oil filtrate is then distilled. The solvent is recovered as an overhead and recycled to the process; the oil fraction, free of the

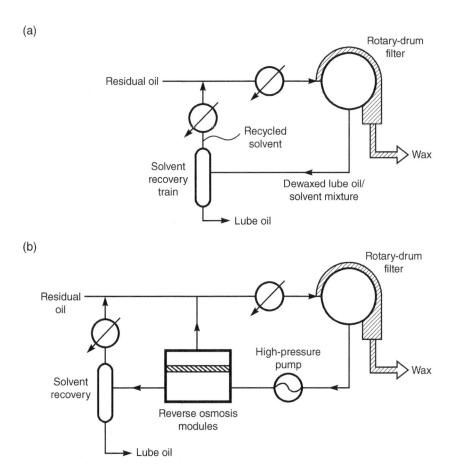

Figure 6.25 *Simplified flow schemes of (a) a conventional and (b) Mobil Oil's membrane solvent dewaxing processes. Refrigeration economizers are not shown. The first 3 million gal/day commercial unit was installed at Mobil's Beaumont refinery in 1998. Polyimide membranes in spiral-wound modules were used [53, 54].*

heavy aromatics, becomes the feed stock for lubricating oil. Cooling the large volumes of diluent solvent and then distilling to recover the solvent is costly and consumes large amounts of energy.

The process developed by Mobil Oil is illustrated in Figure 6.25b. Polyimide nanofiltration membranes formed into spiral-wound modules are used to separate up to 50% of the cold solvent from the dewaxed oil. The membranes have a flux of 20–40 $l/m^2 \cdot h$ at a pressure of 30–40 bar. The solvent filtrate bypasses the distillation step and is recycled, still cold, directly to the incoming oil feed. The net result is a significant reduction in the refrigeration load required to cool the oil. The retentate from the membrane passes as before to the distillation step, but the size and energy consumption of this section are reduced considerably. The plant achieved a payback time of less than one year.

Another refinery application, developed by Shell Oil, is nanofiltration of refinery naphtha fractions contaminated with color bodies and asphaltenes [55]. Silicone rubber composite membranes coated onto polyacrylonitrile microporous supports are used. The process operates at about 20 bar and 80–95% of the naphtha fraction is recovered as a clear filtrate. The black retentate is recycled to the crude distillation unit of the refinery. Shell has installed a number of these plants.

In a different arena, hyperfiltration using various types of membranes has been used to separate vegetable oil/extraction solvent (hexane) mixtures in corn oil and soybean oil plants. The process has reached pilot plant stage, but has yet to take off. Fouling of the membranes by other components in the oil has been an issue.

In the 1990s, several major oil companies, including Exxon, Texaco, and Mobil, had research programs focused on developing membranes and processes for a variety of refinery separations. One application that reached pilot-plant stage was the separation of aromatic/aliphatic mixtures [42]. The aromatics benzene, toluene, and xylene are feed stocks for nine of the top 50 chemicals produced in the United States. Processes to separate these components, together known as BTX, from C_4 to C_{10} hydrocarbons are found in every large refinery; liquid–liquid extraction, extractive distillation, and azeotropic distillation are commonly used. A number of studies have shown that either a membrane process or a hybrid membrane/distillation process is likely to have significantly lower capital and operating costs than traditional complex and energy intensive processes.

Another opportunity within the refinery for membranes that can separate aromatics from aliphatics occurs downstream of the BTX extraction described above. After the raw BTX mix has been extracted, the components are separated and a portion of the toluene product is converted by toluene disproportionation and transalkylation to benzene and xylenes, for which demand exceeds raw supply. A representative process is shown in Figure 6.26. In addition to toluene, the feed usually contains about 5–10% of close boiling aliphatics, such as isooctane. The feed is sent to a catalytic reactor, where a portion of the toluene is converted to xylenes and benzene. The raw product passes to a distillation train. The benzene fraction is removed as the overhead from the first column; the xylene is removed as the bottoms of the second column. Unreacted toluene is recycled to the reactor. The aliphatic fraction of the feed builds up in the recycle loop, so a portion of the recycle stream must be purged to control this build up. The purge stream typically contains 80–90% toluene, with the balance being a mixture of aliphatics. A membrane process able to selectively permeate the toluene can recover a significant amount of toluene that might otherwise be lost [42]. The streams involved are very large, and even a few percent improvement in toluene utilization corresponds to a potentially significant cost savings. Field tests of such designs have been carried out.

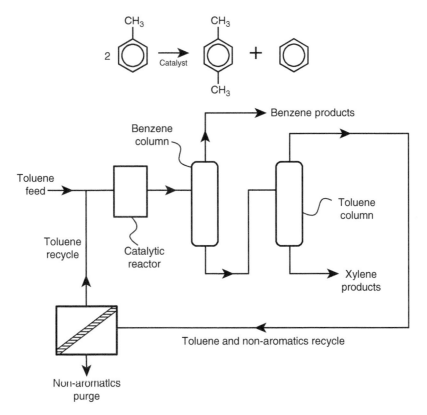

Figure 6.26 *Simplified flow diagram showing use of a toluene/non-aromatics separating membrane to recover toluene from the purge stream of a toluene disproportionation process*

6.8.4.2 *Pharmaceutical and Fine Chemical Applications*

Production of pharmaceuticals and fine chemicals is often a complex, multistep process carried out in small-scale batch operations. Organic nanofiltration membranes have found a place in such processes because, in contrast to distillation, they operate without exposing thermally labile components to high temperatures. Further, unlike liquid extraction or chemical precipitation, they add no components that must subsequently be removed to the streams under treatment. Processes where membranes have found application include

- recovery and recycle of solvents;
- separation and recovery of catalyst from solvents, reactants and products;
- exchanging one solvent for another;
- concentration of dilute reaction mixtures.

An example illustrating the use of membranes in a process involving these operations is shown in Figure 6.27 [51]. The process shown is the separation of a racemic mixture of phenethyl alcohol into two separate enantiomers. This type of separation is common in the production of drugs where only one of the enantiomers of the drug is pharmaceutically active.

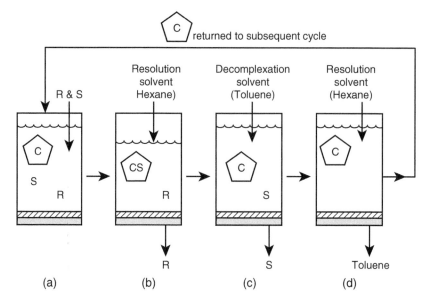

Figure 6.27   *The resolution of the racemic mixture of R and S enantiomers of phenethyl alcohol. Separation is achieved in four discrete steps (a–d) by combining complexing agent (represented in the figure with the pentagon containing 'c') with the S enantiomer, then using organic solvent nanofiltration to recover separate R- and S-fractions. Source: Adapted from [51].*

In the first step (a) of the process, a hexane solution of the R/S racemic mixture is reacted with an optically active complexing agent. The agent reacts exclusively with the S form of the enantiomer to form a bulky complex that is much larger than the unreacted enantiomer. In step (b), fresh hexane is used to flush the relatively small R-enantiomer through a membrane that selectively retains the complexed S form. In step (c), the retained purified S-enantiomer complex is broken by adding toluene as a decomplexation solvent, and essentially pure S-enantiomer is flushed through the membrane as a dilute solution of pure S-enantiomer. In step (d), the remaining complexation agent is solvent exchanged with hexane so that it can be recycled to step (a). In principle, the dilute R and S enantiomer solutions can each be concentrated in further treatment steps by using a tighter membrane to permeate the solvent and retain the enantiomers, thereby facilitating the next steps in drug production and recovering solvent for reuse.

6.8.5   Conclusions and Future Directions

The RO industry is well established. The membrane modules needed for the bulk of the market are provided by four large manufacturers, who between them produce 80% of all modules. These module makers supply a larger number of system builders, also known as OEMs (Original Equipment Manufacturers), who produce relatively small systems for diverse industrial users, and who buy modules as commodities, according to their particular needs. There is a smaller number of large engineering and project management companies that supply equipment for large seawater desalination and municipal tertiary treatment plants. A handful of companies serving various niche markets produce both modules and systems. Total membrane module sales in 2020 were about US$1 billion. Plant equipment and system sales were around US$2–3 billion.

Prospects for future growth are good. Municipalities in arid regions of the world continue to buy seawater desalination units. In addition, the use of RO as a final treatment in water reuse/reclamation projects is becoming increasingly common, and the demand for RO systems to produce ultrapure water for the electronics and pharmaceutical industries is very strong.

The industry is extremely competitive; manufacturers produce similar products and compete mostly on price and service. Many incremental improvements have been made to membrane and module performance over the past 20 years, resulting in steadily decreasing water desalination costs. The result has been to keep the capital cost of seawater desalination at US$500–1000/m³-day of product water. During that same time, the energy cost of the technology has steadily declined as a result of more selective and higher performance membranes, allowing operating pressures to fall. The introduction of energy recovery units on the high-pressure brine reject streams has also been important in reducing energy usage [11]. The impact of these changes on the unit cost of desalted seawater is shown in Figure 6.28. In 1991, the cost of desalination at the 11 000 m³-day Santa Barbara plant was US$1.50/m³. In 2003, the cost at the 150 000 m³/day Toas plant was US $0.50/m³. Much of this three-fold reduction in cost was a result of economies of scale, but reduced power requirements and better membranes also played a role.

The main outstanding technical issue is the limited chlorine resistance of interfacial composite membranes. A number of incremental steps made over the past 10–15 years have improved

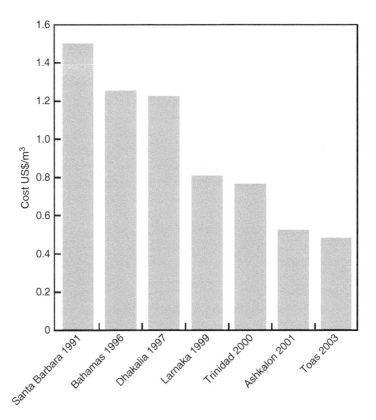

Figure 6.28   *Changes in the cost of seawater desalination from 1991–2003. Current estimate is in the $0.45–$0.5 per m³ range. Source: Adapted from [9].*

resistance, but current chlorine-resistant interfacial composites do not have the rejection and flux of the best conventional membranes. All the major membrane manufacturers are working on this problem, which is slowly being solved. Three other interrelated technical issues are fouling resistance, pretreatment, and membrane cleaning. Current membrane modules are subject to fouling by particulates and scale, which is controlled by feed water pretreatment and regular membrane cleaning. The importance of these problems has declined in recent years as a result of improved membrane modules and process designs. Currently, membrane lifetimes are 7 years or more. Further improvements may be possible but are not going to change costs much.

Development of hyperfiltration membranes to separate organic solvent mixtures is a continuing area of research. The total market for this type of application is still small, probably only a few tens of million dollars a year. However, this is an area where improvements in membrane performance could make a real difference to the economics of the process. Further growth is therefore likely.

References

1. Horvath, A.G. (1931). Water softening. US Patent 1,825,631, issued 29 September 1931.
2. Reid, C.E. and Breton, E.J. (1959). Water and ion flow across cellulosic membranes. *J. Appl. Polym. Sci.* **1**: 133.
3. Loeb, S. and Sourirajan, S. (1963). Sea water demineralization by means of an osmotic membrane. In: *Saline Water Conversion II*, Advances in Chemistry Series Number, vol. **38** (ed. R.F. Gould), 117–132. Washington, DC: American Chemical Society.
4. Westmoreland, J.C. (1968). Spirally wrapped reverse osmosis membrane cell. US Patent 3,367,504, issued 6 February 1968.
5. Bray, D.T. (1968). Reverse osmosis purification apparatus. US Patent 3,417,870, issued 24 December 1968.
6. Cadotte, J.E. (1985). Evolution of composite reverse osmosis membranes. In: *Materials Science of Synthetic Membranes*, ACS Symposium Series Number, vol. **269** (ed. D.R. Lloyd), 273–294. Washington, DC: American Chemical Society.
7. Cadotte, J.E. (1981). Interfacially synthesized reverse osmosis membrane. US Patent 4,277,344, issued 7 July 1981.
8. Côté, P., Liu, M., and Siverns, S. (2008). Water reclamation and desalination by membranes. In: *Advanced Membrane Technology & Applications* (ed. N.N. Li, A.G. Fane, W.S.W. Ho, and T. Matsuura), 171–188. Hoboken, NJ: Wiley.
9. Voutchkov, N. and Semiat, R. (2008). Seawater desalination. In: *Advanced Membrane Technology and Applications* (ed. N.N. Li, A.G. Fane, W.S.W. Ho, and T. Matsuura), 47–86. Hoboken, NJ: Wiley.
10. Greenlee, L.A., Lawler, D.F., Freeman, B.D. et al. (2009). Reverse osmosis desalination: water sources, technology and today's challenges. *Water Res.* **43**: 2317.
11. Fritzmann, C., Löwenberg, J., Wintgens, T., and Merlin, T. (2016). State-of-the-art reverse osmosis desalination. *Desalination* **216**: 1.
12. Schäfer, A.I., Fane, A.G., and White, T.D. (ed.) (2005). *Nanofiltration Principle and Applications*. Amsterdam: Elsevier.
13. Cadotte, J.E., Petersen, R.J., Larson, R.E., and Erickson, E.E. (1980). A new thin film sea water reverse osmosis membrane. Presented at the 5th Seminar on Membrane Separation Technology, *Desalination*, **32**, pp.25–31. Clemson, SC: Clemson University.
14. Petersen, R.J. (1993). Composite reverse osmosis and nanofiltration membranes. *J. Membr. Sci.* **83**: 81.

15. Lonsdale, H.K. (1966). Properties of cellulose acetate membranes. In: *Desalination by Reverse Osmosis* (ed. M. Merten), 93–160. Cambridge, MA: MIT Press.

16. Lonsdale, H.K., Merten, U., and Riley, R.L. (1965). Transport properties of cellulose acetate osmotic membranes. *J. Appl. Polym. Sci.* **9**: 1341.

17. Riley, R.L., Lonsdale, H.K., Lyons, C.R., and Merten, U. (1967). Preparation of ultrathin reverse osmosis membranes and the attainment of theoretical salt rejection. *J. Appl. Polym. Sci.* **11** (11): 2143.

18. Sourirajan, S. (1970). *Reverse Osmosis*. New York: Academic Press.

19. King, W.M., Hoernschemeyer, D.L., and Saltonstall, C.W. Jr. (1972). Cellulose acetate blend membranes. In: *Reverse Osmosis Membrane Research* (ed. H.K. Lonsdale and H.E. Podall), 131–162. New York: Plenum Press.

20. Endoh, R., Tanaka, T., Kurihara, M., and Ikeda, K. (1977). New polymeric materials for reverse osmosis membranes. *Desalination* **21** (35): 26.

21. McKinney, R. and Rhodes, J.H. (1971). Aromatic polyamide membranes for reverse osmosis separations. *Macromolecules* **4**: 633.

22. Richter, J.W. and Hoehn, H.H. (1971). Selective aromatic nitrogen-containing polymeric membranes. US Patent 3,567,632, issued 2 March, 1971.

23. Cadotte, J.E. (1975). Reverse osmosis membrane. US Patent 3,926,798, issued 16 December 1975.

24. Kurihara, M., Harumiya, N., Kannamaru, N. et al. (1981). Development of the PEC-1000 composite membrane for single stage sea water desalination and the concentration of dilute aqueous solutions containing valuable materials. *Desalination* **38**: 449.

25. Cadotte, J.E. (1977). Reverse osmosis membrane. US Patent 4,039,440, issued 2 August 1977.

26. Riley, R.L., Milstead, C.E., Lloyd, A.L. et al. (1977). Spiral-wound thin film composite membrane systems for brackish and seawater desalination by reverse osmosis. *Desalination* **23**: 331.

27. Kamiyama, Y., Yoshioka, N., Matsui, K., and Nakagome, E. (1984). New thin-film composite reverse osmosis membranes and spiral-wound modules. *Desalination* **51**: 79.

28. Antony, A., Fudianto, R., Cox, S., and Leslie, G. (2010). Assessing the oxidative degradation of polyamide reverse osmosis membrane-accelerated ageing with hypochlorite exposure. *J. Membr. Sci.* **347**: 159.

29. Arza, A. and Kucera, J. (2016). Minimize biofouling of RO membranes. *Chem. Eng. Prog.* **112** (9): 60.

30. Kurihara, M., Takeuchi, H., and Ito, Y. (2018). A reliable seawater desalination system based on membrane technology and biotechnology considering reduction of the environmental impact. *Environments* **127**: 5.

31. Matsuura, T. and Sourirajan, S. (1972). Reverse osmosis separation of phenols in aqueous solutions using porous cellulose acetate membranes. *J. Appl. Polym. Sci.* **15**: 2531.

32. Petersen, R.J. and Cadotte, J.E. (1990). Thin film composite reverse osmosis membranes. In: *Handbook of Industrial Membrane Technology* (ed. M.C. Porter), 307–348. Park Ridge, NJ: Noyes Publications.

33. Lonsdale, H.K., Merten, U., and Tagami, M. (1967). Phenol transport in cellulose acetate membranes. *J. Appl. Polym. Sci.* **11**: 1877.

34. Riley, R.L. (1991). Reverse osmosis. In: *Membrane Separation Systems* (ed. R.W. Baker, E.L. Cussler, W. Eykamp, et al.), 276–328. Park Ridge, NJ: Noyes Data Corp.

35. Rahmawati, K., Ghattour, N., Aubry, C., and Amy, G.L. (2012). Boron removal efficiency from red sea water using different SWRO/BWRO membranes. *J. Membr. Sci.* **423**: 522.

36. Van der Bruggen, B., Schaep, J., Wilms, D., and Vandecasteele, C. (1999). Influence of molecular size, polarity and charge on the retention of organic molecules by nanofiltration. *J. Membr. Sci.* **156**: 29.

37. Peters, J.M.M., Boom, J.P., Mulder, M.H.V., and Strathmann, H. (1998). Retention measurements of nanofiltration membranes with electrolyte solutions. *J. Membr. Sci.* **145**: 199.

38. Vandezande, P., Gevers, L.E.M., and Vankelecom, I.F.J. (2008). Solvent resistant nanofiltration separating on a molecular level. *J. Chem. Soc. Rev.* **37**: 365.

39. Tarleton, E.S., Robinson, J.P., Millington, C.R., and Nijmeijer, A. (2005). Non-aqueous nanofiltration: solute rejection in low-polarity binary systems. *J. Membr. Sci.* **252**: 123.

40. Iwama, A. and Kazuse, Y. (1982). New polyimide ultrafiltration for organic use. *J. Membr. Sci.* **11**: 297.

41. White, L.S., Wang, I-F., and Minhas, B.S. (1993). Polyimide membranes for separation of solvents from lube oil. US Patent 5,264,166, issued 23 November 1993.

42. White, L.S. (2006). Development of large-scale applications in organic solvent nanofiltration and pervaporation for chemical and refining processes. *J. Membr. Sci.* **286**: 26.

43. See Toh, Y.H., Loh, X.X., Li, K. et al. (2007). In search of a standard method for the characterization of organic solvent nanofiltration membranes. *J. Membr. Sci.* **291**: 120–125.

44. Schafer A.I., Andritos N, Karabelas A.J., Heok E.M.V., Schneider R., Nystrom M., (2005) *Fouling in Nanofiltration in Nanofiltration Principles and Applications*. (Ed A.I. Schafer, A.G. Fane and T.D. Waite), 169–240. Amsterdam: Elsevier.

45. Ko, A. and Guy, D.B. (1988). Brackish and seawater desalting. In: *Reverse Osmosis Technology* (ed. B.S. Parekh), 141–184. New York: Marcel Dekker.

46. Pittner, G.A. (1993). High purity water production using reverse osmosis technology. In: *Reverse Osmosis* (ed. Z. Amjad). New York: Van Nostrand Reinhold.

47. Penate, B. and Garcia-Rodriguez, L. (2012). Current trends and future prospects in the design of seawater reverse osmosis desalination technology. *Desalination* **284**: 1.

48. Rahardianto, A., Gao, J., Gabelich, C.J. et al. (2007). High recovery membrane desalting of low-salinity brackish water: integration of accelerated precipitation softening with membrane RO. *J. Membr. Sci.* **289**: 123.

49. Strathmann, H. (1991). Electrodialysis in membrane separation systems. In: *Membrane Separation Systems: Recent Developments and Future Directions* (ed. R.W. Baker, E.L. Cussler, W. Eykamp, et al.), 396–420. Park Ridge, NJ: Noyes Data Corp.

50. Nusbaum, J. and Argo, D.C. (1984). Design and operation of a 5-mgd reverse osmosis plant for water reclamation. In: *Synthetic Membrane Processes* (ed. G. Belfort), 378–436. Orlando, FL: Academic Press.

51. Marchetti, P., Jimenez Solomon, M.F., Szekely, G., and Livingston, A.G. (2014). Molecular separation with organic solvent nanofiltration: a critical review. *Chem. Rev.* **114**: 10735.

52. Liu, C., Dong, G., Tsuru, T., and Matsuyama, H. (2021). Organic solvent reverse osmosis membranes for organic liquid mixture separation: a review. *J. Membr. Sci.* **620**: 118882.

53. Bhore, N., Gould, R.M., Jacob, S.M. et al. (1999). New membrane process debottlenecks solvent dewaxing unit. *Oil Gas J.* **97**: 67.

54. White, L.S. and Nitsch, A.R. (2000). Solvent recovery from lube oil filtrates with polyimide membranes. *J. Membr. Sci.* **179**: 267.

55. Cederløf, G. and Geus, E.R. (2008). continuous process to separate colour bodies and/or asphalthenic contaminants from a hydrocarbon mixture. US Patent 7,351,873, issued 1 April 2008.

7

Ultrafiltration

Membrane Technology and Applications, Fourth Edition. Richard W. Baker.
© 2024 John Wiley & Sons Ltd. Published 2024 by John Wiley & Sons Ltd.

7.1 Introduction and History

Ultrafiltration uses finely porous membranes to separate macromolecules and colloids from water and microsolutes. The process can also be used to fractionate macromolecular mixtures. Ultrafiltration membranes are usually anisotropic structures, with a finely porous surface layer or skin, having pore diameters in the 20–200 Å range, supported on a much more open microporous substrate. The surface layer performs the separation; the substrate provides mechanical strength. Components of the feed solution with diameters larger than the diameter of the surface pores are retained at the surface of the membrane and are collected with the non-permeate residue stream.

The first synthetic ultrafiltration membranes were prepared over 100 years ago from collodion (nitrocellulose) by Bechhold [1]. Bechhold was probably the first to measure membrane bubble points, and he coined the term 'ultrafilter'. Other important early workers were Zsigmondy and Bachmann [2], Ferry [3], and Elford [4]. By the mid-1920s, collodion membranes were commercially available for laboratory use, although no industrial applications existed. The crucial breakthrough that expanded the applicability of ultrafiltration was the development of the anisotropic cellulose acetate membrane by Loeb and Sourirajan in 1963 [5]. Their goal was to produce reverse osmosis membranes, but others, particularly Michaels at Amicon, realized the general applicability of the technique. Michaels and his coworkers [6] produced anisotropic ultrafiltration membranes from cellulose acetate, as well as from polyacrylonitrile and its copolymers, and aromatic polyamides, polysulfone and poly(vinylidene fluoride). These materials are still widely used to fabricate ultrafiltration membranes.

The development of large-scale ultrafiltration applications can be divided into three phases. The first was the application of ultrafiltration to industrial and food processing applications. This development began with Abcor (now a division of Koch Industries) who installed the first commercially successful industrial ultrafiltration system [7], to recover electrocoat paint at an automobile assembly plant. The economics were compelling, and within a few years many similar systems were installed. A year later, the first cheese whey ultrafiltration system was installed. Within a decade, 100 similar systems had been sold worldwide. These early systems used tubular or plate-and-frame modules, which are relatively expensive, but lower cost hollow fiber (capillary) modules and spiral-wound modules soon became available, and by the 1990s the industry was well off the ground. All systems installed at that time operated in a pressurized tangential-flow mode, with feed solution circulating under pressure across the membrane surface. The principal problem was membrane fouling. The problem was controlled, but not eliminated, by the use of high feed circulation rates to sweep rejected material from the membrane surface. Regular membrane cleaning cycles were also used to maintain membrane flux, and development of membranes with surface properties designed to minimize fouling helped.

The second phase began in the late 1990s when the U.S. EPA and European regulators introduced new rules requiring that water supplies be treated to control *giardia*, coliform bacteria, and viruses. This opened up a new and large market for ultrafiltration. The water to be treated was clean, so fouling was a much reduced issue, and ultrafiltration provided a cost-efficient treatment option.

An even bigger change at about the same time was the introduction of constant flux operation, a new approach to control membrane fouling. In a constant flux process, the feed solution is circulated across the membrane at a low rate at atmospheric pressure; a constant flow, variable pressure pump is used to withdraw solution from the permeate side. The pump creates a vacuum pressure of 0.1–0.5 bar on the permeate side, thereby sucking permeate at a preset constant rate through the membrane. Over time, as the membrane fouls, the pressure differential required to maintain the constant transmembrane flux increases. At some point, this pressure reaches a set point and the membrane unit is taken offline and cleaned. If the flux through the membrane is set at a low value, the required suction pressure is small and the time between cleanings can be long, especially when constant flux operation is combined with other online fouling control techniques, such as automatic back-flushing and air sparging.

The development of constant flux operation was a major breakthrough. This mode of operation facilitated the development of submersible ultrafiltration modules, particularly for treatment of municipal wastewater. Laboratory- and pilot-scale submerged membrane systems were first tried in Japan by Yamamoto et al. [8] in the late 1980s. Hollow fiber membranes were submerged in a bioreactor tank of raw sewage, and a permeate pump was used to suck clean permeate through the membrane at a low flux. The fibers were maintained clean by backflushing and air sparging. It took several years before companies such as Kubota, Zenon, and Mitsui had solved all the problems involved in making industrial systems, but by the mid-1990s the technology was being used in a number of plants [9, 10].

Membrane bioreactors (MBRs) are now used to treat wastewater of all sorts. Initially, the market was for small systems, with capacities in the 1000 to 10 000 m^3/day range. These units were widely used in Japan to treat water generated from small local communities, sometimes even a single large building. In 2006, it was estimated that there were 2000 systems of this type in operation. There is still a market for such systems, but the bulk of the current market is for plants big enough to treat wastewater from municipal areas of 50 000–200 000 inhabitants. A 2017 review listed more than 30 MBR plants with capacities larger than 100 000 m^3/day [11]. The largest of these plants (Henriksdal, Stockholm) cost US$700 million and was designed to treat 860 000 m^3/day of wastewater. Built by Suez Water Process Technology (originally Zenon), the plant has close to one million m^2 of membrane.

Despite rapid growth, the membrane share of the wastewater treatment market is still less than a few percent of all new wastewater plants. There is general acceptance that the quality of treated water produced by MBR units is significantly better than that produced by conventional technology, making subsequent water reclamation and reuse far easier. The chief barrier to growth is cost. Both capital and operating costs have come down a great deal during the last 15 years and cost reductions should continue. With this trend, an increase in market share is likely.

During the last 20 years, a third major application of ultrafiltration membranes has developed in the processing of biologic drugs. The first biologic drug was recombinant human insulin introduced by Eli Lilly (under license from Genentech) in 1982. Insulin and other early products were highly active hormones that acted as catalysts for complex biological processes; a few thousand kilograms of product per year were enough to satisfy the market. Ultrafiltration and microfiltration membranes were widely used in the manufacturing process, but the total market was small.

In 2003, Humira, the first monoclonal biologic, was launched for the treatment of Crohn's disease and rheumatoid arthritis. Since then, this class of biologic drug has transformed the pharmaceutical industry. Currently (2020), 8 of the 10 biggest selling pharmaceuticals are biologic drugs. These drugs require higher dosage levels than earlier hormone biologics to achieve their effects, so typical production volumes have significantly increased.

The volume of solution to be treated in the production of biological products is still tiny compared to industrial processes, but the product is extremely valuable, and the producers are able to pay premium prices for quality products. Also, most biologic drugs are made by batch processes, so single-use disposable membrane cassettes are widely used. The total membrane market for these products in terms of membrane area is small compared to that of the industrial and water treatment application but is equivalent in terms of dollar value.

Some milestones in the development of ultrafiltration membranes are charted in Figure 7.1.

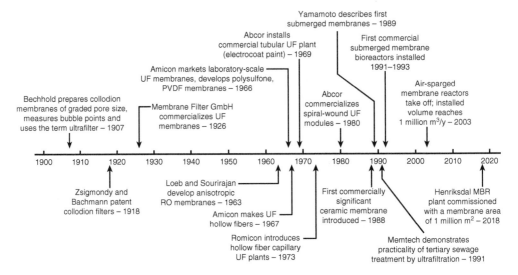

Figure 7.1 *Milestones in the development of ultrafiltration.*

7.2 Characterization of Ultrafiltration Membranes

A common application of ultrafiltration is to remove colloidal or suspended particles with a diameter of 1000 Å or more from solutions that also contain dissolved microsolutes and other small components. The retained components are removed as a concentrated residue (retentate) solution; the permeate is essentially free of the retained material. It might be supposed that microfiltration would be used for this application, since the size of the particles to be filtered is within the microfiltration range. However, microfiltration membranes are generally isoporous, with pores of approximately the same size throughout the membrane. Some of the rejected material is retained at the membrane surface, but much permeates into the pores and is captured at pore constrictions in the membrane interior, causing irreversible fouling. Therefore, use of microfiltration tends to be limited to situations where the solution to be treated has only a small amount of material to be retained, and rapid fouling is less likely to occur. Microfiltration membranes then function like conventional dead-end filters in which all of the feed solution passes through the membrane and is collected as permeate. The retained material is collected on or in the membrane. The two processes are compared in Figure 7.2.

In the processes described above, all of the large components of the feed solution are separated from the microsolutes and water that permeate the membrane. Another type of ultrafiltration application is the selective separation of multiple large components, such as a lower and a higher molecular weight protein, by using a membrane with a pore diameter intermediate between the diameters of the two proteins. In this case, precise control of the surface pore diameter is essential. Such applications provided an important early market for ultrafiltration and continue to grow as new biologic drugs are developed.

In the past, the division between ultrafiltration and microfiltration was made on the basis of the effective pore diameter of the membrane: ultrafiltration if the diameter was less than 1000 Å, microfiltration if the diameter was more than 1000 Å. Because ultrafiltration membranes are now widely used to separate large colloidal particles several microns in diameter, the distinction

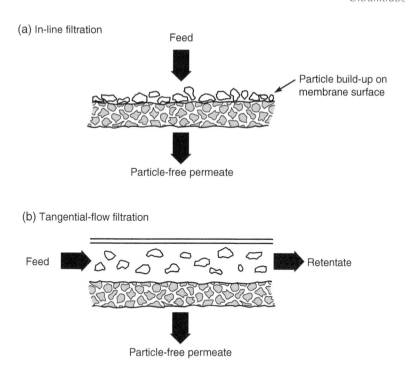

Figure 7.2 *A comparison of microfiltration (a) and ultrafiltration (b) separation processes.*

between the two processes is better based on the type of process used, not the pore diameter of the membrane. Ultrafiltration processes use anisotropic membranes of a range of pore diameters in some form of tangential-flow process, as in Figure 7.2b. The material to be removed is captured as a concentrate in the portion of the feed that does not pass through the membrane. Microfiltration membranes generally use isotropic membranes in a dead end (once through) process, as in Figure 7.2a. The feed solution contains a low level of contaminant that is captured in the membrane. All of the feed solution is collected as a filtrate.

As mentioned above, ultrafiltration membranes are anisotropic structures having surface pore diameters in the 20–200 Å range. To characterize the membranes, laboratory-scale ultrafiltration experiments are usually performed with small, stirred batch cells or flow-through cells in a recirculation system. Diagrams of the two types of systems are shown in Figure 7.3. A fixed pressure is applied to the feed solution and the permeate is collected at atmospheric pressure. Because ultrafiltration experiments are generally performed at pressures below 6 bar, plastic components can be used. Stirred batch cells are often used for quick experiments, but flow-through feed recirculation systems are preferred for systematic work. In feed recirculation systems, the feed solution can be more easily maintained at a constant composition, and the turbulence required to control surface fouling is high and easily reproducible. This allows reliable comparative measurements to be made.

Ultrafiltration membranes are categorized on the basis of their nominal molecular weight cut-off. The cut-off is usually characterized by the molecular weight of a globular protein that has a rejection of 95% by the membrane. Factors other than molecular weight also affect permeation however, and this nominal value should be treated with caution in determining permeation performance under differing conditions.

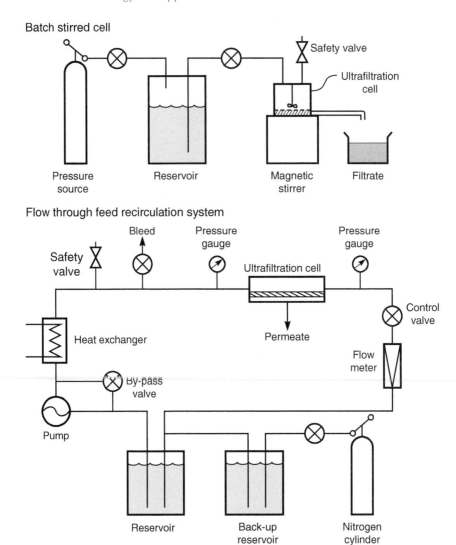

Figure 7.3   *Laboratory tangential-flow ultrafiltration test systems.*

7.3   Membrane Fouling

7.3.1   Constant Pressure and Constant Flux Operation

Until the mid-1990s, most laboratory ultrafiltration systems and all industrial systems operated at an elevated feed pressure, typically 3–10 bar. The feed solution was circulated at this pressure across the surface of the membrane and the permeate was collected at atmospheric pressure. This tangential-flow mode of operation can be called constant pressure/variable flux. A key factor determining the system performance is fouling, due to deposition of retained colloidal and macromolecular material on the membrane surface. A number of reviews have described the process in detail [12–14].

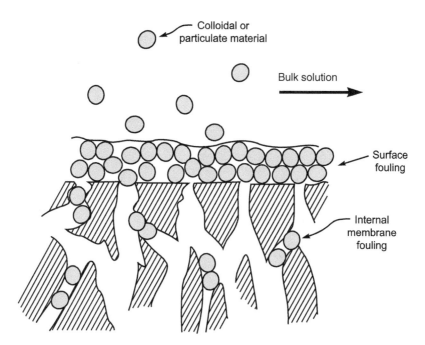

Figure 7.4 *Schematic representation of fouling on an ultrafiltration membrane. Surface fouling is the deposition of solid material on the membrane that consolidates over time. This fouling layer can be controlled by high turbulence, regular cleaning, and using hydrophilic or charged membranes to minimize adhesion to the membrane surface. Surface fouling is generally reversible. Internal fouling is caused by penetration of solid material into the membrane, which results in plugging of the pores. Internal membrane fouling is generally irreversible.*

The pure water flux of ultrafiltration membranes is often very high – greater than 500 l/m$^2 \cdot$ h (350 gal/ft$^2 \cdot$ day). However, when membranes are used to separate macromolecular or colloidal solutions, the flux falls within seconds, typically to about 50 l/m$^2 \cdot$ h. This immediate drop in flux is caused by the formation of a layer of retained colloidal or macromolecular material on the membrane surface. This layer forms a secondary barrier to flow through the membrane, as illustrated in Figure 7.4 and described in detail below. The initial decline in flux is determined by the composition of the feed solution and its hydrodynamics. Sometimes the resulting flux is constant for a prolonged period, and when the membrane is retested with pure water, the flux returns to the original value. More commonly, a further slow decline in flux occurs over a period of hours to weeks, depending on the feed solution. Most of this second decrease in flux is caused by slow consolidation of the secondary layer formed on the membrane surface. Formation of this consolidated layer, called membrane fouling, is difficult to control. Control techniques include regular membrane cleaning, backflushing, or using membranes with surface characteristics that minimize adhesion. More recently, air scrubbing is being used. Operation of the membrane at the lowest practical operating pressure also delays consolidation of the deposited surface layer.

A typical plot illustrating the slow decrease in flux that can result from consolidation of the secondary layer is shown in Figure 7.5 [15]. The pure water flux of this small electrocoating plant was approximately 50 gal/min, but on contact with an electrocoat paint solution containing 10–20% latex, the flux immediately fell to about 10–12 gal/min. This first drop in flux was due to the formation of a layer of latex particles on the membrane surface, as shown in

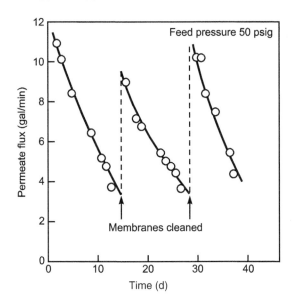

Figure 7.5 *Ultrafiltration flux as a function of time for an electrocoat paint latex solution. Because of fouling, the flux declines over a period of days. Periodic cleaning is required to maintain high fluxes. Source: Reproduced with permission from [15]. Copyright (1982) Gardner Publications, Inc.*

Figure 7.4. Thereafter, the flux steadily declined over a 2-week period. This second drop in flux was caused by slow densification of the fouling layer under the system operating pressure. In this particular example, the densified layer could be removed by periodic cleaning of the membrane. When the cleaned membrane was exposed to the latex solution again, the flux was initially restored to that of a fresh membrane.

If the cleaning cycle shown in Figure 7.5 is repeated many times, the flux no longer returns to the original value when the membrane is cleaned. Part of this slow, permanent loss of flux is due to precipitates on the membrane surface that are not removed by cleaning. A further cause is internal fouling by material that becomes lodged in the interior of the membrane, as illustrated in Figure 7.4. Ultrafiltration membranes are often used to separate relatively large colloids from water and microsolutes. In this case, there is a tendency to use high flux, high molecular-weight cut-off membranes, but the higher fluxes can be transitory, because the membranes are more susceptible to internal fouling. A membrane with smaller pores and a lower molecular-weight cut-off, even though it may have a lower pure water flux, often provides a more sustained flux with actual feed solutions, because less internal fouling occurs.

The process shown in Figure 7.6a is an example of constant pressure/variable flux operation. As the flux declines, the recirculation rate is increased to compensate for the lower flux, or more membrane area is turned on. When the flux declines to an unacceptable value, the system is taken offline, the membranes are cleaned and the process restarted, as was shown in Figure 7.5.

Beginning in about 1995, an alternative type of operation was recognized [16]. This process, called constant flux/variable pressure operation, is compared to constant pressure operation in Figure 7.6b. In constant flux operation, the feed solution is circulated close to atmospheric pressure across the feed side of the membrane, and a sub-atmospheric pressure is created on the permeate side of the membrane with a constant flow/variable pressure pump. The flow through the membrane is maintained at a fixed rate by this permeate pump. As the membrane fouls, the pump has to work harder to draw liquid through the membrane and the pressure on the permeate side of the membrane falls. This increases the transmembrane pressure across the membrane.

(a)

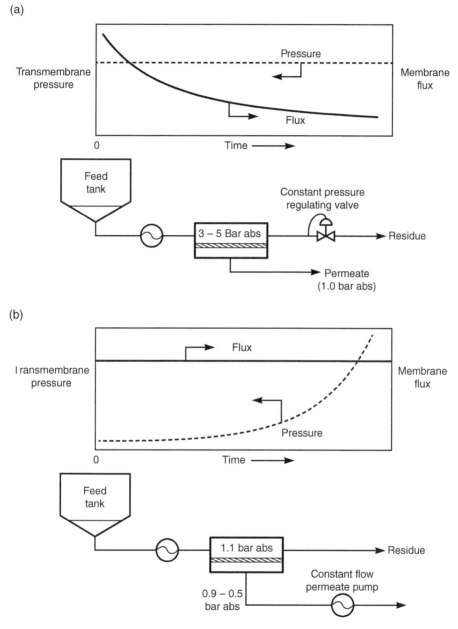

(b)

Figure 7.6 *A comparison of tangential-flow constant pressure/variable flux (a) and constant flux/variable pressure (b) operations in ultrafiltration/microfiltration.*

At some point, the transmembrane pressure increases to a preset cut-off value, at which point the system is taken offline and the membrane is cleaned.

Some results illustrating constant flux/variable pressure operation are shown in Figure 7.7 [17]. In these experiments, the initial pressure difference across the membrane was very small, less than 0.02 bar. As the membrane fouled, the pressure required to maintain the flux increased. When the pressure difference across the membrane reached 0.3 bar, the

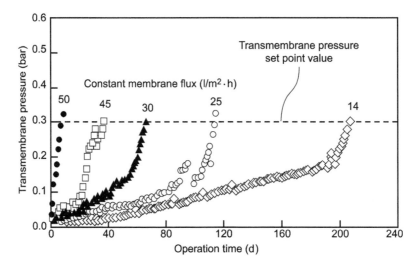

Figure 7.7   Illustration of fixed flux/variable pressure membrane operation. Feed solution is circulated across the membrane surface at ambient pressure. Permeate is removed at a fixed rate by a constant volume permeate pump. Initially, the transmembrane pressure across the membrane is very small. As the membrane fouls, the transmembrane pressure increases. When the transmembrane pressure reaches a preset value of 0.3 bar, the system is taken offline and the membranes are cleaned. As the fixed flux decreases, the time between cleanings increases. Source: Reproduced with permission from [17] Copyright (2006) Elsevier.

membrane was considered fouled and was taken offline for cleaning. When the flux was fixed at a high value, $50 \, \text{l/m}^2 \cdot \text{h}$, the membrane required cleaning after only 5 days. Reducing the flux to $25 \, \text{l/m}^2 \cdot \text{h}$ increased the time between cleanings to 120 days. Reducing the flux to $14 \, \text{l/m}^2 \cdot \text{h}$ increased the time between cleanings to 210 days. There is a steep trade-off in the time between cleaning cycles and membrane flux. Reducing the operating flux means that more membrane area is required to treat the same volume of feed, but the extra capital cost is usually more than offset by the operating cost savings arising from reducing the need for cleaning.

The transmembrane pressure-*versus*-time curves shown in Figure 7.7 have a characteristic form. Initially, the pressure rises at a slow and constant rate for many days, until a critical value is reached. At this critical point, the rate of increase in pressure sharply rises, and within a few days, the pressure reaches the set point value and the system must be taken offline and cleaned.

This behavior is commonly observed. A number of explanations have been given [18]; the most convincing is that the membrane pores are not fouled uniformly. As some pores become blocked, the fixed permeate flux is forced through the remaining open pores. The higher flux that then occurs through these pores causes them to be fouled at a higher rate, forcing the flux to go through even fewer pores, and so on. Thus, once fouling and pore blockage begins to occur, the process rapidly accelerates, leading to complete fouling of the membrane.

Adoption of constant flux/variable (low) pressure membrane operation, combined with improved membrane cleaning methods, has opened up municipal wastewater treatment as a large market for ultrafiltration, especially using submersible membrane systems (described later). The constant flux method can also be used for non-submersible systems, although most suppliers and users continue to favor constant pressure mode, operating at lower pressures than in the past and with better fouling control.

7.3.2 Concentration Polarization

The primary cause of membrane fouling is concentration polarization, which results in the deposition of a layer of material on the membrane surface. The general phenomenon of concentration polarization is described in Chapter 5. In ultrafiltration, solvent and macromolecular or colloidal solutes are carried towards the membrane surface by the solution permeating the membrane. Solvent molecules permeate the membrane, but the larger solutes accumulate at the membrane surface. Because of their size, the rate at which the rejected solute molecules can diffuse from the membrane surface back to the bulk solution is relatively low. Thus, their concentration at the membrane surface is typically 20–50 times higher than the bulk feed solution concentration. These solutes become so concentrated at the membrane surface that they aggregate as a continuous layer of precipitated matter and become a secondary barrier to flow through the membrane. The formation of this aggregated layer on the membrane surface is illustrated in Figure 7.8. The so-called gel layer model of this behavior was developed at the Amicon Corporation in the 1960s [12].

The formation of the layer is easily described mathematically. At any point within the boundary layer shown in Figure 7.8, the convective flux of solute to the membrane surface is given by the volume flux, J_v, of the solution through the membrane multiplied by the concentration of the solute, c_i. At steady state, this convective flux within the laminar boundary layer is balanced by the diffusive flux of retained solute in the opposite direction. This balance is expressed by the equation

$$J_v c_i = D_i \frac{dc_i}{dx} \tag{7.1}$$

where D_i is the diffusion coefficient of the macromolecule in the boundary layer. Once the deposited layer has formed, the concentrations of solute at both surfaces of the boundary layer

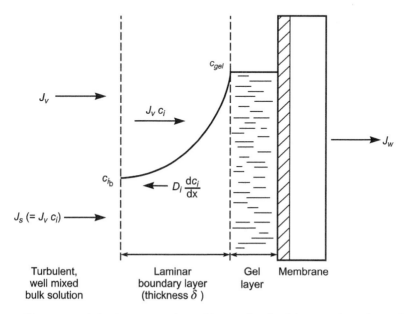

Figure 7.8 *Illustration of the formation of a gel layer of colloidal material on the surface of an ultrafiltration membrane by concentration polarization.*

are fixed. At one surface, the concentration is the bulk feed solution concentration c_{i_b}; at the other surface, it is the concentration of the gel-like deposit (c_{gel}). Integration of Eq. (7.1) over the boundary layer thickness (δ) then gives

$$\frac{c_{gel}}{c_{i_b}} = \exp\left(\frac{J_v \delta}{D_i}\right) \tag{7.2}$$

In any particular ultrafiltration test, the terms c_{i_b}, c_{gel}, D_i and δ in Eq. (7.2) are all fixed because the solution and the operating conditions of the test are fixed. This means that the volume flux, J_v, through the membrane is also fixed, and independent of the intrinsic permeability of the membrane. In physical terms, this is because use of a membrane with a higher intrinsic permeability causes a thicker deposit to form on the surface of the membrane. This lowers the membrane flux until the rate at which solutes are brought toward the membrane surface and the rate at which they are removed are again balanced, as expressed in Eq. (7.1).

The formation of a layer of compacted colloidal material at the ultrafiltration membrane surface produces a limiting or plateau flux that cannot be exceeded. Once the layer has formed, increasing the applied pressure does not increase the flux, but merely increases the thickness of the layer. This is also shown in Eq. (7.2), which contains no term for the applied pressure.

The effect of the deposited layer on the flux through an ultrafiltration membrane at different feed pressures is illustrated in Figure 7.9. At a very low pressure p_1, the flux J_v is low, so the effect of concentration polarization is small, and the layer does not form. The flux is close to the pure water flux of the membrane at the same pressure. As the applied pressure is increased to pressure p_2, the higher flux causes increased concentration polarization, and the concentration of retained material at the membrane surface increases. If the pressure is increased further to p_3, concentration polarization becomes enough for the retained solutes at the membrane surface to reach the critical concentration c_{gel} and the secondary barrier layer begins to form. This is the limiting flux for the membrane with this feed solution and operating conditions. Further increases in pressure to p_4 only increase the thickness of the layer, not the flux.

Experience has shown that the best long-term performance of an ultrafiltration membrane is obtained when the applied pressure is maintained below the pressure p_3 shown in Figure 7.9. We can call this maximum operating pressure the critical pressure. Operating at higher pressures does not increase the membrane flux but does increase the thickness and density of retained material at the membrane surface layer. Over time, material on the membrane surface can become increasingly compacted or precipitate, forming a layer of deposited material that has a lower permeability; the flux then falls from the initial value. When constant flux operation became common in the 1990s, the term critical flux was coined to explain the operating procedures being used by the industrial developers of the technology [16]. The origins of critical flux are shown in Figure 7.9. It is the flux at which the first aggregated layer forms on the membrane surface. The best long-term performance of an ultrafiltration/microfiltration constant-flux system is obtained when the fixed flux is set below the critical flux [19].

One way of determining the critical flux is to develop a tangential-flow flux *versus* applied pressure curve of the type shown in Figure 7.9 and estimate the critical flux value. A second method is illustrated in Figure 7.10, which shows data from Chen et al. [20]. A series of measurements were made with a membrane in a well-controlled tangential-flow test system. The permeate flux was increased in steps from an initial low value. The flux was held constant for 30 minutes at each step and the transmembrane pressure was measured at each step. Figure 7.10a shows the results as the permeate flux was increased stepwise to 130 l/m$^2 \cdot$ h. After

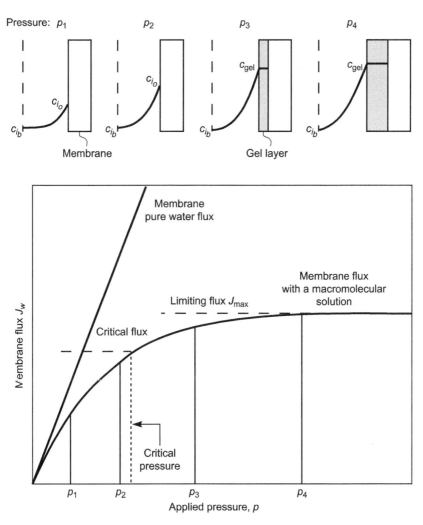

Figure 7.9   *The effect of pressure on ultrafiltration membrane flux and the formation of a secondary gel layer. Ultrafiltration membranes are best operated at pressures below p_3 when the gel layer has not formed. Operation at high pressures such as those above p_3 leads to formation of thick gel layers, which consolidate over time, resulting in permanent fouling of the membrane.*

each incremental change in flux, the transmembrane pressure increased, but stayed constant thereafter until the next change in flux. When the process was reversed, the transmembrane pressure decreased by almost the same amount. All of these measurements were below the critical flux for this membrane at these operating conditions.

Figure 7.10b shows a repeat of this experiment, but this time the permeate flux was increased stepwise to a higher value. Up to 130 l/m$^2 \cdot$ h, the transmembrane pressure remained constant at each step, as before, but when the permeate flux was increased to 140 l/m$^2 \cdot$ h, the transmembrane pressure no longer increased to a steady value. Instead, the pressure began to increase continuously. During the next higher flux increments (155 and 170 l/m$^2 \cdot$ h), the transmembrane pressure also continued to increase steadily, and a constant value was not reached. When the permeate flux decreased, the transmembrane pressure decreased, but did not return to the initial values

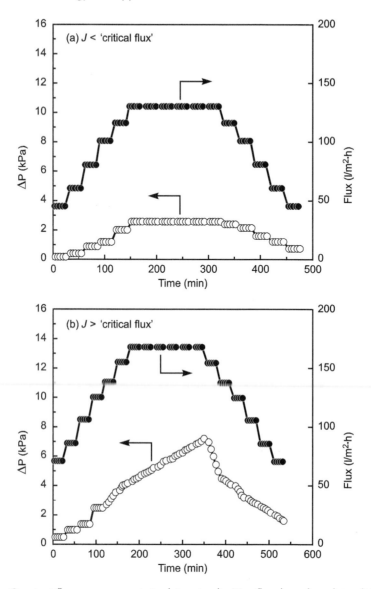

Figure 7.10 *Constant flux measurements to determine limiting flux, based on data of V. Chen et al. The critical flux at which the membrane becomes fouled is between 140 and 150 l/m²h. (a) Maximum flux below critical flux. (b) Maximum flux exceeds critical flux. Source: Reproduced with permission from [20]. Copyright (1997) Elsevier.*

measured when the permeate flux was first increased. The membrane had become permanently fouled. These measurements suggest that the critical flux at which this starts to occur is about 140–150 l/m² · h.

The point at which the foulant layer forms on the membrane can also be determined by constant pressure/variable flux measurements. A series of constant pressure experimental results obtained with latex solutions illustrating the effect of concentration and pressure on flux is shown in Figure 7.11 [13]. The point at which the flux reaches a plateau value depends on the

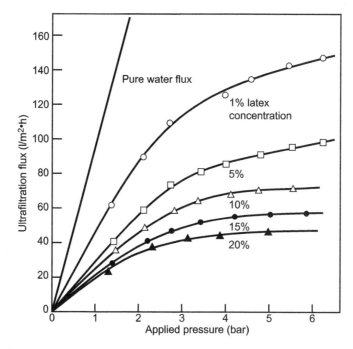

Figure 7.11   *The effect of pressure on membrane flux for styrene-butadiene polymer latex solutions in a high-turbulence, thin-channel test cell. Source. Reproduced with permission from [13]. Copyright (1990) John Wiley and Sons.*

concentration of the latex in the solution: the more concentrated the solution, the lower the limiting flux. The critical flux is about $100 \text{ l/m}^2 \cdot \text{h}$ at 1% latex, but falls to about $35 \text{ l/m}^2 \cdot \text{h}$ at 20% latex. The exact relationship between the maximum flux and solute concentration can be obtained by rearranging Eq. (7.2) to obtain

$$J_{\max} = -\frac{D}{\delta}\left(\ln c_{i_b} - \ln c_{gel}\right) \tag{7.3}$$

where $J_{\max}$ is the plateau or limiting flux through the membrane.

Plots of the limiting flux $J_{\max}$ as a function of solute concentration are shown in Figure 7.12 for a series of latex solutions at various feed solution flow rates. A series of straight-line plots is obtained, and these extrapolate to the latex concentration c_{gel} in the densified fouling layer when the flux drops to zero. The slopes of the plots in Figure 7.12 are proportional to the term D/δ in Eq. (7.3). The increase in flux resulting from an increase in the fluid recirculation rate is caused by the decrease in the boundary layer thickness δ.

Plots of the limiting flux as a function of solute concentration for different solutes using the same membrane under the same conditions are shown in Figure 7.13 [13]. Protein or colloidal solutions, which easily form precipitate layers, have low fluxes that extrapolate to zero at low solids concentrations. Particulate suspensions, pigments, latex particles, and oil-in-water emulsions, which do not easily consolidate, have higher fluxes at the same concentration and operating conditions, and the flux generally extrapolates to zero at higher solids concentrations.

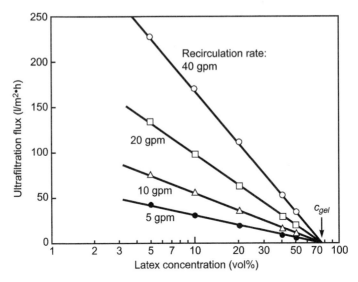

Figure 7.12 *Ultrafiltration flux with a latex solution at an applied pressure of 60 psi (in the limiting flux region) as a function of feed solution latex concentration. These results were obtained in an exceptionally high-turbulence, thin-channel cell. The solution recirculation rate is shown in the figure. Source: Reproduced with permission from [13]. Copyright (1990) John Wiley and Sons.*

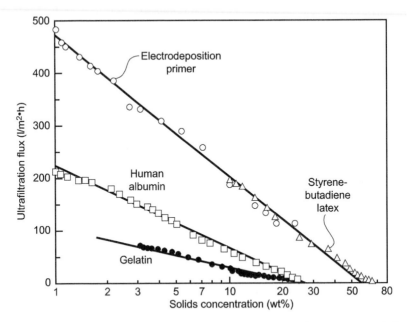

Figure 7.13 *Effect of solute type and concentration on flux through the same type of ultrafiltration membrane operated under the same conditions. Source: Reproduced with permission from [13]. Copyright (1990) John Wiley and Sons.*

The studies of concentration polarization illustrated in Figures 7.11–7.13 were performed in a tangential-flow laboratory test cell, under conditions of high turbulence, yielding unusually high membrane fluxes. Fluxes obtained in industrial processes, operated at less turbulent conditions, are lower, but the same general behavior is seen.

The gel layer model described by Eqs. (7.1)–(7.3) is very appealing and is widely used to rationalize the behavior of ultrafiltration membranes. However, a number of experimental observations cannot be explained by this simple model:

- The flux of many macromolecular colloidal and particulate solutions is too high (sometimes by an order of magnitude) to be rationalized by a reasonable value of the diffusion coefficient and the boundary layer thickness in Eq. (7.2).
- In the plateau region of flux/pressure curves of the type shown in Figure 7.11, different solutes should have fluxes proportional to the value of their diffusion coefficients D in Eq. (7.3). Often, this is not the case, as Figure 7.12 shows. For example, latex and particulate solutes, which have very small diffusion coefficients, typically have higher limiting fluxes than protein solutions measured with the same membranes under the same conditions. This is the opposite of the expected behavior.
- Experiments with different ultrafiltration membranes and the same feed solution often yield very different limiting fluxes. But according to the model shown in Figure 7.8 and represented by Eq. (7.2), the limiting flux is independent of the membrane type.

Contrary to normal experience that falling bread always lands jam-side down, the trend of these observations is that experiment produces a better result than theory predicts. These observational misfits are lumped together and called the flux paradox. The best working model seems to be that, in addition to simple diffusion, solute is also being removed from the membrane surface as undissolved aggregated particles by a scouring action of the tangential flow of the feed fluid. This explains why protein solutions that form tough adherent layers have lower fluxes under the same conditions than pigment and latex solutions that form looser aggregates. The model also explains why increasing the hydrophilicity of the membrane surface or changing the charge on the surface can produce higher limiting fluxes. Decreased adhesion between the foulant and the membrane surface allows the flowing feed solution to remove aggregated particles more easily.

Figure 7.14 illustrates how turbulent eddies, caused by the high velocity of the solution passing through the narrow channel of a spiral-wound module, might remove particles from the membrane surface. The feed solution, flowing at high velocity, has to weave its way through the wales of the feed spacer netting, resulting in highly turbulent flow patterns. Although a relatively laminar boundary layer may form next to the membrane surface, as described by the film model, periodic turbulent eddies may occur. These eddies can dislodge material from the membrane surface, carrying it away with the feed solution.

The most important effect of concentration polarization is to reduce transmembrane flux, but it also affects the retention of macromolecules. Retention data obtained with dextran polysaccharides at various pressures are shown in Figure 7.15 [12]. These data are from stirred cell batch experiments, in which the effect of increased concentration polarization with increased applied pressure is particularly marked. A similar drop in retention with increased pressure is observed with flow-through cells, but the effect is less pronounced, because concentration polarization is better controlled. With macromolecular solutions, the concentration of retained macromolecules at the membrane surface increases with increased pressure, so permeation of the macromolecules also increases, lowering rejection. The effect is particularly noticeable at low pressures, under which conditions increasing the applied pressure produces the largest increase in flux, and hence

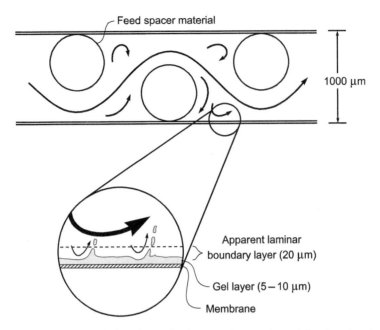

Figure 7.14 *An illustration of the channel of a spiral-wound module showing how periodic turbulent eddies can dislodge deposited gel particles from the surface of ultrafiltration membranes.*

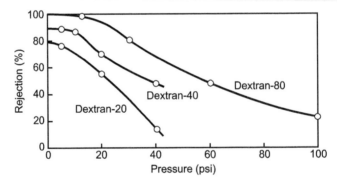

Figure 7.15 *Rejection of 1% dextran solutions as a function of pressure using Dextran 20 (MW 20 000), Dextran 40 (MW 40 000), and Dextran 80 (MW 80 000). Batch cell experiments performed at a constant stirring speed. Source: Reproduced with permission from [12]. Copyright (1970) John Wiley and Sons.*

concentration polarization, at the membrane surface. At high pressure, the change in flux with increased pressure is smaller, so the decrease in rejection by the membrane is less apparent.

The use of the dextran rejection test as a method of providing a nominal molecular weight cutoff for ultrafiltration membranes has a long history. Zydney and Xenopoulos have reviewed the test and its analysis in detail [21].

Concentration polarization can also interfere with the ability of an ultrafiltration membrane to fractionate a mixture of dissolved macromolecules. Figure 7.16 shows the results of experiments using a membrane with a molecular weight cut-off of about 200 000 Da to separate albumin (MW 65 000) from γ-globulin (MW 156 000). Tests with the pure components show that

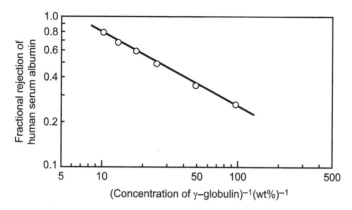

Figure 7.16 *The retention of albumin (MW 65 000) in the presence of varying concentrations of γ-globulin (MW 156 000) by a membrane with a nominal molecular weight cut-off based on one-component protein solutions of MW 200 000. As the concentration of γ-globulin in the solution increases, the membrane water flux decreases, and the albumin rejection increases from 25% at 0.01 wt% (γ-globulin to 80% rejection at 0.1 wt% γ-globulin. Source: Reproduced with permission from [12]. Copyright (1970) John Wiley and Sons.*

albumin passes through the membrane almost completely unhindered, but rejection of γ-globulin is significant. In principle, therefore, the membrane should be able to fractionate the components effectively. However, addition of even a small amount of γ-globulin (less than 0.2 wt%) to the albumin solution causes almost complete rejection of both components. The increased rejection is accompanied by a sharp decrease in membrane flux, suggesting that rejected globulin forms a secondary barrier layer. The secondary layer depresses albumin permeation to almost zero. Hence the process can be used to fractionate the two proteins only at very low γ-globulin concentrations.

Because of the effect of the secondary layer on selectivity, special techniques are needed to use ultrafiltration for fractionation of macromolecular mixtures. This issue is discussed further in the biotechnology applications section. Though these highly specialized applications are growing, most industrial applications today involve processes in which the membrane completely rejects dissolved macromolecular and colloidal material indiscriminately, while completely passing water and dissolved microsolutes.

7.3.3 Fouling Control

As described above, the best way to control fouling is to prevent it, by operating the membrane system at a pressure below the critical flux. Once a membrane is fouled, several cleaning methods can be used. The easiest method is to take the unit offline and circulate an appropriate cleaning solution through the modules for an hour or two. The most common foulants – organic polymer colloids and gelatinous materials – are best treated with alkaline solutions followed by hot detergent solutions. Enzymatic detergents are particularly effective when the fouling layer is proteinaceous. Calcium, magnesium, and silica scales, often a problem in reverse osmosis, are generally not seen in ultrafiltration, because their respective salt ions permeate the membrane freely (ultrafiltration of cheese whey, in which high calcium levels can lead to calcium scaling, is an exception). Because many feed waters contain small amounts of soluble ferrous iron salts, hydrated iron oxide scaling can be a problem. In an ultrafiltration system, these salts are oxidized to ferric iron by entrained air. Ferric iron is insoluble in water, so an insoluble iron hydroxide gel forms and accumulates on the membrane surface. Such deposits are usually removed with a citric or hydrochloric acid wash.

Regular cleaning is required to maintain the performance of all ultrafiltration membranes. The period of the cleaning cycle can vary from daily for food applications, such as ultrafiltration of whey, to once a month or even less often for ultrafiltration membranes used as polishing units in ultrapure water systems. A typical cleaning cycle is as follows:

1. Flush the system several times with hot water at the highest possible circulation rate.
2. Treat the system with an appropriate acid or alkali wash, depending on the nature of the layer.
3. Treat the system with a hot detergent solution.
4. Flush the system thoroughly with water to remove all traces of detergent; measure the pure water flux through the membrane modules under standard test conditions. Even after cleaning, some degree of permanent flux loss over time is expected. If the restoration of flux is less than expected, repeat steps 1–3.

Ultrafiltration systems should never be taken off line without thorough flushing and cleaning. Because membrane modules are normally stored wet, the final rinse solutions should contain a bacteriostat, such as 0.5% formaldehyde, to inhibit bacterial growth.

Backflushing is widely used as a way of controlling membrane fouling. The method is suitable for cleaning capillary and ceramic membrane modules, which can withstand a flow of solution from permeate to feed without damaging the membrane. Backflushing is not usually used for spiral-wound modules, because the membranes are too easily damaged. In a backflushing procedure, a slight over-pressure is applied to the permeate side of the membrane, forcing solution from the permeate side to the feed side of the membrane. The flow of solution lifts deposited materials from the surface. Backflushing must be done carefully to avoid membrane damage. Typical backflushing pressures are not more than 0.5 bar.

In the past, backflushing was performed once every few days or weeks. Nowadays, the procedure is often done much more frequently. In submerged membrane reactors, for example, a typical backflush cycle is 20–30 seconds every 15 minutes, and automatic equipment is used to control the process.

Because of the challenging environments in which ultrafiltration membranes are operated and the frequent cleaning cycles, ultrafiltration membrane lifetimes are usually shorter than those of RO membranes. Although module lifetimes of 5 years are sometimes claimed, typical lifetimes are only 2–3 years, and modules may be replaced annually in cheese whey or electrocoat paint applications. In contrast, reverse osmosis membranes are normally cleaned only a few times per year, and membrane lifetimes of 7 years or more are now routine.

7.4   Membranes

Most ultrafiltration membranes today are made by variations of the Loeb–Sourirajan process (Chapter 4). A limited number of materials are used, primarily polyacrylonitrile, poly(vinyl chloride)-polyacrylonitrile copolymers, polysulfone, poly(ether sulfone), poly(vinylidene fluoride), some aromatic polyamides, and cellulose acetate. In general, hydrophilic membranes are more fouling-resistant than completely hydrophobic membranes. For this reason, water-soluble polymers, such as poly(vinyl pyrrolidone) or poly(vinyl methyl ether), are often added to casting solutions of hydrophobic polymers, such as polysulfone or poly(vinylidene fluoride). During the membrane precipitation step, most of the water-soluble polymer is leached from the membrane, but enough remains to make the membrane surface hydrophilic.

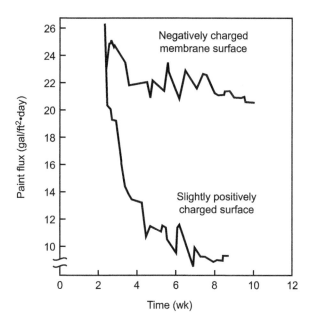

Figure 7.17 *Effect of membrane surface charge on ultrafiltration flux decline. These membranes were used to ultrafilter anodic electrocoat paint, which has a net negative charge. Electrostatic repulsion made the negatively charged membrane significantly more resistant to fouling than the similar positively charged membrane. Source: Reproduced with permission from [13]. Copyright (1990) John Wiley and Sons.*

The charge on the membrane surface is important. Many colloidal materials have a slight negative charge from carboxyl, sulfonic, or other acid groups. If the membrane surface also has a slight negative charge, adhesion of the densified colloidal fouling layer is reduced, which helps to maintain a high flux and inhibit membrane fouling. A slight positive charge on the membrane has the opposite effect. Charge and hydrophilic character can be the result of the chemical structure of the membrane material or can be applied to a preformed membrane surface by chemical grafting or surface treatment. The appropriate treatment depends on the application and the feed solution.

The importance of membrane surface characteristics on performance is illustrated by Figure 7.17. The feed solution in this example was an electrocoat paint solution in which the paint particulates had a net negative charge. As a result, membrane flux declined rapidly with the positively charged membranes. The flux decline with essentially identical membranes that had been treated to give the surface a net negative charge was much slower [13].

7.5 Tangential-Flow Modules and Process Designs

Until the late 1990s, all industrial ultrafiltration processes used constant pressure/variable flux tangential-flow systems, and this operating mode is still widely used to treat industrial process water and most biological applications. Membrane fouling is controlled by recirculation of the feed solution across the membrane surface. The modules, designs, and applications of this type of system are described in this section. The use of constant flux/variable pressure submerged membrane systems is limited to the municipal wastewater treatment market. These systems control membrane fouling by a combination of low-pressure operation, air sparging, and regular backflushing. These designs and operations are covered in the water treatment/membrane bioreactor section.

7.5.1 Modules

The need to control concentration polarization and membrane fouling dominates the design of tangential-flow ultrafiltration modules. In relatively non-fouling applications, such as production of ultrapure water, spiral-wound and capillary modules are universally used. Spiral-wound and capillary modules are also used in some food applications, such as ultrafiltration of cheese whey and clarification of apple juice. Because of their large diameter, and despite their considerably higher cost, tubular modules are used to treat solutions that would rapidly foul other module types. In demanding applications, such as treatment of electrocoat paint, concentration of latex solutions, or separation of oil-water emulsions, the fouling resistance and ease of cleaning of tubular modules outweighs their cost, large footprint, and high energy consumption.

A typical tubular module system contains several 5- to 8-ft-long tubes manifolded in series. The feed solution is circulated through the module array at velocities of 2–6 m/s. This high solution velocity causes a pressure drop of 2–3 psi per tube, or 10–30 psi for a module bank. Because of the high circulation rate and the resulting pressure drop, large pumps are required, so tubular modules have the highest energy consumption of any module design. Tubular ultrafiltration plants often use 10–15 kWh of energy per cubic meter of permeate produced. This corresponds to an energy cost of $1–2/m^3 of permeate, a major cost factor.

Early tubular membrane modules were one inch in diameter. Later, more energy-efficient, higher membrane area modules were produced by nesting four to six smaller-diameter tubes inside a single housing (see Chapter 4). Tubular module costs vary widely but are generally from US\$ 200–400/m^2. Recently, ceramic tubular modules have been introduced; these are more expensive, typically from US\$500–1000/m^2. This high cost limits their use to a few applications that require extreme operating conditions.

Plate-and-frame units compete with tubular units in some applications. These modules are not quite as fouling resistant as tubular modules, but are less expensive. Most consist of a flat membrane envelope with a rubber gasket around the outer edge. The membrane envelope, together with appropriate spacers, forms a plate that is contained in a stack of 20–30 plates. Typical feed channel heights are 0.5–1.0 mm, and the system operates in high-shear conditions. In biotechnology applications, smaller plate-and-frame modules are supplied as sealed steam-sterilizable disposable cassettes. These units have a limited lifetime, but the high value of the separations they perform can support this cost.

Plate-and-frame systems can be operated at higher pressures than tubular or capillary modules – operating pressures up to 10 bar are not uncommon. This can be an advantage in some applications. The compact design, small hold-up volume, and absence of stagnant areas also makes sterilization easy. For these reasons, plate-and-frame units are widely used in pharmaceutical and food industry operations. A photograph of an Alfa Laval plate-and-frame system is shown in Figure 7.18.

Capillary hollow fiber modules were first introduced many years ago by Romicon. A typical capillary module contains 500–2000 fibers with a diameter of 0.5–1.0 mm housed in a 30-in.-long, 3-in.-diameter cartridge. In operation, feed solution is pumped down the bore of the fibers under low operating pressures (normally below 5 bar), to avoid breaking the fibers. The normal feed-to-residue pressure drop of a capillary module is 0.2–0.5 bar. Under these conditions, capillary modules achieve good throughputs with many solutions. High-temperature sanitary systems are available; this, combined with the small hold-up volume and clean flow

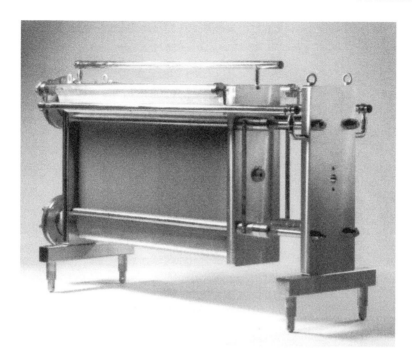

Figure 7.18 *Horizontal M39 plate-and-frame ultrafiltration system. Source: Courtesy of Alfa Laval Nakskov A/S, Nakskov, Denmark.*

path, has encouraged the use of these modules in biotechnology applications in which small volumes of expensive solutions are treated. A major advantage of capillary fiber systems is that the membrane can be cleaned easily by backflushing. With capillary modules it is important to avoid 'blinding' the fibers with particulates caught at the fiber entrance; prefiltration to remove all particulates larger than one-tenth of the fiber inside diameter is essential.

Capillary modules are available in a wide range of sizes. Typical industrial modules contain 5–10 m^2 of membrane, but in pharmaceutical applications, modules containing as little as 0.05–0.1 m^2 of membrane are used. These small modules are often used to process a single batch of high-value material, after which they are discarded.

The use of spiral-wound modules is increasing in industrial applications. This design was first developed for reverse osmosis modules, in which the feed channel spacer is a fine window-screen mesh material. In ultrafiltration, a coarser feed spacer material, often as much as 45 mi thick, is used to prevent particulates from lodging in the spacer corners. However, prefiltration of the feed down to 5–10 μm is still required for long-term operation. In the past, use of spiral-wound modules was limited to clean feed waters, such as occur in preparation of ultrapure water. Development of improved pretreatment and module spacer designs now allows these modules to be used for more fouling solutions, including cheese whey. In food applications, the stagnant volume between the module insert and the module housing is a potentially unsterile area. To eliminate this dead space, the product seal is perforated to allow a small bypass flow to continuously flush this area.

In the last few years, a number of companies have introduced plate-and-frame modules in which the membrane plates are vibrated or rotated, thus controlling concentration polarization by movement of the membrane rather than by movement of the feed solution [22, 23]. Moving the membrane concentrates most of the turbulence right at the membrane surface where it is most needed. The fluxes obtained are high and stable, but vibrating/rotating modules are considerably more expensive than tangential-flow modules, so the first applications have been with high-value, highly fouling feed solutions that are difficult to treat with standard modules.

7.5.2 Process Design

The common forms of tangential-flow constant pressure process designs are shown in Figure 7.19. The simplest process uses a batch unit, shown in Figure 7.19a. In this unit, the feed solution is circulated through the module at a high flow rate. The process continues until the required separation is achieved, after which the concentrate solution is drained from the feed tank, and the unit is ready to treat a second batch of solution. Batch processes are best suited to small-scale operations and are generally used in applications involving small volumes of high value products. Single-use, disposable membrane cassettes are often used in these applications [24]. Batch processes can be adapted to continuous use, but this requires automatic controls, which are expensive and can be unreliable.

In biotechnology applications, batch operations illustrated in Figure 7.19a are used to remove microsolutes and smaller partially retained molecules from a larger DNA or protein, which is the desired product. At the completion of the one-step processes, the concentrated product may still contain significant amounts of undesired impurities. One solution is to redilute the concentrated product and repeat the process. A widely used alternative solution is to use a process called diafiltration, illustrated in Figure 7.19b.

In a diafiltration process, permeate is removed from the recirculating feed solution as in a batch process, but the removed permeate is continuously replenished with an equal amount of fresh solvent from a second diafiltration solvent tank. The process continues until removal of

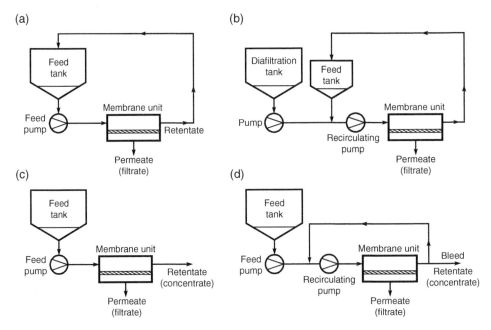

Figure 7.19 *Comparison of tangential-flow operating modes. (a) Batch mode, (b) diafiltration, (c) single-pass, (d) feed-and-bleed. Source: [25]/Taylor & Francis Group.*

impurities with the permeate solution brings the impurity level in the retained product to a low enough level. The product solution is then removed from the feed tank and the process can be repeated with a fresh batch of feed. The amount of diafiltration solvent needed to flush impurities from the retained product can be large, so the process is usually used with membranes that are completely retentive to the product. This limits product loss with the permeate.

The average throughput of biologic drug production has increased significantly in the last decade and, as a result, there is increasing interest in moving to continuous ultrafiltration processes like those shown in Figures 7.19c,d.

A continuous tangential-flow processes is shown in Figure 7.19c. A set of modules is arranged in series to achieve the required separation in a single pass. This type of process is not commonly used, because high feed flow rates across the membrane surface (typically 5–10 times higher than in reverse osmosis) are required to control concentration polarization, and unless a great many modules are used in series, the required degree of contaminant removal cannot be reached. Feed-and-bleed systems of the type shown in Figure 7.19d are used instead. In these systems, a large volume of solution is circulated continuously through a bank of membrane modules. A small volume of feed solution enters the recirculation loop just before the recirculation pump, and an equivalent volume of more concentrated solution is removed (or bled) from the recirculation loop just after the membrane module. The advantage of feed-and-bleed systems is that a high feed solution velocity is easily maintained through a limited number of modules, independent of the volume of solution being treated. In most plants, the flow rate of solution in the recirculation loop is 5–10 times the feed solution flow rate. This high circulation rate means that the concentration of retained material in the circulating solution is close to the concentration of the bleed solution and is significantly higher than the feed solution concentration. Because the flux of ultrafiltration membranes decreases with increasing concentration, more membrane area is required to produce the required separation than in a batch or a once-through continuous system operated at the same feed solution velocity.

To overcome the inefficiency of one-stage feed-and-bleed designs, industrial systems are often divided into multiple stages, as shown in Figure 7.20. By using multiple stages, the difference in concentration between the solution circulating in a stage and the feed solution entering the stage is reduced. A numerical example illustrates this point. In this example, assume the membrane is completely retentive and the object of the separation is to concentrate the feed solution from 1% to 8%. If this is done in a one-stage feed-and-bleed system, the average concentration of the solution circulating through the modules is close to 8%, and the flux is proportionately low. In a more efficient two-stage system, the first stage concentrates the solution from 1% to 3%, and the second stage concentrates the solution from 3% to 8%. Approximately three-quarters of the permeate is removed in the first stage, and the rest in the second stage. Because the modules in the first stage operate at a concentration of 3% rather than 8%, these modules have a higher membrane flux than in the one-stage unit. In fact, the membrane area of each stage is about equal, although the volume of permeate produced by each stage is very different. The two-stage feed-and-bleed design has about 60% of the membrane area of the one-stage system. The three-stage system, which concentrates the solution in three equal-area stages – from 1% to 2% in the first stage, from 2% to 4% in the second stage, and from 4% to 8% in the third stage – is even more efficient. In this case, the total membrane area is about 40% of the area of a one-stage system performing the same separation.

Because of the significantly lower membrane area requirements of multistage systems, large plants may have three to five stages. The limit to the number of stages is reached when the reduction in membrane area does not offset the increase in complexity of the system. Also, because of the high fluid circulation rates involved in feed-and-bleed plants, the cost of

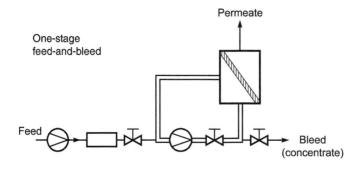

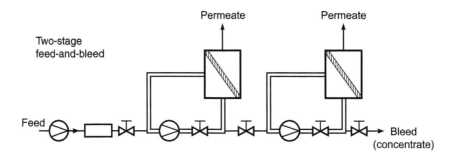

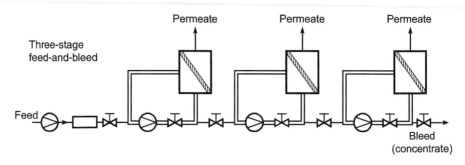

Figure 7.20 *One-, two- and three-stage feed-and-bleed systems. In general, the most efficient design is achieved when all stages have approximately the same membrane area. As the number of stages is increased, the average concentration of the solution circulating through the membrane modules decreases, and the total membrane area of the system is significantly less than for a one-stage design.*

pumps can rise to 30–40% of the total cost of the system. Electricity to power the pumps represents a significant operating expense.

7.6   Applications

Beginning in the early 1970s, the first industrial applications of ultrafiltration were developed. In general, these applications use small plants typically processing one hundred to a few thousand cubic meters of feed solution per day. Operating pressures are typically in the 5–10 bar range, so plastic components and piping are often used.

The current industrial market, excluding water treatment and biotechnology, is in the US$1–2 billion per year range. The market is very fragmented and no single application is more than about US$200 million/year. Each of the diverse applications uses membranes, modules, and system designs tailored to the particular industry served. The result is little product standardization, many custom-built systems, and high costs compared to reverse osmosis.

The economic driving force to install an ultrafiltration process can come from a number of sources:

- *Water recovery.* Depending on the plant location, reduced usage of purchased municipal water can produce savings in the US$0.50–1.0/m^3 range.
- *Heat recovery.* Many process streams are hot, and ultrafiltration usually works better with hot feed solutions. Unlike other treatment techniques, cooling before treatment may not be required. If the hot, clean permeate is recycled at, or close to, the process operating temperature, the energy savings can be considerable. If the water is 50 °C above ambient temperature, for example, the energy savings can amount to about US$1.00/m^3.
- *Avoided wastewater treatment costs.* These costs are process dependent. For a food processing plant, they are likely to be relatively modest – perhaps only US$0.50/m^3 or less – but treating latex emulsion plant effluents (called white water) can cost US$2–3/m^3 or more.
- *Materials recovery value.* If the material concentrated by the ultrafiltration process can be recovered and reused in the plant, this is likely to be the most important credit.

Equipment costs vary widely, depending on plant size, solution to be treated, and separation to be performed. In general, industrial ultrafiltration plants are much smaller than reverse osmosis plants. Typical flow rates are 1000–10 000 m^3/day, one-tenth to one-hundredth that of the average reverse osmosis plant. Rogers [26] compiled the costs shown in Figure 7.21 that, adjusted for inflation, still seem reasonable. These costs are for industrial plants treating wastewater or low value process streams. Operating processes are in the 5–10 bar range and for typical plants treating 1000–10 000 m^3/day of feed solution, the capital cost is in the range US$500–1000/m^3-day capacity. The typical cost breakdown is shown in Table 7.1 [27]. Operating costs are around US$2–3/m^3 of product with module replacement costs about 20–40% of this number, and energy costs for the recirculation pumps 30–40%, depending on

Table 7.1 Typical ultrafiltration capital and operating cost breakdown [27].

Capital costs	%
Pumps	30
Membrane modules	20
Module housings	10
Pipes, valves, frame	20
Controls/other	20
Total	100
Operating costs	
Membrane replacement	20–40
Cleaning costs	10–30
Energy	30–40
Labor	15
Total	100

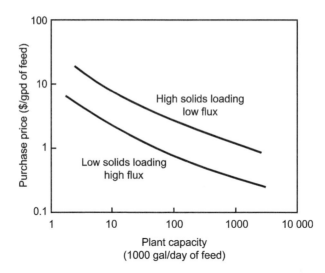

Figure 7.21   *Purchase price in 2010 dollars for ultrafiltration plants as a function of plant capacity. Data of Rogers corrected for inflation. Source: Reproduced with permission from [26]. Copyright (1984) Elsevier.*

the system design. The same type of system built for pharmaceutical applications is all stainless steel, far more instrumented and controlled, and able to be repeatedly sterilized. The cost of these systems is three- to fivefold higher than those shown above. At the other extreme, municipal water and membrane bioreactor systems are 100–1000 fold larger than most industrial plants, so economies of scale kick in and costs are 3–5 fold lower than those shown above.

In the discussion below, a number of the largest industrial applications of ultrafiltration are described. Municipal water treatment and biotechnology applications are practiced on a very different scale and have very different cost and process designs, and so are treated separately in the following section.

The first large successful application was the recovery of electrocoat paint in automobile plants. Later, a number of applications developed in the food processing industry, first in the production of cheese [28], then in the production of apple and other juices [29] and, more recently, in the production of beer and wine [30]. Industrial wastewater and process water treatment is a growing application, but high costs limit growth. Overviews of constant pressure ultrafiltration applications are given in Cheryan's book [14], and Van Reis and Zydney's review [24].

7.6.1   Industrial Applications

7.6.1.1   *Electrocoat Paint Recovery*

Automobile companies use electrodeposition of paint on a large scale. The paint solution is an emulsion of charged paint particles. The metal piece to be coated is immersed in a large tank of paint, where it forms an electrode of opposite charge to the paint particles. When a voltage is applied across the metal part and a second electrode, the charged paint particles migrate towards and are deposited on the metal surface, forming a coating over the entire piece. After electrodeposition, the piece is removed from the tank and rinsed to remove excess paint. The paint is then cured in an oven.

The rinse water from the washing step rapidly becomes contaminated with excess paint. In addition, the stability of the paint emulsion is gradually degraded by ionic impurities carried over

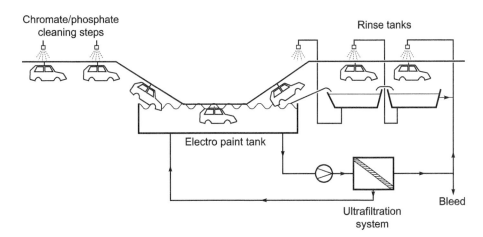

Figure 7.22 *Flow schematic of an electrocoat paint ultrafiltration system. The ultrafiltration system removes ionic impurities from the paint tank carried over from the chromate/phosphate cleaning steps and provides clean rinse water for the counter-current rinsing operation.*

from cleaning steps upstream of the paint tank. Both problems are solved by the ultrafiltration system shown in Figure 7.22. The system takes paint solution, containing 15–20% solids, and produces a permeate free of paint particles but still containing the ionic impurities, and a residue stream in which all of the paint particles have been retained. The residue stream, slightly concentrated in particles compared with the membrane feed solution, is returned to the paint tank; the permeate is passed to the second rinse tank. A portion of the ultrafiltration permeate is bled from the tank and replaced with fresh water to maintain the ionic balance of the process.

Electrocoat paint is a challenging feed solution for an ultrafiltration process. The solids content of the solution is high, typically 15–20 wt%, so a fouling layer easily forms on the membrane, resulting in relatively low fluxes, generally 20–30 l/m^2·h. However, the value of the paint recovered from the rinse water, plus elimination of other rinse-water cleanup steps, made the ultrafiltration process an immediate success when introduced by Abcor. Tubular modules were used in the first plants [7] and are still installed in some electrocoat operations; capillary modules or sometimes spiral-wound modules are used in newer plants. The first electrocoat paint comprised latex emulsion particles carrying a negative charge. These emulsions were best treated with membranes having a slight negative charge to reduce surface adhesion and minimize fouling. Paints carrying a positive charge were introduced later and required development of membranes carrying a slight positive charge.

7.6.1.2 Cheese Production

Ultrafiltration has found a major application in cheesemaking; the technology is now used throughout the dairy industry [28]. During cheese production, milk is coagulated (or curdled) by precipitation of the milk proteins. The solid that forms (curd) is sent to the cheese fermentation plant; the supernatant liquor (whey) presents a disposal problem. The compositions of milk and whey are shown in Table 7.2. Whey contains most of the dissolved salts and sugars present in the original milk and about 25% of the original protein. In the past, whey was often discharged to the sewer because its high salt and lactose content made direct use as a food supplement difficult. Now most whey is processed to recover the protein value and reduce disposal problems.

Table 7.2   *Composition of milk and cheese whey.*

Component (wt%)	Milk	Whey
Total solids	12.3	7.0
Protein	3.3	0.9
Fat	3.7	0.7
Lactose/other carbohydrates	4.6	4.8
Ash	0.7	0.6

The traditional cheese production process and two newer processes using ultrafiltration membranes are shown in Figure 7.23.

The objective of both membrane processes is to increase protein recovery and to reduce the waste disposal problem represented by the whey. In the MMV process, named after the developers Maubois, Mocquot, and Vassal [31], the whole or skimmed milk is treated directly by ultrafiltration. The treatment produces a retentate enriched three- to five-fold in proteins, and this pre-cheese concentrate can be used directly to produce soft cheeses and yogurt. Typically, the total solids level is about 30–35%, and the protein concentration is about 12–17%. This protein content is sufficient to make a number of soft cheeses, such as feta, without further enrichment. When ultrafiltration is used to increase protein utilization, cheese production increases by approximately 10%, so the process has been widely adopted.

The whey separation process on the right of the figure uses both ultrafiltration and reverse osmosis/nanofiltration. A more detailed flow schematic of the operation is shown in Figure 7.24. The goal is to separate the whey into three streams; protein, lactose, and salt. The most valuable stream is the concentrated protein fraction, stripped of salts and lactose. Before this fraction can be used, the lactose content must be reduced by 95% and the protein content increased to at least 60–70% on a dry basis. The objective of the ultrafiltration steps is to concentrate the protein as much as possible (to minimize subsequent evaporator drying costs), and to simultaneously remove the lactose. These two objectives are difficult to meet

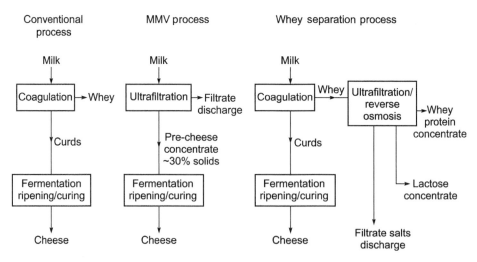

Figure 7.23   *Simplified flow schematic showing the traditional cheese production method, and two new methods using ultrafiltration to increase the production of useful products.*

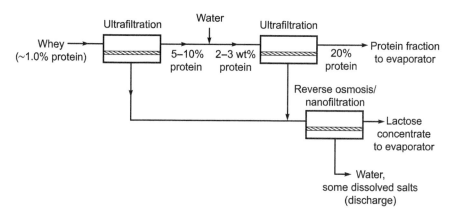

Figure 7.24 *Simplified flow schematic of an ultrafiltration plus reverse osmosis process to extract proteins and lactose from cheese whey. A two-step ultrafiltration unit is used to remove all the lactose and salt from the concentrated protein. The permeate is concentrated by a reverse osmosis unit to produce a lactose concentrate to be sent to an evaporator to make dry protein.*

in a single step, because the very high volume reduction required to achieve sufficient lactose removal results in serious fouling and reduction in flux. Therefore, whey plants commonly use two ultrafiltration steps in series. The first step achieves a 5- to 10-fold volume reduction and removes most of the lactose. The retentate from this step is diluted with water and reconcentrated in a second step, which removes the remaining lactose. The final retentate is sent to an evaporator from which a dry protein-rich solid product is recovered.

The mixed permeate from the two ultrafiltration steps contains lactose and dissolved salts. The lactose fraction has value and is separated from the salts by a nanofiltration membrane that permeates the salt and most of the water, producing a concentrated salt-free lactose retentate. Some plants use an evaporator to recover dry lactose sugar, as shown in Figure 7.24. Other plants ferment the lactose concentrate to make ethanol.

Most whey plants use spiral-wound modules in multistage feed-and-bleed systems. Sanitary spiral-wound module designs are used to eliminate stagnant areas in the module housing, and the entire plant is sterilized daily with hot high- and low-pH cleaning solutions. This harsh cleaning treatment significantly reduces membrane lifetime.

7.6.1.3 Clarification of Fruit Juices

Apple, pear, orange, and grape juices are all clarified by ultrafiltration. Ultrafiltration of apple juice is a particularly successful application. Several hundred plants have been installed, and almost all US apple juice is now clarified by this method. In the traditional process, crude filtration is performed directly after crushing the fruit. Pectinase is added to hydrolyze pectin, facilitating juice release and reducing the juice viscosity. The juice is then passed through a series of decantation and diatomaceous filtration steps to yield clear juice, with a typical yield of about 90%. By replacing the final filtration steps with ultrafiltration, a good-quality, almost-sterile product can be produced, with a yield of almost 97% [29].

Ultrafiltration membranes with a molecular weight cut-off of 10 000–50 000, packaged as tubular or capillary hollow fiber modules, are generally used. The initial feed solution is quite fluid, but in this application almost all of the feed solution is forced through the membrane, and overall concentration factors of 50 are normal. As the concentration of the residue rises, the flux falls significantly. This means that the final residue solution is viscous, so the solution is usually filtered at 50–55 °C to increases flux. Operation at this temperature also reduces bacterial growth.

7.6.1.4   *Industrial Oil/Water Emulsions*

Oil/water emulsions are used in metal machining operations to provide lubrication and cooling. Although recycling of the fluids is widely practiced, spent waste streams are produced. Using ultrafiltration to recover the oil component and allow safe discharge of the water makes good economic sense. In large, automated machining operations, such as in automobile plants, steel rolling mills, and wire mills, a central ultrafiltration system may process up to 500 m³/day of waste emulsion. These are relatively sophisticated plants that operate continuously, using several feed-and-bleed stages in series. At the other end of the scale are small systems dedicated to single machines, which process only a few gallons of emulsion per hour. The principal economic driver for users of small systems is the avoided cost of waste hauling. For larger systems, the value of the recovered oil and associated chemicals is also worthwhile. In both cases, tubular or capillary hollow fiber modules are generally used, because of the high fouling potential and widely variable composition of emulsified oils. A flow diagram of an ultrafiltration system used to treat large volumes of machine oil emulsions is shown in Figure 7.25. The dilute, used emulsion is filtered to remove metal cuttings and is then circulated through a feed-and-bleed system, producing a concentrated emulsion for reuse and a dilute filtrate that can be discharged or reused.

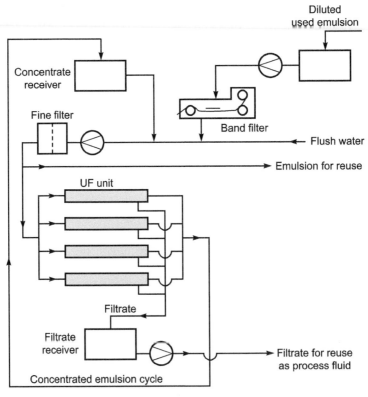

Figure 7.25   *Flow diagram of a feed-and-bleed ultrafiltration unit used to concentrate a dilute oil emulsion.*

7.6.1.5　*Process Water and Materials Recycling*

Ultrafiltration has been applied to a number of process and material recycling operations. Typical applications include cleaning and recycling hot water used in food processing, recovery of latex particles contained in wastewater from paint production, and recovery of poly(vinyl alcohol) sizing agents used as process aids in synthetic fabric weaving operations.

A typical example of a recycling application, shown in Figure 7.26, is the recovery of poly(vinyl alcohol) (PVA) sizing agent [32]. In this application, all the economic drivers listed in the introduction to this section contribute to the total plant economics. The feed stream is produced in fabric weaving when the fiber is dipped into a solution of PVA to increase its strength.

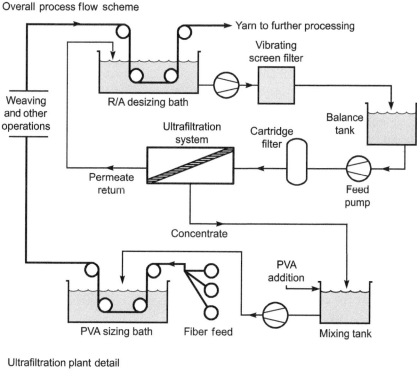

Figure 7.26　*Flow schematic of a three-stage feed-and-bleed ultrafiltration system used to recover polyvinyl alcohol (PVA) sizing agents used in the production of cotton/synthetic blend fabrics [32].*

After weaving, the PVA is removed in a desizing wash bath. The solution produced in this bath is hot (55 °C) and contains 0.5–1.0% PVA. The purpose of the ultrafiltration unit is twofold: to concentrate the PVA, so it can be recycled to the sizing bath, and to send the reclaimed, hot clean permeate back to the desizing bath. After filtration, the PVA solution is relatively particulate-free, so spiral-wound modules are used to reduce costs. For very small plants with flows of less than 20 l/min, batch systems are used. However, most plants are in the 50–500 l/min range and are multistage feed-and-bleed systems, as shown in Figure 7.26. The environment is challenging; the membranes must be cleaned weekly with detergents to remove waxy deposits and with citric acid to remove iron scale. Even so, modules must be replaced every 1–2 years, representing a major operating cost.

7.6.1.6   Drinking Water Sterilization

In 1993, a major *cryptosporidiosis* outbreak in Milwaukee, Wisconsin caused 100 deaths and led the EPA to mandate that better sterilization techniques be used to treat all surface drinking water supplies. European regulators adopted similar rules. When the EPA rule was adopted, it was estimated that about 40 000 small public water works in the United States would be affected. As surface waters from rivers and lakes usually have a low level of contaminants, they are readily amenable to treatment by ultrafiltration and the regulations provided the driver that led to the current large market in this area. Many public water utilities in the US and elsewhere now rely on ultrafiltration to remove bacteria, viruses, and other microorganisms from water supplies. Chlorination is used after ultrafiltration for secondary sterilization and to maintain sterility in the water distribution system.

Many plants are fitted with capillary membrane modules and both shell and lumen side feed modules are used. A photograph of one such plant is shown in Figure 7.27. The goal is to produce a 1000-fold (three-decade) reduction in the level of giardia and a 10 000-fold (four-decade) reduction in the level of virus contamination. This type of reduction is easily achieved by defect-free ultrafiltration membranes. The plants are carefully monitored to detect even a single broken filter that could allow unfiltered bacteria to enter the drinking water supply.

7.6.2   Municipal Water Treatment/Membrane Bioreactors (MBRs)

Until the early 2000s, the ultrafiltration industry was dominated by applications operating in a tangential-flow mode at a constant applied feed pressure. Although constant flux/variable pressure operation was known, better understanding of operating limits and improved fouling control meant that many systems continue to be run in constant pressure mode. The use of constant flux/ variable pressure mode really came into its own with the development of submerged membrane systems, specifically as membrane bioreactors in municipal wastewater treatment plants. These developments had their origin in the late 1980s, when Kazuo Yamamoto and his coworkers began experiments with looped hanks of hollow fibers immersed in sewage treatment tanks [8]. A liquid pump on the permeate side of the fibers generated a vacuum pressure sufficient to draw water through the membrane. Fouling was a serious problem, but Yamamoto showed that regular air sparging went a long way towards controlling fouling. Yamamoto's work was followed up by Kubota and Mitsubishi Rayon in Japan, Zenon in Canada, and Memtech in Australia. The first commercial plants were installed in the 1990s and by the early 2000s the technology was off the ground. Sales have increased steadily since then, and in the last 10 years not only the number of installations but also the size of individual installations has grown considerably. The total membrane wastewater treatment market now comfortably exceeds US$2–3 billion/year and is still growing [11].

(a)

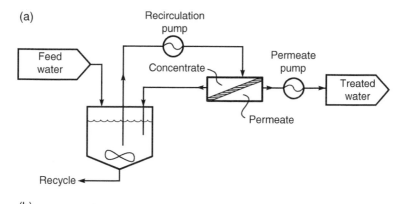

(b)

Figure 7.27 (a) Process configuration for a membrane water sterilization system and (b) photograph of a 25 million gal/day capillary hollow fiber module plant to produce potable water, installed by Norit (X-Flow) in Keldgate, UK. Source: Courtesy of Norit Membrane Technology BV.

7.6.2.1 Modules and System Design

The two most common types of submerged modules are shown in Figure 7.28 [33, 34]. Hollow fiber (capillary) modules were developed by Zenon, Memtech and Mitsubishi; Kubota uses an array of membrane plates. All manufacturers use air sparging to agitate the feed solution and scrub the membrane surface. Periodic pulses of large bubbles seem to be more effective than a continuous stream of small bubbles [35]. In a typical capillary module, the fibers are loosely held between manifolds at each end, so the fibers can sway and bubbles rise between the fiber bundles. This increases the scrubbing action and minimizes channeling. The permeate solution is removed from a manifold at the top of the module connected to a constant flow pump. The pump is set to maintain a flux of 20–30 $l/m^2 \cdot h$, depending on the nature of the feed solution. If the solution is highly fouling, a low flow rate will be set; with less fouling feed solutions, a higher flow rate can be used. The transmembrane pressure is initially less than 0.2 bar, but as the module

(a) Hollow fiber (capillary) modules

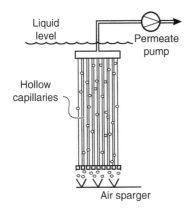

(b) Submerged membrane plates

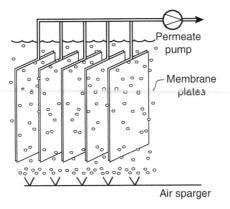
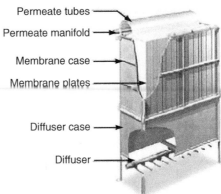

Figure 7.28 *The two most common types of submerged membrane modules. Zenon, Mitsubishi and Memcor make swayable hollow fiber (capillary) modules (a); Kuboto makes submersible plate modules (b). Air sparging is used by all manufacturers to help keep the module surfaces clean.*

is used and slowly fouls, the pressure will rise. Most operators will take the membrane offline for chemical cleaning before the transmembrane pressure reaches about 0.5 bar.

Capillary modules are also maintained clean by automatic short cycle backflushing, typically performed for 20–30 seconds after every 10–20 minutes of operation. The backflushing pressure is small, usually just enough to reverse the flow of fluid and lift off solids accumulating on the membrane surface. Typically, the back flux is 1–3 times the operational flux. Chemical cleaning is usually done every week to once or twice a month, depending on the nature of the feed solution. Operation in this way can achieve membrane lifetimes of several years, even when the membranes are used with highly fouling solutions [36].

Kubota's plate modules are easier to keep clean by air sparging than fiber modules, but can be damaged by repeated backwashing, so backwashing is usually carried out only during periodic chemical cleaning operations, and under a mild pressure.

7.6.2.2 Application to Municipal Wastewater Treatment

A conventional municipal wastewater treatment plant consists of a series of bioreactor operations. The incoming wastewater is first subject to primary waste treatment where a series of screen filters and settling tanks remove large solids and grit. The clarified solution is then sent to a secondary biological treatment process. This may consist of a single aerobic biodigestion step or, as shown in Figure 7.29, a first anaerobic (anoxic) step, followed by a second aerobic step. In the anaerobic tank, a portion of the organic components present are consumed and nitrogen-containing compounds are reduced to ammonia and nitrogen. In the air-sparged aerobic tank that follows, the remaining organic components are removed and ammonia is converted to nitrates. Nitrogen-rich sludge that settles out in this tank is recycled to the anaerobic tank for treatment a second time. The treated water from this secondary biological step is then sent to a larger settler/filter where any suspended solids are removed. The water is then disinfected with chlorine and discharged. It can take a day or more for a unit of water to traverse the whole series of operations, so conventional biological water treatment plants have a large footprint. One of the advantages of membrane bioreactors is their much smaller footprint, because they treat the feed water more rapidly than the conventional process.

The first membrane bioreactors for treating municipal wastewater were developed by Dorr-Oliver working with Amicon in the late 1960s [37]. A flow diagram of the process is shown in Figure 7.30. Liquid was withdrawn from the aerated bioreactor tank and circulated across the surface of the membrane and returned to the tank as a more concentrated residue. The filtrate showed a very significant reduction in BOD and COD levels.

The Dorr-Oliver system was a flat sheet plate-and-frame unit that operated in a tangential-flow, constant pressure/variable flux mode. Membrane fouling was a major problem that could not be solved at that time. Improved side-stream systems have been developed in recent years and they have a small portion of the market, though integrated submerged systems are normally more reliable and cheaper to operate and have become the established preference of the market [38].

The development of air-sparged submerged membranes is what led to today's multi-billion dollar waste water treatment business [39]. A waste water membrane bioreactor process is shown in Figure 7.31.

The membrane units are submerged in the aerobic digestion tank and most of the settler tanks are eliminated. Maintaining the membranes free of fouling is critical for successful operation. All of today's submerged membranes operate below the critical flux in a constant flux/variable

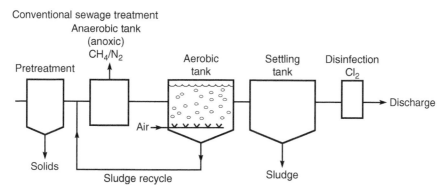

Figure 7.29 *A block diagram of a conventional sewage treatment plant. The conventional plant produces a 10- to 20-fold reduction in BOD and COD levels.*

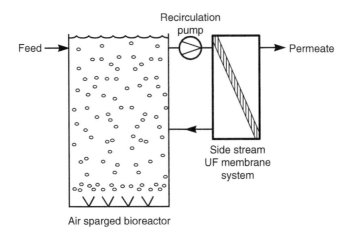

Figure 7.30 *A block diagram of a side-stream membrane bioreactor developed by Dorr-Oliver in the 1960s.*

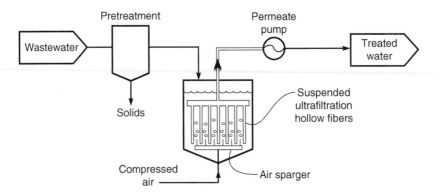

Figure 7.31 *A simplified flow diagram of a membrane bioreactor process. The air-sparged membrane operates in a low-pressure constant flux/variable pressure mode. The membrane can provide as much as a 10^4 to 10^5-fold reduction in BOD and COD levels, far above that provided by the conventional treatment process shown in Figure 7.29.*

pressure mode. Regular automatic backflushing for 10–30 seconds every 15 minutes is generally used. This operation, coupled with air sparging, removes most of the fouling material from the membrane surface. Backflushing delays, but cannot completely eliminate, fouling. Over time, residual fouling remaining after backflushing builds up and the transmembrane pressure required to maintain a constant flux slowly rises. Periodically thereafter, usually every 3–5 months, the system is taken offline and chemically cleaned. This intermittent cleaning cycle removes most but not all foulants, so some residual fouling remains as irrecoverable fouling. It follows that, over time, no amount of backflushing or cleaning will reduce the transmembrane pressure, and after some years of operation, the membranes must be removed and replaced. Membrane lifetime has been increasing. In early plants, modules were replaced after 2–4 years, but newer modern plants can operate for 5–7 years or more [40].

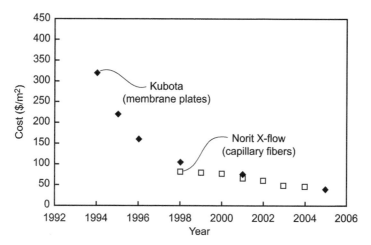

Figure 7.32   *MBR process operating costs (Kubota) versus time. Source: [33] / with permission of Elsevier.*

The cost of submerged membranes has fallen significantly since the first commercial systems were introduced. Data reported by Judd up to 2005 are shown in Figure 7.32 [33]. Between 1992 and 2005, the cost of replacement modules on a per-square-meter-of-membrane basis dropped from US$400 to less than US$50. This trend is continuing, although at a slower pace. The total cost of a membrane bioreactor system is now in the region of US$500/m³-day of treated wastewater capacity, but this number varies significantly with size and design of the plant. Non-membrane parts of the plant such as pumps, pipes, and the associated civil works are usually more costly than the membrane system.

Concurrently with the decrease in system costs, operating costs have also decreased. In part, this is because module replacement costs are lower, but also because process optimization has led to longer module life. The operating cost of new membrane bioreactor plants using Kubota membranes is shown for the time period from 1992 to 2005 in Figure 7.33 [33].

In 1992, membrane replacement was more than half of the total operating cost, while power (mostly used by the air spargers) was less than 10%. By 2005, membrane replacement was less than 10% of the operating cost, and total operating costs had come down more than 10-fold. Currently, the cost of water from new plants is as low as about US$0.20–0.30/m³ of treated water. Power consumption, at about 1–2 kWh/m³ of product water, remains about twice the power required for a conventional bioreactor; as a consequence, improved ways to air sparge and reduce power costs are a research focus for process developers. Despite the power cost, submerged membrane systems continue to gain in acceptance and use, because these extra costs are offset, or more than offset, by the advantages of a much-reduced plant footprint and better quality of the discharged filtrate, which is clear and essentially sterile. In some locations, the bioreactor-treated water is discharged or used as irrigation water. In other locations, the water is used as a supplementary municipal water source. Most of the submerged membrane plants installed to date are aerobic and are sparged with air, but there is increasing interest in anaerobic bioreactors [41]. One advantage of this type of process is increased biogas recovery. The gas can be used as fuel and there is enough produced to make wastewater plants net energy producers. Membrane fouling and the cost of its control is the principal problem slowing down use of the process.

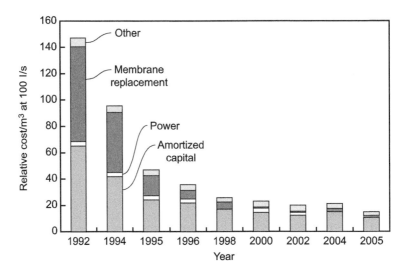

Figure 7.33 Submerged membrane operating costs for Kubota systems. Source: [33] / with permission of Elsevier.

7.6.3 Biotechnology

The application of ultrafiltration to the biotechnology industry has a long history. Developers of biologic drugs were early adopters in laboratory applications to separate and concentrate all types of protein, DNA and polypeptide products. The technology has now moved to the production scale and ultrafiltration/microfiltration membranes are widely used throughout the industry. Early systems used membranes and modules already in use in the food industry; currently, modules specifically developed for biotechnology are preferred. Van Reis and Zydney [24] have written a useful review of the membrane products used in these applications. A comprehensive description of the current industry is given in another review by Zydney [42] and in a monograph by Lutz [43].

In a typical biotechnology application, an ultrafiltration membrane is used to separate a target protein or DNA from other materials created in a recombinant cell culture process. Typical impurities are host cell proteins and debris, a variety of enzymes and virus-like particles. Batch systems are commonly used, and the volume of solution to be treated in each batch is usually only 5–10 m^3, containing a few kilograms of pure product. An alternative approach is to use preparative column chromatography; in practice a combination of ultrafiltration and chromatography is often used. A first separation by ultrafiltration reduces the volume of solution and the mass of impurities; a polishing separation by chromatography yields the pure product.

Biotechnology applications have grown enormously with the development of biologic drugs in the last two decades. A typical biological process flow diagram will contain 5–20 membrane filtration steps, together with the use of multiple sterile filtration steps at every mixing point in the process. The production process used typically involves four main types of operations [42]:

1. Depth filtration for the initial clarification of the cell culture fluid.
2. Virus filtration for removal of adventitious and exogenous viruses.
3. Ultrafiltration/diafiltration for concentration, buffer exchange, and formulation.
4. Sterile filtration for bioburden reduction and final fill–finish operations.

After the initial clarification step, these operations involve processing relatively clean solutions, so fluxes are high. The first generation of biologics typically involved treatment of less than $10\,\text{m}^3$ of solution, so the use of small single-use membrane module cassettes was practical. The second generation of monoclonal antibody biologics is produced on a significantly larger scale, and the size and cost of single-use cassettes has risen. As a result, there is a strong driver toward continuous or semi-continuous manufacturing processes.

In some processes, a particular target protein must be separated from a mixture of similar proteins, and the size difference between the protein to be retained and the protein to be permeated may be as little as a factor of two or less. The ability of the membrane to perform the separation is often conveniently characterized by a term called the protein sieving coefficient, S, where S is simply equal to $1 - \mathbb{R}$, where $\mathbb{R}$ is the protein rejection coefficient. In reverse osmosis, an equivalent term is called the salt passage.

The definition of the rejection coefficient in Eq. (7.4) is

$$\mathbb{R} = \left(1 - \frac{c_p}{c_b} \right) \tag{7.4}$$

Where c_p is the concentration of the protein in the permeate solution and c_b is the concentration in the feed. Hence, it follows that

$$S = \frac{c_p}{c_b} \tag{7.5}$$

The ability of a membrane to separate two components i and j is given by an enrichment factor

$$\gamma_{ij} = \frac{S_i}{S_j} \tag{7.6}$$

Normally, it is difficult to obtain a good separation in a single membrane step, unless components i and j are very different sizes. The trick used to overcome this problem is to choose a membrane that almost completely retains one of the components to be separated, but allows partial passage of the other components. A diafiltration process is then used. As shown in Figure 7.34a, a volume, V, of feed solution, containing the mixed protein components and small dissolved salts, such as buffering agents, is introduced into a processing tank and circulated around a tangential-flow ultrafiltration loop. Permeate that passes through the membrane is replaced by topping up with fresh solvent, hence diluting the contents of the processing tank with an equal amount of fresh, protein-free, salt solution. The result is to wash the smaller, more permeable proteins through the membrane, while still retaining the non-permeable protein. Figure 7.34b shows the result of this process when applied to the separation of bovine serum albumin (BSA; MW = 67 000) from β-lactoglobulin (β-LG; MW = 18 000) and α-lactoglobulin (α-LG; MW = 14 000) [44]. The figure plots the progressively falling normalized protein concentration in the processing tank as a function of the number of feed diafiltration volumes V that have been collected as permeate. The concentration of the most permeable component, α-LG, decreases rapidly and less than 1% is left in the feed solution after 10 diafiltration volumes have passed through the membrane. The β-LG concentration falls less rapidly, but 75% has been removed after 10 diafiltration volumes. In contrast, the BSA concentration stays almost constant and 95% of this component remains in the retentate solution.

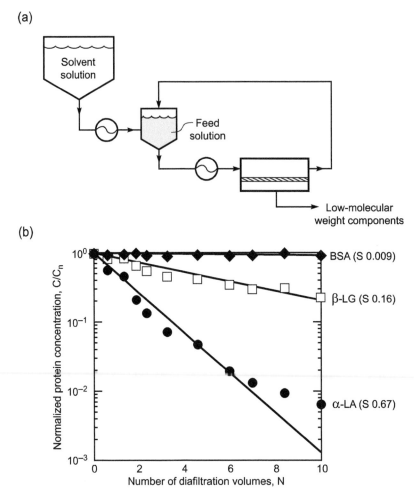

Figure 7.34 *The separation of BSA from β-LG and α-LA by a batch diafiltration process (a). The normalized concentrations (b) of the three proteins in the feed tank are shown as a function of the number of diafiltration volumes passing through the membrane. Source: Adapted from [44].*

The much-diluted permeate solution is reconcentrated by ultrafiltration with a tight membrane that retains all proteins. A second diafiltration operation is then used to separate the α- and β-lactoglobulin components from the reconcentrated permeate solutions. In this case, a membrane is used that is designed to completely retain β-LG while passing a portion of the α-LG. The result of the total separation scheme is shown in Figure 7.35.

7.7 Conclusions and Future Directions

In the last 10–15 years, the ultrafiltration membrane market has grown fivefold. Two very different applications have driven this growth. One is the increasing use of membranes in municipal water treatment and reclamation plants. The second has been the development of biologic drugs

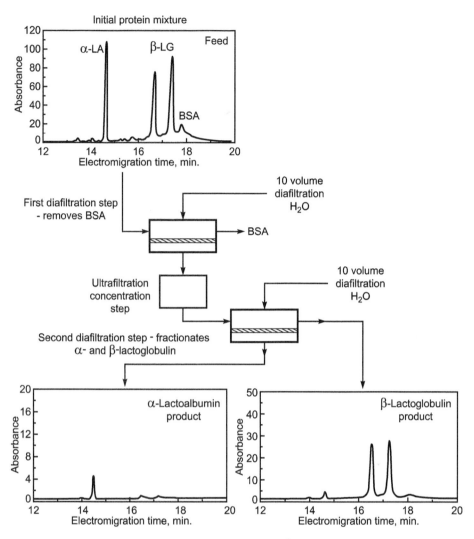

Figure 7.35 *A two-stage membrane ultrafiltration/diafiltration scheme to separate α-LA, β-LG and BSA. Source: Adapted from [44].*

in the pharmaceutical industry. The key breakthrough in the water treatment application was the development of constant flux/variable pressure operating systems, coupled with automatic backflushing and air sparging. These innovations have gone a long way towards solving the membrane fouling problems that previously limited the usability of ultrafiltration. As a consequence, many municipal membrane bioreactor plants treating more than $100\,000\text{ m}^3$/day of wastewater are now online. The largest plants have 1 million m^2 of membrane. The cost of the technology is still coming down, so submerged membrane systems are likely to gain an increasing share of the municipal water and sewage treatment market. Expansion of the technology into industrial water treatment applications is also under way and is likely to continue.

Industrial and municipal wastewater treatment plants can have many hundreds of thousands of square meters of membrane, and can produce treated water at US\$ $0.50/\text{m}^3$ or less. The use of membranes in the preparation and separation of biologic drugs is at the other end of the spectrum.

Many applications only require a few square meters of membrane, but the value of the separation is thousands of dollars per cubic meter of treated solution. The pharmaceutical industry has always been a large user of membrane technology; what has changed in the last few years has been the huge growth in biologic drugs. The drugs are made by fermentation processes which use membranes as an essential component of the production process.

The water treatment and pharmaceutical markets are serviced by different companies but both markets are continuing to grow, so further growth of the ultrafiltration industry seems assured.

References

1. Bechhold, H. (1907). Kolloidstudien mit der Filtrationsmethode. *Z. Physik Chem.* **60**: 257.
2. Zsigmondy, R. and Bachmann, W. (1918). Über Neue filter. *Z. Anorg. Chem.* **103**: 119.
3. Ferry, J.D. (1936). Ultrafilter membranes and ultrafiltration. *Chem. Rev.* **18**: 373.
4. Elford, W.J. (1937). Principles governing the preparation of membranes having graded porosities. The properties of 'Gradocol' membranes as ultrafilters. *Trans. Faraday Soc.* **33**: 1094.
5. Loeb, S. and Sourirajan, S. (1963). Sea water demineralization by means of an osmotic membrane. In: *Saline Water Conversion II*, Advances in Chemistry Series Number 38 (ed. R.F. Gould), 117–132. Washington, DC: American Chemical Society.
6. Michaels, A.S. (1968). High flow membranes. US Patent 3,615,024, issued 26 October 1971.
7. Goldsmith, R.L., deFilippi, R.P., Hossain, S., and Timmins, R.S. (1971). Industrial ultrafiltration. In: *Membrane Processes in Industry and Biomedicine* (ed. M. Bier), 267–300. New York: Plenum Press.
8. Yamamoto, K., Hiasa, M., and Mahmood, T. (1989). Direct solid-liquid separation using hollow fiber membrane in an activated sludge aeration tank. *Water Sci. Technol.* **21**: 43.
9. Ishida, H., Yamada, Y., Izumi, K., and Moro, M. (1993). Apparatus for Treating activated sludge and method of cleaning it. US Patent 5,192,456, issued 9 March 1993.
10. Mahendran, M., Rodrigues, C.F.F., and Pedersen, S.K. (1995). Vertical skein of hollow fiber membranes and method of maintaining clean fiber surfaces while filtering a substrate to withdraw a permeate. US Patent 5,639,373, issued 17 June 1997.
11. Krzeminski, P., Leverette, L., Malamis, S., and Katsou, E. (2017). Membrane bioreactors – a review on recent developments in energy reduction, fouling control novel configurations, LCA and market prospects. *J. Membr. Sci.* **527**: 207.
12. Baker, R.W. and Strathmann, H. (1970). Ultrafiltration of macromolecular solutions with high-flux membranes. *J. Appl. Polym. Sci.* **14**: 1197.
13. Porter, M.C. (1990). Ultrafiltration. In: *Handbook of Industrial Membrane Technology* (ed. M.C. Porter), 136–259. Park Ridge, NJ: Noyes Publications.
14. Cheryan, M. (1998). *Ultrafiltration and Microfiltration Handbook*. Lancaster, PA: Technomic Publishing Co.
15. Walker, R. (1982). Recent developments in ultrafiltration of electrocoat paint. *Electrocoat* **82**: 1.
16. Field, R.W., Wu, D., Howell, J.A., and Gupta, B.B. (1995). Critical flux concept for microfiltration fouling. *J. Membr. Sci.* **100**: 259.
17. Wang, Z., Wu, Z., Yu, G. et al. (2006). Relationship between sludge characteristics and membrane flux determination in submerged membrane bioreactors. *J. Membr. Sci.* **284**: 87.
18. Le-Clech, P., Chen, V., and Fane, A.G. (2006). Fouling in membrane bioreactors used in wastewater treatment. *J. Membr. Sci.* **284**: 17.

19. Bacchin, P., Aimar, P., and Field, R.W. (2006). Critical and sustainable fluxes: theory, experiments, and applications. *J. Membr. Sci.* **281**: 42.

20. Chen, V., Fane, A.G., Madaeni, S., and Wenten, I.G. (1997). Particle deposition during membrane filtration of colloids: transition between concentration polarization and cake formation. *J. Membr. Sci.* **125**: 109.

21. Zydney, A.L. and Xenopoulos, A. (2007). Improving Dextran tests for ultrafiltration membranes: effect of device format. *J. Membr. Sci.* **291**: 180.

22. Culkin, B., Plotkin, A., and Monroe, M. (1998). Solve membrane fouling with high-shear filtration. *Chem. Eng. Prog.* **94**: 29.

23. Jaffrin, M.Y. (2008). Dynamic shear-enhanced membrane filtration: a review of rotating discs, rotating membranes, and vibrating systems. *J. Membr. Sci.* **324**: 7.

24. van Reis, R. and Zydney, A. (2007). Bioprocess membrane technology. *J. Membr. Sci.* **297**: 16.

25. Zeman, L.J. and Zydney, A.L. (1996). *Microfiltration and Ultrafiltration: Principles and Applications*. New York: Marcel Dekker.

26. Rogers, A.N. (1984). Economics of the application of membrane processes. In: *Synthetic Membrane Processes* (ed. G. Belfort), 437–476. Orlando, FL: Academic Press.

27. Eykamp, W. (1995). Microfiltration and ultrafiltration. In: *Membrane Separation Technology: Principles and Applications* (ed. R.D. Noble and S.A. Stern), 1–40. Amsterdam: Elsevier Science.

28. Brans, G., Schroën, C.G.P.H., VanderSman, R.G.M., and Boom, R.M. (2004). Membrane fractionation of milk: state of the art and challenges. *J. Membr. Sci.* **243**: 263.

29. Jiao, B., Cassano, A., and Drioli, E. (2004). Recent advances on membrane process for the concentration of fruit juices: a review. *J. Food Eng.* **63**: 303.

30. El Rayess, Y., Albasi, C., Bacchin, P. et al. (2011). Cross flow filtration applied to oenology: a review. *J. Membr. Sci.* **382**: 1.

31. Maubois, J.L., Mocquot, G., and Vassal, L. (1979). Preparation of cheese using ultrafiltration. US Patent 4,205,080, issued 27 May 1980.

32. Mir, L., Eykamp, W., and Goldsmith, R.L. (1977). Current and developing applications for ultrafiltration. *Ind. Water Eng.* **14**: 1.

33. Judd, S. (2010). *The MBR Book: Principles and Applications of Membrane Bioreactors in Water and Wastewater Treatment*, 2e. London: Butterworth-Heinemann.

34. Carstensen, F., Apel, A., and Wessling, M. (2012). In situ product recovery: submerged membranes vs. external loop membranes. *J. Membr. Sci.* **394**: 1.

35. Cui, Z.F., Chang, S., and Fane, A.G. (2003). The use of gas bubbling to enhance membrane processes. *J. Membr. Sci.* **221**: 1.

36. Yu, H., Li, X., Chang, H. et al. (2020). Performance of hollow fiber ultrafiltration membrane in a full scale drinking water treatment plant in china: a systematic evaluation during 7-year operation. *J. Membr. Sci.* **613**: 118469.

37. Okey, R.W. and Stavenger, P.L. (1966). Reverse osmosis applications in industrial wastewater treatment. In: *Membrane Processes for Industry Proceedings of the Symposium* (19–20 May 1966). Birmingham, AL: Southern Research Institute.

38. Yang, W., Cicek, N., and Ilg, J. (2006). State-of-the-art of membrane bioreactors: worldwide research and commercial applications in North America. *J. Membr. Sci.* **270**: 201.

39. Water Environmental Federation (2012). Membrane bioreactors. In *WEF Manual of Practice No. 36*. Alexandria VA: WEF Press.

40. Drews, A. (2010). Membrane fouling in membrane bioreactors – characterization, contradictions, cause and cures. *J. Membr. Sci.* **363**: 1.

41. Lin, H., Peng, W., Zhang, M. et al. (2013). A reviews of anaerobic membrane bioreactors: applications, membrane fouling and future perspectives. *J. Membr. Sci.* **314**: 169.
42. Zydney, A.L. (2021). New developments in membrane bioprocessing – a review. *J. Membr. Sci.* **620**: 118804.
43. Lutz, H. (2015). *Ultrafiltration for Bioprocessing*. Waltham, MA: Woodhead Publishing.
44. Cheang, B. and Zydney, A.L. (2004). A two-stage ultrafiltration process for fractionation of whey protein isolate. *J. Membr. Sci.* **231**: 159.

8

Microfiltration

8.1 Introduction and History

Microfiltration refers to filtration processes that use porous membranes to separate suspended particles with diameters between 0.1 and 10 μm. Thus, microfiltration applications fall between ultrafiltration and conventional filtration. Like ultrafiltration, microfiltration has its modern origins in the development of collodion (nitrocellulose) membranes in the 1920s and 1930s. In 1926, Membranfilter GmbH was founded and began to produce collodion microfiltration membranes commercially. The market was very small, but by the 1940s, other companies, including Sartorius and Schleicher and Schuell, were producing similar membrane filters.

The first large application of microfiltration was in laboratory tests to culture microorganisms in drinking water; this remains a significant application. The test was developed in Germany during World War II, as a rapid method to monitor the water supply for contamination. Established test methods required water samples to be cultured for at least 96 hours. Mueller and others at Hamburg University devised a method in which a liter of water was filtered through a Sartorius microfiltration membrane. Any bacteria in the water were captured by the filter, which was then placed on a pad of gelled nutrient solution for 24 hours. The nutrients diffused to the trapped bacteria on the membrane surface, allowing them to grow into colonies large enough to be counted easily under a microscope. After the war, there was no US supplier of these membranes, so, in 1947, the US Army sponsored a program by Goetz at CalTech to duplicate the Sartorius technology. The membranes developed there were made from a blend of cellulose acetate and nitrocellulose and were formed by controlled precipitation of a film of casting solution with water from the vapor phase. This technology was passed to the Lowell Chemical Company, which in 1954 became the Millipore Corporation, producing the Goetz membranes on a commercial scale. Millipore remains one of the largest microfiltration membrane producers. Membranes made from a number of non-cellulosic materials, including poly(vinylidene fluoride), polyamides, polyolefins, and poly(tetrafluoroethylene), have been developed by Millipore and others. Nonetheless, the cellulose acetate/cellulose nitrate blend membrane remains a widely used microfilter.

For many years, the use of microfiltration membranes was confined to laboratory applications. The introduction of pleated membrane cartridges by Gelman in the 1970s was an important step forward and made industrial applications possible. The first of these were in pharmaceutical and microelectronics manufacturing. The production of low-cost, single-use, disposable cartridges for such users now represents a major part of the microfiltration industry. In these applications, trace amounts of particles are removed from solutions that are already very clean. The membranes used have a generally uniform, fine sponge-like structure with pores in the 1–5 µm range. Large particulates in the feed solution are captured at the membrane surface; the rest are captured at pore constrictions in the interior of the membrane. Pharmaceutical products are often made in a batch process where single-use disposable cassettes or cartridges are used. A semiconductor factory uses large quantities of ultrapure water (water purified down to ppb levels or below of contaminants) to rinse wafers and other components during their manufacture. Microfiltration cartridges are used to maintain the rinse water particle- and bacteria-free. These filters have a longer lifetime. As particulates accumulate on and in the membrane, the pressure needed to maintain permeate water flow is increased. When the required flux can no longer be maintained under acceptable pressures, the membrane cartridge is replaced.

In the 1970s, tangential-flow filtration using newly developed anisotropic ultrafiltration membranes began to be used, especially with solutions containing high concentrations of particulates. These membranes and the related filtration processes are described in the ultrafiltration chapter (Chapter 7). The present chapter is focused on in-line filtration processes, as shown in Figure 8.1.

Some important milestones in the development of microfiltration are charted in Figure 8.2.

8.2   Background

8.2.1   Types of Membrane

Microfiltration membranes are used in applications where the filtrate must be completely free of all microorganisms and particles above a certain diameter. By convention, membranes are classified by the size of the smallest particle that is completely retained by the membrane. Figure 8.3 shows the top surfaces and cross sections of two types of microfiltration membrane able to produce a sterile filtrate in a challenge test with bacteria of 0.45 µm diameter. Thus, both

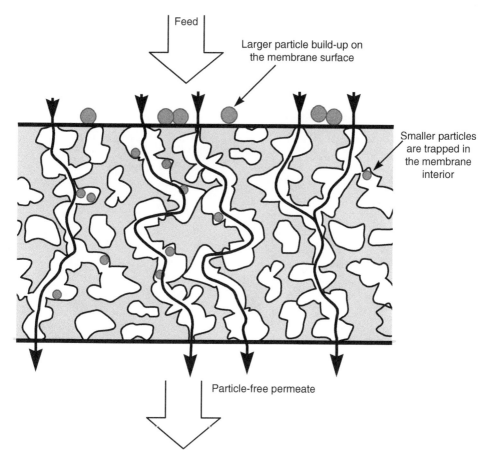

Feed

Larger particle build-up on
the membrane surface

Smaller particles
are trapped in
the membrane
interior

Particle-free permeate

Figure 8.1 *Schematic representation of an in-line microfiltration membrane. Large particles are captured at the membrane surface; smaller ones are trapped within the membrane.*

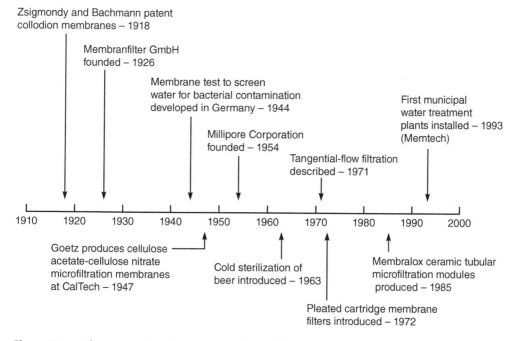

Zsigmondy and Bachmann patent
collodion membranes – 1918

Membranfilter GmbH
founded – 1926

Membrane test to screen
water for bacterial contamination
developed in Germany – 1944

First municipal
water treatment
plants installed – 1993
(Memtech)

Millipore Corporation
founded – 1954

Tangential-flow filtration
described – 1971

| 1910 | 1920 | 1930 | 1940 | 1950 | 1960 | 1970 | 1980 | 1990 | 2000 |

Goetz produces cellulose
acetate-cellulose nitrate
microfiltration membranes
at CalTech – 1947

Cold sterilization of
beer introduced – 1963

Membralox ceramic tubular
microfiltration modules
produced – 1985

Pleated cartridge membrane
filters introduced – 1972

Figure 8.2 *Milestones in the development of microfiltration.*

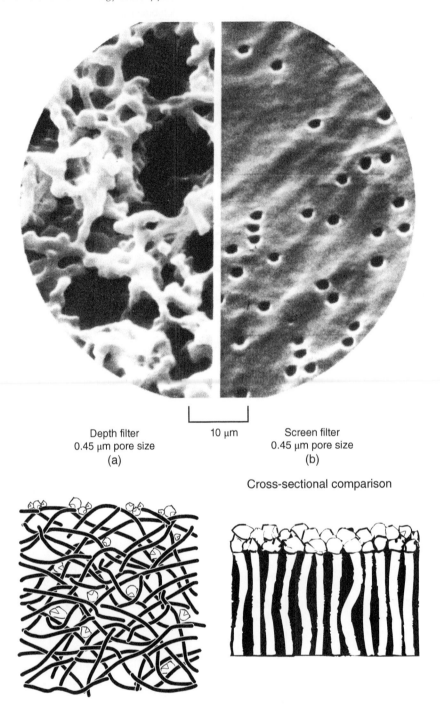

Depth filter
0.45 μm pore size
(a)

10 μm

Screen filter
0.45 μm pore size
(b)

Cross-sectional comparison

Figure 8.3 *Surface scanning electron micrograph and schematic comparison of nominal 0.45 μm screen and depth filters. The depth filter pores (a) are almost 5–10 times larger than the screen filter equivalent. A few large 5–10 μm particles are captured on the surface of the membrane, but all particles larger than 0.45 μm are captured by adsorption in the membrane interior. In contrast, the screen filter pores (b) are uniform and small and capture the retained particles on the membrane surface.*

membranes have an *effective* pore diameter of 0.45 μm and are classified as 0.45 μm membranes. As the photo shows, however, the membrane of Figure 8.3a clearly has an average *geometric* pore size that is in the 5–10 μm range, an order of magnitude larger than its nominal 0.45 μm *effective* rating. The discrepancy arises from the capture mechanism. In use, some bacteria are retained on the surface, but most are captured in the membrane interior, either at pore entanglements or by adsorption on the interior pore surfaces. Such a membrane is known as a depth filter. Figure 8.3b shows a nucleation track membrane in which uniform pores with a diameter of about 0.4 μm traverse the membrane. This type of membrane is a screen filter, in which bacteria larger than 0.4 μm are captured on the surface. Screen filters are not much used in microfiltration processes, because they are rapidly fouled by the captured material on the surface. They have, however, found application in some analytical applications to determine the type and concentration of bacteria in a test solution. For these analyses, a known volume of solution is filtered through the membrane, after which material captured on the membrane surface can be inspected, counted and classified under a microscope.

Until the 1980s, most microfiltration membranes had the finely microporous sponge structure shown in Figure 8.3a. The membranes were 200–300 μm thick. Depending on the way the membranes were made, the pores tapered from smaller pores at one surface to larger pores at the other surface, but the degree of tapering or anisotropy was not large. Later, when variants of the Loeb–Sourirajan technique were used, membranes with much more noticeable anisotropy were produced. These membranes were sometimes used with the biggest pores facing the feed solution. Larger particulates in the feed were trapped in the top surface layer and smaller particles were trapped in the underlying layers. In this way, pore plugging caused by captured particulates was more uniformly spread over the whole membrane, extending the useful life of the membrane.

More recently, multi-layered membrane structures have been developed by several membrane companies. Two or more casting operations are performed in series to form a layered casting solution, which is precipitated in the usual way by water vapor inhibition or direct water precipitation. The process is described in Chapter 4 and can produce a variety of layered membranes having different distributions of pores in each layer. The objective of the process is to extend the useful operating life of the membrane by forming a more open surface layer on top of a more finely porous substrate. The change in pore size from the top to the bottom of the filter is called the taper angle. The optimal taper angle depends on the material being filtered. If the taper angle is too small, the upper layers of the filter will become blocked before the lower layers. If the taper angle is too large, the bottom layers will become blocked before the upper layers. Attempts have been made to put the calculation of the optimum taper angle on a mathematical basis [1]; membrane producers, however, use an empirical approach.

8.2.2 Membrane Characterization

Microfiltration membranes are often used in applications for which penetration of even one particle or bacterium through the membrane can be critical. Therefore, membrane integrity, that is, the absence of membrane defects or oversized pores, is extremely important. Several tests are used to characterize membrane pore size, pore size distribution, and membrane defects. Modern techniques for measuring pore size distribution are reviewed by Tanis-Kanbur et al. [2] and Giglia et al. [3].

The most important test is the bacterial challenge test. The ability of a membrane to filter bacteria from solutions depends on the pore size of the membrane, the size of the bacterium being filtered, and the number of organisms used to challenge the membrane. Some results of Elford that illustrate these effects with the bacterium *B. prodigiosus* are shown in Figure 8.4 [4]. Elford prepared a series of membranes with a range of pore sizes. He found that membranes with

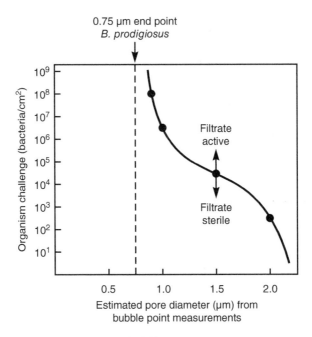

Figure 8.4 *Membrane pore diameter from bubble point measurements versus* Bacillus prodigiosus *concentration. Source: Reproduced with permission from [4]. Copyright (1933) The Royal Society, London, UK.*

relatively large pores could completely filter bacteria from the challenge solution to produce a sterile filtrate, providing the challenge concentration was low. If the organism concentration was increased beyond a certain value, breakthrough of bacteria to the filtrate occurred. If the membrane pore size was small enough, however, a point was reached at which no breakthrough of bacteria to the filtrate occurred, no matter how concentrated the challenge solution. This point is taken to be the effective pore size of the membrane. As mentioned above, this pore size is much smaller than the average pore diameter seen with an electron microscope.

The industry has adopted two bacterial challenge tests to measure pore size and membrane integrity [5]. The tests are based on *Serrata marcescens*, a bacterium originally thought to have a diameter of 0.45 μm, and *Pseudomonas diminuta*, originally thought to have a diameter of 0.22 μm. In fact, both organisms are ellipsoids with aspect ratios of about 1.5:1. The tests have changed several times over the years, but by convention a membrane is designated as 0.45 μm pore size if it is completely retentive when challenged with 10^7 *S. marcescens* organisms per cm^2 and 0.22 μm pore size if it is completely retentive when challenged with 10^7 *P. diminuta* organisms per cm^2. Most commercial microfiltration membranes are categorized as 0.22 or 0.45 μm diameter pore size based on these tests.

Currently, most bacterial challenge tests are performed with *P. diminuta*. This organism has an average size of 0.3–0.4 μm, although the size varies significantly with the culture conditions. In a rich culture medium, the cells can form larger clumps. Thus, to obtain consistent results, the culture characteristics must be carefully monitored, and control experiments must be performed with already certified 0.45 and 0.22 μm filters to confirm that no clumping has occurred. A detailed ASTM procedure is available [5]. Factors affecting this test are discussed in detail by Brock [6] and Meltzer [7].

The performance of membranes in bacterial challenge tests is often quantified by a log reduction value (LRV), defined as

$$\text{LRV} = \log_{10}\left(\frac{c_f}{c_p}\right) \tag{8.1}$$

where c_f is the concentration of bacteria in the challenge solution and c_p is the concentration in the permeate. It follows that at 99% rejection, c_f/c_p is 100 and the LRV is 2; at 99.9% rejection, the LRV is 3; and so on. In pharmaceutical and electronic applications, an LRV of 7 or 8 is usually required. In municipal water filtration, an LRV of 4 or 5 is the target.

Bacterial challenge tests require careful, sterile laboratory techniques and an incubation period of several days before the results are available. For this reason, a number of secondary tests based on filtration of suspensions of latex particles or on bubble point measurements are used. These quicker, easier tests are used as quality control tools in membrane production facilities. The accuracy of the secondary test can be checked by applying it to similar membranes that have already been characterized by the bacterial challenge test.

8.2.2.1 Latex Challenge Tests

The first generation of latex challenge tests used a suspension of latex particles of uniform size. The solution was filtered through the membrane. The number of particles in the permeate was determined by filtering the permeate solution a second time with a tight screen filter that captured the particles for easy counting. This test was used in fundamental studies but was too time consuming to be widely adopted by membrane producers and users.

The second generation of latex challenge tests is more instrumented and easier to use, and has found a place in industrial processes. A multistandard latex suspension containing up to 10 latex components of graded size from 0.1 to 1.5 μm is used. A dilute solution of the standard is drawn through the test filter. Submicron particle size analyzers are then used to measure the distribution of latex particles in the permeate, which is then compared to the distribution in the feed. Calculation of the percent capture as a function of latex diameter gives a good measure of the effective pore size distribution of the filter. The type of data that can be obtained is shown in Figure 8.5.

8.2.2.2 Bubble Point Test

The bubble point test is simple, quick, and reliable and is by far the most widely used method of characterizing flat-sheet microfiltration membranes. The membrane is first wetted with a suitable test liquid, usually water for hydrophilic membranes and an organic liquid for hydrophobic membranes. The membrane is then placed in a holder with a layer of the test liquid on the top surface. Air is fed to the bottom of the membrane, and the pressure is slowly increased until the first continuous string of air bubbles at the membrane surface is observed. This pressure is called the bubble point pressure and is a characteristic measure of the diameter of the largest pore in the membrane. Obtaining reliable and consistent results with the bubble point test requires care. It is essential, for example, that the membrane be completely wetted with the test liquid; this may be difficult to determine. Because this test is so widely used by membrane manufacturers, a great deal of work has been devoted to developing a reliable test procedure. The use of this test is reviewed in Meltzer's book [7].

The bubble point pressure can, in principle, be correlated to the membrane pore diameter, r, by the equation

$$\Delta p = \frac{2\gamma \cos\theta}{r} \tag{8.2}$$

where Δp is the bubble point pressure, γ is the fluid surface tension, and θ is the liquid–solid contact angle. For completely wetting solutions, θ is 0°, so $\cos\theta$ equals 1. Properties of liquids commonly used in bubble point measurements are given in Table 8.1.

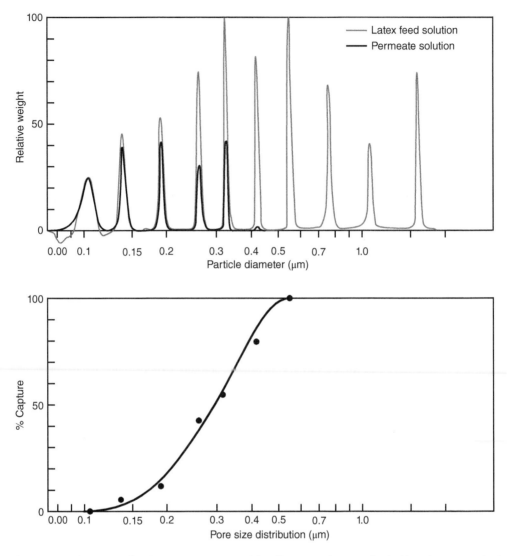

Figure 8.5 *Pore size distribution as measured by filtration of a standard dilute latex standard solution. Source: Adapted from [8].*

Table 8.1 *Properties of liquids commonly used in bubble point measurements. The conversion factor divided by the bubble pressure (in psi) gives the maximum pore size (in μm).*

Wetting liquid	Surface tension (dyn/cm)	Conversion factor
Water	72	42
Kerosene	30	17
Isopropanol	21.3	12
Silicone fluid[a]	18.7	11
Fluorocarbon fluid[b]	16	9

[a]Dow Corning 200 fluid, 2.0 centistoke.
[b]3M Company, Fluorochemical FC-43.

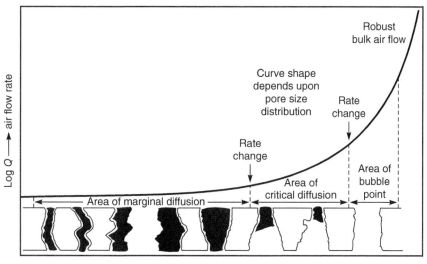

Figure 8.6 Schematic of the effect of applied gas pressure on gas flow through a wetted microporous membrane in a bubble pressure test. Source: Reproduced with permission from [7]. Copyright (1987) Elsevier.

Microfiltration membranes are heterogeneous structures having a distribution of pore sizes. The effect of the applied gas pressure on the liquid in a bubble test is illustrated schematically in Figure 8.6. At pressures well below the bubble point, all pores are completely filled with liquid; gas can only pass through the membrane by diffusion through the liquid film. Just below the bubble point pressure, liquid begins to be forced out of the largest membrane pores. The diffusion rate then starts to increase as the pressure is raised, until the liquid is completely forced out of the largest pore. Bubbles of gas form on the membrane surface; this is the bubble point and the pressure at which it occurs is the bubble point pressure. As the gas pressure is increased further, liquid is forced out of more pores, and general convective flow of gas through the membrane takes place. This is sometimes called the 'foam-all-over pressure', and is a measure of the average pore size of the membrane.

Bubble point measurements are subjective, and different operators can obtain different results. Nonetheless, the test is quick, simple, and an effective quality control technique, and companies that make and use the filters will have the simple manual apparatus. Nowadays automated test equipment is also available and widely used for routine measurements. The use of this equipment is reviewed by Tanis-Kanbur et al. [2].

Although bubble point measurements can be used to determine the pore diameter of membranes using Eq. (8.2), the results do not equate with the filtration capability even approximately, unless the filter is strictly a tight screen filter. Based on Eq. (8.2), a 0.22 μm pore diameter membrane should have a bubble point of about 15 bar. In fact, a membrane that has a nominal 0.22 μm pore diameter, based on the bacterial challenge test, typically has a bubble point pressure of only 3–4 bar, indicating an actual geometric pore diameter of about 1 μm.

Figure 8.7 shows typical results comparing microbial challenge tests with membrane bubble point measurements. The tests were conducted on a series of related membranes and used *P. diminuta* as the challenge organism [9]. A membrane with a microbial reduction factor of

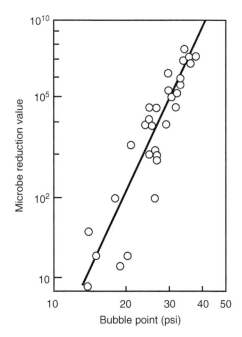

Figure 8.7   *Correlation of* P. diminuta *microbial challenge and bubble point test data for a series of related membranes. Reproduced with permission from [9]. Copyright (1987) / Taylor & Francis.*

10^8–10^9 has, according to the convention mentioned earlier, a pore diameter of about 0.22 μm. According to the figure, such a membrane has a bubble point pressure of only about 35 psi (2.3 bar), far below the theoretical value of 15 bar from Eq. (8.2). A correction factor is sometimes included in the equation to account for the shape of the membrane pores, but no reasonable shape factor can account for the more than fivefold discrepancy. This is because retention by surface pores is only one of the capture mechanisms of a microfiltration membrane. Adsorption and electrostatic attraction also remove bacteria, even when the pore diameter is much larger than the organism diameter. These mechanisms account for the observed performance of a depth filter, albeit qualitatively.

8.2.2.3   Module Integrity Tests

The bubble point test is a good quality control technique when applied to the manufacture or use of individual membrane discs or cartridges, but for microfiltration plants a more sensitive test is required. In a 50–100 m² plant that is required to produce water of ultra-high purity, even a single pinhole defect or damaged fiber may be enough to compromise the operation. For this type of critical application, an air diffusion test may be used [10, 11].

The apparatus involved is shown in Figure 8.8 [11]. The feed and permeate sides of the module are first filled with water. The feed side of the membrane is then completely drained and replaced with air at a pressure of a few psig. Under these conditions, there is a small flow of air into the permeate side produced by diffusion of air through the water-filled membrane pores. The air flow in defect-free modules is quite small and is often measured by collecting air displaced from a tube connecting the module to an inverted water-filled burette as shown in Figure 8.8. The test is

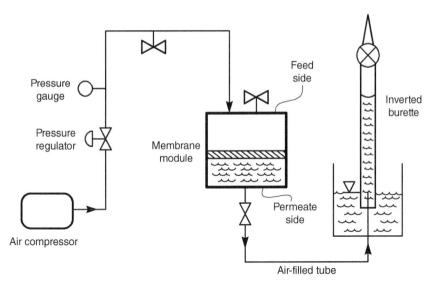

Figure 8.8 *Schematic diagram illustrating the apparatus used to measure diffusive air flow rates. The diagram shows a single module under test; the same arrangement can be used to test several modules linked together. Source: Adapted from [11].*

highly accurate; even a very small defect is immediately apparent as an unexpectedly high displaced water flow rate. In Figure 8.8, the test is shown as applied to a single module, but the test can be applied in like manner to an entire system consisting of a number of modules manifolded together.

8.2.3 Microfiltration Membranes and Modules

The first major application of microfiltration was biological testing of water. This remains an important laboratory application in microbiology and biotechnology. For this application, the early cellulose acetate/cellulose nitrate phase separation membranes made by vapor-phase precipitation with water are still used. Other membrane materials with improved mechanical properties and chemical stability include polyacrylonitrile–poly(vinyl chloride) copolymers, poly(vinylidene fluoride), polysulfone, cellulose triacetate, various nylons, and poly(tetrafluoroethylene). Most cartridge filters today use membranes made from these materials.

During the early decades of commercial microfiltration, the in-line plate-and-frame module was the only type available. These units contained between 1 and 20 separate membrane envelopes sealed by gaskets. In most operations, all the membrane envelopes were changed after each use; the labor involved in disassembly and reassembly of the module was a significant drawback. Nonetheless, these systems are still used to process small volumes of solution. A typical plate-and-frame filtration system is shown in Figure 8.9.

A variety of cartridges that allow a much larger area of membrane to be incorporated into a disposable unit are now available. Disposable plate-and-frame cassette modules have been produced, but a larger portion of the market is for pleated cartridges. These cartridges have evolved over the years. A simple cut-away view of a basic cartridge design is shown in Figure 8.10.

The first pleated cartridges were based on the standard straight concertina fold shown in Figure 8.10. Later, designs in which the pleats are slightly oversized and curved into a crescent,

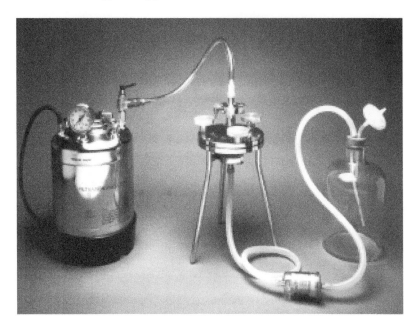

Figure 8.9   *Sterile filtration of a small-volume pharmaceutical solution with a 142 mm plate-and-frame filter used as a prefilter in front of a small disposable cartridge final filter.*

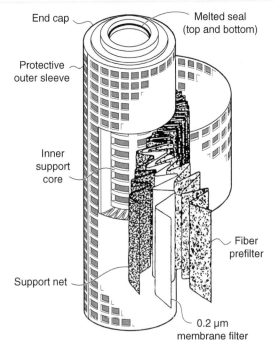

Figure 8.10   *Cut-away view of a simple pleated cartridge filter. By folding the membrane, a large surface area can be contacted with the feed solution, producing a high particle loading capacity. Source: From Membrana product literature.*

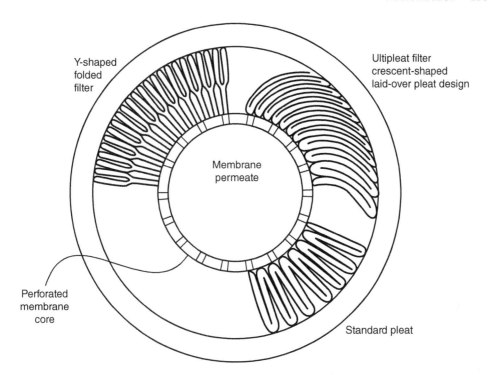

Figure 8.11 *Optional configurations for a pleated cartridge filter. The first cartridges were based on a concertina fold structure called a standard pleat. Later, to pack more membrane area into the same size cartridge, crescent-shaped folded filters were developed. Y-shaped filters are also made by some manufacturers.*

or folded into a Y-shape, were developed, as shown in Figure 8.11. Such arrangements pack more membrane area into the same module volume [12]. A typical cartridge is 25 cm long, has a diameter of 5–6 cm, and contains about 0.3–0.5 m^2 of membrane. Often the membrane consists of several layers: an outer prefilter facing the solution to be filtered, followed by a finer polishing filter.

Irrespective of the pleating details, the membrane is folded around a perforated permeate core. The cartridge fits inside a specially designed housing into which the feed solution enters at a pressure of 1–10 bar. Pleated membrane cartridges, which are fabricated with high-speed automated equipment, are cheap, disposable, reliable, and hard to beat if the solution to be filtered has a relatively low particle level. Ideal applications are production of aseptic solutions in the pharmaceutical industry or ultrapure water for wafer manufacture in the electronics industry, where the low feed load allows the cartridges to filter large volumes of solution before needing replacement. Manufacturers produce cartridge holders that allow a number of cartridges to be connected in series or in parallel to handle large solution flows.

In-line cartridge filters are unsuitable for microfiltration of highly contaminated feed streams because the high feed load results in an impractically short lifetime. Such streams are treated by ultrafiltration, using an ultrafiltration membrane as a screen filter. The design of appropriate membranes, modules, and processes is covered under ultrafiltration (Chapter 7).

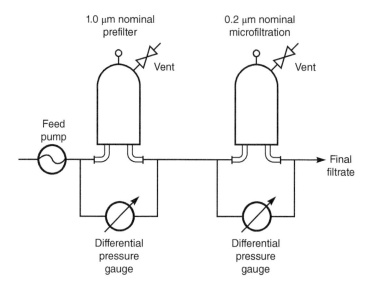

Figure 8.12 *Typical in-line filtration operation using two cartridge filters in series. The prefilter removes all of the large particles and some of the smaller ones. The final polishing filter removes the remaining small particles.*

8.2.4   Process Design

A typical microfiltration process arrangement is illustrated in Figure 8.12. A pump forces liquid through the filter, and the pressure across the filter is measured by a pressure gauge. Initially, the pressure difference measured by the gauge is small, but as retained particles block the filter, the pressure difference increases until a predetermined limiting pressure is reached, and the filter is changed.

To extend its life, a microfiltration cartridge may contain two or more membrane filters in series, or, as shown in Figure 8.12, a coarse prefilter cartridge may be used before the microfiltration cartridge. The prefilter captures the largest particles, allowing smaller particles to pass and be captured by the following microfiltration membrane. The use of a prefilter extends the life of the microfiltration cartridge significantly; without a prefilter, the fine microfiltration membrane would be rapidly blinded by accumulation of large particles on the membrane surface. The correct combination of prefilter and final membrane must be determined for each application. This can be done by placing the prefilter on top of the required final filter membrane in a small test cell, or better yet, with two test cells in series. With two test cells, the pressure drop across each filter can be measured separately. Results obtained from such tests for a given microfilter with different prefilters are shown in Figure 8.13 [7]. Figure 8.13a shows the rate of pressure rise across the fine filter alone. The limited dirt-holding capacity of this filter means that it is rapidly plugged by a surface layer of large particles. Figure 8.13b shows the case when a prefilter that is too coarse is used. In this case, the pressure difference across the prefilter remains small, whereas the pressure difference across the final filter increases almost as rapidly as before, because of plugging by particles passing the prefilter. Little improvement in performance is obtained. Figure 8.13c shows the case where the prefilter is too fine. This situation is the opposite of Figure 8.13a – the pressure difference across the prefilter increases rapidly, and the lifetime of the combination filter is limited by this filter. Figure 8.13d shows the optimum combination, in which the pressure difference is uniformly distributed across the prefilter and final filter. This condition maximizes the lifetime of the filter combination.

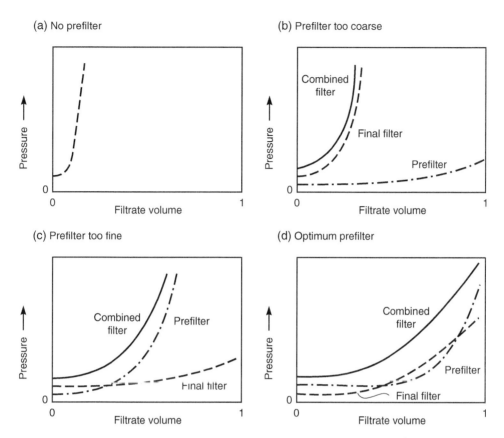

Figure 8.13 *The pressure difference across the prefilter, the final filter and the combined filters for various combinations (a–d) of prefilter and final filter. The optimum prefilter distributes the particle load evenly between the two filters so both filters reach their maximum particle load at the same time. This maximizes the useful life of the combination.*

Cartridge microfiltration is a stable area of membrane technology – few changes in cartridge design or use have occurred in the past 20 years. Most changes have focused on improving membrane resistance to high temperatures, solvents, and extremes of pH, to allow application of the technology to more challenging environments.

8.3 Applications

The first use of microfiltration membranes was in microbiological and water quality testing laboratories. The original cellulose acetate/nitrocellulose membrane used by Mueller to test water in 1945 is still used. But as the biological drug industry has grown, so has the market for membranes. As a result, laboratory applications of all types remain a significant application even as the industrial applications described below have been developed and grown.

One feature of the microfiltration market differs significantly from that of other industrial membrane processes: membrane lifetimes are often measured in hours. In a few completely passive applications, such as treating sterile air vents, membranes may last several months; in general, the market is dominated by single-use cartridges designed to filter a relatively small

Table 8.2   *Approximate volume of fluid that can be filtered by 1 m²
of a 5-µm membrane before fouling [13].*

Solution	Volume filtered (m³/m²)
Solvents	500
Tap water	200
Wine	50
Pharmaceuticals for ampoules	50
20% glucose solution	20
Vitamin solutions	10
Parenterals	10
Peanut oil	5
Fruit juice concentrate	2
Serum (7% protein)	0.6

mass of particles from a solution. The volume of solution that can be treated by a microfiltration membrane is directly proportional to the particle level in the feed. As a rough rule of thumb, the particle-holding capacity of a cartridge filter in a noncritical use is between 100 and 300 g/m² of membrane area. Thus, the volume of fluid that can be treated may be large if the microfilter is a final safety filter for an ultrapure water system, but much smaller if treating contaminated surface water or a food processing stream. The approximate volume of various solutions that can be filtered by a 5-µm filter before the filter is completely plugged is given in Table 8.2 [13].

Despite the limited volumes that can be treated before a filter must be replaced, microfiltration is economical, because the cost of disposable cartridges is low. Currently, a 25 cm long pleated cartridge costs between US$10 and US$20 and contains 0.3–0.5 m² of active membrane area. The low cost reflects the large numbers that are produced.

As stated previously, the primary markets for the disposable cartridge are the pharmaceutical and semiconductor industries. In both industries, the cost of microfiltration is small compared to the value of the products, so these markets have driven the microfiltration industry for the past 20 years.

8.3.1   Sterile Filtration of Pharmaceuticals

Microfiltration is used widely in the pharmaceutical industry to produce injectable drug solutions. Regulating agencies require rigid adherence to standard preparation procedures to ensure a consistent, safe, and sterile product. Microfiltration needs to remove particles, and more importantly, all viable bacteria, so a 0.22 µm rated filter is usually used. Because the cost of validating membrane suppliers is substantial, users usually develop long-term relationships with individual suppliers.

A microfilter for this industry is considered a sterile filter if it achieves a log reduction factor of better than 7. This means that if 10^7 bacteria/cm² are placed on the filter, none appears in the filtrate. Even though virus and bacteria removal through these filters is extremely good, the use of a second redundant filter (two filters in series) is often used as a safety measure [14, 15].

Microfiltration cartridges produced for this market are often sterilized directly after manufacture and again just prior to use. Live steam, autoclaving at 120 °C, or ethylene oxide sterilization may be used, depending on the application. A flow schematic of an ampoule-filling station (after material by Schleicher and Schuell) is shown in Figure 8.14.

The process starts with a pretreatment train, which includes a deionization system utilizing reverse osmosis, followed by mixed bed ion exchange, and a 5 µm microfiltration step. Water

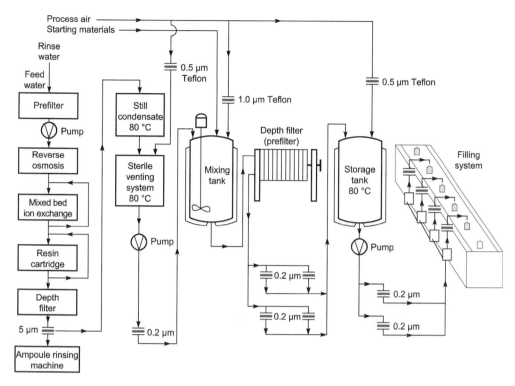

Figure 8.14 Flow diagram illustrating the use of microfiltration sterilization filters in a production line used to prepare ampoules of injectable drug solutions.

for injection need not be ultrapure, so the treatment system used to prepare the feed water is relatively straightforward. After pretreatment, the water is sterilized with a 0.2 μm microfilter, mixed with the drug solution, then sent to a storage tank for the ampoule-filling station. Figure 8.14 shows several redundant sterile steps as the drug solution moves to the ampoule filling system. These extra operations are safety measures to ensure the stability of the final packaged product. Before use, the solution is filtered at least twice more with 0.2 μm filters to ensure sterility. Because pharmaceuticals are produced by a batch process, all filters are replaced at the end of each batch.

8.3.2 Microfiltration in the Electronics Industry

Reverse osmosis and microfiltration for the production of ultrapure water are enabling technologies for the semiconductor industry. The size of the microdevices on a silicon wafer has decreased 100-fold over the last two decades. As a result of the higher density and finer structure of these components, the volume of ultrapure water required per cm^2 for wafer cleaning has increased significantly. More importantly, the quality of the water used has also increased. A general description of an ultrapure water plant was given in Chapter 6 on Reverse Osmosis. Cartridge microfiltration units are used in several locations in the ultrapure water production process and as a final-point-of-use filter. Although fine filters with 0.1 μm pore diameter are used, lifetimes are relatively long.

The electronics industry also uses a variety of reactive gases and solvents which must be particle free. Teflon® microfilters are widely used to treat these materials.

8.3.3 Sterilization of Wine and Beer

Cold sterilization of beer using microfiltration was introduced on a commercial scale in 1963. The process was not broadly adopted at that time, but has since become more common. Sterilization of beer and wine is much less stringent than pharmaceutical sterilization. The main objective is to remove yeast cells, which are quite large, so that the product is clear and bright, although bacterial removal is also desirable. The industry has found that 1 μm filters can remove essentially all the yeast, as well as provide a 10^6 reduction in the commonly found bacteria. Because the cost structure of beer and wine production is very different from that of pharmaceuticals, the filtration system typically involves one or more prefilters to extend the life of the final polishing filter.

8.4 Conclusions and Future Directions

The main microfiltration market is for in-line disposable cartridge filters. These cartridges are sold into two growing industries – microelectronics and pharmaceuticals – so prospects for continued market growth of the industry are very good. In addition to these markets, a significant market exists for microfiltration in bacterial control of drinking water. This application is discussed along with other municipal water treatment applications in Chapter 7.

References

1. Griffiths, I.M., Kumar, A., and Stewart, P.S. (2016). Designing asymmetric multilayered filters with improved performance. *J. Membr. Sci.* **511**: 108.
2. Tanis-Kanbur, M.B., Peinador, R.I., Calvo, J.I. et al. (2021). Porosimetric membrane characterization techniques: a review. *J. Membr. Sci.* **619**: 118750.
3. Giglia, S., Bohonak, D., Greenhalgh, P., and Leahy, A. (2015). Measurement of pore size distribution and prediction of membrane filter virus retention using liquid-liquid porometry. *J. Membr. Sci.* **476**: 399.
4. Elford, W.J. (1933). The principles of ultrafiltration as applied in biological studies. *Proc. R. Soc. London, Ser. B* **112**: 384.
5. ASTM F838-83 (1983). *Determining Bacterial Retention of Membrane Filters Utilized for Liquid Filtration*. Philadelphia: American Society for Testing and Materials.
6. Brock, T.D. (1983). *Membrane Filtration: A User's Guide and Reference*. Madison, WI: Science Tech.
7. Meltzer, T.H. (1987). *Filtration in the Pharmaceutical Industry*. New York: Marcel Dekker.
8. Rideal, A. (2009). What's new in filter testing. *Filtration News* **28**: 20.
9. Leahy, T.J. and Sullivan, M.J. (1978). Validation of bacterial retention capabilities of membrane filters. *Pharm. Technol.* **2**: 65.
10. Farahbakhsh, K. and Smith, D.W. (2004). Estimated air diffusion contribution to pressure decay integrity tests. *J. Membr. Sci.* **237**: 203.
11. Farahbakhsh, K., Adham, S.S., and Smith, D.W. (2003). Monitoring the integrity of low-pressure membranes. *J. AWWA* **95**: 95.
12. Van Reis, R. and Zydney, A. (2007). Bioprocess membrane technology. *J. Membr. Sci* **297**: 16.
13. Hein, W. (1980). Mikrofiltration. Verfahren fur Kritisch Trenn-unde Reinigungsprobleme bei Flussigkeiter und Gasen, Chem. Produkt. November.
14. Parenteral Drug Association (1988). Sterilizing filtration of liquids. Technical report no. 26. *J. Pharm. Sci. Technol.* **58** (Suppl 1): 1–31.
15. Food and Drug Administration (2004). *Guidance for Industry: Sterile Drug Products Produced by Aseptic Processing, Current Good Manufacturing Practice*. FDA.

9

Gas Separation

9.1   Introduction and History

Gas separation has become a major industrial application of membrane technology in the last 30–40 years. The study of gas permeation through membranes has a much longer history, however. Systematic studies began in the 1860s with Thomas Graham who, over a period of 20 years, measured the permeation rates of all the gases then known, through every diaphragm available to him [1]. This was no small task because his experiments had to start with synthesis of the gas. Graham gave the first description of the solution-diffusion model, and his work on porous membranes led to Graham's law of diffusion. Through the remainder of the nineteenth and the early twentieth centuries, the ability of gases to permeate membranes at different rates had no industrial or commercial use. The concept of the perfectly selective membrane was, however, used as a theoretical tool to develop physical and chemical theories, such as Maxwell's kinetic theory of gases.

From 1943 to 1945, Graham's law of diffusion was exploited for the first time, to separate $U^{235}F_6$ from $U^{238}F_6$ as part of the Manhattan Project. Finely microporous metal membranes were used. The separation plant, constructed in Knoxville, Tennessee, represented the first large-scale use of gas separation membranes and remained the largest membrane separation plant in the world for the next 40 years. However, this application was unique and so secret that it had essentially no impact on the long-term development of gas separation technology.

In the 1940s–1950s, Barrer [2], van Amerongen [3], Stern [4], Meares [5], and others laid the foundation of modern studies of gas permeation. The solution-diffusion model of gas permeation developed then is still the accepted model for gas transport through membranes. However, despite the availability of interesting polymer materials, membrane fabrication technology was not sufficiently advanced to make useful membrane systems.

The development of high-flux anisotropic membranes and large-surface-area membrane modules for reverse osmosis (RO), which took place in the late 1960s and early 1970s, provided the needed technology. The first company to establish a commercial presence was Monsanto, which launched its hydrogen-separating Prism® membrane in 1980 [6]. Monsanto had the advantage of being a large chemical company with ample opportunities to test pilot- and demonstration-scale systems in its own plants. The economics were compelling, especially for the separation of hydrogen from ammonia plant purge-gas streams. Within a few years, Prism systems were installed in many such plants [7].

Monsanto's success encouraged other companies to pursue their own membrane applications. By the mid-1980s, Cynara, Separex, and Grace Membrane Systems were all producing membrane plants to remove carbon dioxide from methane in natural gas. This application has grown significantly over the years. At about the same time, Dow launched Generon®, the first commercial membrane system for nitrogen separation from air. Initially, membrane-produced nitrogen was cost-competitive in only a few niche areas, but the development by Dow, Ube, and DuPont/ Air Liquide of materials with higher selectivities has since improved the competitiveness of membrane separation, which now represents more than half of the industrial nitrogen-production market. In about 2010, a large new market developed for aircraft fuel tank inerting systems, and all large jets are now fitted with membrane systems to produce nitrogen for inerting. Gas separation membranes are also being used for a wide variety of other applications ranging from dehydration of air and natural gas to organic vapor removal from air and nitrogen streams to helium recovery in natural gas processing plants. The largest of the new potential applications is the capture and sequestration of CO_2 from power plants and various industrial processes, as a partial solution to global warming. This application, now at the demonstration plant scale, has the potential to become very large. Figure 9.1 summarizes the development history of the technology.

Essentially all commercial gas separation applications to date use dense polymeric membranes that operate by the solution-diffusion process described in Chapter 2. Depending on the pore size

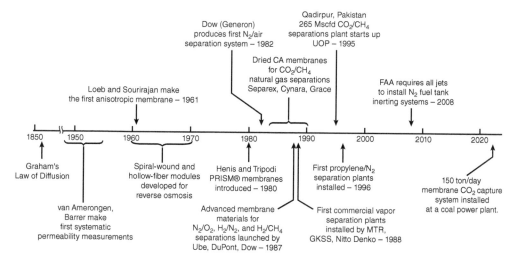

Figure 9.1   *Milestones in the development of gas separation.*

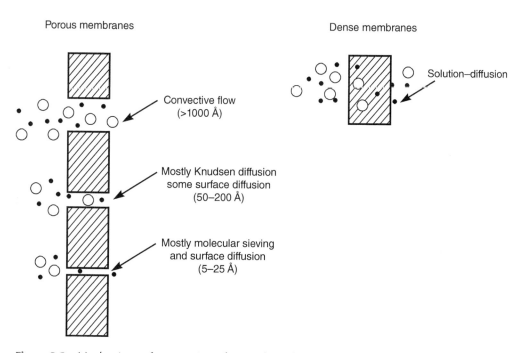

Figure 9.2   *Mechanisms of permeation of gases through porous and dense membranes.*

and structure, microporous membranes also have the capacity to separate gases. Figure 9.2 compares the mechanisms of gas permeation through dense and porous membranes. Three types of porous membranes, differing in pore size, are shown. If the pores are relatively large – from 0.1 to 10 µm – gases permeate by convective flow, and no separation occurs. If the pores are smaller than 0.1 µm (1000 Å), the pore diameter is smaller than the mean free path of most gases at atmospheric pressure. Knudsen diffusion then begins to affect permeation. Membranes with pores below 200 Å are comfortably in the Knudsen diffusion region and the transport rate of a gas is inversely proportional to the square root of the molecular weight. Finally, if the membrane pores are extremely small, on the order of 5–25 Å, then surface diffusion and molecular sieving become factors affecting membrane permeation. Transport through this type of very-small-pore membrane is complex and includes both diffusion in the gas phase and diffusion of adsorbed species on the surface of the pores (surface diffusion), as described in Chapter 3. In the last decade, there has been a surge of interest in such very-small-pore membranes, which often have much higher permeabilities than their dense polymer equivalents, and sometimes good selectivities for important gas separation applications. But the difficulty of producing thin, defect-free, stable membranes on a large scale is keeping them in the laboratory for now.

The theoretical background and structural requirements for dense polymeric membranes and for finely microporous membranes are discussed separately below.

9.2   Dense Polymeric Membranes

9.2.1   Theoretical Background

In Chapter 2, it was shown that gas transport through a dense polymer membrane is governed by the expression

$$J_i = \frac{D_i K_i^G \left(p_{i_o} - p_{i_\ell}\right)}{\ell} \tag{9.1}$$

where J_i is the mass flux of component i (g/cm^2 · s), p_{i_o} and p_{i_ℓ} are the partial pressures of the component i on either side of the membrane, ℓ is the membrane thickness, D_i is the permeant diffusion coefficient, and K_i^G is the Henry's law sorption coefficient (g/cm^3 · pressure). In gas permeation, it is much easier to measure the volume flux through the membrane than the mass flux, and so Eq. (9.1) is usually recast as

$$j_i = \frac{D_i K_i \left(p_{i_o} - p_{i_\ell}\right)}{\ell} \tag{9.2}$$

where j_i is the volume (molar) flux of component i expressed as cm^3(STP)/cm^2 · s and K_i is a volume sorption coefficient with units cm^3(STP) gas/cm^3 of polymer. The product $D_i K_i$ can be written as P_i, which is called the membrane permeability, a measure of ability to permeate gas, normalized for pressure driving force and membrane thickness.[1] A measure of the ability of a membrane to separate two gases, i and j, is the ratio of their permeabilities, α_{ij}, called the membrane selectivity

$$\alpha_{ij} = \frac{P_i}{P_j} \tag{9.3}$$

---

[1] The permeability of gases through membranes is most commonly measured in Barrer, defined as 10^{-10} cm^3(STP) · cm/cm^2 · s · cmHg and named after R.M. Barrer, a pioneer in gas permeability measurements. The term $j_i/\left(p_{i_o} - p_{i_\ell}\right)$, best called the permeance or pressure-normalized flux, is often measured in terms of gas permeation units (gpu), where 1 gpu is defined as 10^{-6} cm^3(STP) · cm/cm^2 · s · cmHg. One gpu is therefore one Barrer/μ. Occasional academic purists write permeability in terms of mol · m/m^2 · s · Pa (1 Barrer = 0.33×10^{-15} mol · m/m^2 · s · Pa), but fortunately this has been slow to catch on.

Since permeability P is the product of the diffusion coefficient, reflecting the mobility of the individual molecules in the membrane material, and the gas sorption coefficient, reflecting the number of molecules dissolved in the membrane material, Eq. (9.3) can also be written as

$$\alpha_{ij} = \left[\frac{D_i}{D_j}\right]\left[\frac{K_i}{K_j}\right] \tag{9.4}$$

The ratio D_i/D_j is the ratio of the diffusion coefficients of the two gases to be separated and can be viewed as the mobility selectivity, reflecting the different sizes of the two molecules. The ratio K_i/K_j is the ratio of the sorption coefficients of the two gases and can be viewed as the solubility selectivity, reflecting the relative solubilities of the two gases.

In all polymers, the diffusion coefficient decreases with increasing permeant molecular size, because large molecules interact with more segments of the polymer chain than do small molecules. Hence, the mobility selectivity always favors the passage of small molecules over large ones. However, the magnitude of the mobility selectivity term depends greatly on whether the membrane material is above or below its glass transition temperature (T_g). If the material is below the glass transition temperature, the polymer chains are essentially fixed and segmental motion is limited. The material is then called a glassy polymer and is tough and rigid. Above the glass transition temperature, segments of the polymer chains have sufficient thermal energy to allow limited rotation around the chain backbone. This motion changes the mechanical properties of the polymer, and it becomes a rubber. The transition is sharp, occurring over a temperature change of just a few degrees. As characterized by their diffusion coefficients, the relative mobilities of gases in rubber and glasses differ significantly, as illustrated in Figure 9.3 [8]. Diffusion coefficients in glassy materials are small and decrease much more rapidly with increasing permeant size than do diffusion coefficients in rubbers. This means the mobility selectivity of rubbery membranes is smaller than that of glassy membranes, sometimes by several orders of magnitude. For example, the mobility selectivity for nitrogen over pentane in natural rubber is approximately 10, whereas in poly(vinyl chloride), a rigid, glassy polymer, it is more than 100 000.

The sorption coefficient of gases and vapors increases with increasing condensability of the permeant. This dependence on condensability means that the sorption coefficient increases with molecular diameter, because large molecules are normally more condensable than smaller ones. The gas sorption coefficient can, therefore, be plotted against boiling point or molar volume. As shown in Figure 9.4 [9], sorption selectivity favors larger, more condensable molecules, such as hydrocarbon vapors, over permanent gases, such as oxygen and nitrogen. However, the difference between the sorption coefficients of permeants in rubbery and glassy polymers is far less marked than the difference in the diffusion coefficients.

It follows from the discussion above that the balance between mobility selectivity and sorption selectivity is different for glassy and rubbery polymers. This difference is illustrated by the data in Figure 9.5. In glassy polymers, the mobility term is usually dominant, permeability falls with increasing permeant size, and small molecules permeate preferentially. Therefore, when used to separate organic vapors from nitrogen for example, glassy membranes permeate nitrogen preferentially. In rubbery polymers, the sorption selectivity term is usually dominant, permeability increases with increasing permeant size, and larger molecules permeate preferentially. Therefore, when used to separate organic vapor from nitrogen, rubbery membranes permeate the organic vapor preferentially. The separation properties of polymer membranes for a number of the most important gas separation applications have been summarized by Robeson [11]. Reviews of structure/property relations have been given by Stern [12] and Park and Lee [13]. Gas permeabilities for some representative and widely used membrane materials are summarized in Table 9.1.

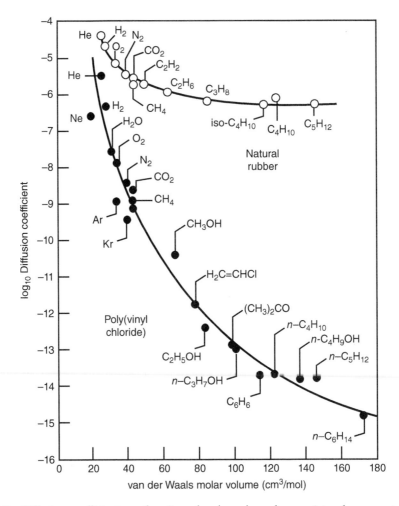

Figure 9.3 *Diffusion coefficient as a function of molar volume for a variety of permeants in natural rubber and in poly(vinyl chloride), a glassy polymer. Source: Adapted from [8].*

An important tool to rationalize the properties of different membrane materials is the plot of selectivity versus permeability popularized by Robeson [11, 14]. A Robeson plot for the separation of oxygen and nitrogen is shown in Figure 9.6. The selectivity obtained from the ratio of the pure gas permeabilities of oxygen and nitrogen is plotted against the oxygen pure gas permeability. Each point on the figure represents a different membrane material. A wide range of selectivity/permeability combinations is provided by different polymers; for a gas separation, the most permeable polymers at a particular selectivity are the ones of most interest. The line linking these polymers is called the upper bound, beyond which no better material is currently known. There is a strong inverse relationship between permeability and selectivity. The most permeable membranes with a selectivity of 6 to 7 have 1% of the permeability of membranes with a selectivity of 2 to 3. The relative positions of the upper bound in 2008 and in 1991 show the progress that was made in producing polymers specifically tailored for this separation. Development of better materials goes on, and the position of the upper bound has continued

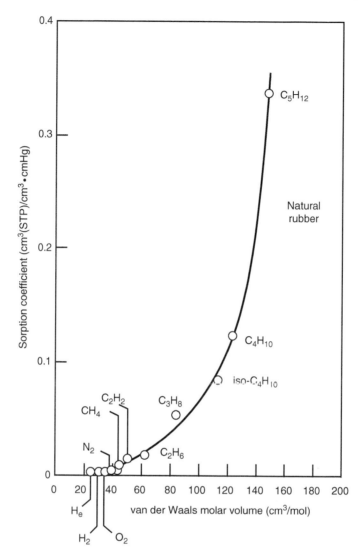

Figure 9.4 *Gas sorption coefficient as a function of molar volume for natural rubber membranes. Larger permeants are more condensable and have higher sorption coefficients. Source: Adapted from [9].*

to move, albeit slowly. What has changed is the development of very finely porous films and membranes from MOFs, carbonized polymers, TR polymers and PIMs, some of which have permeabilities well beyond the upper bound line in the figure.

Robeson plots have been created for other gas pairs, and the upper bound lines for pairs of commercial interest are shown in Figure 9.7. For some separations, it is possible to switch the selectivity by the choice of membrane material. The separation of nitrogen/methane gas mixtures, which is of interest in the processing of natural gas, is an example [15]. The mobility selectivity term D_{N_2}/D_{CH_4} favors permeation of the small molecule, nitrogen (kinetic diameter 3.64 Å), over the larger methane (kinetic diameter 3.80 Å). On the other hand, the sorption selectivity K_{N_2}/K_{CH_4} favors sorption of the more condensable gas, methane (boiling point

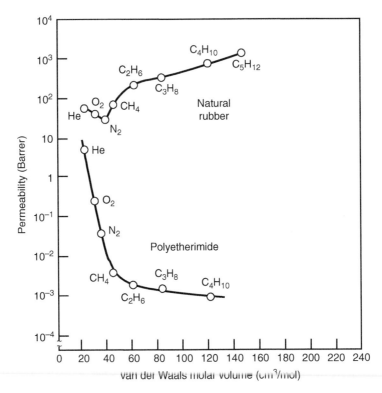

Figure 9.5 Permeability as a function of molar volume for a rubbery and a glassy polymer, illustrating the different balance between sorption and diffusion in these polymer types. The natural rubber membrane is highly permeable; permeability increases rapidly with increasing permeant size because sorption dominates. The glassy polyetherimide membrane is much less permeable; the permeability decreases with increasing permeant size because diffusion dominates. Source: Reproduced with permission from [10]. Copyright (1989) The American Institute of Chemical Engineers.

Table 9.1 Pure-gas permeabilities (Barrer (10^{-10} cm^3(STP) · cm/cm^2 · s · cmHg)) of widely used polymers.

Gas	Rubbers		Glasses		
	Silicone rubber at 25 °C T_g −129 °C	Natural rubber at 30 °C T_g −73 °C	Cellulose acetate at 25 °C T_g 124 °C	Polysulfone at 35 °C T_g 186 °C	Polyimide (Ube Industries) at 60 °C T_g > 250 °C
H_2	550	41	24	14	50
He	300	31	33	13	40
O_2	500	23	1.6	1.4	3
N_2	250	9.4	0.33	0.25	0.6
CO_2	2700	153	10	5.6	13
CH_4	800	30	0.36	0.25	0.4
C_2H_6	2100	—	0.20	—	0.08
C_3H_8	3400	168	0.13	—	0.015
C_4H_{10}	7500	—	0.10	—	—

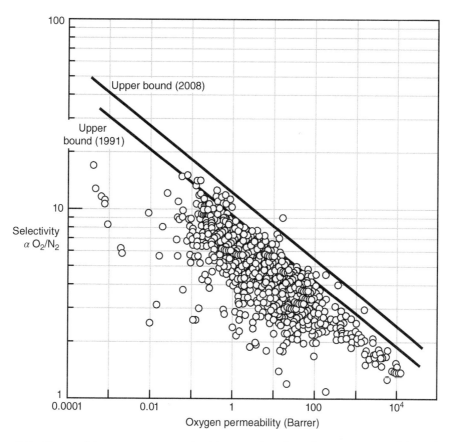

Figure 9.6 *Oxygen/nitrogen pure gas selectivity as a function of oxygen permeability. The plot shows the wide range of selectivity/permeability combinations achieved by polymer materials. Source: Reproduced with permission from [14]. Copyright (2008) Elsevier.*

111 K), over the less condensable gas nitrogen (boiling point 77 K). It follows that the effects of these terms in Eq. (9.4) are opposed. Glassy polymers generally have low permeability and permeate nitrogen preferentially ($\alpha_{N_2/CH_4} > 1$) because the diffusion mobility selectivity term is dominant. Rubbery polymers have higher permeabilities and preferentially permeate methane ($\alpha_{N_2/CH_4} < 1$) because the sorption selectivity term is dominant.

The separation of CO_2 from hydrogen mixtures is another example. Polar rubbery membranes have a CO_2/H_2 solubility selectivity strongly favoring CO_2 and membranes with a CO_2/H_2 selectivity of 10 have been made. Glassy polymers with a very high T_g have exceptionally high diffusion selectivities in favor of hydrogen, and membranes with H_2/CO_2 selectivities of 10 have also been reported.

The upper bound lines shown in Figure 9.7 can be expressed mathematically as

$$\ln \alpha_{A/B} = \ln \beta_{A/B} - \lambda_{A/B} \ln P_A \tag{9.5}$$

or

$$\alpha_{A/B} = \beta_{A/B} / P_A^{\lambda_{A/B}} \tag{9.6}$$

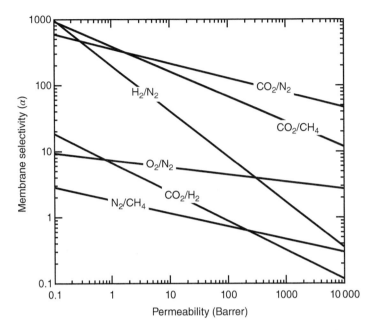

Figure 9.7 *Upper bound lines for commercially important gas separations. This figure allows the trade-off between selectivity and permeability to be estimated for the best available membrane materials.*

where A and B represent the two gases, $\lambda_{A/B}$ is the slope of the line in Figure 9.7, and $\ln\beta_{A/B}$ is the intercept at $\ln P_A = 0$. Freeman [16] has shown that these parameters have physical significance and can be calculated with reasonable accuracy from first principles. The slope $\lambda_{A/B}$ depends on the relative sizes of the gases in the pair, and $\beta_{A/B}$ depends on the gas condensabilities.

Applying the data shown above to actual gas separation problems is not straightforward. Permeabilities used to prepare Robeson plots are measured with pure gases; the selectivity obtained from the ratio of pure gas permeabilities gives the ideal membrane selectivity. However, gas separation processes are performed with gas mixtures. If the gases in a mixture do not interact with the membrane material, the pure gas selectivity and permeability and the mixed gas selectivity and permeability will be similar. This is usually the case for mixtures of oxygen and nitrogen, for example. In many other cases, such as the separation of carbon dioxide/methane mixtures, one of the components (carbon dioxide) is sufficiently sorbed by the membrane to affect the permeability of the other component (methane). The selectivity measured with a gas mixture may then be much less (half or even less) of the selectivity calculated from pure gas measurements. Pure gas selectivities are much more commonly reported in the literature than gas mixture data because they are easier to measure, but neglecting the difference between ideal and real values can lead to a serious overestimate of the ability of a membrane to separate a target gas mixture. Figure 9.8 [17] shows selected data for CO_2/methane separation using cellulose acetate membranes. The calculated pure gas selectivity is very good, but in gas mixtures, enough carbon dioxide dissolves in the membrane to increase the methane permeability far above the pure gas value. As a result, the selectivities measured with gas mixtures are much lower than those calculated from pure gas data.

For mixed gas experiments in the laboratory, a flow of helium sweep gas or a vacuum is sometimes applied to the permeate side of the membrane. Results of this type of experiment

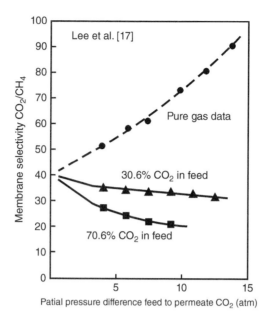

Figure 9.8 *The difference between selectivities calculated from pure gas measurements and selectivities measured with gas mixtures can be large. Data of Lee et al. [17] for carbon dioxide/ methane separation with cellulose acetate films.*

are also poor predictors of industrial performance. Either technique maintains a permeate partial pressure far below that of the permeate gas in an industrial system. The result is that permeance and selectivity calculated from such laboratory data are far higher than will be found in a real industrial process.

9.2.2 Structural Features and Considerations

To achieve economical fluxes, most gas separation processes require that the selective membrane layer be extremely thin. Typical membrane thicknesses are less than 0.5 μm and sometimes as low as 0.1 μm. Early gas separation membranes were made by modifying membranes produced by the Loeb–Sourirajan phase separation process. It is difficult to make these membranes completely defect free, which is unfortunate because gas separation membranes are far more sensitive to minor defects layer than are membranes used in reverse osmosis or ultrafiltration. Even a few tiny defects in the selective layer can allow several percent of the feed to pass unseparated through the membrane, diminishing the separation performance substantially. This sensitivity posed a serious problem to early developers, as generation of a few defects is difficult to avoid during membrane preparation and module formation.

In the late 1970s, Henis and Tripodi [6] at Monsanto devised an ingenious solution. The Monsanto group made polysulfone hollow fiber membranes by the Loeb–Sourirajan technique, then coated the selective skin side of the membranes with a thin layer of silicone rubber. Silicone rubber was chosen because it is much more permeable but much less selective than polysulfone, and hence did not adversely change the selectivity or flux of the defect-free portions of the selective layer. The coating layer did, however, plug defects in the layer very effectively, thereby eliminating convective flow through them. The coating also protected the membrane during handling. The development of silicone-rubber-sealed anisotropic membranes was a critical step

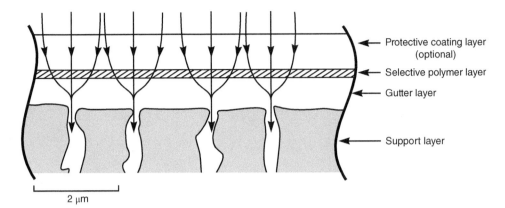

Figure 9.9 *Multilayer composite membrane formed by coating thin layers of polymer onto a microporous support membrane.*

in the production of the first successful gas separation membrane for hydrogen/nitrogen separations by Monsanto.

Another type of gas separation membrane has the multilayer composite structure shown in Figure 9.9. The preparation of these membranes is described in Chapter 4. The base layer that provides the mechanical strength is a finely microporous support membrane. This support is then coated with a series of thin polymer coatings. A gutter layer is often applied first to provide a defect-free, smooth surface, onto which the ultrathin selective layer is deposited. A final protective layer of silicone rubber or other highly permeable polymer seals any defects. It is difficult to make composite membranes with glassy selective layers as thin as good-quality Loeb–Sourirajan membranes, and every layer in the multilayer structure has to be optimized if the final membrane is to have the theoretical permeation properties of the material chosen for the selective layer. However, composite membranes can be made from a much wider range of materials than the Loeb–Sourirajan process allows, and each layer can be made from a material most suited to the functions it must perform. For example, it is possible to make composite membranes that use a rubbery soft polymer as the selective layer, with a tough, glassy polymer as the support. Rubbery composite membranes of this type can withstand pressure differentials of 100 bar or more.

Following the installation of the first membrane hydrogen separation system in 1980, new membranes were developed for different gas separation applications. By the year 2000, membrane gas separation was an established unit operation. There were four large industrial applications; hydrogen from nitrogen or methane in petrochemical and refinery operations, CO_2 from natural gas, nitrogen from air, and light hydrocarbons from nitrogen in various purge streams in petrochemical plants. Approximately 10–15 different polymer membranes were produced by various companies for these applications.

Since 2000, the introduction of new materials and applications has slowed greatly. The four applications commercial in 2000 still represent greater than 90% of the market, with, for the most part, only small modifications to the membranes. Hundreds, even thousands, of polymers have been evaluated in the laboratory and reported in the literature as potentially better membrane materials. None has been shown to be significantly better than those in use under real conditions, nor has any large new application reached the industrial stage.

The lack of new applications is not from a shortage of good targets. Examples of unsolved problems include the separation of propylene/propane mixtures, ethylene/ethane mixtures, CO_2 from CO_2/H_2 mixtures, C_1–C_4 hydrocarbons from hydrogen, and oxygen from air. It is unlikely that further development of conventional polymers will lead to a better result for any of these; new approaches are required. One possibility is to use polymer chemistries sufficiently different from conventional materials that a different result might be possible. For example, ultra-rigid polybenzimidazole-type materials require operation at elevated temperatures to get useful permeances, but can then give impressive selectivities. Fully fluorinated polymers can sometimes have very different separation properties from their partially or non-fluorinated counterparts, owing to the odd sorption properties that result from the all-fluorine chemistry [18, 19]

Facilitated transport membranes offer another possible approach. In these membranes, normal permeation is enhanced by incorporating into the polymer matrix chemical groups that react reversibly with one of the permeating species. Facilitated transport membranes have been around for 50 years, and spectacular separations have been achieved in the laboratory. The unsolved problem is membrane stability. However, a breakthrough is still possible. These membranes are discussed in Chapter 12.

The most promising of the new approaches involves the use of finely microporous membranes. Since these membranes have yet to become commercial for any major gas separation application, they are mentioned only in brief outline in this chapter. Their composition and structure are key to understanding their properties, so these attributes are covered along with the transport mechanism, insofar as it is currently understood, in Chapter 3. Though yet to be proven fitting for industrial separations, they should not be overlooked, and may offer a route to solve some separation problems where conventional membranes have failed.

9.3 Microporous Membranes

The four main categories of microporous membranes applicable to gas separation processes are shown in Figure 9.10. The average pore diameter and the main permeating mechanism are also shown. In brief, the first category is membrane materials with the smallest pores, in the range of

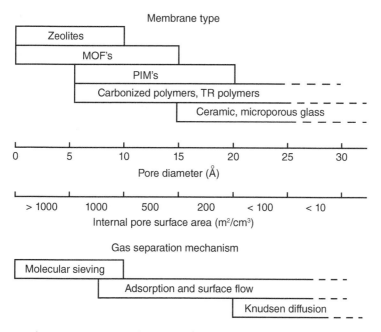

Figure 9.10 *Finely microporous membrane types that can be used for used gas separation.*

3–10 Å in diameter, which are made from zeolites or metal organic frameworks (MOFs), rigid crystal structures that contain a regular network of tiny, connected cavities [20–23]. The cavities are linked by even smaller windows that determine the size of the largest permeating component. Membranes made from these materials separate gas mixtures in the manner of a filter by molecular sieving.

The second category consists of membranes made from polymers of intrinsic microporosity (PIMs). The first membrane of this type was made from poly(1-trimethylsilyl-1-propyne) (PTMSP) [24]. The polymer backbone of PTMSP is extremely sterically hindered, which prevents efficient packing of the polymer chains. The spaces between the polymer chains have diameters of 5–20 Å; there are enough spaces to form a continuous connected network, and almost all gas permeation takes place through the network. Later, a series of polybenzodioxane polymers was synthesized that were even more sterically hindered; these have similar properties to PTMSP [25]. Membranes made from these and other PIMs have exceptionally high permeabilities, sometimes coupled with potentially useful selectivities for gas mixtures. Unfortunately, the sterically hindered movement of the polymer chains that gives the membranes these interesting properties is not completely prohibited. Over time, the polymer chains rearrange and the membranes lose their porosity, and with this, many of their useful high flux properties [19, 26].

The third category is microporous carbon membranes. These membranes were first developed and characterized during the 1950s and 1960s [27]. Membranes were made by compressing high-surface-area carbon powder in a small tube, to form a plug of compressed microporous carbon. The average pore diameter of the plug could be controlled by changing the surface area/g of the carbon powder used. Typical pore diameters were in the 30–50 Å range. Later, Koresh and Soffer [28] produced membranes with a similar structure by carbonizing polymeric films in an inert atmosphere or in vacuo to 500–800 °C. At these temperatures, most polymers lose their hetero atoms, leaving behind a graphitic network that contains pores in the 5–20 Å range. These membranes have high permeances and exceptional selectivities for some gas mixtures; air separation membranes with selectivities for oxygen over nitrogen of between 10 and 12 have been reported, for example.

Carbon membranes are difficult to make and scale-up, and are brittle and easily fouled by trace contaminants in the feed. To better control their mechanical and permeation properties, some recent developers have only partially carbonized the base polymer. The resulting membranes are called TR, or thermally rearranged, membranes. Because the matrices of carbon and TR polymers are made of fused graphite molecules, the membranes are not subject to the densification problems that have hindered the application of PIM membranes [29, 30].

The final category is ceramic and glass membranes, made by various specialized preparation techniques. Like carbon membranes, the membranes are mechanically stable but are easily fouled by contaminants in the feed. Typical membrane pore diameters are greater than about 20 Å. These membranes, though usable for some gas separation applications, are more usually used in ultrafiltration.

The reason for the interest in finely microporous membranes is illustrated in Figure 9.11, where the CO_2/CH_4 permeability and selectivity of most of the polymeric materials that have been measured are compared with a group of microporous TR polymer membranes on a Robeson trade-off plot [14]. Some of the TR polymers have twice the selectivity at equivalent permeabilities of the best polymeric membranes. Similar results are obtained with TR polymers, as well as with MOFs, for a number of other gas mixtures, most notably ethylene/ethane and propylene/propane. The properties of finely microporous membranes are described in more detail in Chapter 3.

Despite these outstanding permeation properties, finely microporous membranes have not yet been brought to the industrial scale. The difficulty of making defect-free thin membranes on a

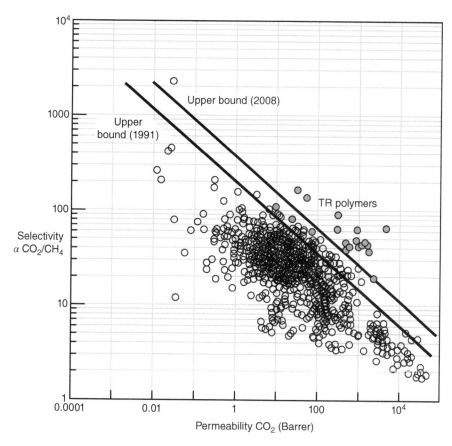

Figure 9.11 *A Robeson trade-off plot for CO$_2$/CH$_4$ separation membranes. The performances of polymeric membranes and TR polymer membranes are compared.*

large scale is an issue, but this problem can probably be overcome, for example by incorporating the microporous material in dispersed form in a polymer matrix, to form a mixed matrix membrane. A more fundamental issue is adsorption of higher boiling components in the feed gas mixture on the interior surface of the membrane pores. This adsorption can result in partial or complete blocking of the pores at constrictions in the porous network. Pore blocking, which significantly affects permeation and selectivity, may be due to one of the major components in the feed gas (carbon dioxide for example in the separation of CO$_2$/CH$_4$ mixtures), but it can also be caused by low levels of C$_{6+}$ hydrocarbons in the CO$_2$/CH$_4$ mixture.

9.4   Membrane Modules

Gas separation membranes are most commonly packaged into spiral-wound or hollow fiber modules. Because the membrane is underpinned by a tough fabric web and a microporous support structure, spiral-wound modules are usually better suited than hollow fibers modules to high-pressure applications. This is especially so when the separation involves condensable components, which could sorb into a hollow fiber and cause the unsupported polymer layer to collapse under pressure. On the other hand, it is possible to pack hollow fibers more densely

Table 9.2 *Module designs used for the most important gas separation applications.*

Application	Typical membrane material	Selectivity (α)	Pressure-normalized flux of most permeable component (gpu)	Typical feed pressure (bar)	Commonly used module designs
N$_2$/Air	Polyimide Polysulfone Brominated Polycarbonate Poly(phenylene oxide)	5–7	O$_2$ 10–30	5–10	Almost exclusively hollow fiber, bore-side feed
H$_2$/N$_2$/CH$_4$	Polyimide Polysulfone	50–100	H$_2$ 100–200	30–60	Generally hollow fiber, shell-side feed
CO$_2$/CH$_4$ (natural gas)	Cellulose acetate	15	CO$_2$ 100–200	30–60	Spirals have 90% market share of the high-pressure market. Hollow fibers with shell-side feed are used in lower pressure applications
Other natural gas applications H$_2$S, He, N$_2$	Various	—	He, H$_2$S 200+	30–60	Mostly spirals
C$_3^+$, VOC, N$_2$, CH$_4$	Silicone rubber	10–30	C$_3^+$, VOC 500–2000	5–20	Mostly spirals, some plate-and-frame

than flat-sheet membranes, thus making available for use a much higher surface area of membrane per unit volume of module. This can be important, since the metal housing of a module is often more costly than the membranes it contains. Finally, hollow fiber membranes can be adapted to a counter-flow process design, a useful attribute in some separations. The balance of these factors means there is no overall preferred module design – rather, different module types are used for different applications, as shown in Table 9.2.

9.5 Process Design

The three factors that determine the gas separation performance of a membrane are illustrated in Figure 9.12. The role of membrane selectivity is obvious; not so obvious are the importance of the ratio of total feed pressure (p_o) to total permeate pressure (p_ℓ), usually called the pressure ratio, φ, and defined as

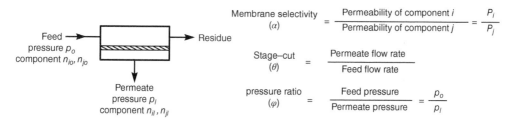

Figure 9.12 *Parameters affecting membrane gas separation performance.*

$$\varphi = \frac{p_o}{p_\ell} \tag{9.7}$$

and of the stage-cut, θ, which is the fraction of the feed gas that permeates the membrane, defined as

$$\theta = \frac{\text{permeate flow}}{\text{feed flow}}. \tag{9.8}$$

In the section that follows, analytical solutions first derived by Weller and Steiner [31] and refined by Pan and Habgood [32] are used to link pressure ratio, stage-cut, and membrane selectivity for a simple binary gas mixture. These analytical solutions can be used to generate the plots that follow; nowadays plots are usually generated using a differential element membrane code incorporated into a computer process simulation package (ChemCad, Aspen, HySys, etc.).

9.5.1 Pressure Ratio

The importance of pressure ratio on separation performance is best explained by a concrete example – the separation of carbon dioxide from a nitrogen/CO_2 mixture, shown in Figure 9.13. The calculation assumes that the membrane unit preferentially permeates only a small portion of CO_2 from a feed gas stream containing 10% CO_2, and that the pressure ratio is 5 (5 bar feed/1 bar permeate). Permeation of CO_2 takes place because the feed gas has a higher CO_2 partial pressure than the permeate, which inequality can be manipulated as shown on the

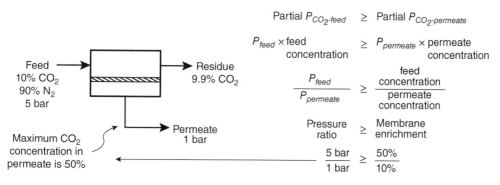

Figure 9.13 *The separation of carbon dioxide from a nitrogen/carbon dioxide mixture. The pressure ratio limits the carbon dioxide concentration in the permeate.*

right of the figure, leading to the result that the ratio of the CO_2 concentration permeate/feed, cannot be greater than the pressure ratio. That is

$$\frac{p_{feed}}{p_{permeate}} \geq \frac{\text{Permeate concentration}}{\text{Feed concentration}} \qquad (9.9)$$

Since the pressure ratio is 5, the maximum concentration of the permeate gas cannot be greater than $5 \times 10\% = 50\%$ CO_2, no matter how selective the membrane. In other words, at least half of the permeate must be the slower permeating component, nitrogen. As a result, the nitrogen permeance, not the CO_2 permeance, determines how much membrane area is needed to perform the separation. This leads to the counterintuitive result that an infinitely selective membrane, which would permeate no nitrogen, would require an infinite membrane area.

The restraining impact of pressure ratio on separation is shown in the graphs of Figure 9.14, which assume the same conditions (10% feed CO_2 and pressure ratio of 5) as Figure 9.13. Figure 9.14a shows the permeate concentration as a function of increasing membrane selectivity. At a selectivity of 1 (no separation), the permeate concentration is the same as the feed. As the selectivity increases, the permeate concentration increases, reaching about 40% CO_2 at a selectivity of 40. At this point, the separation is already significantly constrained by the low pressure ratio. As the selectivity increases further, the permeate concentration asymptotically approaches the limiting value of 50%.

Figure 9.14b shows the relative membrane area required to permeate a fixed amount of CO_2 at the same conditions as Figure 9.14a. This plot has an exponential form; the required membrane area increases rapidly as the selectivity increases. It follows that at a pressure ratio of 5, the optimum selectivity under the chosen conditions is between 20 and 40. A higher selectivity produces a slight increase in permeate CO_2 concentration but at the expense of a very large increase in membrane area (and cost).

The relationship between pressure ratio and membrane selectivity can be obtained as an analytical equation from the Fick's law expressions for the fluxes of components i and j

$$j_i = \frac{P_i \left(p_{i_o} - p_{i_\ell} \right)}{\ell} \qquad (9.10)$$

and

$$j_j = \frac{P_j \left(p_{j_o} - p_{j_\ell} \right)}{\ell}. \qquad (9.11)$$

The total gas pressures on the feed and permeate side are the sum of the partial pressures. For the feed side

$$p_o = p_{i_o} + p_{j_o} \qquad (9.12)$$

and for the permeate side

$$p_\ell = p_{i_\ell} + p_{j_\ell} \qquad (9.13)$$

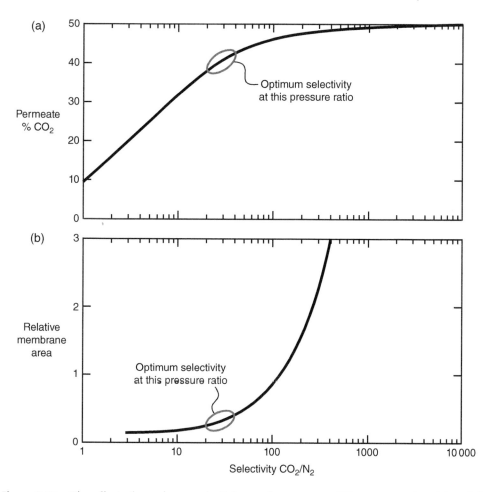

Figure 9.14 *The effect of membrane selectivity on the permeate CO_2 concentration (a) and the relative membrane area (b) for the separation process illustrated in Figure 9.13. In these calculations, the CO_2 permeance is fixed and the selectivity is varied by changing the nitrogen permeance. The feed CO_2 concentration is 10%, as in Figure 9.13. The optimum trade-off between permeate concentrations and required membrane area for this separation occurs around a selectivity of 20–40.*

The volume fractions of components i and j on the feed and permeate side are also related to partial pressures. For the feed side

$$n_{i_o} = \frac{p_{i_o}}{p_o} \quad n_{j_o} = \frac{p_{j_o}}{p_o} \tag{9.14}$$

and for the permeate side

$$n_{i_\ell} = \frac{p_{i_\ell}}{p_\ell} \quad n_{j_\ell} = \frac{p_{j_\ell}}{p_\ell} \tag{9.15}$$

From mass balance considerations

$$\frac{j_i}{j_j} = \frac{n_{i_\ell}}{n_{j_\ell}} = \frac{n_{i_\ell}}{1 - n_{i_\ell}} = \frac{1 - n_{j_\ell}}{n_{j_\ell}} \tag{9.16}$$

Dividing Eq. (9.10) by Eq. (9.11) and using the definition of α in Eq. (9.3), Eqs. (9.14) through (9.16) lead to

$$n_{i_\ell} = \frac{1}{2} \cdot \frac{n_{i_o} + \frac{1}{\varphi} + \frac{1}{\alpha-1} - \sqrt{\left(n_{i_o} + \frac{1}{\varphi} + \frac{1}{\alpha-1}\right)^2 - \frac{4 \cdot \alpha \cdot n_{i_o}}{(\alpha-1) \cdot \varphi}}}{\frac{1}{\varphi}} \tag{9.17}$$

This awkward expression breaks down into two limiting cases, depending on the relative magnitudes of the pressure ratio and the membrane selectivity. First, if the membrane selectivity is very much larger than the pressure ratio, that is,

$$\alpha \gg \varphi \tag{9.18}$$

Then Eq. (9.17) becomes

$$n_{i_\ell} = n_{i_o} \varphi \tag{9.19}$$

The performance is determined by the pressure ratio across the membrane and is independent of the membrane selectivity; this region is the pressure-ratio-limited region.

If the membrane selectivity is very much smaller than the pressure ratio, that is,

$$\alpha \ll \varphi \tag{9.20}$$

Then Eq. (9.17) becomes (after some manipulation and application of the rule of L'Hôpital)

$$n_{i_\ell} = \frac{\alpha \cdot n_{i_o}}{n_{i_o} \cdot (\alpha-1) + 1} \tag{9.21}$$

The performance is determined only by the membrane selectivity and is independent of the pressure ratio; this is the selectivity-limited region. There is, of course, an intermediate region between these two limiting cases, in which both the pressure ratio and the selectivity affect the system performance. These three regions are illustrated in Figure 9.15, in which the calculated permeate concentration (n_{i_ℓ}) is plotted versus pressure ratio for a membrane with a selectivity of 30 [33]. At a pressure ratio of 1, the feed pressure is equal to the permeate pressure, and no separation is achieved. As the difference between the feed and permeate pressures increases, the concentration of the more permeable component in the permeate gas begins to increase, first according to Eq. (9.19) and then, when the pressure ratio and membrane selectivity are comparable, according to Eq. (9.17). At very high pressure ratios, that is, when the pressure ratio is four to five times higher than the membrane selectivity, the process enters the selectivity-controlled region. In this region, the permeate concentration reaches the limiting value given by Eq. (9.21).

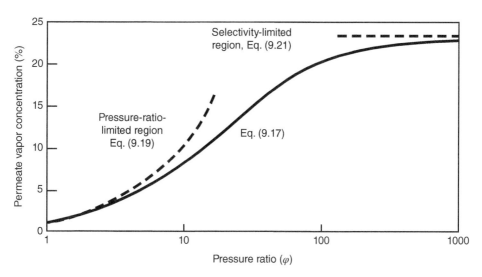

Figure 9.15 *Calculated permeate vapor concentration for a vapor-permeable membrane with a vapor/nitrogen selectivity of 30 as a function of pressure ratio. The feed vapor concentration is 1%. Below pressure ratios of about 10, separation is limited by the pressure ratio across the membrane. At pressure ratios above about 100, separation is limited by the membrane selectivity [33] / Taylor & Francis.*

The relationship between pressure ratio and selectivity is important because pressure ratio is not a freely adjustable parameter in practical separation processes. Compressing the feed stream to a high pressure or drawing a vacuum on the permeate side of the membrane to achieve a large pressure ratio requires both large amounts of energy and expensive equipment. As a result, typical practical pressure ratios are in the range 5–20. Even at these pressures, the capital cost of the compression equipment often surpasses that of the membrane unit. Likewise, the power cost to run compressors and pumps easily exceeds the module replacement costs and other operating expenses.

Because the attainable pressure ratio is limited, the benefit of highly selective membranes is often less than might be expected. This was seen in Figure 9.14a and is apparent again from Figure 9.16, where the calculation is based on the data for the separation of a typical organic vapor from nitrogen. The feed vapor concentration is assumed to be 1% and the pressure ratio is fixed at 20. Increasing the membrane selectivity from 10 to 20 will significantly improve system performance, but a much smaller incremental improvement results from increasing the selectivity from 20 to 40. Increases in selectivity above 100 will produce negligible improvements but, by the reasoning above, will produce a significant increase in the membrane area required to perform the separation. A selectivity of 100 is five times the pressure ratio of 20, placing the system in the pressure-ratio-limited region. Huang et al. have given a detailed description of the effect of pressure ratio on a number of practical applications [34].

9.5.2 Stage-Cut

The ideal target of most separation processes is a complete separation; creation of a permeate stream containing all of the most permeable component but none of the less permeable component, and concurrently a residue stream containing all of the less permeable components but none of the more permeable component; that is, 100% component purity and 100% component recovery. These two goals cannot be met simultaneously; a trade-off must be made. The process attribute that controls this trade-off is the stage-cut.

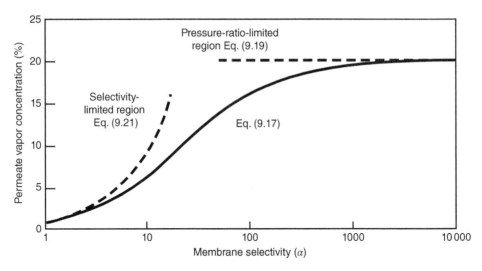

Figure 9.16 *Calculated permeate vapor concentration as a function of selectivity. The feed vapor concentration is 1%; the pressure ratio is fixed at 20. Below a vapor/nitrogen selectivity of about 10, separation is limited by the low membrane selectivity; at selectivities above about 100, separation is limited by the low pressure ratio.*

Analytical expressions linking stage cut with performance have been developed for simple binary mixtures. These expressions are clumsy for routine use; all industrial membrane producers have developed their own simulation software. The techniques used to create these programs have been described in the literature [35, 36].

The effect of stage-cut on module performance calculated with a differential element software for different stage-cuts is shown in Figure 9.17.

In the example of Figure 9.17, the feed gas contains 50% of a permeable gas A and 50% of a relatively impermeable gas B. Under the assumed operating conditions of the system (pressure ratio 20, membrane selectivity 20), it is possible, by keeping the stage-cut close to zero, to produce a permeate stream containing almost the maximum possible 95% of component A. But the permeate stream is small, meaning the recovery of A is very low and the residue stream is still close to the feed gas concentration. As the fraction of the feed gas permeating the membrane (that is the stage-cut) is increased by increasing the membrane area, the concentration of the permeable component in the residue stream falls. At a stage-cut of 25%, a quarter of the total feed gas has permeated the membrane, the permeate gas concentration of A remains high at 93.1%, but only about 46% of component A is in the permeate. Increasing the stage-cut to 50% by adding more membrane area produces a residue stream containing 88.2% B and a permeate that contains 88.2% A. If the stage-cut is increased further to 75%, by adding even more membrane area, the concentration of A in the residue stream is reduced to only 0.04% and the residue is almost pure component B. But the recovery of B in the residue is only 49.5%. In comparison, the recovery of component A in the permeate is almost 100% but the concentration of A has only been enriched to 66.7%.

The calculations in Figure 9.17 also show that increasing the stage-cut to produce a pure residue stream requires a disproportionate increase in membrane area. As the feed gas is stripped of the more permeable component, the average permeation rate through the membrane falls toward the permeation rate of the slow gas. In the example shown, permeating the first 25%

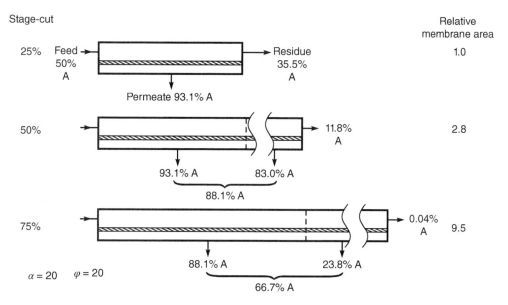

Figure 9.17 *The effect of stage-cut on the separation of a 50/50 feed gas mixture of components A and B. Pressure ratio and selectivity are both set to 20. At low stage-cuts, a permeate product with a high concentration of A can be obtained, but only modest removal of A from the feed is achieved. At high stage-cuts, an essentially A-free residue is obtained, but the permeate is only slightly more enriched in A than the original feed. Required membrane area increases disproportionately with increasing stage-cut.*

of the feed gas requires a relative membrane area of 1. For the next 25%, nearly twice as much membrane area is needed. The third 25% uses nearly 7 times the membrane area of the first 25%, so to run at 75% stage-cut uses about 10 times as much membrane area as to run at 25%.

9.5.3   Multistep and Multistage System Designs

The solution to the trade-off problem is to use a multistep or multistage process. By way of example, process designs for treatment of a nitrogen feed gas containing 1% VOC (volatile organic compound) are compared below. All calculations assume a VOC/nitrogen selectivity of 20 (typical of silicone rubber) and a pressure ratio of 20, achieved by compressing the feed gas to 20 bar.

Figure 9.18, the one-stage option, shows that when 90% of the VOC in the feed stream is removed, the permeate stream will contain approximately 4% of the more permeable component.

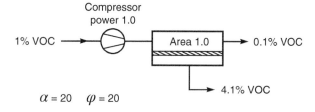

Figure 9.18 *One-stage vapor separation. Performance was calculated for a cross-flow module using a VOC/nitrogen selectivity of 20 and a pressure ratio of 20.*

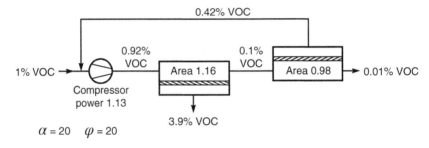

Figure 9.19 *A two-step process achieves 99% VOC removal from the feed stream, but increases the power consumption and doubles the membrane area.*

With this simple design, it may well be that either the residue stream or the permeate stream, or both, will fail to meet specifications for discharge or reuse.

If the main problem is insufficient VOC removal from the feed stream, a two-step system, as shown in Figure 9.19, can be used. The residue stream from the first membrane unit is passed to a second unit, where the VOC concentration is reduced by a further factor of 10, from 0.1% to 0.01%. Because the concentration of VOC in the feed to the second membrane unit is low, the permeate stream is relatively dilute and is recirculated to the feed stream.

A multistep design of this type can achieve almost complete removal of the permeable component from the feed stream to the membrane unit. However, greater removal of the permeable component is achieved at the expense of increases in membrane area and power consumption by the compressor. As a rule of thumb, the membrane area required to remove the last 9% of a component from the feed equals the membrane area required to remove the first 90%.

Sometimes, 90% removal of the permeable component from the feed stream is acceptable, but a higher concentration is needed to make the permeate gas usable. In this situation, a two-stage system of the type shown in Figure 9.20 is used. In a two-stage design, the permeate from the first membrane unit is recompressed and sent to a second membrane unit, where a further separation is performed. The final permeate is then twice enriched. In the most efficient two-stage designs, the residue stream from the second stage is reduced to about the same concentration as the original feed gas, with which it is mixed. In the example shown in Figure 9.20, the permeate stream,

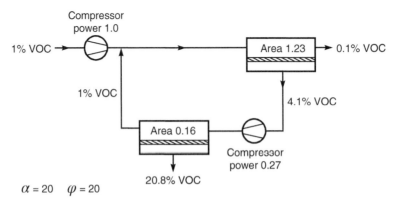

Figure 9.20 *A two-stage system produces a highly concentrated permeate stream with moderate increases in membrane area and power consumption.*

concentrated a further fivefold, leaves the system at a concentration of 21%. Because the volume of gas treated by the second stage membrane unit is much smaller than in the first stage, the membrane area of the second stage is relatively small. Incorporation of a second stage increases the overall membrane area by about 40% and the power consumption by about 30%.

Designs with cascades of stages or steps are possible, but are seldom used in commercial systems because their complexity makes them uncompetitive with alternative technologies. Relatively simple recycle designs, limited to a couple of stages and/or steps, are more practical.

9.5.4 Recycle Designs

A recycle design proposed by Wijmans [37] combining two stages and two steps (sometimes called a two-and-a-half-stage system), is shown in Figure 9.21, again assuming selectivity and pressure ratio of 20. In this design, the permeate from the first stage is recompressed and sent to a second stage, whence a portion of the gas is removed as an enriched permeate product, as in the two-stage design of Figure 9.20. The residue gas from this stage passes to a second step, which brings the gas concentration close to the original feed value. The permeate from this stage is mixed with the first stage permeate, forming a recycle loop. By controlling the relative size of the two second stages, any desired concentration of the more permeable component can be achieved in the product. In the example shown, the permeable component is concentrated to 50% in the permeate. The increased performance is achieved at the expense of a slightly larger second stage compressor and more membrane area. Normally, however, this design is preferable to a more complex three-stage system, which would require a third compressor.

Figure 9.22 shows another type of recycle design, in which a recycle loop increases the concentration of the permeable component to the point at which it can be removed by a second process, most commonly condensation [38]. Process parameters are as above. After compression to 20 bar, the feed gas passes through a condenser at 30 °C, but the VOC content is still below the condensation concentration at this temperature. The membrane unit separates the gas into a VOC-depleted residue stream, which is bled from the loop, and a vapor-enriched permeate stream, which is recirculated to the front of the compressor. Because the bulk of the vapor is recirculated, the concentration of vapor in the loop increases rapidly until the pressurized gas entering the condenser exceeds the vapor dew point of 6.1%. At this point, the system is at steady

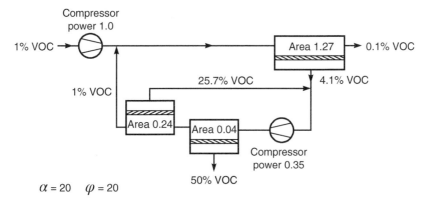

Figure 9.21 *Two-and-one-half-stage system. By forming a recycle loop around the second stage, a small, 50-fold concentrated product stream is created. Source: Adapted from [37].*

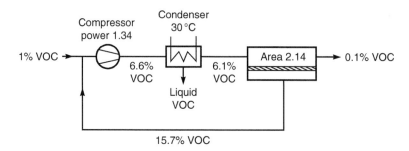

Figure 9.22 *Recycle design using one membrane stage, preceded by a compressor and condenser. The VOC is assumed to be pentane, and the loop serves to build up the pentane concentration in the feed gas until condensation occurs.*

state; the mass of VOC entering the recirculation loop is equal to the mass discharged in the residue stream plus the mass removed as liquid condensate.

Recycle designs of this type are useful in applications in which the components of the gas mixture, if sufficiently concentrated, can be separated from the gas by some other technique. With organic vapors, condensation is often possible; adsorption, chemical scrubbing, or absorption can also be used. The process shown in Figure 9.22 is used to separate VOCs from nitrogen and air, or to separate propane, butane, pentane, and higher hydrocarbons from natural gas (methane).

The examples of Figure 9.18 through Figure 9.22 assume that the modules operate in cross flow; the improved separation performance achieved by counter-flow does not normally compensate for the extra cost of fabrication and use of this type of module. However, some special cases do exist where this type of module can offer real benefits. The dehydration of air and natural gas with counter-flow sweep modules, discussed in Chapter 4, are examples.

9.6   Applications

The membrane gas separation industry is growing and changing. Two of the largest industrial gas companies now have membrane affiliates: Air Products (Permea) and Air Liquide (Medal). The affiliates focus mainly on producing membrane systems to separate nitrogen from air, but also produce hydrogen separation systems. Ube (Japan) and Aquillo (The Netherlands) are active in these same markets. Another group of companies – UOP (GMS/Separex), Schlumberger (Cynara), Evonik, and MTR – produce membrane systems for natural gas separations. Systems for hydrocarbon recovery from petrochemical vent gases are offered by MTR and Borsig. The following section covers the major current applications. Overview articles can be found in several recent reviews [39–42].

9.6.1   Hydrogen Separation

The first large-scale commercial application of membrane gas separation was to treat ammonia plant purge gas streams, to remove inert contaminants from the reactor loop. The process, launched in 1980 by Monsanto, was soon followed by others, such as hydrogen/methane separation in refinery off-gases and hydrogen/carbon monoxide ratio adjustment in oxo chemical synthesis plants [43]. Hydrogen is a small, noncondensable gas, to which most polymers, especially glassy polymers, are highly permeable; hydrogen fluxes, and selectivities obtained with

Table 9.3 Hydrogen separation membranes.

Membrane (Developer)	Selectivity			Hydrogen pressure-normalized flux [10^{-6} cm^3(STP)/cm$^2 \cdot$ s $\cdot$ cmHg]
	H$_2$/CO	H$_2$/CH$_4$	H$_2$/N$_2$	
Polyaramide (Medal)	100	>200	>200	100
Polysulfone (Permea)	40	80	80	100
Cellulose acetate (Separex)	30–40	60–80	60–80	200
Polyimide (Ube)	50	100–200	100–200	80–200

representative glassy membranes are shown in Table 9.3. With fluxes and selectivities as high as these, it is easy to understand why hydrogen separation was the first commercial gas separation process to be developed. Early plants used polysulfone or cellulose acetate membranes, but now specifically synthesized materials, such as polyimides (Ube) and polyaramide (Medal) are used.

A typical flow scheme for recovery of hydrogen from an ammonia plant purge gas stream is shown in Figure 9.23. A photograph of such a system is shown in Figure 9.24. During the

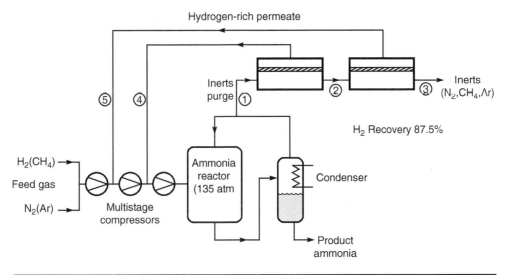

	Stream composition (%)				
	Membrane feed ①	Intermediate feed ②	Membrane vent ③	High-pressure permeate ④	Low-pressure permeate ⑤
Hydrogen	62	44.9	21	87.3	84.8
Nitrogen	21	30.9	44	7.1	8.4
Methane	11	16.2	23	3.6	4.3
Argon	6	8.9	13	2.0	2.5
Pressure (atm)	135	133	132	70	28
Flow (scfm)	2000	1170	740	830	430

Figure 9.23 Simplified flow schematic of the PRISM® membrane system to recover hydrogen from an ammonia reactor purge stream. A two-step membrane system is used to reduce permeate compression costs.

Figure 9.24 *Photograph of an Air Products and Chemicals, Inc. PRISM® system installed at an ammonia plant. The membrane modules are mounted vertically.*

production of ammonia from nitrogen and hydrogen, argon enters the high-pressure ammonia reactor as an impurity with the nitrogen stream, and methane enters the reactor as an impurity with the hydrogen. Ammonia produced in the reactor is removed by condensation, so the argon and methane impurities accumulate until they represent as much as 15% of the gas in the reactor. To control the concentration of these components, the reactor must be continuously purged. The hydrogen lost with this purge gas can represent 2–4% of the total hydrogen consumed. Ammonia plants are very large, so recovery of purged hydrogen for recycle to the reactor is economically worthwhile.

In the process shown in Figure 9.23, a two-step membrane design is used to reduce the cost of recompressing the hydrogen permeate stream to the very high pressures of ammonia reactors. In the first step, the feed gas to the membrane is maintained at the reactor pressure of 135 atm, and the permeate is maintained at 70 atm, a pressure ratio of 1.9. The hydrogen concentration in the feed to this first step is about 62%, high enough that even at this low pressure ratio, the initial permeate contains more than 90% hydrogen, good enough to return to the reactor. However, as hydrogen is removed from the feed gas, the feed concentration falls and, consequently, so does the permeate concentration. When the feed hydrogen concentration is reduced to about 45%, hydrogen, the permeate is no longer high enough for recycle to the reactor. This remaining hydrogen is recovered in a second membrane step operated at a lower permeate pressure of 28 atm; the pressure ratio is then 4.7. The increased pressure ratio increases the hydrogen concentration in the permeate significantly. By dividing the process into two steps operating

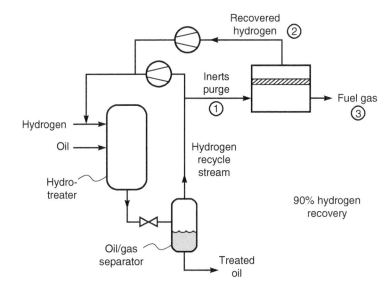

	Stream composition (mole%)		
	Untreated purge ①	Recovered hydrogen ②	Treated purge ③
Hydrogen	82	96.5	34.8
Methane	12	2.6	43.3
Ethane	4.6	0.7	17.1
Propane	1.2	0.2	4.8
Pressure (psig)	1800	450	1450
Flow (MMscfd)	18.9	14.5	4.4

Figure 9.25 *Hydrogen recovery from a hydrotreater used to lower the molecular weight of a refinery oil stream. The calculation assumes the use of PRISM® membranes [43]/ US Department Energy.*

at different pressure ratios, maximum hydrogen recovery is achieved at minimum permeate hydrogen recompression costs.

A second major application is recovery of hydrogen from off gases produced in refinery operations [7, 43, 44]. A typical separation – treatment of the high-pressure purge gas from a hydrotreater – is shown in Figure 9.25. The process is designed to recycle hydrogen to the hydrotreater. As in the case of the ammonia plant, there is a trade-off between the concentration of hydrogen in the permeate and the permeate pressure and subsequent cost of recompression. In the example shown, a permeate of 96.5% hydrogen, achievable at a pressure ratio of 4, is considered adequate.

9.6.2 Air Separation

A major use of gas separation today is in the production of nitrogen from air. The first membranes used for this process were based on poly(4-methyl-1-pentene) (TPX) or ethyl cellulose. These materials have oxygen/nitrogen selectivities of only 4, making the economics marginal. The second-generation materials now used have selectivities of about 6–7, providing favorable

Table 9.4   *Permeabilities and selectivities of polymers of interest in air separation.*

Polymer	Oxygen permeability (Barrer)	Nitrogen permeability (Barrer)	Oxygen/ nitrogen selectivity
Poly(1-trimethylsilyl-1-propyne) (PTMSP)	7600	5400	1.4
Teflon AF 2400	1300	760	1.7
Silicone rubber	600	280	2.2
Poly(4-methyl-1-pentene) (TPX)	30	7.1	4.2
Poly(phenylene oxide) (PPO)	16.8	3.8	4.4
Ethyl cellulose	11.2	3.3	3.4
6FDA-DAF (polyimide)	7.9	1.3	6.2
Polyaramide	3.1	0.46	6.8
Tetrabromobisphenol A polycarbonate	1.4	0.18	7.5
Polysulfone	1.1	0.18	6.2

economics, especially for small plants, producing 5–500 scfm of nitrogen. In this range, membranes are the low-cost process, and most new plants use membrane systems.

Table 9.4 lists the permeabilities and selectivities of some materials that are used or have been used for this separation. There is a sharp trade-off between permeability and selectivity. This trade-off was illustrated in Figure 9.6.

The effect of improved membrane selectivities on the efficiency of nitrogen production from air is illustrated in Figure 9.26. This figure shows the fraction of nitrogen in the feed gas recovered as nitrogen product gas as a function of the nitrogen concentration in the product gas. All oxygen-selective membranes, even membranes with an oxygen/nitrogen selectivity as low as 2, can produce better than 99% nitrogen, albeit at very low recoveries. The figure also shows the significant improvement in nitrogen recovery that results from an increase in oxygen/ nitrogen selectivity from 2 to 6, and that would result if membranes with even higher selectivities could be used.

The first systems, using TPX membranes with a selectivity of about 4, were incorporated into one-stage designs as shown in the figure to produce 95% nitrogen for blanketing storage tanks containing flammable liquids, but they were soon displaced by second generation membranes with selectivities of 6–7. In 2008, a significant new market for inerting systems was created when the Federal Aviation Authority (FAA) ruled that the fuel tanks of all large passenger jets must be fitted with such systems. The ruling had its origins in the crash of a Boeing 747 jet over Long Island, NY in 1996, the cause of which was an explosion in one of the fuel tanks of the plane. Blanketing with nitrogen reduces the oxygen level of air in the tanks to less than the lower explosive limit (12% oxygen). Hollow fiber membrane units are light and compact, and almost all modern jets are fitted with them. Since there are about 30 000 jets in current use, the total market is large and now exceeds the industrial nitrogen-membrane market in dollar value.

Inerting requires only a moderate reduction in oxygen content, such as could be achieved with early membranes and a simple one-step design. As membranes improved, more complex process designs, of the type shown in Figure 9.27, started to be used to produce higher purity gas. The first improvement was the two-step process. As oxygen is removed from the air passing through the membrane modules, the oxygen concentration in the permeating gas falls. At some point, the oxygen concentration in the permeate gas is less than the concentration in normal ambient feed air. Mixing this oxygen-depleted gas permeate with the incoming air then becomes worthwhile. The improvement is most marked when the system is used to produce high-quality nitrogen

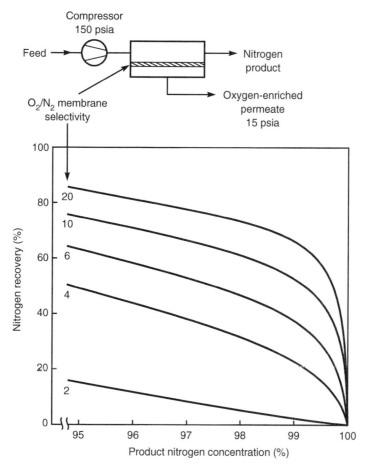

Figure 9.26 *Nitrogen recovery as a function of product nitrogen concentration for membranes with an O₂/N₂ selectivity between 2 and 20.*

containing less than 1% oxygen. In the example shown in Figure 9.27, the second-step permeate gas contains 12.5% oxygen, and recycling this gas to the incoming feed air reduces the membrane area and compressor load by about 6%. This relatively small saving is worthwhile because it is achieved at essentially no cost, by making a simple piping change to the system. In the two-step design, the 12.5% oxygen permeate recycle stream is mixed with ambient air containing 21% oxygen. A more efficient design would be to adjust the recycle gas composition to have about the same oxygen content as the incoming air, as in the three-step design of the figure. This saves a further 2% in membrane area and compressor power, but requires two separate compressors, so it not generally used. A discussion of factors affecting the design of nitrogen plants is given by Prasad et al. [45, 46].

Membranes compete with other technologies for nitrogen production. Competitive ranges for the various methods are shown in Figure 9.28. Very small users usually purchase gas cylinders or delivered liquid nitrogen, but once consumption exceeds 5000 scfd, membranes become the low-cost process. This is particularly true if the required nitrogen purity is between 95% and 99% nitrogen. Membrane systems can still be used if high quality nitrogen (up to 99.9%) is

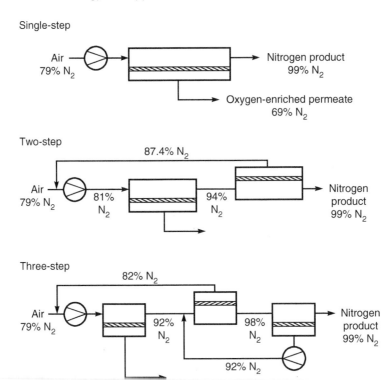

Single-step

Air
79% N$_2$ → Nitrogen product
99% N$_2$

Oxygen-enriched permeate
69% N$_2$

Two-step

87.4% N$_2$

Air
79% N$_2$ 81% N$_2$ 94% N$_2$ Nitrogen product 99% N$_2$

Three-step

82% N$_2$

Air
79% N$_2$ 92% N$_2$ 98% N$_2$ Nitrogen product 99% N$_2$

92% N$_2$

Design	Relative membrane area	Relative compressor HP
One-step	1.0	1.0
Two-step	0.94	0.94
Three-step	0.92	0.92

Figure 9.27   *Single and multistep designs for nitrogen production from air.*

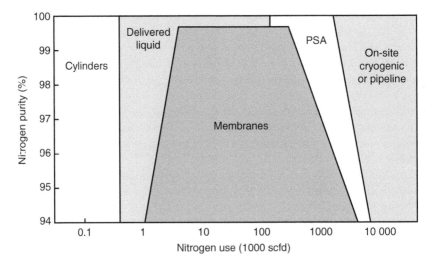

Figure 9.28   *Competitive ranges for nitrogen supply options.*

required, but the cost of the system increases significantly. Very large nitrogen users – above 10 MMscfd of gas – generally use pipeline gas or on-site cryogenic systems. Pressure swing adsorption (PSA) systems are also used in the 1–10 MMscfd range. The ranges shown are good general indicators, but site-specific factors may trump them on a case-by-case basis.

A membrane process to separate nitrogen from air inevitably produces oxygen-enriched air as a by-product. Sometimes this by-product gas, containing about 35% oxygen, can be used beneficially, but usually it is vented. Markets for pure oxygen and oxygen-enriched air exist, but because oxygen is produced as the permeate stream, it is more difficult to produce high-purity oxygen than high-purity nitrogen with membrane systems. Figure 9.29 shows the maximum permeate oxygen concentration that can be produced by a one-step membrane process, using membranes of various selectivities. Even at zero stage-cut and an infinite pressure ratio, the best currently available membrane, with an oxygen/nitrogen selectivity of 8, can only produce 68% oxygen. At useful stage-cuts and affordable pressure ratios this concentration falls. These constraints limit membrane systems to the production of oxygen-enriched air in the 30–40% oxygen range.

Oxygen-enriched air is used in the chemical industry, in refineries, and in various fermentation and biological digestion processes, but it must be produced very cheaply for these applications. The competitive technology is pure oxygen produced cryogenically, then diluted with atmospheric air. The quantity of pure oxygen that must be blended with air to produce the desired oxygen enrichment determines the cost. This means that in membrane systems producing oxygen-enriched air, only the fraction of the oxygen above 21% can be counted as a credit. This fraction is called the equivalent pure oxygen (EPO_2) basis. A comparison of the cost of oxygen-enriched air produced by membranes and by cryogenic separation shows that current

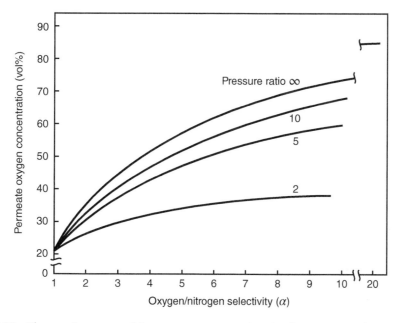

Figure 9.29 *The maximum possible oxygen concentration in the permeate from a one-step membrane process with membranes of various selectivities. The best current membrane materials, with a selectivity of 8, can only produce 68% oxygen, even in the idealized case of zero stage-cut and infinite pressure ratio.*

membranes are generally uncompetitive. The only exception is for very small users in isolated locations, where the logistics of transporting liquid oxygen to the site increase the oxygen cost to US$80–100/ton.

Development of better membranes for producing oxygen-enriched air has been, and continues to be, an area of research, because of the potential application of the gas in combustion processes. When methane, oil, and other fuels are burned with air, a large amount of nitrogen passes as an inert diluent through the burners and is discarded as hot exhaust gas. If oxygen-enriched air were used, the energy lost with the hot exhaust gas would decrease considerably and fuel consumption would decrease. The useful energy that can be extracted from the same amount of fuel increases significantly even if air is enriched only to 25–35% oxygen.

To make these oxygen enrichment applications viable economically, the fuel savings achieved must offset the cost of the oxygen-enriched air. Calculations show that the process would be cost-effective for some applications at an EPO$_2$ cost as high as US$60/ton and, for many applications, at an EPO$_2$ cost of US$30–40/ton [47]. For membranes to produce oxygen at these costs requires optimistic assumptions that are at the outer limit of current technology, but could be feasible with improvements in the technology.

9.6.3 Natural Gas Separations

US production of natural gas is about 20 trillion scf/year; total worldwide production is about 100 trillion scf/year. Before it can be delivered to the pipeline, all of this gas requires some treatment, if only to remove water, and approximately 20% requires extensive treatment. As a result, equipment worth several billions of dollars is installed annually worldwide. The current membrane market share is about 5–10%, essentially all for carbon dioxide removal. This fraction is slowly increasing as better carbon-dioxide-selective membranes are developed and the application of membranes to other separations in the natural gas processing industry becomes more widespread [48, 49].

Raw natural gas varies substantially in composition from source to source. Methane is always the major component, typically 75–90% of the total. Natural gas also contains significant amounts of ethane, some propane and butane, and 1–3% of other higher hydrocarbons. These C$_2$–C$_8$ hydrocarbons are usually more valuable as separated natural gas liquids (NGLs) rather than being left in the gas as a fuel. Natural gas is also the world source of helium, typically present at very low levels (less than 100 ppm) but occurring up to as much as 0.5–1.0% in a few exceptional fields. In addition, the gas contains undesirable impurities: water, carbon dioxide, nitrogen, and hydrogen sulfide. Although raw natural gas has a wide range of compositions, the composition of gas delivered to the pipeline is tightly controlled. Typical US natural gas specifications are shown in Table 9.5. The opportunity for membranes lies in the processing of gas to meet these specifications.

Natural gas is usually produced and transported to the gas processing plant at high pressure, in the range 30–90 bar. To minimize recompression costs, the membrane process must remove

Table 9.5 *Composition of natural gas required for delivery to the US national pipeline grid.*

Component	Specification	Fraction of raw gas out of specification
CO$_2$	<2%	10–15%
H$_2$O	<120 ppm	all
H$_2$S	<4 ppm	about 15%
C$_{3+}$ content	950–1050 Btu/scf Dew point, −20 °C	lots
Nitrogen	<4%	about 10%

impurities from the gas into the permeate stream, leaving the methane, ethane, and other hydrocarbons in the high-pressure residue gas. This requirement determines the type of membranes that can be used. Figure 9.30 shows the molecular size and condensability of components of natural gas. Both parameters affect selection of membranes for natural gas separations.

As Figure 9.30 shows, water is small and condensable; therefore, it is easily separated from methane by both rubbery and glassy polymer membranes. Both rubbery and glassy membranes can also separate carbon dioxide and hydrogen sulfide from natural gas. However, in practice, carbon dioxide is best separated by glassy membranes (utilizing size selectivity), whereas hydrogen sulfide, which is larger and more condensable than carbon dioxide, is best separated by rubbery membranes (utilizing sorption selectivity). Nitrogen can be separated from methane by glassy membranes utilizing the difference in size, or rubbery membranes using the difference in sorption. In both cases, the differences are small, so the membrane selectivities are low. Finally, propane and other hydrocarbons, because of their condensability, are best separated from methane by rubbery membranes. Table 9.6 shows typical membrane materials and the selectivities that can be obtained with good-quality membranes under normal natural gas processing conditions.

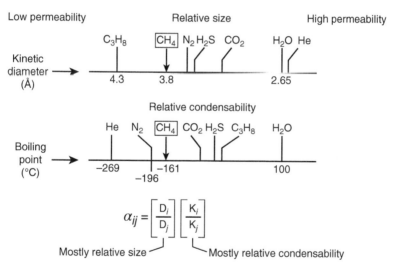

Figure 9.30 *The relative size and condensability (boiling point) of the principal components of natural gas. Glassy membranes generally separate by differences in size; rubbery membranes separate by differences in condensability.*

Table 9.6 *Membrane materials and selectivities for separation of impurities from natural gas under normal operating conditions.*

Component to be permeated	Category of preferred polymer material	Typical polymer used	Typical selectivity over methane
CO_2	Glass	Cellulose acetate, polyimide	10–20
H_2S	Rubber	Ether-amide block copolymer	20–30
N_2	Glass	Polyimide, perfluoro polymers	2–3
N_2	Rubber	Silicone rubber	0.35
H_2O	Rubber or glass	Many	>200
Butane	Rubber	Silicone rubber	7–10

9.6.3.1   Carbon Dioxide Separation

Removal of carbon dioxide is the only membrane-based natural gas separation process currently practiced on a large scale; several hundred plants have been installed, some very large. Most were installed by Grace, Separex (UOP), and Cynara, and all use cellulose acetate membranes in hollow fiber or spiral-wound module form. More recently, hollow fiber polyaramide and polyimide membranes have been introduced by Evonik, Air Liquide, and Fuji, but their use has been slow to take off.

The designs of two carbon dioxide removal plants are illustrated in Figure 9.31. One-stage plants are simple, contain no rotating equipment, and require minimal maintenance, but methane loss to the permeate is often 10–15%. Unless there is a large fuel use for this gas, it must be flared, which represents a significant revenue loss.

For all but a few small plants, the methane loss from a one-stage system is unacceptable, and the permeate gas must be recompressed and passed through a second membrane stage, reducing methane loss to a few percent. However, the cost of recompression is considerable, and may make the membrane option uncompetitive compared with amine absorption, which remains the dominant technology. In general, membrane systems have proven to be most competitive in applications where amine systems have problems – small-scale operations and those with high CO_2 concentrations in the feed gas, for example – or for systems to be used on offshore platforms, where system footprint and weight are prime concerns. A plot illustrating the relative competitive positions of membrane and amine technology is shown in Figure 9.32 [48]. As with Figure 9.28, the ranges are guidelines, not hard and fast rules. For those seeking more detail, White [49] has reviewed the competitive position of membrane systems for this application.

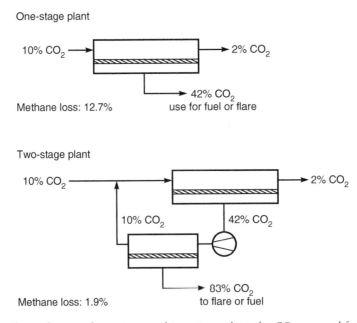

Figure 9.31   *Flow schemes of one-stage and two-stage plants for CO_2 removal from natural gas. Because the one-stage design has no moving parts, it is simple and cheap, but can only be used if there is a use for the low-pressure permeate gas. Two-stage processes are more expensive because a compressor is required, but methane loss with the fuel gas is much reduced.*

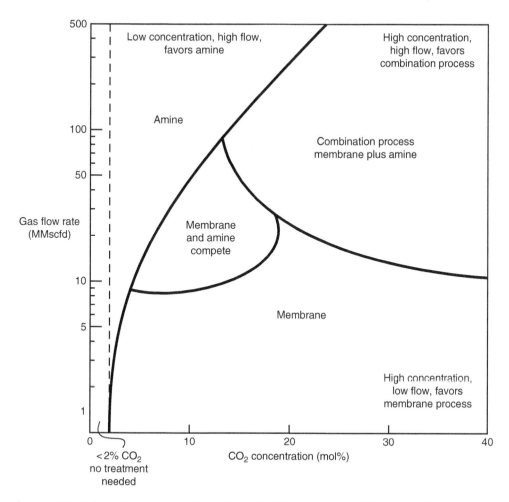

Figure 9.32 *Schematic plot indicating options for CO₂ removal at different combinations of gas flow and CO₂ content. Source: Reprinted with permission from [48]. Copyright (2008) American Chemical Society.*

In principle, the combination of membranes for bulk CO_2 removal with amine units as polishing systems offers a low-cost alternative to all-amine plants for many streams. However, this approach has not been generally used, because the savings in capital cost are offset by the increased complexity of the plant, which now contains two distinct processes. One exception has been in enhanced oil recovery projects, in which carbon dioxide is injected into an oil formation to lower the viscosity of the oil. Water, oil, and gas are removed from the formation; the carbon dioxide is separated and reinjected. In these projects, the composition and volume of the gas changes significantly over the lifetime of the project. The modular nature of membrane units allows easy retrofitting to an existing amine plant, allowing the performance of the plant to be adjusted to meet the changing separation needs. Also, the capital cost of the separation system can be spread more evenly over the project lifetime. An example of a membrane/amine plant design is shown in Figure 9.33. In this example, the membrane unit removes two-thirds of the carbon dioxide, and the amine plant removes the remainder. The combined plant is less expensive than an all-amine or all-membrane plant.

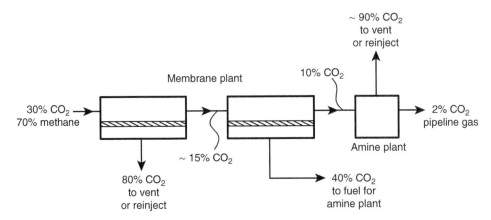

Figure 9.33 *A typical membrane/amine absorption plant for the treatment of associated gas produced in enhanced oil projects. The permeate gas from the second membrane step can be used as a fuel for the amine plant.*

9.6.3.2   Dehydration

All natural gas must be dried before entering the national distribution pipeline, both to control corrosion of the pipeline and to prevent formation of solid hydrocarbon/water hydrates, which can choke valves. Glycol dehydrators are widely used; approximately 50 000 units are in service in the United States. They are not well suited for use on small gas streams or on offshore platforms, however, and these are increasingly common sources of natural gas. In addition, glycol units coextract the trace contaminant benzene, a known carcinogen, which is then released to the atmosphere. The Environmental Protection Agency (EPA) now requires that glycol units be fitted with benzene emission control systems.

Membrane processes offer an alternative approach to natural gas dehydration. Two possible process designs are shown in Figure 9.34. In the first design, a small one-stage system removes 90% of the water in the feed gas, producing a low-pressure permeate gas representing about 5% of the initial gas flow. Selectivities for water over methane are very high, typically in the hundreds, so the process is pressure ratio limited. If the permeate gas can be used as low-pressure fuel at the site, this design is economically competitive with glycol dehydration. In the second design, the wet, low-pressure permeate gas is recompressed and cooled, and the water vapor condenses and is removed as liquid water. The natural gas that permeates the membrane is then recovered in a recycle loop as shown. However, the need for the compressor doubles the capital cost of the system, and such designs are only useful in special situations where glycol dehydration has problems.

9.6.3.3   Dew Point Adjustment, NGL Recovery

Natural gas usually contains varying amounts of C_{3+} hydrocarbons. The gas is often close to saturation with respect to some of these components, which means liquids will condense at cold spots in the pipeline transmission system. To avoid such problems, the hydrocarbon dew point of US natural gas is lowered to about $-20\,°C$ before delivery to the pipeline. For safety reasons, the Btu rating of the pipeline gas is also usually controlled within a narrow range, typically 950–1050 Btu/scf. Because the Btu values of ethane, propane, and pentane are higher than that of methane, natural gas that contains significant amounts of these hydrocarbons may have an excessive Btu value, requiring their removal. Of equal importance, the C_{3+} hydrocarbons are generally more valuable as recovered NGLs than for their fuel value. For all of these reasons, almost all natural gas is treated to control the C_{3+} hydrocarbon content.

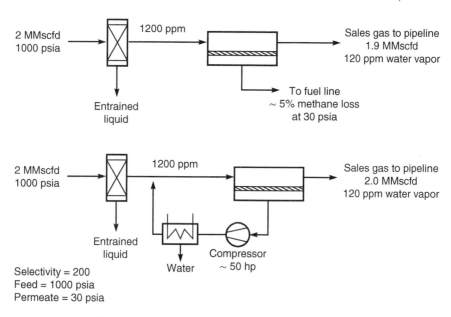

Figure 9.34 *Two possible process designs for natural gas dehydration. Dehydration of natural gas is easily performed by membranes, but cost often limits its scope to niche applications.*

The current treatment is condensation, shown as the upper scheme in Figure 9.35. The natural gas stream is cooled by refrigeration or expansion to between −20 and −40 °C. The resulting condensed liquids, which include the higher hydrocarbons and water, are removed and subjected to fractional distillation to recover the individual components. Because refrigeration is capital-intensive and uses large amounts of energy, there is interest in alternative techniques.

The lower scheme of Figure 9.35 shows a membrane-based option that obviates the need for refrigeration. Gas is fed to a first stage containing rubbery membranes selective for C_{3+} hydrocarbons over methane. This stage produces a hydrocarbon-lean residue that meets pipeline dewpoint specification and a C_{3+} hydrocarbon-enriched permeate. The permeate is recompressed and cooled by simple heat exchange against cold water, yielding a condensed NGL product. The non-condensed bleed stream from the condenser remains saturated with heavy hydrocarbons under the condensation conditions, and is fed a second set of membrane modules. The permeate streams from the two sets of modules are combined, creating a recirculation loop around the condenser, which continuously concentrates the higher hydrocarbons.

The silicone-rubber-based membranes of today are insufficiently selective to be widely used to treat raw gas destined for the main gas pipeline. However, they have found an application in treating gas used as fuel at remote compressor stations. If left untreated, this hydrocarbon rich fuel gas causes knocking and frequent shutdowns. By removing the heavy hydrocarbons, the octane number is substantially improved at little cost. Several hundred of these fuel conditioning units have been installed at large compression stations around the world.

9.6.3.4 Nitrogen Removal

The US pipeline specification for natural gas requires the total inert content – predominantly nitrogen – to be less than 4%. Fourteen percent of known US gas reserves do not meet this requirement. Most commonly, such gas is simply diluted with low-nitrogen gas, but if dilution is not practical, a nitrogen removal unit must be installed. Cryogenic distillation is currently used;

Current technology

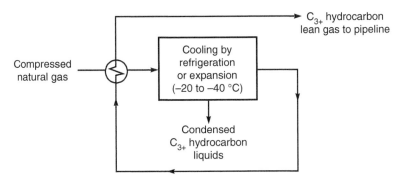

Membrane system using C$_{3+}$ hydrocarbon-selective membranes

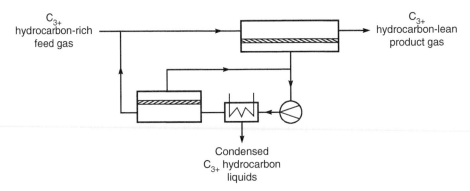

Figure 9.35   *Recovery of C$_{3+}$ hydrocarbons from natural gas.*

as of 1999, 26 cryogenic nitrogen removal plants were in operation in the United States. These plants are complex and expensive, and are only really suited to large gas fields, where production of 50–500 million scfd of gas is likely to continue for 10–20 years, enabling the high capital costs to be defrayed over a long period. Many small gas wells are shut in for lack of suitable nitrogen separation technology. One technology that has been tried with limited success is pressure-swing adsorption (PSA), using molecular sieves that preferentially adsorb nitrogen. Another technology is membrane separation [15]. Membranes that selectively permeate methane over nitrogen are available, albeit the selectivities are low, necessitating multistep, or multistage system designs. To date, most plants have used silicone rubber membranes with a methane/nitrogen selectivity of 3.

A typical unit is illustrated in Figure 9.36 [15]. The operator was producing 12 MMscfd of gas that contained up to 16% nitrogen, and had a heating value of about 900 Btu/scf. The pipeline company was ready to accept the gas for dilution with low nitrogen gas, provided the nitrogen content was less than 10% and heating value was more than 970 Btu/scf. To reach this target, the feed gas, at a pressure of 65 bar, was passed through three sets of modules in series. The permeate from the front set of modules was preferentially enriched in methane, ethane, and the C$_{3+}$ hydrocarbons, and the nitrogen content was reduced to 9% nitrogen. These changes raised the heating value of the gas to 990 Btu/scf. This gas was compressed and sent to the pipeline.

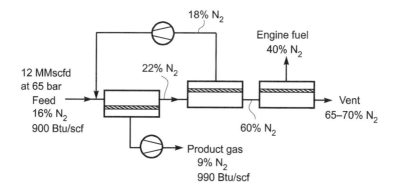

Figure 9.36 *Flow diagram for a three-step membrane process to reduce the nitrogen content of natural gas. The photograph shows of one of the two membrane skids at an installation in California that uses this design.*

The residue gas from the first set of modules contained 22% nitrogen and was sent to a second membrane step, where it was concentrated to 60% nitrogen. The permeate from the second step contained 18% nitrogen and was recycled to mix with the feed gas. The second-step residue gas was sent to a third and final small membrane system to be fractionated. The permeate gas – containing 40% nitrogen – was used as fuel for the compressor engines. The final residue contained 65–70% nitrogen and was essentially stripped of all C_{3+} hydrocarbons; it was flared.

The unit operated for 5 years until the gas field was depleted. Overall, the unit recovered 95% of the hydrocarbon values for delivery to the pipeline, 2% of the hydrocarbons were used as compressor fuel and the final 3% were flared with the final residue nitrogen.

9.6.3.5 Helium Recovery

Natural gas is the sole source of helium on Earth. In the past, a few fields in Texas and Poland containing gas with high helium concentrations (where high means in the 1% range) were enough to supply the market. In these fields, the gas is cooled to condense the hydrocarbons and the helium is recovered as a concentrate with the non-condensable inert gas fraction. More recently, demand from the semiconductor industry and for medical applications has made it worthwhile to recover the gas from relatively low helium concentration gas streams containing as little as 0.1–0.2% helium. Membrane units are generally used first, to concentrate the gas about 5-fold to 10-fold. The concentrate is then sent to a cryogenic plant from which helium is recovered as an overhead product. A flow scheme and photograph of such a plant are shown in Figure 9.37. Membranes of the type normally used for hydrogen separation and listed in Table 9.3 are used. The membrane selectivity is more than sufficient for the separation, so modified versions of the membrane with higher permeances but lower selectivities are used to reduce the system size.

Another new and large source of helium is as a by-product of the production of liquified natural gas (LNG) in Qatar and Algeria. Gas from these sources contains very little helium, but liquefaction of the hydrocarbons to make the LNG product produces a small

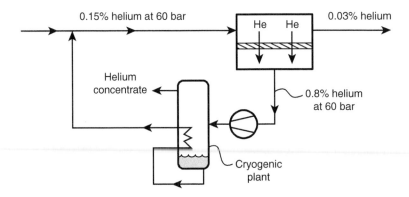

Figure 9.37 *A helium recovery plant using a combination of membrane and cryogenic separation. This configuration can recover about 80% of the helium from the raw feed. The photo shows an installation in Kościan Poland. Each of the two large vessels contains 20 spiral-wound modules. The permeate compressors are contained in the buildings in the background.*

non-condensable off-gas stream containing mostly helium and nitrogen. The helium content of the gas is generally less than 5%, but further concentration to 70–80% helium, using any of the membranes listed in Table 9.3, is straightforward and can be followed by a cryogenic or PSA process to make high purity gas.

9.6.4   Organic Vapor/Gas Separations

In the separation of vapor/gas mixtures, rubbery membranes can be used to permeate the more condensable vapor, or glassy membranes can be used to permeate the smaller gas. Most installed plants use silicone rubber vapor-permeable membranes, often in conjunction with a second process such as condensation [38] or absorption [50]. The first plants, installed in the early 1990s, were used to recover vapors from gasoline terminal vent gases or to remove chlorofluorocarbon (CFC) vapors from the vents of industrial refrigeration plants. The biggest current applications are in the petrochemical industry, where membranes are used to recover valuable feedstocks and solvents.

One of the most successful petrochemical applications is in polyolefin manufacturing plants [51, 52]. Olefin monomer, catalyst, solvents, and other co-reactants are fed at high pressure into the polymerization reactor. The polymer product (resin) is removed from the reactor and separated from excess monomer in a flash-separation step. The recovered monomer is recycled to the reactor. Residual monomer is removed from the resin by stripping with nitrogen in a purge bin, a process that results in a vent stream typically comprising 20–50% of mixed hydrocarbon monomers in nitrogen. The monomer content represents about 1% of the hydrocarbon feedstock entering the plant. This amount might seem small, but plants are large, and the recovery value of the hydrocarbons can be as much as US$5 million/year.

Various membrane-based processes can be used to treat the vent gas from the nitrogen purge, but the most common is the hybrid membrane/condensation design, shown as it applies to propylene recovery, in Figure 9.38. In this design, the compressed vent gas is sent to a condenser, from which a portion of the propylene content is recovered as a condensed liquid. The remaining, uncondensed propylene is removed by the membrane separation system to produce a 99% nitrogen stream, which can be reused in the purge bin. The permeate gas is recycled to mix with the incoming feed gas.

Because the gas sent to the membrane stage is cold, the solubility of propylene in the membrane is enhanced, and the selectivity of the membrane unit is increased. The propylene condensate contains some dissolved nitrogen, so the liquid is flashed at low pressure to remove

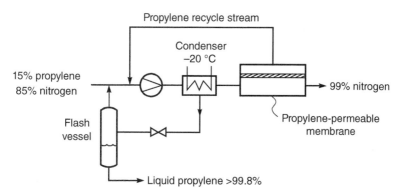

Figure 9.38 *A hybrid process combining condensation and membrane separation to recover propylene and nitrogen from a propylene/nitrogen mixture.*

Figure 9.39   *Photograph of a membrane unit used to recover nitrogen and propylene from a polypropylene plant vent gas.*

this gas, producing a better than 99.5% pure hydrocarbon product. A photograph of a treatment system is shown in Figure 9.39.

9.6.5   To-Be-Developed Applications

The section above has described existing commercial applications of membranes. This section reviews a number of potential next-generation applications. The first two, carbon dioxide/nitrogen and carbon dioxide/hydrogen separations, are related to global warming and to CO_2 capture and sequestration. Both are areas of considerable current research, but development to full-scale depends on government regulations and incentives. Another global-warming-related application is the separation of oxygen from air. With existing membranes, this is a small niche application (described in Section 9.6.2), but if higher permeance membranes with reasonable selectivities could be made, such membranes would facilitate the production of oxygen-enriched air for many combustion processes [47]. The resulting market would be substantial.

Finally there is the general area of organic/organic separations. Such separations can be done in the liquid phase (pervaporation) or vapor phase (vapor permeation). This topic as a whole is discussed in Chapter 10. Very light hydrocarbon mixtures, specifically propylene/propane and ethylene/ethane mixtures, are better treated as gas separation problems, however, and because of this and their industrial importance, they are treated in this section.

9.6.5.1   Carbon Dioxide/Nitrogen Separation (Carbon Capture)

The most important developing gas separation application is the capture of carbon dioxide from emissions produced by power plants and industrial processes. Current carbon dioxide atmospheric emissions are about 35 billion tons/year and are growing at 2% per year. There is a general consensus that CO_2 emissions have already raised global temperatures by 1 °C and that, by 2050, continued emissions are likely to raise temperatures by another 1–2 °C. This change in the global climate will significantly affect life on this planet.

One approach to controlling CO_2 emissions is to abandon coal, oil and natural gas, and switch all energy production to renewable sources. A second approach is to capture the CO_2 produced by fossil fuel combustion and sequester it underground. These are not either/or scenarios. The technology to implement an all-renewable energy economy is not in place and will take many years to fully deploy. Even with widespread use of renewable energy, industrial processes where CO_2 is produced regardless of energy source, such as steel and cement production, will continue to generate CO_2 emissions. Carbon dioxide capture plants can reduce emissions from these persistent industrial sources, as well as from the large installed base of fossil fuel energy plants until they are replaced by renewable alternatives. Membrane-based capture plants could also treat emissions from bioenergy production processes, thereby rendering such processes carbon dioxide negative, which could be a valuable contribution to achieving current net zero emissions targets [53].

Over the past decade, many nations have invested billions of dollars in research funding for capture technology. Currently, amine absorption, which has been used for many years to remove CO_2 from natural gas, is the most developed. Advances in amine chemistry and process design have improved performance, but amine capture from power plant flue gas is likely to be costly. Current best estimates (2022) are in the range $70–$100/ton of CO_2 captured. Three full-scale amine capture plants have been built and are in successful operation, two at coal-fired power plants, another at a cement plant. More plants are being built, but there are multiple problems: the high cost of capture, the large footprint of the plant, the complex flow scheme, and the potential for chemical emissions resulting from the use of enormous volumes of hazardous amine solvents.

Since this is a book about membranes, it will not surprise the reader to learn that membrane technology has a number of advantages over amine absorption. These include potential lower capital and operating costs, an absence of hazardous emissions, a relatively simple flow scheme, a smaller footprint, and the opportunity for modular construction. In addition, the membrane process is driven by electricity, so no changes to the power plant steam cycle are required. The polar rubbery membranes currently being considered for this application offer a CO_2 permeance of about 1000–1500 gpu and a CO_2/N_2 mixed gas selectivity of about 30–40 [54]. Facilitated transport membranes have also been developed for this application, mostly using amine carriers [55–57]. These membranes are described in Chapter 12. The membranes have comparable permeances to existing solution-diffusion membranes but selectivities are better; up to 200 has been claimed. To obtain this performance, however, the membranes must be operated with a hot humidified feed gas, which significantly affects the process feasibility.

Although a significant effort is under way to develop improved membranes, the properties of current membranes are good enough to make the technology viable. Process designs need to be different from other industrial gas separation processes, however, as these almost always use compression of the feed gas to generate the driving force for permeation. The huge gas flows involved (an average sized coal-fired power plant produces 1300 MMscfd of flue gas) mean that the energy required to compress this gas even modestly from the 1 bar at which it is produced can represent a significant fraction of the electricity generated by the power plant. The alternative is to generate a partial vacuum on the permeate side of the membrane, so that energy is used only to treat the much smaller permeate gas stream [54].

Analyses have shown that an improvement in membrane selectivity above the current range produces little improvement in the separation cost, because increasing the pressure ratio above about 10 increases the power consumption to unacceptable values. The pressure ratio limitation means that high permeance membranes are needed. A full-sized power plant will need more than 1 million m^2 of membrane. These membrane plants will be big, on a scale comparable with the largest plants already built for reverse osmosis.

A design for a CO_2 capture plant fitted on a coal-fired power plant is shown in Figure 9.40. The flue gas is at 80–100 °C and contains 13% CO_2, 15–20% water and 10–30 ppm SO_2 and NO_x. The gas is cooled and scrubbed in a pretreatment system to remove the water and other impurities.

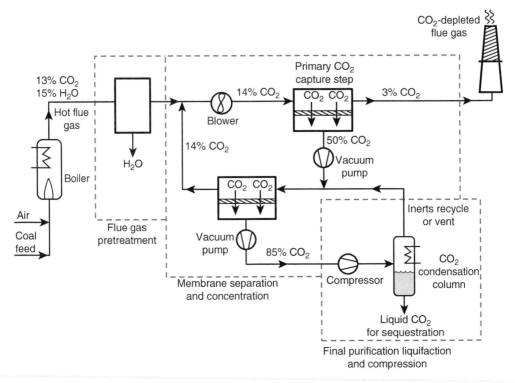

Figure 9.40 *Process diagram of a two-stage membrane CO_2 capture system. The system can be configured to capture 70–80% of the CO_2 emitted from the power plant.*

A blower is used to push the gas through the first membrane separation step. The permeate from this step contains 45–55% CO_2 and is further concentrated to 85–95% in a second stage membrane unit. If the gas is to be sequestered in a nearby salt aquifer, simple compression to the injection pressure of 100–150 bar is all that is required. If the CO_2 is to be transported a significant distance for use in enhanced oil recovery, further purification to remove oxygen, nitrogen, and other inerts is required as shown in the figure.

9.6.5.2   *CO_2/H_2, H_2/CO_2 Separations*

One partial solution to the global warming issue is the so-called hydrogen economy. The idea is to use hydrogen as an energy carrier to produce carbon-free power using fuel cells or gas turbines. The hydrogen would be produced at a central plant by electrolysis of water using electricity from renewable sources, or by steam reforming natural gas (CH_4) combined with CO_2 capture and sequestration.

Currently, almost all the world supply of hydrogen is made by steam methane reforming (SMR), so this technology is likely to remain the main source of hydrogen for some time. There are more than 700 refinery hydrogen plants around the world and nearly 90% of these plants are SMR based.

A block flow diagram of a steam reformer is shown in Figure 9.41. Pipeline natural gas is treated in ZnO beds to remove trace amounts of H_2S and mercaptans. The gas is then mixed with steam, heated to 500 °C and passed through external fired tubes filled with a nickel catalyst. Methane is converted to CO, CO_2, and H_2 by the reactions

$$CH_4 + H_2O \rightarrow CO + 3H_2$$

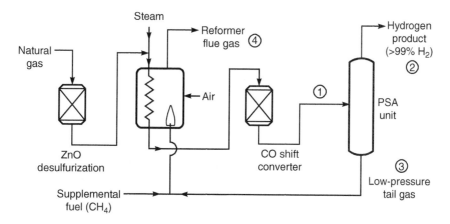

Parameter	① PSA feed	② H₂ product	③ PSA tail gas	④ Reformer flue gas
Composition (vol%)				
H_2	78.0	99.9	36.1	0
CH_4	1.0	0.1	3.0	0
CO	2.0	–	6.1	0
CO_2	19.0	–	54.8	23.1
H_2O	–	–	–	15.2
O_2	–	–	–	5.2
N_2	–	–	–	56.5
Flow rate (MMscfd)	100	67.2	32.8	90.9
Pressure (bar)	30	30	5	1-2

Figure 9.41 Block flow diagram of a steam methane reformer (SMR) using pressure swing adsorption (PSA) to purify the product hydrogen.

and

$$CH_4 + 2H_2O \rightarrow CO_2 + 4H_2$$

The gas leaving the reactor contains up to 10% carbon monoxide and 1–2% of unreacted methane. The gas is cooled and passed to a second catalytic step, the shift converter, where most of the remaining CO is converted to CO_2 and more hydrogen is produced by the reaction

$$CO + H_2O \rightarrow CO_2 + H_2$$

The reformed gas, consisting primarily of CO_2 and hydrogen, is sent to a PSA system that produces essentially pure hydrogen still at pressure, plus a low-pressure tail gas, which contains CO_2, CO, and CH_4, and about 10–15% of the hydrogen produced by the reformer. For the last 30 years, almost all SMR plants have used PSA as the hydrogen purification system.

The PSA feed from the shift converter (stream 1 in the figure) contains about 20% carbon dioxide and the tail gas from the PSA unit (stream 3 in the figure) is about 50% CO_2. Membranes could be used to capture CO_2 from either of these streams, as shown in Figure 9.42. Figure 9.42a

(a) Hydrogen selective membranes

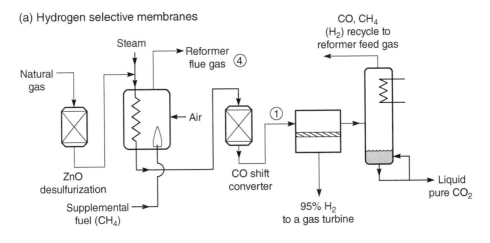

(b) CO$_2$ selective membranes

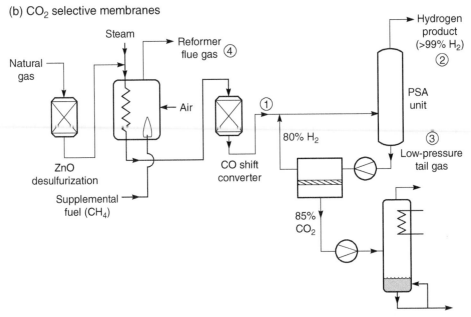

Figure 9.42   *The use of hydrogen selective (a) and CO$_2$ selective (b) membranes to capture CO$_2$ in methane steam reformer plants.*

shows the case where membrane separation is used to treat stream 1. Hydrogen-selective membranes are used, thus maintaining the carbon-dioxide-enriched residue stream at pressure to facilitate subsequent liquefaction. At room temperatures, H$_2$/CO$_2$ selectivities as high as 10 are possible with glassy membranes, such as polybenzimidazole membranes. Heating to 100 °C improves the hydrogen permeance and selectivity [58]. The membrane unit produces a permeate containing 95% hydrogen, usable without further purification as gas turbine fuel. The retentate stream, still at approximately 30 bar, is cooled to produce liquid CO$_2$ for sequestration. The non-condensed carbon monoxide, methane, and residual hydrogen are recycled to the reformer and mixed with the natural gas feed to make more hydrogen.

Figure 9.42b shows an option for treating the PSA tail gas, stream 3. In this case, CO_2-selective membranes are used. The CO_2-rich permeate, containing 80–90% CO_2, is compressed and liquefied for sequestration, and the hydrogen-rich retentate is recycled to the PSA feed. This design has the benefit that the value of the additional hydrogen recovered as product from the tail gas helps to offset the cost of sequestration.

9.6.5.3   Olefin/Paraffin Separations

The separations of ethylene from ethane and of propylene from propane underpin the petrochemical industry and consume vast amounts of energy. Currently, these separations are performed by distillation, but these are close boiling mixtures (ethylene bp = −110 °C and ethane bp = −89 °C; propylene bp = −47 °C; and propane bp = −43 °C), so tall towers with high reflux ratios are needed to produce a good separation. Separation of these mixtures has been a target of membrane developers for many years, and the installation of commercial plants has been predicted to be just around the corner a number of times.

A useful membrane does not require very high selectivities. A stable membrane with a propylene/propane selectivity of 6–10 and a propylene permeance 20–40 gpu could be used to recover and recycle olefins from the reactor purge gas of petrochemical processes. Olefin crackers would be a much larger target, but membranes with selectivities in the 15–20 range would then be needed.

Glassy polymeric membranes have been tried for propylene/propane separation, and promising results are sometimes reported. However, this data is often overly optimistic. Some authors report selectivities based on the ratio of the permeabilities of the pure gases; others use a hard vacuum or sweep gas on the permeate side of the membrane, as described earlier in this chapter. Both procedures produce unrealistically high selectivities. In an industrial plant, the feed gas will be at 5–20 bar and a temperature sufficient to maintain the gas in the vapor phase; the permeate gas will be at a pressure of about 2–5 bar. Under these conditions, plasticization and loss of selectivity occur with even the most rigid polymer membranes. Few conventional polymers maintain mixed-gas high-pressure selectivities of more than 5. Some rigid or crosslinked polyimides can give good selectivities, but the permeances are then low. Air Liquide, for example, has reported selectivities of 10–12 with P84 polyamide hollow fiber membranes, but the propylene permeances are only 1–2 gpu – too low for an industrial process. Burns and Koros have reviewed these results [59].

In the past few years, some promising olefin/paraffin selective membrane results using mixed-matrix membranes have been reported. The first generation of mixed-matrix membranes consisted of small inorganic zeolite crystals dispersed in a polymer matrix phase. More recently, MOF and COF nanocrystals have been used. Some of these results are described in more detail in Chapter 3. Obtaining reproducible data and overcoming the problems of scale-up has proved difficult. Nonetheless this approach has hope.

Another approach is to use facilitated transport membranes, described in Chapter 12. These membranes have the selectivity and permeability required for an olefin/paraffin separation process and have been a subject of research since the work of Steigeleman and Hughes at Standard Oil in the 1970s. Unfortunately, no one has solved the stability problem that plagues these membranes.

9.7   Conclusions and Future Directions

The market for gas separation membranes has grown significantly since the installation of the first industrial plants in the early 1980s. The current status is summarized in Table 9.7, in which the processes are divided into three groups: established processes, developing processes, and

Table 9.7   Status of membrane gas separation processes.

Process	Application	Comments
Established processes		
Oxygen/nitrogen	Nitrogen from air	Processes are all well developed. Only incremental improvements in performance and market share expected
Hydrogen/methane; hydrogen/nitrogen; hydrogen/carbon monoxide	Hydrogen recovery; ammonia plants and refineries	
Water/air	Drying compressed air	—
VOCs and light hydrocarbons from nitrogen or hydrogen	Reactor purge gas, petrochemical process streams, refinery waste gas	Several newer applications being developed; e.g., gasoline stations and terminals, but costs inhibit growth
VOC/air	Air pollution control applications	—
Carbon dioxide/methane	Carbon dioxide from natural gas	Many plants installed, but better membranes would change economics and increase market share
Developing processes		
C$_{3+}$ hydrocarbons/ methane	NGL recovery from natural gas	Processes used for fuel gas conditioning, but NGL recovery requires better economics
Nitrogen Hydrogen sulfide, water/methane Helium recovery	Natural gas treatment	Niche applications, difficult for membranes to compete with existing technology for large flows
To-be-developed processes		
Oxygen/nitrogen	Oxygen-enriched air	Requires better membranes to become commercial. Size of ultimate market will depend on properties of membranes developed. Could be large
Carbon dioxide/nitrogen	Carbon dioxide capture and sequestration	Potential application is enormous and technically feasible but requires government regulation of CO$_2$ emissions. Also has stiff competition from absorption technology
Carbon dioxide/ hydrogen	Hydrogen production in refineries and IGCC plants	Could be big, but also depends on adoption of government regulations for CO$_2$ recovery
Olefin/paraffin mixtures	Separation of mixtures in refineries and petrochemical plants	Requires better membranes and modules. Potential size of application is large

to-be-developed processes. The first group consists of nitrogen production from air, hydrogen recovery, natural gas processing, treatment of petrochemical purge gas, and air drying. These processes represent more than 90% of the current market. All have been used on a large commercial scale for 20 years, and improvements in membrane properties and process design have been made during that time. For most of these applications, the technology has reached a point at which, barring a completely unexpected breakthrough, further changes in productivity are likely to be the result of small incremental changes. The one exception is the removal of CO_2 from natural gas, where membranes have a small market share. Development of more selective membranes (not an impossible dream) could improve the competitiveness of the membrane process for this separation.

Developing processes include recovery of light hydrocarbons from refinery and petrochemical plant purge gases, and separation of C_{3+} hydrocarbons, hydrogen sulfide, nitrogen, and water from natural gas. All of these processes are performed on a commercial scale. Expansion in these applications, driven by the development of better membranes and process designs, is occurring.

The 'to-be-developed' processes represent the future expansion of membrane gas separation technology. The production of oxygen-enriched air is a large potential application for membranes. The market size depends completely on the properties of the membranes that can be produced. Improvements in flux by a factor of two at current oxygen/nitrogen selectivities would probably produce a limited membrane market; improvements by a factor of 5–10 would make the use of oxygen-enriched air in natural gas combustion processes attractive. In the latter case, the market could be very large indeed. The separations of carbon dioxide from nitrogen at electric power plants and of carbon dioxide from hydrogen in steam methane reformers are two applications linked to global warming. If governments decide that carbon dioxide separation and sequestration will be carried out, these applications could be huge. The final application listed in Table 9.7 is the separation of olefin/paraffin mixtures (for example, propylene/propane mixtures) using membranes in competition, or perhaps in combination, with distillation. Ten years ago, plants for these separations seemed to be just around the corner. Today, they do not look so near. Membranes that retain their properties at high temperature and in the presence of high concentrations of organic vapors are required. This may be a separation for which ceramic membranes finally find an application.

References

1. Graham, T. (1866). On the adsorption and dialytic separation of gases by colloid septa. *Philos. Mag.* **32**: 401.
2. Barrer, R.M. (1951). *Diffusion in and Through Solids*. London: Cambridge University Press.
3. van Amerongen, G.J. (1950). Influence of structure of elastomers on their permeability to gases. *J. Appl. Polym. Sci.* **5**: 307.
4. Stern, S.A. (1966). Industrial applications of membrane processes: the separation of gas mixtures. In: *Membrane Processes for Industry, Proceedings of the Symposium*, 196–217. Birmingham, AL: Southern Research Institute.
5. Meares, P. (1954). Diffusion of gases through polyvinyl acetate. *J. Am. Chem. Soc.* **76**: 3415.
6. Henis, J.M.S. and Tripodi, M.K. (1980). A novel approach to gas separation using composite hollow fiber membranes. *Sep. Sci. Technol.* **15**: 1059.

7. MacLean, D.L., Bollinger, W.A., King, D.E. and Narayan, R.S. (1986) Gas separation design with membranes, in *Recent Developments in Separation Science*, (eds. N.N. Li and J.M. Calo), p. 9, Boca Raton, FL: CRC Press.

8. Gruen, F. (1947). Diffusionmessungen an Kautschuk (Diffusion in rubber). *Experimenta* **3**: 490.

9. van Amerongen, G.J. (1946). The permeability of different rubbers to gases and its relation to diffusivity and solubility. *J. Appl. Phys.* **17**: 972.

10. Behling, R.D., Ohlrogge, K., Peinemann, K.-V., and Kyburz, E. (1989). The separation of hydrocarbons from waste vapor streams. In: *Membrane Separations in Chemical Engineering*, AIChE Symposium Series Number 272, vol. **85** (ed. A.E. Fouda, J.D. Hazlett, T. Matsuura, and J. Johnson), 68. New York: AIChE.

11. Robeson, L.M. (1991). Correlation of separation factor versus permeability for polymeric membranes. *J. Membr. Sci.* **62**: 165.

12. Stern, S.A. (1994). Polymers for gas separation: the next decade. *J. Membr. Sci.* **94**: 1.

13. Park, H.B. and Lee, Y.M. (2008). Polymeric membrane materials and potential use in gas separation. In: *Advanced Membrane Technology and Applications* (ed. N.N. Li, A.G. Fane, W.S. Ho, and T. Matsuura), 633–670. Hoboken, NJ: Wiley.

14. Robeson, L.M. (2008). The upper bound revisited. *J. Membr. Sci.* **320**: 390.

15. Lokhandwala, K.A., Pinnau, I., He, Z. et al. (2010). Membrane separation of nitrogen from natural gas: a case study from membrane synthesis to commercial deployment. *J. Membr. Sci.* **346**: 270.

16. Freeman, B.D. (1999). Basis of permeability/selectivity tradeoff relations in polymeric gas separation membranes. *Macromolecules* **32** (2): 375.

17. Lee, S. Y., Minhas, B.S. and Donohue, M.D. (1988). Effect of gas composition and pressure on permeation through cellulose acetate membranes, in *New Membrane Materials and Processes for Separation*, AIChE Symposium Series Number 261, **84**, 93–101 (eds. K. K. Sirkar and D.R. Lloyd), New York: AIChE.

18. Okamoto, Y., Zhang, H.Z., Mikes, F. et al. (2014). New perfluoro-dioxolane based membranes for gas separations. *J. Membr. Sci.* **471**: 412.

19. Tiwari, R.P., Smith, Z.P.S., Lin, H. et al. (2014). Gas permeation in thin films of high free volume glassy perfluoro polymers: part 1. Physical aging. *Polymer* **55**: 5788.

20. Zhang, C., Wu, B.-H., Ma, M.-Q. et al. (2019). Ultrathin metal/covalent- organic framework membranes towards ultimate separation. *Chem. Soc. Rev* **48**: 3811.

21. Qiao, Z., Zhao, S., Sheng, M. et al. (2019). Metal-induced ordered microporous polymers for fabricating large-area gas separation membranes. *Nat. Mater.* **18**: 163.

22. Qian, Q., Asinger, P.A., Lee, M.J. et al. (2020). MOF-based membranes for gas separations. *Chem. Rev.* **120**: 8161.

23. Kalaj, M., Bentz, K.C., Ayala, S. et al. (2020). MOF-polymer hybrid material: from simple composites to tailored architectures. *Chem. Rev* **120**: 8267.

24. Nagai, K., Masuda, T., Nakagawa, T. et al. (2001). Poly[1-(trimethylsilyl)-1-propyne] and related polymers: synthesis, properties and functions. *Prog. Polym. Sci.* **26**: 721.

25. Budd, P.M., Ghanem, B.S., Makhseed, S. et al. (2004). Polymers of intrinsic microporosity (PIMs): robust, solution-processable, organic nanoporous materials. *Chem. Comm.* **2**: 230–231.

26. Low, Z.-X., Budd, P.M., McKeown, N.B., and Patterson, D.A. (2018). Gas permeation properties, physical aging and its mitigation in high free volume glassy polymers. *Chem. Rev.* **118**: 5871.

27. Aylmore, L.A.G. and Barrer, R.M. (1966). Surface and volume flow of simple gases and of binary gas mixtures in a microporous carbon membrane. *Proc. Roy. Soc.* **290**: 466.

28. Koresh, J.E. and Sofer, A. (1983). Molecular sieve carbon selective membrane. *Sep. Sci. Technol.* **18**: 723.

29. Liu, Q., Borjigin, H., Paul, D.R. et al. (2016). Gas permeation properties of thermally rearranged (TR) isomers and their aromatic polyimide precursors. *J. Membr. Sci.* **518**: 88.

30. Bum Park, H., Han, S.H., Jung, C.H. et al. (2010). Thermally rearranged polymer membranes for CO_2 separation. *J. Membr. Sci.* **359**: 11.

31. Weller, S. and Steiner, W.A. (1950). Separation of gases by fractional permeation. *J. Appl. Phys.* **21**: 279.

32. Pan, C.Y. and Habgood, H.W. (1978). Gas separation by permeation, part 1: calculation method and parametric analysis. *Can. J. Chem. Eng.* **56**: 197.

33. Baker, R.W. and Wijmans, J.G. (1994). Membrane separation of organic vapors from gas streams. In: *Polymeric Gas Separation Membranes* (ed. D.R. Paul and Y.P. Yampolskii), 353–398. Boca Raton, FL: CRC Press.

34. Huang, Y., Merkel, T.C., and Baker, R.W. (2014). Pressure ratio and its impact on membrane gas separation processes. *J. Membr. Sci.* **463**: 33.

35. Coker, D.T., Freeman, B.D., and Fleming, G.K. (1998). Modeling multicomponent gas separation using hollow fiber membrane contactors. *AIChE J.* **44** (6): 1289.

36. Stern, S.A. and Wang, S.C. (1978). Countercurrent and Cocurrent gas separation in a permeation stage: comparison of computation methods. *J. Membr. Sci.* **4**: 141.

37. Baker, R.W. and Wijmans, J.G. (1993). Two-stage membrane process and apparatus, US Patents 5,256,295 and 5,256,296, issued 26 October 1993.

38. Wijmans, J.G. (1992). Process for removing condensable components from gas streams, US Patent 5,199, 962, issued 6 April 1993, and 5,089, 033, issued 18 February 1992.

39. Paul, D.R. and Yampol'skii, Y.P. (ed.) (1994). *Polymeric Gas Separation Membranes*. Boca Raton, FL: CRC Press.

40. Koros, W.J. and Fleming, G.K. (1993). Membrane based gas separation. *J. Membr. Sci.* **83**: 1.

41. Baker, R.W. (2002). Future Directions of membrane gas separation technology. *Ind. Eng. Chem. Res.* **41**: 1393.

42. Yampol'skii, Y. and Freeman, B.D. (ed.) (2010). *Membrane Gas Separation*. Chichester: Wiley.

43. Bollinger, W.A., MacLean, D.L., and Narayan, R.S. (1982). Separation systems for oil refining and production. *Chem. Eng. Prog.* **78**: 27.

44. Henis, J.M.S. (1994). Commercial and practical aspects of gas separation membranes. In: *Polymeric Gas Separation Membranes* (ed. D.R. Paul and Y.P. Yampol'skii), 441–530. Boca Raton: FL, CRC Press.

45. Prasad, R., Shaner, R.L., and Doshi, K.J. (1994). Comparison of membranes with other gas separation technologies. In: *Polymeric Gas Separation Membranes* (ed. D.R. Paul and Y.P. Yampol'skii), 531–614. Boca Raton, FL: CRC Press.

46. Prasad, R., Notaro, F., and Thompson, D.R. (1994). Evolution of membranes in commercial air separation. *J. Membr. Sci.* **94**: 225.

47. Lin, H., Zhou, M., Ly, J. et al. (2013). Membrane-based oxygen enriched combustion. *I&E Res.* **52**: 10820.

48. Baker, R.W. and Lokhandwala, K.A. (2008). Natural gas processing with membranes: an overview. *Ind. Eng. Chem. Res.* **47**: 2108.

49. White, L.S. (2010). Evolution of natural gas treatment with membrane systems. In: *Membrane Gas Separation* (ed. Y. Yampolskii and B.D. Freeman), 313–332. Chichester, UK: Wiley.

50. Ohlrogge, K., Wind, J., and Behling, R.D. (1995). Off gas purification by means of membrane vapor separation systems. *Sep. Sci. Technol.* **30**: 1625.

51. Baker, R.W. and Jacobs, M.L. (1996). Improve monomer recovery from polyolefin resin degassing. *Hydrocarbon Process* **75**: 49.

52. Baker, R.W., Lokhandwala, K.A., Jacobs, M.L., and Gottschlich, D.E. (2000). Recover feedstock and product from reactor vent streams. *Chem. Eng. Prog.* **96**: 51.

53. The International Energy Agency Special Report (2021). Net Zero by 2050: A Roadmap for the Global Energy Sector.

54. Merkel, T.C., Lin, H., Wei, X., and Baker, R.W. (2010). Power plant post-combustion carbon dioxide capture: an opportunity for membranes. *J. Membr. Sci.* **359**: 126.

55. Taniguchi, I., Duan, S., Kazama, S., and Fujioka, Y. (2008). Facile fabrication of a novel high performance CO_2 separation membrane: immobilization of poly(amidoamine) dendrimers in poly(ethylene glycol) networks. *J. Membr. Sci.* **322**: 277.

56. Chen, Y. and Ho, W.S.W. (2016). High-molecular-weight polyvinylamine/piperazine glycinate membranes for CO_2 capture from flue gas. *J. Membr. Sci.* **514**: 376.

57. Dang, L., Kim, T.-J., and Hägg, M.B. (2009). Facilitated transport of CO_2 in novel PVAm, PVA blend membranes. *J. Membr. Sci.* **359**: 154.

58. Merkel, T.C., Zhou, M., and Baker, R.W. (2012). Carbon dioxide capture with membranes at an IGCC power plant. *J. Membr. Sci.* **389**: 441.

59. Burns, R.L. and Koros, W.J. (2003). Defining the challenges for C_3H_6/C_3H_8 separations using polymeric membranes. *J. Membr. Sci.* **211**: 299.

10

Pervaporation/Vapor Permeation

10.1   Introduction and History

As its name suggests, pervaporation is a combination of two separation processes: evaporation and membrane permeation. A simple pervaporation process to separate liquid mixtures is illustrated in Figure 10.1. A heated liquid feed mixture contacts one side of a membrane;

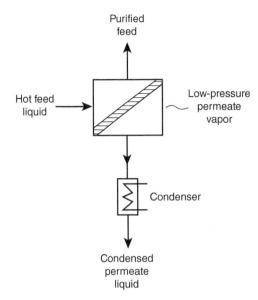

Figure 10.1 In a pervaporation process, a hot liquid mixture contacts a selective membrane. The permeate is withdrawn as a vapor mixture. The vapor, enriched in the more permeable component, is cooled and condensed, spontaneously generating the vacuum that drives the process.

permeate vapor is removed from the other side. Transport through the membrane is driven by the vapor pressure difference between the heated feed and the low-pressure permeate vapor. This vapor pressure difference can be maintained in several ways. In the laboratory, a vacuum pump is used to draw a vacuum on the permeate side of the system. Industrially, the permeate vacuum is most economically generated by cooling the permeate vapor, causing it to condense; condensation spontaneously creates a partial vacuum.

This chapter treats both pervaporation processes, where the feed is a hot liquid mixture, and vapor permeation processes, where the feed is a hot vapor mixture. The permeating components and membranes used are similar, and the permeate collection and condensation systems are the same. In the laboratory, pervaporation tends to be more frequently studied because it is easier to work with hot liquids than vapor mixtures above their boiling point. Companies that supply industrial equipment produce either type of system depending on the particular application.

The origins of pervaporation can be traced to the nineteenth century, but the word itself was coined by Kober in 1917 [1]. The process was first studied in a systematic fashion by Binning and co-workers at American Oil in the 1950s [2, 3]. Binning was interested in applying the process to the separation of organic mixtures. Although this work was pursued at the laboratory and bench scale for a number of years and several patents were obtained, the process was not commercialized. The technology of the time could not produce the high-performance membranes and modules required for a commercially competitive process. The process was picked up in the 1970s by Eli Perry and others at Monsanto. More than a dozen patents assigned to Monsanto were issued from 1973 to 1980, and cover a wide variety of pervaporation applications [4], but none of this work led to a commercial process. Academic research on pervaporation was also carried out over a period of years by Neel, Aptel, and others at the University of Toulouse [5, 6]. By the 1980s, advances in membrane technology finally made it possible to construct economically viable systems.

GFT Membrane Systems was the first successful industrial developer and installed a pilot plant in 1982 [7]. The application was the removal of water from organic solvent solutions, most importantly ethanol solutions. The ethanol feed to the membrane contained 5–10 wt% water and was close to the azeotropic composition. Pervaporation removed the water as permeate, producing an ethanol residue containing less than 1% water. All the problems of azeotropic distillation were avoided. But pervaporation had its own problems. Pervaporation plants run at about 100 °C and pressures of 2–5 bar; finding membranes that can operate with ethanol/water mixtures at these conditions was not easy. Developing membrane modules to package the membranes was even harder. GFT's solution was to use an all-metal plate-and-frame module, with porous metal as the permeate support and stainless-steel head spacers [8]. A 50 m^2 module, equivalent in membrane area to a standard 12-in. diameter spiral-wound module, was a steel block weighing close to a ton. Module lifetimes were in the 1–3 year range, and changing out the membranes required special tools, so modules had to be returned to the factory for refitting.

In 1988, GFT built a 5 tons/h demonstration plant, which was installed at Bétheniville, France. The plant operated for 1–2 years on an intermittent basis, but had multiple problems. This was a pity, because a huge opportunity for pervaporation/vapor permeation was to appear a few years later. From 2000 to 2008, approximately 400 corn-to-ethanol plants, each producing an average of 300–1000 tons of bioethanol per day, were built in the US in response to a government mandate that all gasoline contain 5–15% bioethanol. An equivalent number of smaller plants were built in Brazil. But the membrane technology was not ready, and almost all US plants installed molecular sieve adsorption units for the final dehydration step; no membrane plants were installed.

For a number of years, it was thought that a second generation of cellulose-to-ethanol plants would be built [9]. Alas, despite a significant development effort from 2000 to 2010 by a number of oil companies, as well as by Dupont in the United States, Abengoa in Spain, and Iogen in Canada, these plants were never built. The economics for cellulose-to-ethanol conversion were just not there. Vaperma, MTR, and other companies that had been developing polymeric membranes in anticipation of these opportunities were left with a solution in search of a problem.

Since 2010, the most significant development has been the use of zeolite membranes formed on ceramic supports [10, 11]. This effort has been centered in Japan (Mitsui, Hitachi-Zosen) and China (Jiangsu Nine Heaven). The membranes are too expensive to compete with molecular sieves or distillation for large plants, but can compete for smaller applications in the 10–50 tons/day range. In China in particular, a large number of such systems has been installed. Most operate with the feed in the vapor phase rather than as a liquid, but the membranes used and the process designs are essentially the same.

The current market for pervaporation and vapor permeation is essentially all for small-scale solvent dehydration, and probably does not exceed US$50 million per year. There is some new interest in cellulose-to-ethanol production as an outgrowth of concern over global warming and the search for non-fossil fuels. However, over the long term, the big opportunity for pervaporation is to replace distillation in organic/organic separations (aromatic/aliphatic; linear/branched hydrocarbons; xylene isomers). These are all areas of research interest, but currently remain in the laboratory. A timeline illustrating some of the key milestones in development of pervaporation is shown in Figure 10.2.

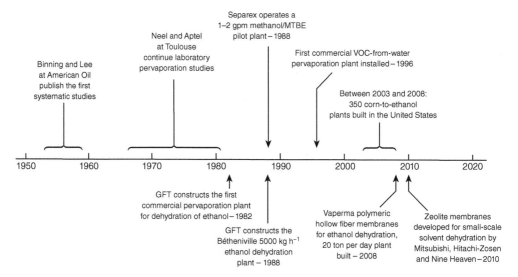

Figure 10.2   *Milestones in the development of pervaporation/vapor permeation.*

10.2   Theoretical Background

The link between pervaporation and vapor permeation is illustrated by Figure 10.3. The right-hand side of the figure shows a pervaporation process. A portion of the heated feed liquid is circulated from the large feed tank past the pervaporation membrane. The membrane removes a small amount of the feed as a low pressure permeate vapor. After cooling and condensing, this vapor forms the condensed permeate.

The left-hand side of Figure 10.3 shows a vapor permeation process. A portion of feed vapor is circulated past the feed side of a vapor permeation membrane. The membrane removes a small amount of the feed as a low pressure permeate vapor. After cooling and condensing, this vapor also forms a condensed permeate.

The membranes, assumed to be identical and of thickness ℓ, are in contact with different feed fluids, but those fluids are in equilibrium, and hence the driving force for permeation through both membranes is the same [12, 13]. This means that the transmembrane flux produced by that driving force is also the same in both cases, as is shown by the equations derived in Chapter 2. For the vapor permeation case described by Eq. (2.83), the flux of component i (J_i) through the membrane is

$$J_i = \frac{D_i K_i^G \left(p_{i_o} - p_{i_\ell} \right)}{\ell} = \frac{P_i^{\,G}}{\ell} \left(p_{i_o} - p_{i_\ell} \right) \tag{10.1}$$

where p_{i_o} is the partial vapor pressure of component i in equilibrium with the feed liquid, p_{i_ℓ} is the partial pressure of the permeating vapor on the permeate side of the membrane, D_i is the diffusion coefficient for component i in the membrane, K_i^G is the gas phase sorption coefficient of component i in the membrane and $P_i^{\,G}$ is the permeability, and the product of the diffusion and sorption coefficients.

Vapor permeation   Pervaporation

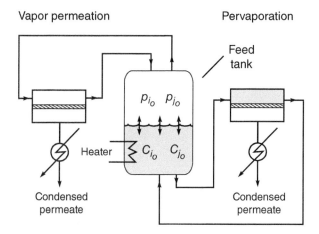

Figure 10.3 *A schematic illustration of the equivalence of vapor permeation and pervaporation. The vapor feed to the vapor permeation membrane and the liquid feed to the pervaporation membrane are in equilibrium with each other. This means the driving force across both membranes is equivalent and identical membranes will produce identical permeation fluxes and separations. Source: Adapted from [12].*

For the pervaporation case, the mass flux J_i of component i through the membrane is described by Eq. (2.78), derived in Chapter 2 as

$$J_i = \frac{D_i\left(K_i^L c_{i_o} - K_i^G p_{i_\ell}\right)}{\ell} \tag{10.2}$$

where K_i^L is the liquid phase sorption coefficient of component i in the membrane and c_{i_o} is the concentration of component i in the feed liquid. In Chapter 2, it was shown that Eq. (10.2) can be rearranged (Eqs. (2.78)–(2.82)) to yield

$$J_i = \frac{D_i K_i^G \left(p_{i_o} - p_{i_\ell}\right)}{\ell} = \frac{\mathcal{P}_i^G}{\ell}\left(p_{i_o} - p_{i_\ell}\right) \tag{10.3}$$

which is identical to the gas Eq. (10.1). Equations (10.1) and (10.3) can also be written in molar form

$$j_i = \frac{\mathcal{P}_i^G}{\ell}\left(p_{i_o} - p_{i_\ell}\right) \tag{10.4}$$

where j_i is the molar flux of i, ℓ is the membrane thickness and $\mathcal{P}_i^G$ is the gas separation permeability coefficient. A similar equation can be written for component j

$$j_j = \frac{\mathcal{P}_j^G}{\ell}\left(p_{j_o} - p_{j_\ell}\right) \tag{10.5}$$

Having obtained the driving forces and measured the fluxes, the permeability $\mathcal{P}_i^G$ for each component can be calculated from the equations above, and the membrane selectivity, defined as the ratio of the molar permeabilities or permeances, can then be determined:

$$\alpha_{ij}^G = \frac{\mathcal{P}_i^G}{\mathcal{P}_j^G} = \frac{\mathcal{P}_i^G/\ell}{\mathcal{P}_j^G/\ell} \tag{10.6}$$

Expressing the permeability and selectivity in molar terms, as in Eqs. (10.4) and (10.5), allows direct comparison of pervaporation data with the larger body of gas permeation data.

The above equations are easily applied to vapor permeation, since the partial pressures on the feed and permeate sides of the membrane are readily calculated from the measured total pressures and the feed and permeate concentrations. In pervaporation separations, calculating the vapor pressure driving force is not easy. Determining the permeate side vapor pressure for each component is straightforward; the total permeate pressure and the permeate gas composition are all that is required. The problem is to calculate the feed side vapor pressure. For this, conversion of the measured liquid phase composition and temperature into component vapor pressures is required. This process used to require the accumulation of extensive data, and creation from the same of VLE diagrams. Today, process simulator software (ChemCad, HYSYS, ProSim) delivers instant answers, provided the correct equation of state is selected. Most simulators do a good job of selecting the appropriate equation of state and the activity coefficients required for simple two-component mixtures. The vapor pressures calculated this way are normally within a few percent of the experimental values. However, obtaining reliable vapor pressures for multicomponent mixtures is more difficult and may not be possible if one of the components is an electrolyte. Fortunately, the vast bulk of laboratory pervaporation data involve only two-component mixtures.

Describing pervaporation in terms of membrane permeabilities, permeances, and selectivities is the preferred method of reporting the data, since it links pervaporation to the related processes of gas and vapor separation, where the results are normally reported in this way. In much of the published literature, however, the separation performed by a pervaporation process is characterized in terms of a separation factor β_{pervap} defined as the ratio of component concentrations on the permeate and feed sides:

$$\beta_{pervap} = \frac{c_{i_\ell}/c_{j_\ell}}{c_{i_o}/c_{j_o}} \tag{10.7}$$

or, since the ratio of the permeate side component partial vapor pressures is also the ratio of the component concentrations

$$\beta_{pervap} = \frac{p_{i_\ell}/p_{j_\ell}}{c_{i_o}/c_{j_o}} \tag{10.8}$$

The factors that determine β_{pervap} can be illustrated by considering the thermodynamically equivalent vapor permeation process shown in Figure 10.3. Pervaporation can be divided into two steps. The first step is evaporation, governed by the vapor–liquid equilibrium of the liquid mixture. Evaporation (distillation) produces a separation because of the different volatilities of the components of the feed liquid. The evaporative separation can be defined as β_{evap}, the ratio of the component i and j concentrations in the feed vapor to their concentrations in the feed liquid. Expressing the vapor ratio as partial pressures

$$\beta_{evap} = \frac{p_{i_o}/p_{j_o}}{c_{i_o}/c_{j_o}} \tag{10.9}$$

The second step is permeation of components i and j through the membrane; this step is equivalent to conventional gas permeation, where the driving force for permeation is the difference in the vapor pressures of the components in the feed and permeate vapors. The separation achieved in this step, β_{mem}, is the ratio of the components in the permeate vapor to the ratio of the components in the feed vapor

$$\beta_{mem} = \frac{p_{i_\ell}/p_{j_\ell}}{p_{i_o}/p_{j_o}} \tag{10.10}$$

It follows by combining Eqs. (10.9) and (10.10) that the separation achieved in pervaporation is the simple product of the evaporative separation and the separation achieved by selective vapor permeation through the membrane.[1]

$$\beta_{pervap} = \beta_{evap} \cdot \beta_{mem} \tag{10.11}$$

Equation (10.11) is useful in showing the two processes that contribute to the total pervaporation permeation process. The first step is evaporation of the feed liquid to form a saturated vapor in contact with the membrane; the second step is diffusion of this vapor through the membrane to the low-pressure permeate side. It is the presence of this second step that allows pervaporation to achieve useful separation of close-boiling or azeotropic mixtures.

Equation (10.11) also illustrates the problem with reporting data in terms of fluxes and separation factors. These values are not only a function of the intrinsic membrane properties, but also depend on the vapor–liquid equilibrium of the feed solution and the operating conditions of the experiments (feed concentration, permeate pressure, feed temperature); change the operating conditions and all the numbers change [13]. Using flux and separation factors makes comparison of pervaporation data sets obtained under different operating conditions difficult. Combining the contributing terms into a single parameter masks the individual effect of each term.

The benefit of reporting pervaporation data in terms of molar permeances (vapor-pressure normalized fluxes) and membrane selectivities is apparent from the experimental data shown in Figures 10.4 and 10.5. Figure 10.4 shows plots of membrane flux and separation factor as a function of feed water concentration for a series of pervaporation experiments performed with ethanol/water mixtures and a hydrophilic cellulose ester membrane. At a low water concentration

---

[1] Figure 10.3 illustrates the concept of permeation from a saturated vapor phase in equilibrium with the feed liquid as a tool to obtain Eq. (10.11). A number of workers have experimentally compared vapor permeation and pervaporation separations and have sometimes shown that permeation from the liquid is faster and less selective than permeation from the equilibrium vapor. This is an experimental artifact. In vapor permeation experiments, the vapor in contact with the membrane is often not completely saturated. This means that the activities of the feed components in vapor permeation experiments are less than their activity in pervaporation experiments. Because sorption by the membrane in this range is extremely sensitive to activity, the vapor permeation fluxes are lower than pervaporation fluxes. Kataoka et al. [14] have illustrated this point in a series of careful experiments.

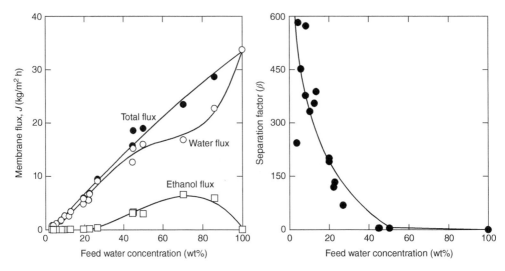

Figure 10.4 *Dehydration of ethanol/water solutions with a commercial cellulose ester membrane (75 °C, permeate pressure < 5 Torr), reported as membrane flux and water/ethanol separation factor data. Source: Reproduced with permission from [13]. Copyright (2010) Elsevier.*

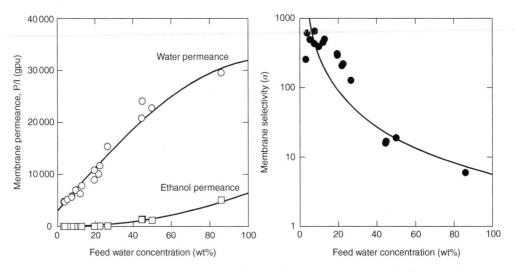

Figure 10.5 *Data from Figure 10.4, replotted as membrane permeance and selectivity data. As a reminder, $1\,gpu = 1 \times 10^{-6}\ cm^3(STP)/cm^2 \cdot s \cdot cmHg$. Source: Reproduced with permission from [13]. Copyright (2010) Elsevier.*

in the feed, the membrane is very selective for water from ethanol and the separation factor is high (right-hand graph), but the flux of water and ethanol through the membrane is low (left-hand graph). As the water concentration in the feed is increased, both the ethanol and water fluxes increase and the membrane separation factor falls. At a water concentration of about 70 wt%, the ethanol flux reaches a maximum value. Beyond this maximum, the water flux increases sharply, reaching about 34 kg/m$^2 \cdot$ h at 100 wt% water, while the ethanol flux falls to zero as the ethanol concentration drops. Something is going on, probably related to plasticization of the membrane by water, but it is hard to sort this out from the membrane flux and separation factor data presented in Figure 10.4.

The same data are replotted as membrane permeances and selectivities versus feed concentration in Figure 10.5. The situation is now clearer. At low water concentrations in the feed, the membrane is extremely selective; for example, at a water concentration of less than 5 wt%, the membrane has a water permeance of about 5000 gpu and the ethanol permeance is below 10 gpu. As the water concentration increases, the ethanol and water permeances both increase. This increase is probably due to swelling of the membrane by water, leading to plasticization. Plasticization increases the water permeance from 5000 gpu at less than 5 wt% water to 32 000 gpu at 100 wt% water, but has a much larger relative effect on ethanol permeance. The corresponding ethanol permeance increases from less than 10 gpu at less than 5 wt% water to more than 6000 gpu at close to 100 wt% water, an almost 600-fold increase. As a consequence, the water/ethanol selectivity falls to a selectivity of only 5 at very high water concentrations.

The data in Figure 10.5 highlight the effect of sorption on the behavior of polymeric membranes in general. Absorption of a few wt% of the feed is necessary to ensure useful permeances, but excessive sorption (more than 10%) results in plasticization of the polymer and a loss in selectivity. Since pervaporation and vapor permeation processes involve contacting the membrane with a feed at close to its boiling point, controlling sorption can be tricky. The composition of the membrane material must be adjusted to achieve the correct balance of adequate flux and selectivity; typically this means maintaining permeant sorption between 3% and 10%.

Membrane selectivities and permeances are intrinsic properties of the membrane, whereas the membrane separation factor, β_{mem}, is affected by the properties of the membrane, the vapor–liquid equilibrium diagram of the feed solution and the process operating conditions. The term β_{mem} in Eq. (10.11) can be replaced with the intrinsic properties of the membrane. This leads to a more complicated equation, but untangles the respective contributions of membrane properties and operating conditions to the separation process [12]. By combining Eqs. (10.4) and (10.5), we can write

$$\frac{j_i}{j_j} = \frac{\mathcal{P}_i^G}{\mathcal{P}_j^G} \frac{\left(p_{i_o} - p_{i_\ell}\right)}{\left(p_{j_o} - p_{j_\ell}\right)} = \alpha_{mem} \frac{\left(p_{i_o} - p_{i_\ell}\right)}{\left(p_{j_o} - p_{j_\ell}\right)} \tag{10.12}$$

The ratio of the molar fluxes is also the same as the ratio of the permeate partial pressures, and so

$$\frac{j_i}{j_j} = \frac{p_{i_\ell}}{p_{j_\ell}} \tag{10.13}$$

And the expression for β_{mem} from Eq. (10.10) can then be written

$$\beta_{mem} = \frac{\alpha_{mem}^G \left(p_{i_o} - p_{i_\ell}\right)}{\left(p_{j_o} - p_{j_\ell}\right)} \cdot \frac{1}{\left(p_{i_o} / p_{j_o}\right)} \tag{10.14}$$

And Eq. (10.11) becomes

$$\beta_{pervap} = \frac{\beta_{evap}\alpha_{mem}^G\left(p_{i_o} - p_{i_\ell}\right)}{\left(p_{j_o} - p_{j_\ell}\right)\left(p_{i_o}/p_{j_o}\right)} \tag{10.15}$$

Equation (10.15) identifies the three factors that determine the separation achieved by a pervaporation process. The first factor, β_{evap}, is the vapor–liquid equilibrium, determined by the feed liquid composition and temperature; the second is the membrane selectivity, α_{mem}^G, an intrinsic property of the membrane material; and the third includes the feed and permeate vapor pressures, reflecting the effect of process operating parameters on performance.

As in gas separation, the separation achieved by pervaporation is determined both by the membrane selectivity and by the transmembrane pressure ratio, and there are two limiting cases in which one factor dominates the separation. The first limiting case is when the membrane selectivity is very large compared to the vapor pressure ratio:

$$\alpha_{mem}^G \gg \frac{p_o}{p_\ell} \tag{10.16}$$

This means that for a membrane with infinite selectivity for component i, the permeate vapor pressure of component i will equal the feed partial vapor pressure of i. That is,

$$p_{i_\ell} = p_{i_o} \tag{10.17}$$

Equation (10.17) combined with Eq. (10.10) gives

$$\beta_{mem} = \frac{p_{j_o}}{p_{j_\ell}} \tag{10.18}$$

which, combined with Eq. (10.11), leads to the limiting case

$$\beta_{pervap} = \beta_{evap} \cdot \frac{p_{j_o}}{p_{j_\ell}} \quad \text{when} \quad \alpha_{mem}^G \gg \frac{p_o}{p_\ell} \tag{10.19}$$

Similarly, in the case of a very large membrane selectivity in favor of component j,

$$\beta_{pervap} = \beta_{evap} \cdot \frac{p_{i_o}}{p_{i_\ell}} \tag{10.20}$$

For the special case in which component i is the minor component in the feed liquid, p_{j_o} approaches p_o, p_{j_ℓ} approaches p_ℓ, and Eq. (10.19) reverts to

$$\beta_{pervap} = \beta_{evap} \frac{p_o}{p_\ell} \tag{10.21}$$

where p_o/p_ℓ is the feed-to-permeate ratio of the total vapor pressures. In other words, membrane separation is completely determined by the pressure ratio, and intrinsic membrane selectivity becomes irrelevant.

The second limiting case occurs when the vapor pressure ratio is very large compared to the membrane selectivity. This means that the permeate partial pressure is much smaller than the feed partial vapor pressures, and p_{i_ℓ} and $p_{j_\ell} \to 0$. Eq. (10.14) then becomes

$$\beta_{pervap} = \beta_{evap}\alpha_{mem}^G \qquad \text{when} \qquad \alpha_{mem}^G \ll \frac{p_o}{p_\ell} \qquad (10.22)$$

In other words, the full intrinsic selectivity of the membrane is available, and the pressure ratio is irrelevant.

The relationship between the three separation factors, β_{pervap}, β_{evap}, and β_{mem}, is illustrated in Figure 10.6. This type of plot was introduced by Shelden and Thompson [15] to illustrate the effect of permeate pressure on pervaporation separation, and is a convenient method to represent the pervaporation process graphically. When the permeate pressure, $p_\ell = p_{i_\ell} + p_{j_\ell}$, approaches the feed vapor pressure, $p_o = p_{i_o} + p_{j_o}$, the vapor pressure ratio across the membrane approaches unity. No separation is produced by the membrane, and the composition of the permeate vapor approaches the composition obtained by simple evaporation of the feed liquid. This composition is shown by the line labeled β_{evap} in the figure. As the permeate pressure decreases to below the feed vapor pressure, the vapor pressure ratio across the membrane increases. The overall separation obtained, β_{pervap}, is then the product of the separation due to evaporation of the feed liquid, β_{evap}, and the separation due to permeation through the membrane, β_{mem}. The line labeled

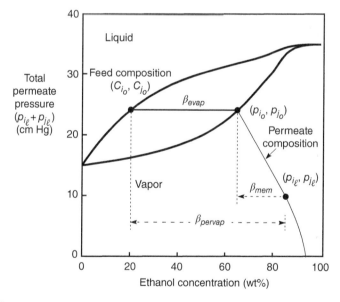

Figure 10.6 *The permeate composition of an ethanol-selective membrane used to separate ethanol from an ethanol/water feed solution containing 20 wt% ethanol at 60°C. The total separation, β_{pervap}, is made up of two contributions: β_{evap}, due to the evaporative VLE contribution, and β_{mem}, due to selective permeation of ethanol through the membrane. The line labeled permeate composition is calculated from Eq. (10.14) and shows that the membrane performance improves as the permeate pressure decreases [12, 15].*

'permeate composition' in Figure 10.6 is calculated from Eq. (10.14). As the permeate pressure decreases, the feed-to-permeate pressure ratio across the membrane increases and a better separation is obtained.

10.3 Membrane Materials and Modules

10.3.1 Membrane Characterization

The experimental set-up used to characterize membrane performance in the laboratory is shown in Figure 10.7. The upper diagram shows a pervaporation experiment, which is the more straightforward to perform. Hot liquid feed is circulated across the surface of a membrane or small module, and a vacuum pump maintains the permeate side of the apparatus under a pressure of 10–20 millibar. A dual liquid nitrogen trap system condenses the permeate vapor. The permeate stream can be switched from one condenser to the other, allowing samples to be withdrawn for analysis without interrupting operation of the unit.

The equipment for vapor permeation experiments, shown in the lower diagram, is a little more complex [16]. A charge of feed solution is placed inside a steel feed vessel. A hot plate/stirrer is used to boil the liquid vigorously. Because the apparatus is jacketed, the rising vapor soon brings the temperature of the entire enclosed apparatus to the feed boiling point. The membrane cell is designed so that a flow of vapor passes across the membrane feed surface to the water-cooled vapor condenser. The condensed residue or retentate stream flows back to the feed vessel.

As with the pervaporation experiment, a combination of a vacuum pump and two liquid nitrogen traps is used to collect the membrane permeate at a pressure of 10–20 millibar. The feed vapor pressure is one bar, so the pressure ratio across the membrane is high. The permeate system can be switched from one condenser to the other, allowing multiple permeate samples to be collected.

10.3.2 Membrane Materials

As with the other processes in which permeant transport takes place in accordance with the solution-diffusion model, the membrane selectivity in pervaporation is the product of the mobility (diffusion) selectivity and the sorption selectivity. In gas separation, the total sorption of gases by the membrane material is often low, such as less than 2–3 wt%, and the membrane structure is not much affected by the sorbed gas. In consequence, the selectivity measured with gas mixtures is often close to the selectivity calculated from the ratio of the pure gas permeabilities. By contrast, in pervaporation the membrane is in contact with the feed liquid at close to its boiling point and sorption is much higher, typically 5–30 wt%. Sorption of this much permeant can cause substantial plasticization and swelling of the membrane, changing the permeation properties correspondingly. As a rule of thumb, the total sorption of the feed liquid by the membrane should be in the range of 3–10 wt%. Below 3 wt% sorption, the membrane selectivity may be good, but the flux through the relatively low sorption material will be low. Above 10 wt% sorption, fluxes will be high, but the swollen membrane will permit freer entry and passage of all components, and selectivity will be lost; indeed, at very high total sorption, the sorption selectivity tends toward unity. To control sorption, polymeric pervaporation materials are usually crosslinked.

For any given separation, membranes may be predominantly sorption selective or diffusion selective, depending on the choice of membrane material, and large differences in performance are possible. The range of results that can be obtained with the same liquid mixture is illustrated

Pervaporation test system

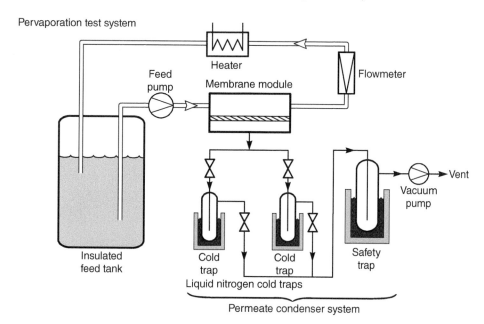

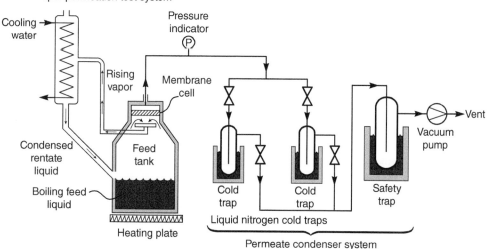

Figure 10.7 *A process flow diagram of laboratory apparatus for pervaporation and vapor permeation membrane characterization experiments [12, 16].*

in Figure 10.8 for the separation of acetone from water [17]. The figure shows the concentration of acetone in the permeate as a function of the concentration in the feed. The silicone rubber membrane, made from a hydrophobic rubbery material, preferentially sorbs acetone, the more hydrophobic organic compound. For rubbery materials, the diffusion selectivity term, which would favor permeation of the smaller component (water), is small compared to the sorption selectivity. Therefore, the silicone rubber membrane is sorption selectivity controlled and preferentially permeates acetone. In contrast, the poly(vinyl alcohol) (PVA) membrane is made from a hydrophilic, rigid, crosslinked material. Because poly(vinyl alcohol) is hydrophilic, the

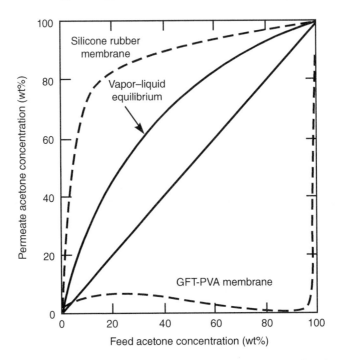

Figure 10.8 *Pervaporation separation of acetone-water mixtures achieved with a water-selective membrane (poly(vinyl alcohol) [PVA]) and an acetone-selective membrane (silicone rubber), compared to the acetone/water VLE. Source: Reproduced with permission from [17]. Copyright (1993) Taylor and Francis.*

sorption selectivity favors permeation of water, the more hydrophilic polar component. Also, because poly(vinyl alcohol) is glassy and crosslinked, the diffusion selectivity favoring the smaller water molecules over the larger acetone molecules is substantial. As a result, PVA membranes permeate water several hundred times faster than acetone.

In most membrane processes, it is desirable for the minor components to permeate the membrane (less membrane area is needed and the permeate recovery/disposition is simpler), so the acetone-selective silicone rubber membrane is best used to treat dilute acetone feed streams, thereby concentrating most of the acetone in a small volume of permeate. The water-selective poly(vinyl alcohol) membrane is best used to treat concentrated acetone feed streams, concentrating most of the water in a small volume of permeate. In their preferred operating ranges, both membranes outperform distillation, which relies on the vapor–liquid equilibrium to achieve separation. Table 10.1 summarizes the membrane materials used for the main applications of pervaporation and vapor permeation.

10.3.2.1 Dehydration Membranes

A huge body of pervaporation data describing the separation of ethanol-water mixtures has been published and reviewed elsewhere [18, 29, 30]. This is an easy separation, and many materials have high permeabilities and water/ethanol selectivities of more than 100. The only real issue, usually bypassed in laboratory studies, is the long-term stability of membranes and modules under real world operating conditions. Crosslinked poly(vinyl alcohol) was the first

Table 10.1 *Widely used pervaporation membrane materials.*

Dehydration of organics Water/ethanol Water/isopropanol Water/glycol, and so on	Microporous polyacrylonitrile support membrane coated with a 5–20 µm layer of crosslinked poly(vinyl alcohol) is the most commonly used commercial material [18]
	Chitosan [19] and polyelectrolyte membranes such as Nafion® [20, 21] have equivalent properties
	Zeolite membranes are also used commercially [10]
	Perfluorinated polymers have been used for some aggressive solvent mixtures such as acetic acid/water [22]
VOC/water separation Toluene/water Trichloroethylene/water Methylene chloride/water	Membranes comprising silicone rubber coated onto polyimides, polyacrylonitrile or other microporous supports are widely used [13, 23]
	Other rubbers, such as ethylene-propylene terpolymers, have been reported to have good properties [24]
	Polyamide-polyether block copolymers have been used for pervaporation of some polar VOCs [25]
Organic/organic separation	Depends on the nature of the organics. Poly(vinyl alcohol) and cellulose acetate [26] have been used to separate alcohols from ethers
	Polyurethane-polyimide block copolymers have been used for aromatic/aliphatic separations [27, 28]

membrane material used commercially by GFT and it remains widely used [7]. Chitosan and Nafion® provide equivalent performance. Various cellulose esters and ethers have also been used. Typical module lifetimes for ethanol-water pervaporation membranes are 1–3 years. The same membranes used in vapor permeation generally do better and can have lifetimes in the 3–4 year range.

Perfluoro polymers have been tried for some dehydration applications. The membranes are chemically stable even in aggressive solvents such as acetic acid. Most perfluoro polymers are glassy, so diffusion selectivity strongly favors permeation of water, but this is offset by the low water sorption of the hydrophobic polymer. The net result is membranes with good stability but modest water permeances and water/solvent selectivities [22, 31].

In recent years, there has been increasing interest in tubular zeolite-coated ceramic membranes [10, 11, 32–34]. The Bussan Research Institute brought tubular NaY and NaA zeolite membranes to the commercial stage and a demonstration plant was built. Membrane water/ethanol selectivities are high; at temperatures of 100–150 °C, high fluxes have also been obtained. The membranes can be used for pervaporation or vapor permeation. Ceramic membranes are expensive, but are competitive with distillation for small applications (less than 50 tons/day). Jiangsu Nine Heaven in China has installed more than a hundred small plants using tubular zeolite-coated ceramic tubes. A ceramic membrane module developed to separate a particular solvent/water mixture has the advantage that it can often be applied without change to other mixtures. The membrane polymers, glues, and spacers that make up a polymeric membrane module are less adaptable, and changing the solvent to be treated may require significant changes to some or all of these components. However, for large volume applications, for example water/ethanol separations in bioethanol plants, the cost differential between ceramic and polymeric membrane modules is enough to make the development of polymeric modules worthwhile.

10.3.2.2   *Organic/Water Separation Membranes*

Several types of membrane have been used to separate VOCs from water and are discussed in the literature [35, 36]. The membranes are usually made from rubbery polymers, such as silicone rubber, polybutadiene, and polyamide-polyether copolymers (Pebax®). Rubbery membranes are remarkably effective at separating hydrophobic organic solutes from dilute aqueous solutions; the concentration of VOCs such as toluene or trichloroethylene (TCE) in the condensed permeate is typically more than 1000 times that in the feed solution. For example, a feed solution containing 100 ppm VOC can yield a permeate vapor containing 10–20% VOC. This concentration is well above the saturation limit, so condensation produces a two-phase permeate. This permeate comprises an essentially pure condensed organic phase and an aqueous phase containing a small amount of VOC, which can be recycled to the aqueous feed. The separations achieved with moderately hydrophobic VOCs, such as ethyl acetate, methylene chloride and butanol, are still impressive, typically providing at least 100-fold enrichment in the permeate. However, the enrichment obtained with hydrophilic solvents, such as methanol, acetic acid and ethylene glycol, is usually much lower, at 5 or below. Typical enrichments obtained for common VOCs with silicone rubber are shown in Table 10.2.

A number of academic studies have identified rubbery hydrophobic membrane materials with higher selectivities than silicone rubber. For example, Nijhuis et al. [24] measured the separation factors of various rubbery membranes with dilute toluene and trichloroethylene solutions. The separation factor of silicone rubber was in the 4000–5000 range, but other materials had separation factors as high as 40 000. However, in practice, an increase in membrane separation factor beyond about 200–500 provides very little additional benefit. Once a separation factor of this magnitude is obtained, other factors, such as ease of manufacture, mechanical strength, chemical stability, and control of concentration polarization, become more important. For this reason, silicone rubber remains the prevalent membrane material.

Membranes with improved separation factors would be useful for hydrophilic VOCs, for which the separation factor of silicone rubber is 5–10, a little less than the separation produced by simple evaporation. As yet, no good replacement for silicone rubber has been developed. The most promising results to date have been obtained with mixed matrix membranes containing dispersed zeolite particles in a silicone rubber matrix [37]. Ethanol preferentially permeates the pores of the zeolite particles; membranes have been produced in the laboratory with ethanol/water separation factors

Table 10.2   *Typical separation factors for VOC removal from water using silicone rubber membrane modules.*

Separation factor for VOC over water	Volatile organic compound (VOC)
200–1000	Benzene, toluene, ethylbenzene, xylenes, TCE, chloroform, vinyl chloride, ethylene dichloride, methylene chloride, perchlorofluorocarbons, hexane
20–200	Ethyl acetate, propanols, butanols, MEK, aniline, amyl alcohol
5–20	Methanol, ethanol, phenol, acetaldehyde
1–5	Acetic acid, ethylene glycol, DMF, DMAC

of 40 or more. The evaporative separation factor (β_{evap}) is 12–15, so from Eq. (10.11), separation factors of 40 imply a membrane selectivity of about 3. Membranes with these properties could be applied in fermentation processes and solvent recovery. Unfortunately, the membranes made to date are not stable when used with industrial solutions. Small amounts of esters or long-chain alcohols present in the solution penetrate the zeolite pores, where they are adsorbed and block further ethanol transport.

Polyamide-polyether block copolymers (Pebax®, Elf Atochem, Inc., Philadelphia, PA) have been used successfully with polar organics, such as phenol and aniline [25]. The separation factors obtained are greater than 100, far higher than those obtained with silicone rubber. The improved selectivity reflects the greater sorption of the polar organic in the polar membrane. On the other hand, toluene separation factors are below those measured with silicone rubber.

10.3.2.3 Organic/Organic Separation Membranes

The membranes used for organic/organic separations are very varied, to suit the needs of each individual case. However, the general requirement is that the sorption of the feed mixture should be limited. The membranes will typically be exposed to a feed liquid at an applied pressure of 2–5 bar and at a temperature above its atmospheric boiling point. Under these conditions, many polymers will sorb large amounts of the feed, becoming highly plasticized. Both sorption and diffusion components of the selectivity will diminish rapidly, and the overall selectivity will drop correspondingly or disappear altogether.

A number of strategies are used to control plasticization. The most used approach is to crosslink the selective membrane layer; another is use of rigid backbone polymers (for example, the Matrimid® polyimides developed by Grace). However, the permeability of these materials is often low. ExxonMobil [27, 28] has used block copolymers consisting of rigid polyimide segments, which provide a strong network, and softer segments formed from more flexible polymers, through which permeant transport occurs. The permeation properties of the polymer can be controlled by tailoring the size and chemistry of the two blocks. Amorphous, but polar, rubbery or glassy polymers have also been used. Because of their polar nature, sorption of hydrophobic organics is limited.

10.3.3 Membrane Modules

The first commercial pervaporation modules were made by GFT and used a plate-and-frame design. A photograph of one of these units is shown in Figure 10.9 [38]. All of the membrane plates were arranged in parallel, so multiple modules with interstage heating were required for a complete plant. Two modules were normally stacked inside the vacuum chamber of the system shown in the figure.

In later years, GFT, then a division of Sulzer (working with W.R. Grace) developed tubular modules having a shell-and-tube (heat exchanger) design as shown in Figure 10.10 [39]. The tubular modules were mounted inside perforated steel tubes. Permeate passing through the perforated tubes collected in the low-pressure central shell of the unit. The U-tube manifolds at each end of the tubes were contained inside steam-heated end chambers separated from the central vacuum chamber by a tube sheet. As the feed circulated through these manifolds, it was reheated to the set temperature.

Tubular and plate-and-frame modules of the type described above have been used in many small plants (plants using tens to a few hundred square meters of membrane), but are too expensive to be used in large plants, such as those for dehydration of bioethanol. These plants require several thousand square meters of membrane, so more economical module designs

Figure 10.9   *Photographs of a 50-m² GFT plate-and-frame module and an ethanol dehydration system fitted with this type of module. The module is contained in the large vacuum chamber on the left-hand-side of the pervaporation system. Source: Reproduced with permission from [36]. Copyright (1992) Elsevier.*

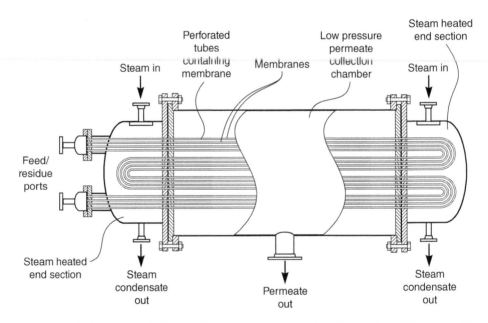

Figure 10.10   *Sulzer/Grace shell-and-tube pervaporation module design, from U.S. Patent 7,655, 141 [38]. The feed liquid pumped through the tubular membranes is reheated every time it passes into the steam-heated end sections of the module.*

are needed. It was in response to this need that Ube in Japan and Vaperma in the United States developed hollow fiber polyimide membranes, and MTR developed spiral-wound modules. Developing a stable membrane was relatively straightforward; developing modules with potting compounds, glues, seals, and spacer that would withstand prolonged exposure to boiling ethanol was difficult.

10.4　Process Design

10.4.1　Basic Principles

In pervaporation/vapor permeation processes, the feed can be a hot liquid or a vapor (usually at 10–30 °C above its boiling point). The driving force in either case is a vapor pressure gradient across the membrane. Figure 10.11 illustrates two ways to achieve this.

In the laboratory, the low vapor pressure required on the permeate side is often produced with a vacuum pump, as shown in Figure 10.11a. In a commercial-scale system, however, the vacuum pump requirement would be impossibly large. (In the early days of pervaporation research, the calculated vacuum pump size was sometimes used as proof that pervaporation would never be commercially viable.) An alternative used in all industrial plants, and illustrated in Figure 10.11b, is to cool the permeate vapor below the boiling point of the constituents; condensation of the liquid spontaneously generates the permeate side vacuum. In this process, the driving force is the difference in vapor pressure between the hot feed solution and the cold permeate liquid at the temperature of the condenser. The cost of providing the required cooling and heating is much less than the cost of a vacuum pump, and the process is operationally more reliable. A small vacuum pump is still included to remove any non-condensable gas that leaks into the system on the low-pressure side.

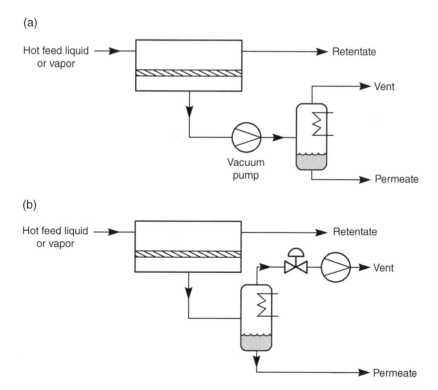

Figure 10.11　*Techniques for maintaining driving force in pervaporation/vapor permeation processes; (a) using a vacuum pump, (b) cooling to condense the permeate and generate a spontaneous partial vacuum.*

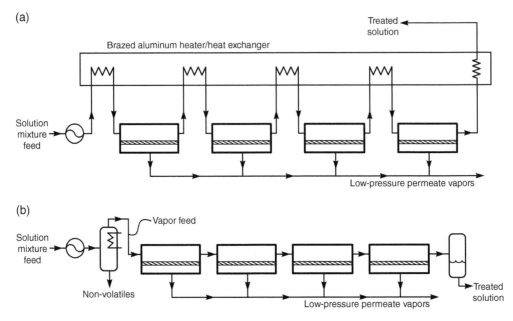

Figure 10.12 *Simplified flow schemes of (a) a pervaporation process and (b) a vapor permeation process.*

A major distinction between pervaporation and vapor permeation is the need to maintain the feed temperature during separation. In pervaporation, for permeate vapor to be formed, a portion of the feed liquid must vaporize. The latent heat of vaporization is extracted from the feed liquid, which cools as a result; a feed solution temperature drop of about 3–5 °C for every 1% of the feed that permeates the membrane is characteristic. In industrial processes, an average of 5–10% of the feed permeates the membrane per module, and the corresponding temperature drop is a substantial 20–50 °C. This temperature drop is reversed by heaters or heat exchangers between modules, as shown in Figure 10.12a, a simplified scheme in which four pervaporation/heating cycles are shown. The investment in the necessary piping, valves, and flanges can represent a significant fraction of the total cost in an industrial plant. In contrast, the vapor permeation process shown in Figure 10.12b uses an initial feed solution boiler and does not require interstage reheating, but the energy consumption of the evaporator is high because all of the feed solution is vaporized.

In the process flow schemes shown in Figure 10.11, a simple low-temperature condenser, operating as a stand-alone unit after the membrane separation step, is used to condense the permeate. However, depending on the vapor–liquid equilibrium diagram of the particular vapor mixture, the potential for an additional separation step is possible, as shown in Figure 10.13. When the vapor–liquid equilibrium is favorable, significant improvements to the overall separation processes are possible if the configuration of Figure 10.11 (repeated for comparison in 10.13a) is replaced by the fractional condensation arrangement of Figure 10.13b [40], or even better, if a fractionating condenser is used, as in Figure 10.13c [41]. In each case, the feed is assumed to be 6 wt% ethanol in water, and to be at a temperature of 80 °C. The design of Figure 10.13a can produce a permeate enriched to 20 wt% ethanol and a residue reduced to 0.5 wt% ethanol. Since the condenser operates as a stand-alone unit, it must run at a temperature

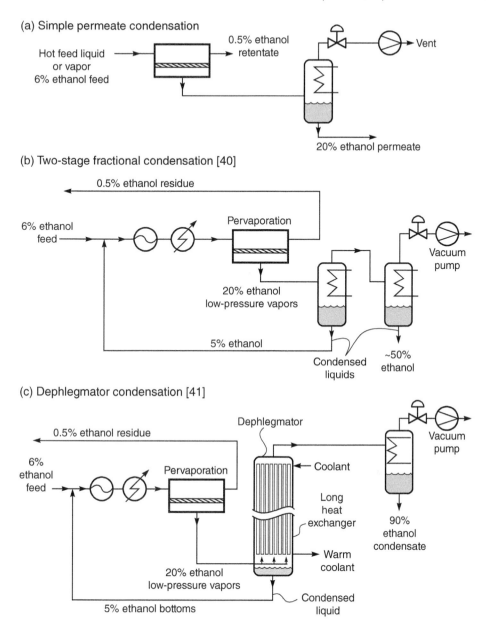

Figure 10.13 *The use of permeate vapor fractional condensation systems to improve the separation achieved in pervaporation of dilute ethanol solutions: (a) stand-alone pervaporation, followed by a single condenser to liquify the permeate; (b) pervaporation integrated with two-stage condensation [40]; (c) pervaporation integrated with multistage condensation in a dephlegmator. Source: Adapted from [41].*

low enough to condense all of the permeate vapor. Figure 10.13b shows the improvement that results from using two-stage fractional condensation [40]. The first (higher temperature) condenser separates the permeate into a first condensate containing about 5 wt% ethanol, which is recycled to the incoming feed, and an ethanol-enriched overhead vapor. The second (lower temperature) condenser condenses the remaining vapor to produce an ethanol product stream containing about 50 wt% ethanol. Integrating the unit operations by returning the first condensate within the process more than doubles the ethanol content of the product.

As shown in Figure 10.13c, an even better separation can be obtained if a vertical heat exchanger (sometimes called a dephlegmator) is used as the permeate condenser [41]. Warm, low-pressure permeate vapor from the pervaporation unit enters the heat exchanger at the bottom. As the vapor rises up the heat exchange tubes, some condenses on the cold tube walls. The resultant liquid flows downward countercurrent to the rising feed vapor. Mass transfer between the liquid and vapor enriches the liquid in the less volatile components as the more volatile components are re-vaporized. As a result, three or four theoretical stages of separation are achieved at little cost. The separation achieved with this approach depends on the VLE of the mixture being separated, but can be impressive for mixtures like ethanol and water. In the example shown, the 20 wt% ethanol vapor is separated into a 5 wt% ethanol bottoms, which is recycled to the pervaporation unit, and a 90–95 wt% overhead ethanol product stream.

The designs shown in Figure 10.13 can all operate with the feed in either the liquid or the vapor phase. The choice as to which to adopt is usually determined by how the membrane unit needs to be, or can be, integrated with other unit operations, or by the overall energy balance of the plant where the unit is to be installed. The following sections provides some examples of integrated operations.

10.4.2   Hybrid Distillation/Membrane Processes

The use of membrane separation in conjunction with distillation has been studied by many authors [22, 42–44]. The most promising applications are those in which distillation alone has a problem, for example, the separation of azeotropes or of close-boiling mixtures. Three example flow diagrams that highlight combinations of distillation and membrane permeation are shown in Figure 10.14.

Each drawing in the figure consists of two parts; the left side shows the VLE diagram for a two-component mixture and the membrane permeation performance curve; the right side shows a hybrid scheme for separating that mixture. The preferred operating ranges for the unit operations can be seen from the diagrams. For example, Figure 10.14a shows that ethanol/water mixtures containing up to 75 wt% ethanol can be easily separated into pure water and an ethanol-rich overhead by simple distillation. Above 75 wt% ethanol, it becomes increasingly difficult and energy intensive to produce a higher concentration ethanol overhead as the composition approaches the azeotrope at 95 wt% ethanol. The composition diagram also shows the separation that can be achieved with a PVA membrane. Feed compositions above 75 wt% ethanol are best separated by the membrane, which exhibits a very high selectivity in favor of water over ethanol under those conditions [8].

An example of a hybrid process design that takes advantage of the features identified by the VLE diagram is shown on the right-hand side of Figure 10.14a. Distillation produces a pure water distillation bottom stream and an overhead vapor containing about 75 wt% ethanol. This vapor stream is sent to the membrane unit, which produces a retentate stream consisting of 99% ethanol, and an aqueous permeate stream containing about 20 wt% ethanol that is recycled back to the distillation step.

The acetic acid/water VLE diagram in Figure 10.14b has no azeotrope, but the proximity of the equilibrium curve to the diagonal indicates a close-boiling mixture. The VLE diagram shows that distillation can perform the total separation, but most trays in the column will be at the top, removing the last 10–15% of the acetic acid from water, and at the bottom, removing the last

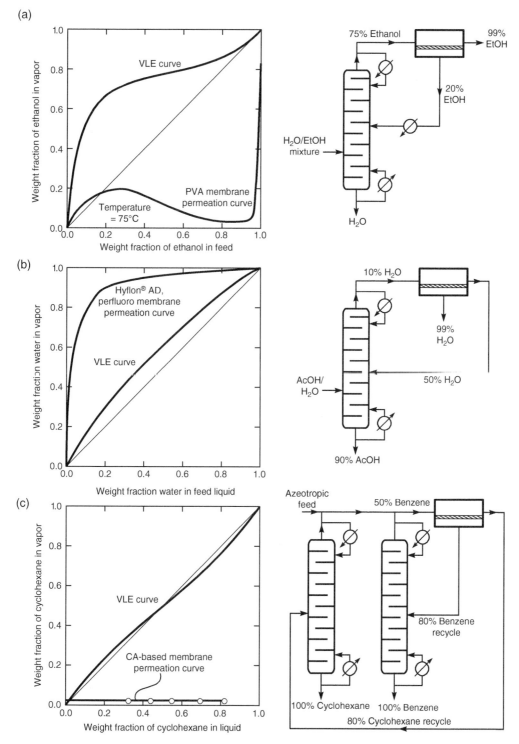

Figure 10.14 *Three hybrid flow schemes (a)–(c), for separating azeotropes and close-boiling mixtures. The membrane separation step improves the separation, reduces energy usage, or both.*

10–15% of water from the acetic acid. Membranes that preferentially permeate water, such as the perfluoro Hyflon® AD membrane shown, can be useful at the top or bottom of the column, reducing the size and energy consumption of the distillation processes [22]. Removal of the last fraction of acetic acid from water at the top of the column has been most studied because the acetic acid recovered is a valuable product, and because disposal of water contaminated with acetic acid is expensive.

Separation of benzene/cyclohexane mixtures, illustrated in Figure 10.14c, is more complicated. The mixture forms an azeotrope at about 50/50 benzene/cyclohexane. This means two distillation columns must be used: one producing pure benzene, the other producing pure cyclohexane. The overheads of the columns are mixed and sent to a membrane unit that preferentially permeates benzene over cyclohexane. The membrane unit splits the azeotropic mixture; the benzene-enriched permeate is reintroduced at an appropriate composition point into the benzene column; the cyclohexane-enriched retentate is recycled in like manner to the cyclohexane column [45].

Based on the membrane performance indicated on the VLE diagrams for cases (b) and (c), it might be concluded that dispensing with distillation entirely and using membrane separation alone is the way to go. This would indeed be an option for a new plant, but for improving the performance and energy usage of an existing column the hybrid designs would be beneficial.

In the processes shown in Figure 10.14, the membrane treating the column overhead could operate in either pervaporation or vapor permeation mode. In the examples shown, however, energy has already been used to create the overhead vapor, so vapor permeation would normally be preferred. On the other hand, if the membrane unit were to be used to treat the hot liquid bottoms stream from a column, pervaporation would be favored, to avoid the cost of evaporating the entire bottoms stream. In the acetic acid/water case, for example, pervaporation mode could be used to dehydrate the bottoms stream, which still contains 10 wt% water.

The drawings of Figure 10.14 do not show heat integration, but in an industrial process this would be used. For example, in ethanol dehydration, a turbo fan vapor compressor can be used to increase the pressure of the ethanol vapor stream to be sent to the membrane. Such a design is shown in Figure 10.15. Increasing the pressure of the vapor sent to the membrane unit reduces the size of the system, but more importantly, it raises the condensation temperature of the ethanol-rich retentate vapor. This vapor can then be condensed in the reboiler of the distillation unit, thereby recovering its latent heat content and significantly reducing the energy consumption of the total process [22]. A number of related process flow schemes of this type are described in the literature. These processes have much in common with conventional mechanical vapor compression processes.

10.5   Applications

There are three main applications of pervaporation/vapor permeation. The most important, and currently the only significant commercial application, is dehydration of organic solvents, primarily bioethanol, but also isopropanol, butanol, acetone, and acetonitrile. Other applications that have been the subject of research, but have yet to be used on any real commercial scale, include removal of organics from water for pollution control, and organics from food processing streams for flavor recovery. A long list of organic/organic separations has been tried, including removal of sulfur compounds from diesel, aromatic/aliphatic separation, linear/branch hydrocarbon separation and separation of xylene isomers. We will briefly review these applications below.

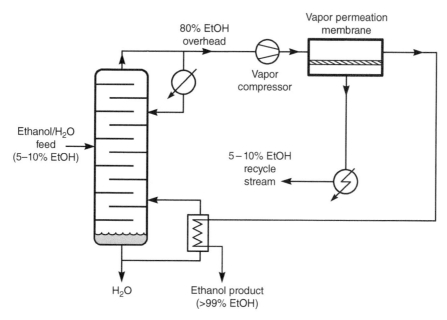

Figure 10.15 *Flow schematic of a hybrid distillation/vapor permeation process to produce dry ethanol.*

10.5.1   Bioethanol and Solvent Dehydration

10.5.1.1   Bioethanol

The separation of ethanol/water mixtures using pervaporation has been the target of continuous research for more than 40 years. Many small plants have been built, but of the 25 billion gallons of bioethanol produced annually, less than 1% is made by processes that include membranes.

In the US, ethanol is made by fermentation of corn or sorghum; in Brazil, the other major producer, sugar is used. Between 2006 and 2012, there was a surge of research and government interest in using cellulose as a feedstock. The potential was huge; studies suggested that a billion tons per year of cellulosic biomass (corn stover, wood chips, municipal solid waste, etc.) were available in the US alone. Conversion to ethanol at the expected yield of 100 gal/ton implied that 30% of US gasoline could be replaced with ethanol [9]. Unfortunately, this was not to be; cellulose conversion to ethanol is possible, but is not cheap enough to displace oil. A small market for pervaporation technology continues to exist in debottlenecking operations, or where an ethanol stream is produced as a byproduct of another process, but the anticipated major opportunities for membranes have not yet materialized. However, interest in the use of cellulose-to-ethanol conversion as a source of renewable liquid fuel and a potential replacement for oil has received a new boost as a contributor to the mitigation of global warming. Research interest in the technology is on the upswing.

A process flow diagram for a membrane-based process suitable for a cellulose-to-ethanol plant is shown in Figure 10.16. The beer feed from the fermentation step is sent to a beer still, which produces an overhead vapor containing about 50% ethanol. This vapor is compressed to 1.5 bar by a turbo fan compressor. The compressed hot vapor is sent to a two-step vapor permeation unit. The first membrane step reduces the water content of the vapor to about 10% water.

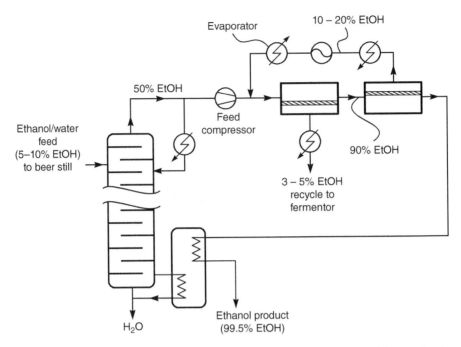

Figure 10.16 *Flow diagram of a two-step vapor permeation unit proposed for use in the next generation of cellulose-to-ethanol bioethanol plants. Source: Adapted from [22].*

The water-rich permeate is condensed and recycled to the fermentation step. The second vapor permeation step removes the remaining water, producing dry ethanol (99% EtOH) and a condensed permeate containing ethanol. This permeate is evaporated and recycled to the intake of the first membrane step. The dry ethanol product is condensed in the beer still reboiler, recovering its latent heat and reducing the heat duty of the unit.

10.5.1.2　Solvent Dehydration

Although most of the published data have focused on ethanol dehydration, there is a market for dehydration of other solvents [46]. Water forms azeotropes with a number of commonly used solvents, so a membrane process that could separate these mixtures, alone or in combination with distillation, could be widely used. The vapor–liquid equilibrium diagrams for some of the potential candidate applications are shown in Figure 10.17.

Tubular zeolite-coated ceramic membranes have been developed for this application. A process flow diagram and photograph of a small system are shown in Figure 10.18. Although the unit shown is for ethanol dehydration, similar systems have been built to dehydrate other solvents. These systems are competitive with extractive distillation or molecular sieve dehydration for small plants.

10.5.2　VOC/Water Separations

The applications above involve dehydration of an organic product. It is also possible to use pervaporation to treat water streams that contain small amounts of organics, with the goal of purifying the water. Examples include removal of VOCs from industrial process waters and

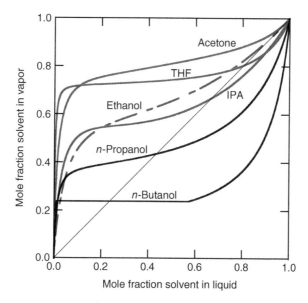

Figure 10.17 *Vapor–liquid equilibrium diagrams for solvent/water mixtures that are potential candidates for separation by water-selective pervaporation. As can be seen, all except acetone form an azeotrope with water, so that complete separations are not possible by simple distillation. Source: Adapted from [46].*

wastewaters, and treatment of contaminated groundwater. Though some interesting process designs have been proposed, including hybrid designs operating under conceptually similar principles to the distillation/membrane hybrid designs shown in Figure 10.14, commercial development has been slower than many predicted. The main problem is that membrane separation has to compete with the well-established conventional technologies of steam stripping, air stripping and carbon adsorption.

Another set of applications with more promise is flavor recovery in the food industry. Evaporator condensate and processing streams are generated in the production of concentrated orange juice, tomato paste, apple juice and the like. These streams often contain a mixture of alcohols, esters, and ketones that are the flavor elements of the juice. Steam distillation can be used to recover these elements, but the high temperatures involved damage the product. Pervaporation at low temperatures (50–70 °C) recovers essentially all of these components, producing a concentrated, high-value oil without damaging the flavor elements [47–49]. Figure 10.19 shows gas chromatography (GC) traces of the feed and permeate streams produced by pervaporation of an orange juice evaporator condensate stream.

10.5.3 Separation of Organic Mixtures

Approximately 90% of all separations carried out in the United States petrochemical and refining industries are performed by distillation of organic mixtures. In 2005, this was estimated to use about 4.5 quads of energy per year. This high energy consumption occurs because distillation involves a phase change. For example, distillation of a mixture of hydrocarbons having a relative volatility of three, a relatively easy mixture to separate, has a thermodynamic efficiency of only 6%. Mixtures with lower relative volatilities have even lower thermodynamic efficiencies, making the separation of aromatic from aliphatic hydrocarbons, where the relative volatilities

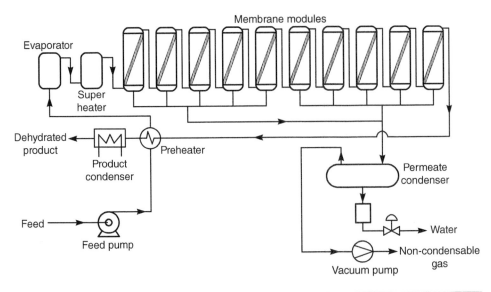

Figure 10.18 *Flow schematic of a vapor permeation dehydration system fitted with zeolite A tubular ceramic membranes. The membrane modules are arranged in a row on top of the skid. The feed evaporator is on the left-hand end and the permeate collection system is on the right. More than 100 of these 5–50 ton/day systems have been installed, mostly in China by Jiangsu Nine Heaven.*

are usually less than two, among the most energy consuming processes in the petrochemical industry. Aromatic hydrocarbons (benzene, toluene, ethylbenzene, and xylenes) and C_4–C_{10} aliphatic hydrocarbons are often separated by liquid extraction, extractive distillation, or azeotropic distillation, depending on the aromatic content.

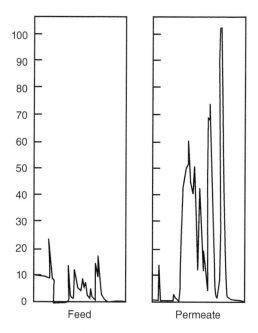

Figure 10.19 *GC traces showing the increase in concentration of flavor and aroma compounds after treatment of orange juice evaporator condensate by pervaporation.*

The use of membranes as a low-energy alternative to distillation has been proposed for more than 30 years [50]. In the early 1990s in particular, major oil companies such as Exxon, Texaco, and Mobil had research programs focused on developing membrane technology for refinery operations. These programs have since been abandoned or scaled back. The problem is not the lack of suitably selective materials, but the difficulty of making membranes and modules that can operate reliably under refinery conditions, which involve prolonged exposure to hydrocarbon mixtures at temperatures well above 100 °C. Most of the academic literature has sidestepped this problem by operating at less than 60 °C, where selectivities are often good but permeabilities are too low for an industrial process. Also, many published studies relate to simple two component mixtures that are far from the multi-component feeds of industrial interest.

One of the first refinery target separations was the removal of thiophenes during gasoline production [51, 52]. Over a 10-year period beginning around 2000, regulators throughout the developed world required refineries to reduce the sulfur content of gasoline from the then current levels of 300–1000 ppm to an average of less than 30 ppm. Some representative sulfur compounds found in refinery hydrocarbon streams are shown in Figure 10.20. Crosslinked rubbery polymers and charged polymers such as Nafion can selectively remove most of the smaller thiophenes, but benzothiophenes are not well separated. The results of a laboratory test of sulfur removal from a light naphtha feed are shown in Figure 10.21 [50]. The selectivities obtained are modest, but were good enough to be taken to the large pilot scale by Sulzer and W.R. Grace, using a tubular module configuration of the type shown in Figure 10.10.

The type of process that was being developed is shown in Figure 10.22 [52]. The sulfur-rich hydrocarbon feed is sent to the pervaporation unit which is able to remove 80–90% of the sulfur components. A membrane stage-cut of 30–35% is required to achieve this removal. The sulfur rich permeate is then treated with a conventional hydrotreater, where hydrogen is used to convert

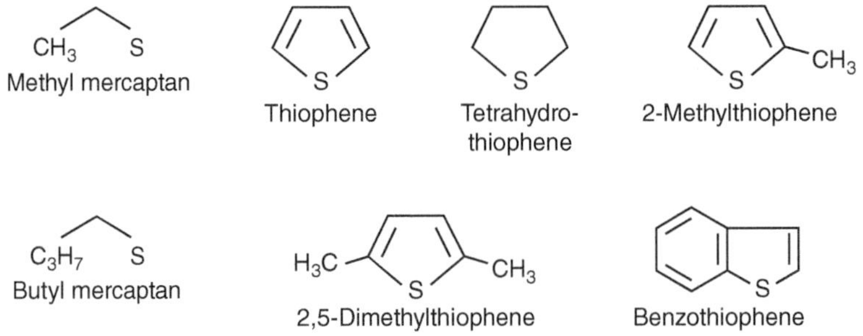

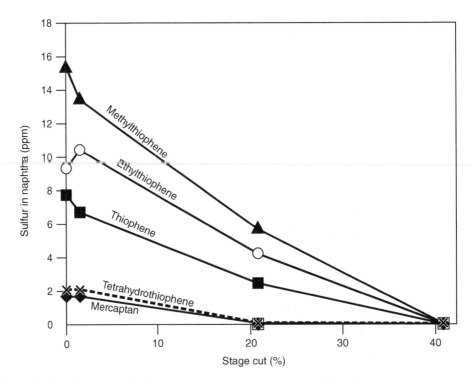

Figure 10.20   *Sulfur compounds found in gasoline feedstocks.*

Figure 10.21   *Removal of sulfur compounds from a light naphtha hydrocarbon stream in a laboratory pervaporation test unit (102 °C, hard permeate vacuum). At a stage-cut of about 30%, the bulk of the sulfur compounds are removed. Source: Adapted from [50].*

the sulfur components to hydrogen sulfide. The sulfur-free hydrotreated oil and the sulfur-lean pervaporation retentate are then mixed and sent to the gasoline pool.

A hydrotreater could have been used to treat the total initial high sulfur hydrocarbon stream but hydrotreating, as well as removing sulfur compounds, also hydrogenates olefins and aromatics in the feed. Hydrogenation of these components reduces the octane number of the gasoline product. By using a pervaporation process, approximately 70% of the feed olefins

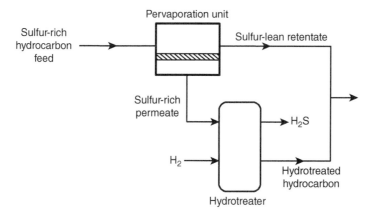

Figure 10.22   *Flow scheme of a hybrid pervaporation/hydrotreater process to produce low sulfur hydrocarbon fuel.*

and aromatics bypass the hydrotreater step, so the process impact on the octane number is much reduced.

The pilot plant results of the process were good but unfortunately, despite good results from pilot systems, the process was not adopted. The industry was under severe regulatory pressure, and it was quicker and easier to adapt conventional hydrotreating technology by using improved selective catalysts.

A second group of refinery separations that has attracted significant industrial and academic interest is the separation of C_6–C_{10} aromatics from a variety of light to medium naphtha streams. Benzene, toluene and xylene are feedstocks for nine of the top 50 chemicals produced in the United States, and are produced at a rate of about 35 million tons per year. A complete separation to recover these components is not required. For example, a process that could recover a 70% aromatics stream from a feed containing 30% aromatics and 70% paraffins and saturated cyclic hydrocarbons would significantly reduce the size and cost of downstream purification.

Refinery feeds are a complex mixture of many hydrocarbons ranging from C_6 to C_{12} components. The separation obtained by membranes is partially determined by size and partially by aromatic character. In general, the component permeance is in the order

saturated hydrocarbons < unsaturated hydrocarbons < aromatics

Saturated hydrocarbons of approximately the same boiling range but different configurations permeate in the order of increasing permeance:

branched chain < cyclic-chain < straight chain

Many patents on membranes and potential applications have issued over the years but none of the suggested processes have become commercial. White [50] and Mahdi et al. [53] have reviewed some of this work.

If the problems of membrane and module scale-up and stability are solved for mixed aromatic/ aliphatic streams, the technology could be expanded to important, but more challenging, separations such as benzene/ethylbenzene, benzene/toluene and xylene isomer separations. Laboratory studies of all of these separations have been reported but none has reached industrial use.

10.6 Conclusions and Future Directions

The use of pervaporation or vapor permeation as a low-energy alternative to distillation has been proposed for decades, yet the current market for this technology is less than US$50 million/year, almost all for the separation of water from ethanol and other hydrophilic solvents. In the early 1990s, major oil companies such as Exxon, Texaco, and Mobil had research programs focused on developing membrane technology for refinery separations, but most of these have now fallen by the wayside. The problem was the difficulty of making membranes and modules that could hold up under pervaporation conditions, while not being prohibitively expensive. Furthermore, early developers often linked the membrane systems and distillation as a simple series of unit operations. This overlooked the very substantial energy reductions achievable when heat integrated process designs are used. Significant progress has been made in solving some of these problems, and the pervaporation/vapor permeation market is slowly growing. Most of the growth is in small applications that use ceramic/zeolite membranes. However, the high cost of these membranes is likely to limit their use to small and niche applications. For pervaporation to make serious inroads into areas now dominated by distillation, it will be necessary to develop polymeric membranes and modules that will be able to hold up under refinery conditions.

The big target opportunity is to replace or supplement distillation in petrochemical and refining applications. This is not an impossible dream, but will require a long-term commitment by developers with deep pockets and vision, attributes not easily found in industry today.

References

1. Kober, P.A. (1917). Pervaporation, perstillation, and percrystallization. *J. Am. Chem. Soc.* **944**: 944–948.
2. Binning, R.C. and Stuckey, J.M. (1954). Method of separating hydrocarbons using ethyl cellulose selective membrane. US Patent 2,958,657, issued 1 November 1960 (see also US Patent 2,970,106 (1961), US Patent 3,035,060 (1962) and many others.)
3. Binning, R.C., Lee, R.J., Jennings, J.F., and Martin, E.C. (1961). Separation of liquid mixtures by permeation. *Ind. Eng. Chem.* **53**: 45.
4. Strazik, W.F. and Perry, E. (1972). Process for removing water from organic chemicals. US Patent 3,776,970, issued 4 December 1973 (see also US Patents 4,067,805 and 4,218,312 and many others).
5. Neel, J. (1995). Pervaporation. In: *Membrane Separations Technology, Principles and Applications* (ed. R.D. Nobel and S.A. Stern), 143–204. Elsevier.
6. Neel, J., Nguyen, Q.T., Clement, R., and Le Blanc, L. (1983). Fractionation of a binary liquid mixture by continuous pervaporation. *J. Membr. Sci.* **15**: 43.
7. Ballweg, A.H., Brüschke, H.E.A., Schneider, W.H. et al. (1982). Pervaporation membranes. In: *Proceedings of the Fifth International Alcohol Fuel Technology Symposium*, Auckland, New Zealand (May 1982), 97–106.
8. Brüschke, H.E.A. (1988). State of art of pervaporation. In: *Proceedings of the Third International Conference on Pervaporation Processes in the Chemical Industry* (ed. R. Bakish), 2–11. Englewood, NJ: Bakish Materials Corp.
9. Perlack, R.D., Wright, L., Turhollow, A. et al. (2005). Biomass as Feedstock for Bioenergy and Bioproducts: The Technical Feasibility of a Billion Ton Annual Supply. US Dept. of Energy/Dept. of Agriculture report. http://www.osti.gov/bridge.

10. Okamoto, K.-I., Kita, H., Horii, K., Tanaka, K. And Kondo, M. (2001). Zeolite NaA membranes: pervaporation, single gas permeation, and pervaporation and vapor permeation of water/organic liquid mixtures, *Ind. Eng. Chem. Res.* **40**, 163.

11. Bowen, T.C., Noble, R.D., and Falconer, J.L. (2004). Fundamentals and applications of pervaporation through zeolite membranes. *J. Membr. Sci.* **245**: 1.

12. Wijmans, J.G. and Baker, R.W. (1993). A simple predictive treatment of the permeation process in pervaporation. *J. Membr. Sci.* **79**: 101.

13. Baker, R.W., Wijmans, J.G., and Huang, Y. (2010). Permeability, permeance and selectivity: a preferred way of reporting pervaporation performance data. *J. Membr. Sci.* **348** (1, 2): 346–352.

14. Kataoka, T., Tsuro, T., Nakao, S.-I., and Kimura, S. (1991). Permeation equations developed for prediction of membrane performance in pervaporation, vapor permeation and reverse osmosis based on the solution–diffusion model. *J. Chem. Eng. Jpn.* **24**: 326–333.

15. Shelden, R.A. and Thompson, E.V. (1978). Dependence of diffusive permeation rates on upstream and downstream pressures. *J. Membr. Sci.* **4**: 115.

16. Bell, G.M., Huang, I., Zhou, M. et al. (2014). A vapor permeation process for the separation of aromatic compounds from aliphatic compounds. *Sep. Sci. Tech.* **49**: 2271.

17. Hollein, M.E., Hammond, M., and Slater, C.S. (1993). Concentration of dilute acetone-water solutions using pervaporation. *Sep. Sci. Technol.* **28**: 1043.

18. Chapman, P.D., Oliveira, T., Livingston, A.G., and Li, K. (2008). Membranes for the dehydration of solvents by pervaporation. *J. Membr. Sci.* **318**: 5.

19. Watanabe, K. and Kyo, S. (1992). Pervaporation performance of hollow fiber chitosan-polyacrylonitrile composite membrane in dehydration of ethanol. *J. Chem. Eng. Jpn.* **25**: 17.

20. Wenzlatt, A., Böddeker, K.W., and Hattenbach, K. (1985). Pervaporation of water-ethanol through ion exchange membranes. *J. Membr. Sci.* **22**: 333.

21. Cabasso, I. and Liu, Z.-Z. (1985). The permselectivity of ion-exchange membranes for non-electrolyte liquid mixtures. *J. Membr. Sci.* **24**: 101.

22. Huang, Y., Baker, R.W., and Vane, L.M. (2010). Low-energy distillation-membrane separation process. *Ind. Eng. Chem. Res.* **49**: 3760–3768.

23. Athayde, A.L., Baker, R.W., Daniels, R. et al. (1997). Pervaporation for wastewater treatment. *Chemtech* **27**: 34.

24. Nijhuis, H.H., Mulder, M.V.H., and Smolders, C.A. (1988). Selection of elastomeric membranes for the removal of volatile organic components from water. In: *Proceedings of the Third International Conference on Pervaporation Processes in the Chemical Industry* (ed. R. Bakish), 239–251. Englewood, NJ: Bakish Materials Corp.

25. Böddeker, K.W. and Bengtson, G. (1991). Selective pervaporation of organics from water. In: *Pervaporation Membrane Separation Processes* (ed. R.Y.M. Huang), 437–460. Amsterdam: Elsevier.

26. Chen, M.S.K., Eng, R.M., Glazer, J.L., and Wensley, C.G. (1987). Pervaporation process for separating alcohols from ethers. US Patent 4,774,365, issued 27 September 1988.

27. Schucker, R.C. (1989). Highly aromatic polyurea/urethane membranes and their use for the separation of aromatics from non-aromatics. US Patent 5,063,186, issued 5 November 1991 (see also US Patent 5,055,632 (October 1991) and many others).

28. Schucker, R.C. (1990). Isocyanurate crosslinked polyurethane membranes and their use for the separation of aromatics from non-aromatics. US Patent 4,983,338, issued 8 January 1991.

29. Vane, L.M. (2005). A review of pervaporation for product recovery from biomass fermentation processes. *J. Chem. Technol. Biotechnol.* **80**: 603.

30. Vane, L.M. (2019). Review: Membrane materials for removal of water from industrial solvents by pervaporation and vapor permeation. *J. Chem. Technol. Biotechnol.* **94**: 343.

31. Huang, Y., Ly, J., Nguyen, D., and Baker, R.W. (2010). Ethanol dehydration using hydrophobic and hydrophilic polymer membranes. *Ind. Eng. Chem. Res.* **49**: 12067.

32. Sato, K., Sugimoto, K., and Nakane, T. (2008). Synthesis of industrial scale NaY zeolite membranes and ethanol permeating performance in pervaporation and vapor permeation up to 130 °C and 570 kPa. *J. Membr. Sci.* **310** (1, 2): 161–173.

33. Okamoto, K., Kita, H., Horii, K., and Tanaka, K. (2001). Zeolite NaA membrane: preparation, simple gas permeation, and pervaporation and vapor permeation of water/organic liquid mixtures. *Ind. Eng. Chem. Res.* **40**: 163.

34. Wee, S.-L., Tye, C.-T., and Bhatia, S. (2008). Membrane separation process – pervaporation through zeolite membranes. *Sep. Purif. Tech.* **63**: 500.

35. Baker, R.W., Wijmans, J.G., Athayde, A.L. et al. (1998). Separation of volatile organic compounds from water by pervaporation. *J. Membr. Sci.* **137**: 159.

36. Vane, L.M. (2008). Separation technologies for the recovery and dehydration of alcohols from fermentation broths. *Biofuels, Bioprod. Biorefin.* **2**: 553.

37. Vane, L.M., Namboodiri, V.V., and Bowen, T.C. (2008). Hydrophobic zeolite-silicone rubber mixed matrix membranes for ethanol-water separation: effect of zeolite and silicone component selection on pervaporation performance. *J. Membr. Sci.* **308**: 230.

38. Abouchar, R. and Brüschke, H. (1992). Long-term experience with industrial pervaporation plants. In: *Proceedings of the Sixth International Conference on Pervaporation Processes in the Chemical Industry* (ed. R. Bakish), 494–502. Englewood, NJ: Bakish Materials Corp.

39. Brüschke, H.E.A., Wynn, N., and Marggraff, F.-K. (2004). Membrane pipe module, US Patent 7,655,141, issued 2 February 2010.

40. Kaschemekat, J., Barbknecht, B., and Böddeker, K.W. (1986). Konzentrierung von Ethanol durch pervaporation. *Chem. Ing. Tech.* **58**: 740.

41. Vane, L.M., Alvarez, F.R., Mairal, A.P., and Baker, R.W. (2004). Separation of vapor-phase alcohol/water mixtures via fractional condensation using a pilot-scale dephlegmator: enhancement of the pervaporation process separation factor. *Ind. Eng. Chem. Res.* **43**: 173.

42. Stephan, W., Noble, R.D., and Koval, C.A. (1995). Design methodology for a membrane/distillation column hybrid. *J. Membr. Sci.* **99**: 259.

43. Fontalvo, J., Cuellar, P., Timmer, J.M.K. et al. (2005). Comparing pervaporation and vapor permeation hybrid distillation process. *Ind. Eng. Chem. Res.* **44**: 5259.

44. Zong, C., Guo, Q., Shen, B. et al. (2021). Heat integrated pervaporation-distillation hybrid system for the separation of methyl acetate-methanol azeotropes. *Ind. Eng. Chem. Res.* **60**: 10327.

45. Cabasso, I. (1983). Organic liquid mixtures separation by permselective polymer membranes. 1. Selection and characteristics of dense isotropic membranes employed in the pervaporation process. *Ind. Eng. Chem. Prod. Res. Dev.* **22**: 313.

46. Wynn, N. (2001). Pervaporation comes of age. *Chem. Eng. Prog.* **97**: 66.

47. Rajagopalan, N. and Cheryan, M. (1995). Pervaporation of grape juice aroma. *J. Membr. Sci.* **104**: 243.

48. Karlsson, H.O.E. and Trägårdth, G. (1996). Applications of pervaporation in food processing. *Trends Food Sci. Technol* **7**: 78.

49. Pereira, C.C., Ribeiro, C.P., Nobrega, R., and Borges, C.P. (2006). Pervaporation recovery of volatile aroma compounds from fruit juice. *J. Membr. Sci.* **274**: 1.

50. White, L.S. (2006). Development of large-scale applications in organic solvent nanofiltration and pervaporation for chemical and refining processes. *J. Membr. Sci.* **286**: 26.

51. Saxton, R.J and Minhas, B.S. (2001). Ionic membranes for organic sulfur separations from liquid hydrocarbon solutions. US Patent 6,702,945, issued 9 March, 2004.
52. Balko, J.W. (2003). Method of reducing sulfur in hydrocarbon feedstock using a membrane separation zone. US Patent 7,267,761, issued 11 September, 2007.
53. Mahdi, T., Almad, A., Nasef, M.M., and Ripin, A. (2015). State-of-the-art technologies for separation of azeotropic mixtures. *Sep. Purif. Rev.* **44**: 308.

11

Ion Exchange Membrane Processes

Membrane Technology and Applications, Fourth Edition. Richard W. Baker.
© 2024 John Wiley & Sons Ltd. Published 2024 by John Wiley & Sons Ltd.

11.1 Introduction and History

Ion exchange membranes are used in electrodialysis stacks, chlor-alkali cells, fuel cells, and batteries. They are characterized by charged groups attached to the polymer backbone of the membrane material; these fixed groups partially or completely repel ions of the same charge from the membrane. An anion exchange membrane has fixed positive groups and excludes positive ions but is freely permeable to negative anions. Similarly, a cation exchange membrane has fixed negative groups and excludes negative ions but is freely permeable to positive cations, as illustrated in Figure 11.1.

Experiments with ion exchange membranes were described as early as 1890 by Ostwald [1]. Work by Donnan [2] a few years later led to development of the concept of membrane potential and to the quantitative analysis of the phenomenon now referred to as Donnan exclusion. The membranes were made from natural materials or chemically treated collodion membranes, and their mechanical and chemical properties were very poor. Nonetheless, as early as 1939, Manegold and Kalauch [3] suggested their application to separate ions from water, and within another year Meyer and Strauss [4] described the concept of a multicell arrangement of ion exchange membranes between a single pair of electrodes, the first electrodialysis stack. The advances in polymer chemistry during and immediately after World War II led to the production of much better membranes by Kressman [5], Murphy et al. [6], and Juda and McRae [7] at Ionics. With the development of these membranes, electrodialysis became practical. Ionics was the principal early developer and installed the first successful plant in 1952; by 1956 eight plants had been installed.

In the United States, electrodialysis was developed primarily for desalination of water, with Ionics, now a division of Suez, being the industry leader. In Japan, Asahi Glass, Asahi Chemical (a different company), and Tokuyama Soda developed the process along opposite lines, as a way to concentrate seawater [8]. This application of electrodialysis is confined to Japan, which has no domestic salt sources. Electrodialysis membranes concentrate the salt in seawater to about 20% solids, after which the brine is further concentrated by evaporation, and the salt recovered by crystallization.

Electrodialysis is now a mature technology, with Ionics remaining an industry leader except in Japan. Desalting of brackish water and the production of boiler feed water and industrial process

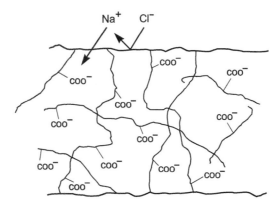

Figure 11.1 *A cationic membrane with fixed carboxylic acid groups is permeable to cations such as sodium but is impermeable to anions such as chloride.*

water were the main applications until the 1990s, when electrodialysis began to lose market share, due to stiff competition from improved reverse osmosis (RO) membranes. Beginning about that time, electrodeionization, a process combining electrodialysis and ion exchange, was developed and began to be used to achieve very good salt removal in ultrapure water plants. This is now a major use of electrodialysis. Other applications are control of ionic impurities from industrial effluent streams, water softening, and desalting certain foods, particularly milk whey [9–11].

Over the last 30 years, a number of other uses of ion exchange membranes have been found. An important application in chlor-alkali cells was enabled by the development of perfluorinated membranes with exceptional chemical stability by Asahi, Dow and DuPont [12]. These membranes were introduced in 1979. Since then, cells containing a total of more than 1 million square meters of ion exchange membranes have been installed, and membrane processes have displaced the older mercury amalgam and asbestos diaphragm processes.

Another growing use of ion exchange membranes is in fuel cells, where the membrane regulates ion transport [13–15]. The operating principle of the fuel cell was described by Sir William Grove (and independently by the Swiss scientist Christian Shoenbein) in 1839. Grove's device consisted of an electrochemical cell with two electrodes. Hydrogen was bubbled over the surface of one electrode, oxygen over the other. As long as the gas supply was maintained, a current flowed through a wire connecting the two electrodes. Grove's cell was a scientific curiosity until the 1940s, when Francis Bacon, at Cambridge and Kings College, London, started to develop practical devices. By the late 1950s, he had made a 6 kW fuel cell. The technology took off with the US space program in the 1960s. Spacecraft needed far more electric power than could be stored in the batteries, and fuel cells are an efficient lightweight method of converting chemical energy into electric power. The Gemini, Apollo, and all subsequent space programs were equipped with fuel cells. Since then, a huge effort has been spent developing fuel cells for more mundane applications, with some limited success.

Ion exchange membranes have also found a market in electrochemical processes of various types. One application that has received a good deal of attention is the use of bipolar membranes, which are laminates of anion and cation exchange membranes, to produce acids and alkalis by electrolysis of salts. The first practical bipolar membranes were developed by K.J. Liu and others at Allied Chemicals in about 1977 [16]; they were later employed in Allied's Aquatech acid/base production process [17]. Ion exchange membranes are also being increasingly used in batteries and other energy storage devices.

A timeline illustrating the major milestones in the development of ion exchange membranes is shown in Figure 11.2.

11.2   Theoretical Background

Electrodialysis was the first industrial process to use ion exchange membranes on a large scale. In an electrodialysis system, anion and cation membranes are formed into a multicell arrangement built on the plate-and-frame principle, to form up to 100 cell pairs in a stack. The membranes are arranged in an alternating pattern between the anode and cathode. Each set of anion and cation membranes forms a cell pair. Salt solution is pumped through the cells, while an electrical potential is maintained across the electrodes. The positively charged cations in the solution migrate toward the cathode and the negatively charged anions migrate toward the anode. Cations pass easily through the cation exchange membrane but are retained by the anion exchange membrane. Similarly, anions pass through the anion exchange membrane but are retained by

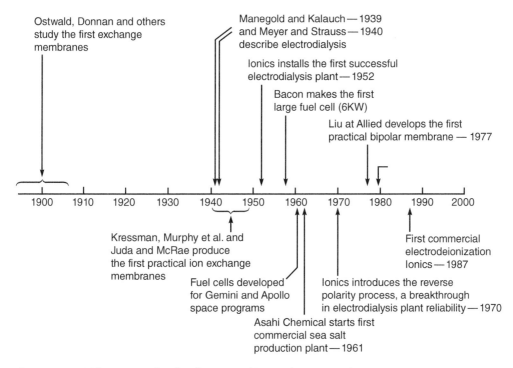

Figure 11.2   Milestones in the development of ion exchange membrane processes.

the cation exchange membrane. The overall result is that one cell of the pair becomes depleted of ions while the adjacent cell becomes enriched in ions. The process, which is widely used to remove dissolved ions from water, is illustrated in Figure 11.3.

11.2.1   Transport Through Ion Exchange Membranes

In electrodialysis and the other separation processes using ion exchange membranes, transport of components generally occurs under two driving forces: a concentration gradient and an electric potential difference (voltage). However, because anions and cations move in opposite directions when an external voltage is applied, ion exchange membrane processes are often more easily evaluated in terms of the amount of charge transported than the amount of material transported. Consider, for example, a simple univalent–univalent electrolyte such as sodium chloride, which can be considered to be completely ionized in dilute solutions. The total concentration, c, of the solution is made up of the concentration of sodium cations c^+, and the concentration of chloride anions c^-. In a simple equimolar salt solution, c, c^+, and c^- are equal. The velocity of the cations in an externally applied field of strength, is u (cm/s), and the velocity of the anions measured in the same direction is $-v$ (cm/s). Each cation carries the protonic charge $+e$, and each anion the electronic charge of $-e$, so the total amount of charge transported per second across a plane of 1 cm^2 area is

$$\frac{I}{F} = c^+\,(u)(+e) + c^-\,(-v)(-e) = ce(u + v) \tag{11.1}$$

where I is the current and F is the Faraday constant to convert transport of electric charge to a current flow in amps. This equation links the electric current with the transport of ions.

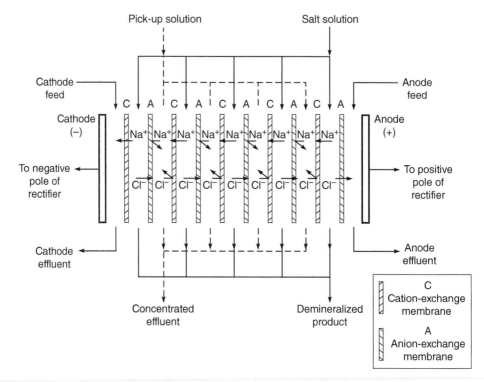

Figure 11.3 *Schematic diagram of a plate-and-frame electrodialysis stack in operation. Alternating cation and anion exchange membranes are arranged in a stack of up to 100 cell pairs.*

The fractions of the current carried by the anions and cations do not necessarily have to be equal. The fraction of the total current carried by any particular ion is known as the transport number of that ion. If the transport number for the cations is t^+, and for anions is t^-, then

$$t^+ + t^- = 1 \tag{11.2}$$

Combining Eqs. (11.1) and (11.2), the transport number of the cations in the univalent–univalent electrolyte described above is given as

$$t^+ = \frac{c^+ ue}{ce(u + v)} = \frac{u}{u + v} \tag{11.3}$$

and similarly for the anion

$$t^- = \frac{v}{u + v} \tag{11.4}$$

Transport numbers for different ions, even in aqueous solutions, can vary over a wide range, reflecting the different sizes of the ions. Ions with the same charge as the fixed charge groups in an ion exchange membrane are excluded from the membrane and, therefore, carry a very small fraction of the current through the membrane. The transport number of the excluded ions is very small, normally between 0 and 0.05. Counter ions with a charge opposite to the fixed charged groups permeate the membrane freely and carry almost all of the current through the membrane. The transport numbers of these ions are between 0.95 and 1.0. This difference in transport number makes the membrane selective for ions of the opposite charge and allows separations to be achieved with ion exchange membranes.

Equation (11.1) shows that, as in other transport processes, the flux of the permeating component is the product of a mobility term (u or v) and a concentration term (c^+ or c^-). In ion exchange transport processes, most of the separation is achieved by manipulating the concentration terms.

The ability of ion exchange membranes to discriminate between oppositely charged ions was put on a mathematical basis by Donnan in 1911 [2]. Figure 11.4 shows the distribution of ions between a salt solution and an ion exchange membrane containing fixed negative charges, R^-.

The equilibrium between the ions in the membrane (m) and the surrounding solution (s) can be expressed as

$$c^+_{(m)} \cdot c^-_{(m)} = k \, c^+_{(s)} \cdot c^-_{(s)} \tag{11.5}$$

where k is an equilibrium constant. Likewise, in equilibrium, the charges within the membrane are in balance

$$c^+_{(m)} = c^-_{(m)} + c_{R^-_{(m)}} \tag{11.6}$$

where $c_{R^-_{(m)}}$ is the concentration of fixed negative charges in the membrane. For a fully dissolved salt, such as sodium chloride, the total molar concentration of the salt $c_{(s)}$ is equal to the concentration of each of the ions, so

$$c_{(s)} = c^+_{(s)} = c^-_{(s)} \tag{11.7}$$

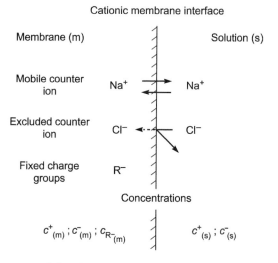

Figure 11.4 *An illustration of the distribution of ions between a cation exchange membrane (fixed negative R⁻ groups) and the adjacent salt solution.*

Combining these three equations and rearranging gives the expression

$$\frac{c^+_{(m)}}{c^-_{(m)}} = \frac{\left[c^-_{(m)} + c_{R^-_{(m)}}\right]^2}{k\left[c_{(s)}\right]^2} \tag{11.8}$$

Because the membrane has fixed negative charges, the concentration of negative counter-ions in the membrane will be small compared to the concentration of fixed charges, that is,

$$c^-_{(m)} \ll c_{R^-_{(m)}} \tag{11.9}$$

so it can be assumed that

$$c^-_{(m)} + c_{R^-_{(m)}} \approx c_{R^-_{(m)}} \tag{11.10}$$

Equation (11.8) can then be written

$$\frac{c^+_{(m)}}{c^-_{(m)}} = \frac{1}{k}\left(\frac{c_{R^-_{(m)}}}{c_{(s)}}\right)^2 \tag{11.11}$$

This expression shows that the ratio of sodium to chloride ions in the membrane $(c^+_{(m)}/c^-_{(m)})$ is strongly dependent on the ratio of the fixed charge groups in the membrane to the salt concentration in the adjacent solution $\left(c_{R^-_{(m)}}/c_{(s)}\right)$. In commonly used membranes, the fixed ion concentration in the membrane is very high, typically at least 3–4 milliequivalents per gram (meq/g). Figure 11.5 shows a plot of the sodium-to-chloride concentration ratio in a cation exchange membrane, calculated using Eq. (11.11). The membrane is assumed to have a fixed negative charge concentration of 3 meq/g. The plot shows that, at salt solution concentrations of less than 0.2 meq/g (~1 wt% sodium chloride), chloride ions are almost completely excluded from the membrane. This means that in this concentration range, the transport number for sodium is close to one, and for chloride it is close to zero. Only at high salt concentrations – above about 0.6 meq/g (3 wt% sodium chloride) – does the ratio of sodium to chloride ions in the membrane fall below 30, and the membrane becomes measurably permeable to chloride ions.

11.3   Chemistry of Ion Exchange Membranes

A wide variety of ion exchange membrane chemistries has been developed. Most early work was aimed at electrodialysis applications, and each manufacturer developed its own tailored membrane. Many have been kept as trade secrets, or are only described in the patent literature. Korngold [18] gives a description of early ion exchange membrane development. More recently, most membrane development has focused on membranes for fuel cells. Summaries of these developments can be found in a book by Strathmann [10] and reviews by Nunes and Peinemann [13] and Ran et al. [11].

Ion exchange membranes contain a high concentration of fixed ionic groups, which tend to absorb water; charge repulsion of the groups can then cause the membrane to swell excessively in aqueous environments. Most ion exchange membranes are highly crosslinked to limit

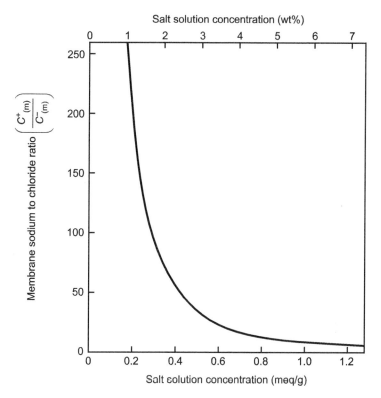

Figure 11.5 *The sodium-to-chloride ion concentration ratio inside a cation exchange membrane containing a concentration of fixed negative groups of 3 meq/g as a function of salt concentration. At salt concentrations in the surrounding solution of less than about 1 wt% sodium chloride (0.2 meq/g), chloride ions are almost completely excluded from the membrane.*

swelling, but high crosslinking densities make polymers brittle; the membranes are usually stored and handled wet to allow absorbed water to plasticize the membrane. Membranes are usually produced as homogenous films 50–200 μm thick, and often reinforced by casting the film onto a net or fabric to maintain the shape and to minimize swelling.

Ion exchange membranes fall into two broad categories: homogeneous and heterogeneous. In homogeneous membranes, the charged groups are uniformly distributed through the membrane matrix. These membranes swell relatively uniformly when exposed to water, the extent of swelling being controlled by their crosslinking density. In heterogeneous membranes, the ion exchange groups are contained in small domains distributed throughout an inert support matrix, which provides mechanical strength. Heterogeneous membranes can be made, for example, by dispersing finely ground ion exchange particles in a polymer support matrix. In recent years, finely dispersed heterogeneous membranes have been made by casting membrane films from ABA block copolymers. The A and B blocks phase separate as the casting solvent evaporates, forming two domain structures. The membrane film is then chemically treated to introduce fixed charges into one of the phase separated domains. This domain forms the ion conducting path through the membrane. Because of the difference in the degree of swelling between the ion exchange portion and the inert portion of heterogeneous membranes, mechanical failure, leading to leaks at the boundary between the two domains, can be a problem.

11.3.1   Homogeneous Membranes

A number of early homogeneous membranes were made by simple condensation reactions of suitable monomers, such as the phenol–formaldehyde condensation reaction shown in Figure 11.6. The mechanical stability and ion exchange capacity of these condensation resins were modest. It was found that a better approach is to prepare a suitable crosslinked base membrane, which can be converted to a charged form in a subsequent reaction. Ionics used this method to make many of their membranes. In a typical preparation procedure, a 60:40 mixture of styrene and divinylbenzene is cast onto a fabric web, sandwiched between two plates and heated in an oven to form the membrane matrix. The membrane is then sulfonated with 98% sulfuric acid or a concentrated sulfur trioxide solution. The degree of crosslinking in the final membrane is controlled by varying the divinylbenzene concentration in the initial mix. The degree of sulfonation can also be varied. The chemistry of the process is shown in Figure 11.7.

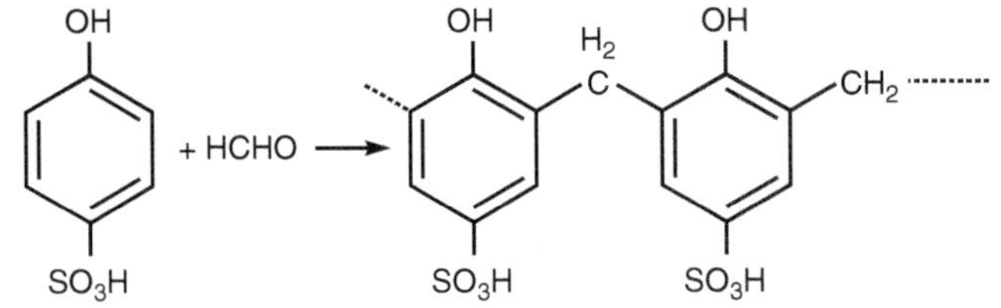

Figure 11.6   *Phenol-formaldehyde condensation reaction used to make early cationic membranes.*

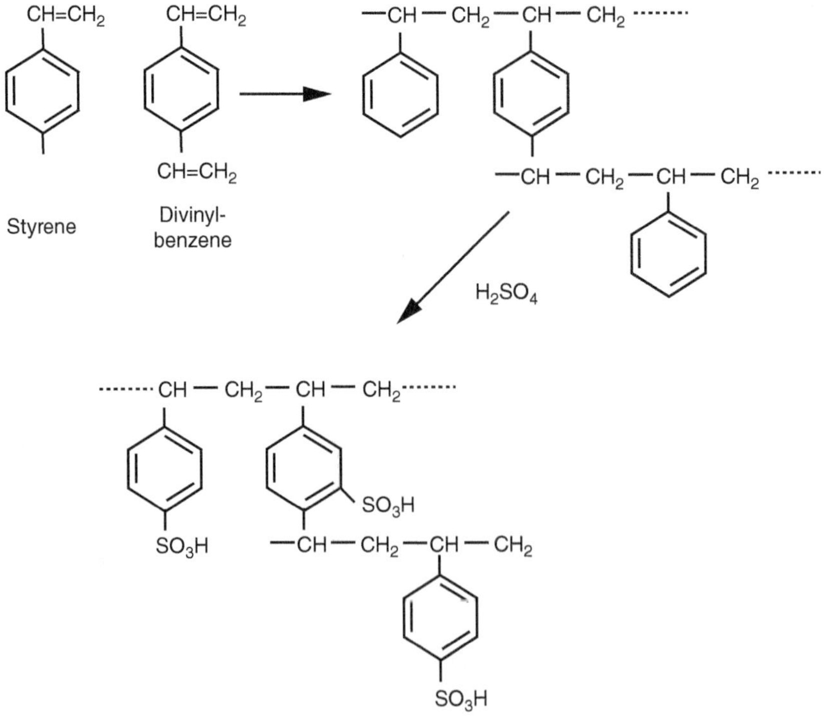

Figure 11.7   *Reactions involved in preparation of a cation exchange membrane.*

Anion exchange membranes can be made from the same crosslinked polystyrene membrane base by post-treatment with monochloromethyl ether and aluminum chloride to introduce chloromethyl groups into the benzene ring, followed by formation of quaternary amines with trimethylamine (Figure 11.8).

A particularly important category of ion exchange polymers is the perfluorocarbon type made by DuPont under the trade name Nafion® [19]. The base polymer is made by polymerization of a sulfinol fluoride vinyl ether with tetrafluoroethylene. The copolymer formed is extruded as films about 120 μm thick, after which the sulfinol fluoride groups are hydrolyzed to form sulfonic acid groups (Figure 11.9).

Asahi Chemical [20] and Tokuyama Soda [21] have developed similar chemistries in which the $-CF_2SO_3H$ groups are replaced by carboxylic acid groups. Membranes made from these polymers have better selectivity than the perfluorosulfonic membranes. In all these perfluoro polymers, the backbone is extremely hydrophobic, whereas the charged acid groups are strongly polar. Because the polymers are not crosslinked, some phase separation into different domains takes place. The hydrophobic perfluoropolymer domains provide a nonswelling matrix, ensuring the integrity of the membrane. The ionic hydrophilic domains absorb water and form connected water-filled channels throughout the supporting matrix. This configuration, illustrated in Figure 11.10, minimizes both the interaction of ions and

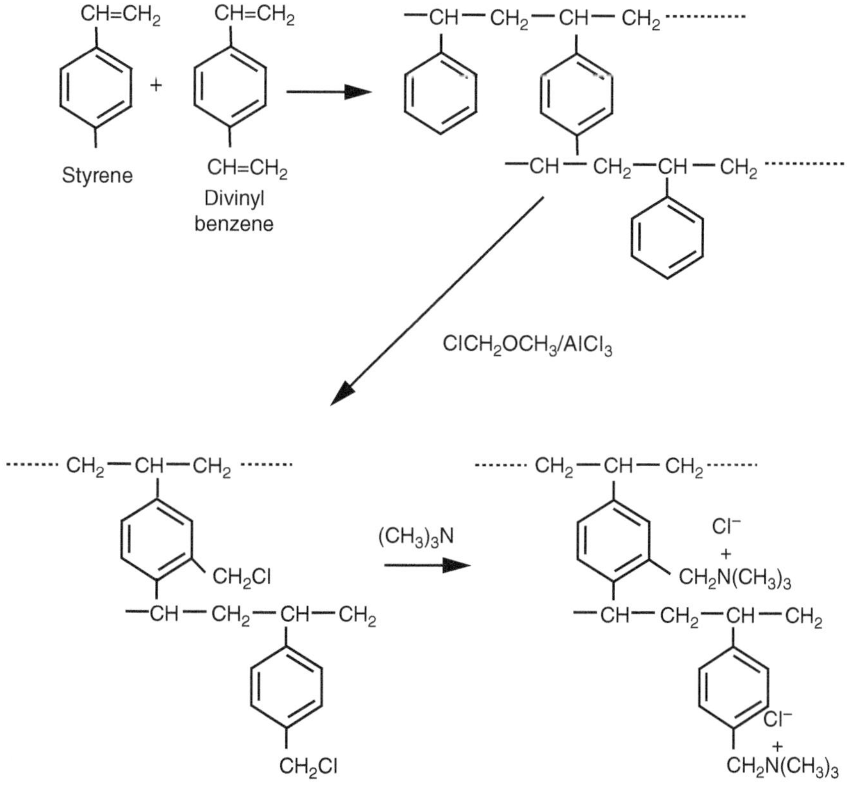

Figure 11.8 *Reactions involved in preparation of a crosslinked anion exchange membrane.*

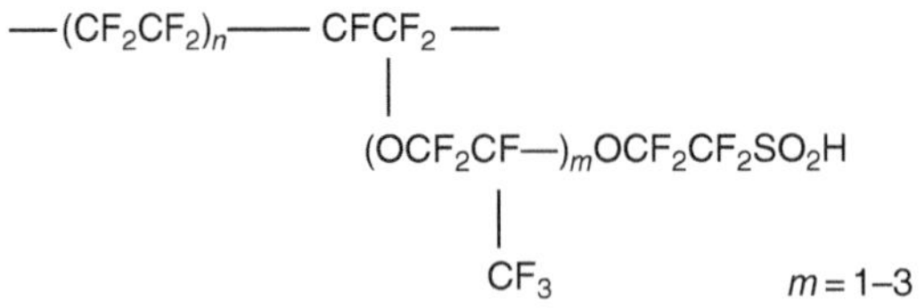

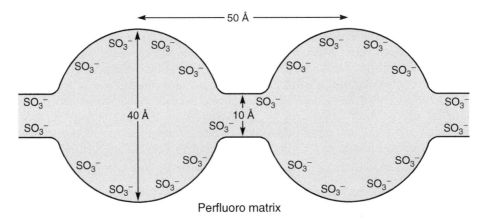

$$—(CF_2CF_2)_n——CFCF_2—$$
$$(OCF_2CF—)_mOCF_2CF_2SO_2H$$
$$CF_3 \qquad m=1\text{--}3$$

Figure 11.9   *DuPont's Nafion® perfluorocarbon cationic exchange membrane.*

Figure 11.10   *Schematic of the cluster model used to describe the distribution of sulfonate groups in perfluorocarbon-type cation exchange membranes such as Nafion. Source: [19] / American Chemical Society.*

water with the backbone and the electrostatic repulsion of close sulfonate groups. These perfluorocarbon membranes are completely inert to concentrated sodium hydroxide solutions and have been widely used in membrane electrochemical cells in the chlor-alkali industry and as polymer electrolyte membranes in fuel cells. One drawback of the membranes is their high cost: US$100/m² of membrane.

11.3.2   Heterogeneous Membranes

Heterogeneous membranes have been produced by a number of Japanese manufacturers. The simplest form has very finely powdered cation or anion exchange particles uniformly dispersed in polypropylene. A film of the material is extruded to form the membrane. The mechanical properties of these membranes are often poor because of swelling of the relatively large – 10–20 μm diameter – ion exchange particles. A much finer heterogeneous dispersion of ion exchange particles, and consequently a more stable membrane, can be made with a poly(vinyl chloride) (PVC) plastisol. A plastisol of approximately equal parts PVC, styrene monomer, and crosslinking agent in a dioctyl phthalate plasticizing solvent is prepared, then cast and polymerized as a film. The PVC and polystyrene polymers form an interconnected domain structure. The styrene groups are then sulfonated by treatment with concentrated sulfuric acid or sulfur trioxide to form a very finely dispersed but heterogeneous structure of sulfonated polystyrene in a PVC matrix, which provides toughness and strength.

11.4 Electrodialysis

11.4.1 Concentration Polarization and Limiting Current Density

Transport of permeating species in electrodialysis cell is driven by an applied voltage as well as a concentration gradient. This makes the inevitable issue of concentration polarization both more interesting and more challenging than in some other processes. Transport of ions in an electrodialysis cell in which the salt solutions in the chambers are very well stirred is shown in Figure 11.11. In this example, negative chloride ions migrating left toward the anode permeate the anionic membranes (labeled A) and are stopped by the cation exchange membranes (labeled C). Similarly, sodium ions migrating to the cathode permeate the cation exchange membranes, but are stopped by the anion exchange membranes. The overall result is an increased salt concentration in alternating compartments, and corresponding salt depletion in the other compartments.

The drawing indicates that the voltage drop within the apparatus takes place entirely across the membranes. This is the case for a well-stirred cell, in which the solutions in the compartments are completely turbulent. In this case, the flux of ions across the membranes, and hence the productivity of the electrodialysis system, can be increased without limit by increasing the voltage across the stack. In practice, however, the resistance of the membrane is often small in proportion to the resistance of the water-filled compartments, particularly the dilute compartments, where the concentration of ions carrying the current is low. As the voltage is raised, the dilute compartments become increasingly ion-depleted, and hence resistant to current flow, increasing the energy consumption per unit transport of charge and making the process progressively more inefficient. Furthermore, as explained below, the formation of a severely ion-depleted boundary layer within the dilute cells adjacent to the membrane places an upper limit on the current that can flow by electrodialysis. It is ion transport through this ion-depleted aqueous boundary layer that generally controls system performance.

The formation of concentration gradients caused by the flow of ions through a single cationic membrane in the electrodialysis cell stack of Figure 11.11 is shown in

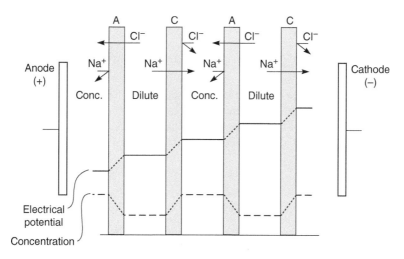

Figure 11.11 *Schematic of the concentration and potential gradients in a well-stirred electrodialysis membrane stack.*

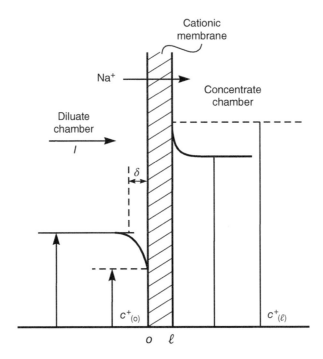

Figure 11.12   *Schematic of the concentration gradients adjacent to a single cationic membrane in an electrodialysis stack. The concentration of sodium ions in the bulk solution and at the membrane interface are shown.*

Figure 11.12. As in the treatment of concentration polarization in other membrane processes, the resistance of the aqueous solution is modeled as a thin boundary layer of unstirred solution separating the membrane surface from the well-stirred bulk solution. In electrodialysis, the thickness δ of this unstirred layer is generally 20–50 μm. Concentration gradients form in this layer because only one of the ionic species is transported through the membrane. This species is depleted in the boundary layer on the feed side and enriched in the boundary layer on the permeate side.

Figure 11.12 shows the concentration gradient of univalent sodium ions next to a cation membrane (Exactly equivalent gradients of chloride ions form adjacent to the anion exchange membranes in the stack, and can be modeled in the same way). Transport of sodium ions across the boundary layer to the membrane surface takes place by a combination of Fickian diffusion and current flow in response to an applied voltage. The rate of diffusion J_{diff}^{+} of cations through the boundary layer to the surface by concentration-driven diffusion according to Fick's law is given by:

$$J_{diff}^{+} = D^{+} \frac{\left(c^{+} - c_{(o)}^{+} \right)}{\delta} \tag{11.12}$$

where D^{+} is the diffusion coefficient of the cation in water, c^{+} is the bulk concentration of the cation in the diluate chamber solution, and $c_{(o)}^{+}$ is the concentration of the cation in the solution adjacent to the membrane surface (o).

The flux of cations through the boundary layer by electric current flow is t^+I/F. It follows that the total flux of sodium ions through the boundary layer to the membrane surface, J_b^+, is the sum of these two terms. That is

$$J_b^+ = \frac{D^+ \left(c^+ - c_{(o)}^+ \right)}{\delta} + \frac{t^+ I}{F} \tag{11.13}$$

Likewise, transport through the membrane itself, J_m^+, is also the sum of two terms, one due to the voltage, the other due to the difference in ion concentrations on each side of the membrane. Thus, the ion flux through the membrane can be written

$$J_m^+ = \frac{t_{(m)}^+ I}{F} + \frac{P^+ \left(c_{(o)}^+ - c_{(\ell)}^+ \right)}{\ell} \tag{11.14}$$

where P^+ is the permeability of the sodium ions in a membrane of thickness ℓ. The quantity $P^+ \left(c_{(o)}^+ - c_{(\ell)}^+ \right)/\ell$ is much smaller than transport due to the voltage gradient, and the flux through the boundary layer and the membrane must be equal in equilibrium, so Eqs. (11.13) and (11.14) can be combined and simplified to

$$D^+ \frac{\left(c^+ - c_{(o)}^+ \right)}{\delta} + \frac{t^+ I}{F} = \frac{t_{(m)}^+ I}{F} \tag{11.15}$$

For a cation exchange membrane that excludes anions, $t_{(m)}^+ \approx 1$, and Eq. (11.15) can be further simplified to

$$I = \frac{F}{1 - t^+} \cdot \frac{D^+}{\delta} \left(c^+ - c_{(o)}^+ \right) \tag{11.16}$$

As the solution next to the membrane surface becomes increasingly depleted of permeating ions, the concentration gradient across the boundary layer becomes steeper. A point can be reached at which the ion concentration at the membrane surface, $c_{(o)}^+$, is approaching zero and the current, I, is approaching a maximum. In the limit Eq. (11.16) reduces to

$$I_{\lim} = \frac{D^+ F c^+}{\delta (1 - t^+)} \tag{11.17}$$

This limiting current, $I_{\lim}$, is the maximum current that can be carried by electrodialysis, and is generally normalized to unit membrane area and referred to as the limiting current density, expressed in mA/cm^2. Equation (11.17) shows that the performance of electrodialysis processes is limited by the boundary layer thickness, that is by concentration polarization, rather than by membrane permeation. If the applied voltage across the stack is increased beyond that required to produce the limiting current, the extra current will be carried by other processes, first by transport of anions through the cation exchange membrane and, at higher potentials, by hydrogen and hydroxyl ions formed by dissociation of water. Both of these undesirable processes consume power without producing any separation. This decreases the current efficiency of the process, that is, the separation achieved per unit of energy consumed. A more detailed discussion of the effect of the limiting current density on electrodialysis performance is given by Krol et al. [22].

The limiting current can be determined experimentally by plotting the electrical resistance across the membrane stack against the reciprocal electric current. This is called a Cowan–Brown plot after its original developers [23]; Figure 11.13 shows an example for a laboratory cell [24]. At a reciprocal current of 0.1 A^{-1}, the resistance has a minimum value. When the limiting current is exceeded, the excess current is not used to transport ions. Instead, the current causes water to dissociate into protons and hydroxyl ions. The pH of the solutions in the cell chambers undergoes an abrupt and pronounced change, reflecting this water splitting. This change in pH, also shown in Figure 11.13, can also be used to determine the value of the limiting current density.

In industrial-scale electrodialysis systems, determining the limiting current is not so easy. In large membrane stacks, the boundary layer thickness will vary from place to place across the membrane surface. The limiting current where the boundary layer is relatively thick, because of poor fluid flow distribution, will be lower than where the boundary layer is thinner. Thus,

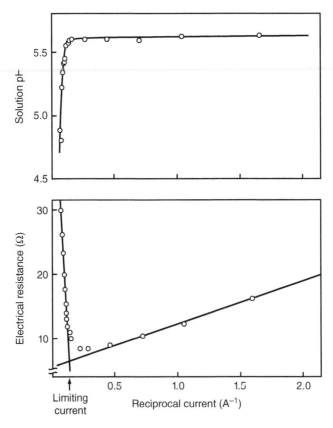

Figure 11.13   *Cowan–Brown plots showing how the limiting current density can be determined by measuring the stack resistance or the pH of the dilute solution as a function of current.*
Source: Redrawn from data in [24].

the measured limiting current may be only an average value. In practice, systems are operated at currents well below the limiting value.

The limiting current density for an electrodialysis system operated at a constant solution flow rate is a function of the feed solution salt concentration, as shown in Eq. (11.17). As the salt concentration in the solution increases, more ions are available to transport current in the boundary layer, so the limiting current density also increases. For this reason, large electrodialysis systems with several electrodialysis stacks in series will operate with different current densities in each stack, reflecting the change in the feed water concentration as salt is removed.

As we have seen, once the limiting current is reached, further changes in applied voltage do not increase the current through the membrane, and energy is dissipated without achieving an increase in separation. However, at very high applied potential, an increase in current does occur. The appearance of this so-called overlimiting current, shown in Figure 11.14, is a phenomenon the origins of which have been debated for a number of years. Recent thinking is that surface heterogeneity leads to charge separation and electroconvection in the ion-depleted boundary layer of the membrane [25]. At high applied voltages, tiny cells are created by charge separation, and these cells bring ions to the membrane surface by convection.

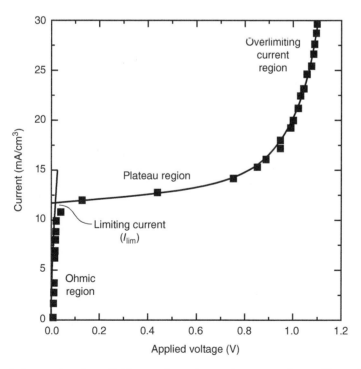

Figure 11.14 *Full current–voltage (I–V) curve for an ion exchange membrane, illustrating the ohmic region controlled by the diffusion in the membrane boundary layer; the plateau region above the limiting current, where the diffusion gradient has reached its maximum value; and the overlimiting current region, where electroconvection supplements diffusion in transporting ions through the membrane boundary layer [25] / American Chemical Society.*

11.4.2 Current Efficiency and Power Consumption

A key factor determining the overall efficiency of an electrodialysis process is the energy consumed to perform the separation. Power consumption E in kilowatts, is linked to the current I through the stack and the resistance R of the stack by the expression

$$E = I^2 R \tag{11.18}$$

The theoretical electric current I_{theor} is directly proportional to the number of charges transported across the ion exchange membrane and is given by the expression

$$I_{theor} = z\Delta C F Q \tag{11.19}$$

where Q is the feed flow rate, ΔC is the difference in molar concentration between the feed and the dilute solutions, z is the valence of the salt, and F is the Faraday constant. Thus the theoretical power consumption E_{theor} to achieve a given separation is given by substituting Eq. (11.19) into Eq. (11.18) to give:

$$E_{theor} = IRz\Delta C Q F \tag{11.20}$$

or

$$E_{theor} = Vz\Delta C Q F \tag{11.21}$$

where V is the theoretical voltage drop across the stack. In the absence of concentration polarization and any resistance losses in the membrane or solution compartments, the theoretical energy required to desalinate seawater (~35 000 ppm salt) is about 2 kWh/m^3 of product. Energy consumption is inversely proportional to the salt concentration, so the theoretical energy consumption for desalination of dilute brackish water feeds in the 2000–3000 ppm range is only about 0.2 kWh/m^3.

The actual energy consumed in desalination is higher than the theoretical value for two reasons. First, owing to the concentration polarization effects discussed above, the actual voltage applied across the stack has to be higher than the theoretical voltage of Eq. (11.21), because much of the voltage is dissipated in overcoming resistance in the dilute compartments, and especially in the boundary layers. The result is to increase the actual energy consumption to well above the theoretical minimum value. In commercial electrodialysis plants, concentration polarization is controlled by circulating the solutions through the stack at a high rate. Various feed spacer designs are used to maximize turbulence in the cells. Because electric power is used to power the feed and product solution circulation pumps, a trade-off exists between the power saved because of the increased efficiency of the electrodialysis stack and the power consumed by the pumps. In modern electrodialysis systems, the circulation pumps can consume one-quarter of the total power. Even under these conditions, concentration polarization is not fully controlled and actual energy consumption is substantially higher than the theoretical value.

Although most inefficiencies in electrodialysis systems are related to the difficulty in controlling concentration polarization, a number of minor current utilization losses arise from the following factors [10]:

1. Ion exchange membranes are not completely semipermeable; some leakage of co-ions of the same charge as the membrane can occur. This effect is generally negligible at low feed

solution concentrations, but can be serious with concentrated solutions, such as when seawater is treated for salt production.

2. Ions permeating the membrane carry solvating water molecules in their hydration shell. Also, osmotic transport of water from the dilute to the concentrated chambers can occur.

3. A portion of the electric current can be carried by the stack manifold, bypassing the membrane cell. Modern electrodialysis stack designs generally make losses due to this effect negligible.

11.4.3   System Design

An electrodialysis plant consists of several elements:

- a feed pretreatment system;
- the membrane stack;
- the power supply and process control unit;
- the solution pumping system.

Many small plants use a single electrodialysis stack, as shown in Figure 11.15. Manifolding may be used to allow the feed and brine solutions to pass through several cell pairs in series, but the entire procedure is performed in a single stack.

In large systems, using several separate electrodialysis stacks in series to perform the separation is more efficient [10]. Multiple stacks are used because the current density of the first stack is higher than the current density of the last stack, which is operating on a more dilute feed solution. As in the single-stack system, the feed solution may pass through several cell pairs in each stack in series. Because concentration polarization worsens as the solution becomes more dilute, the solution velocity in the stacks is increased progressively to counter this effect. The

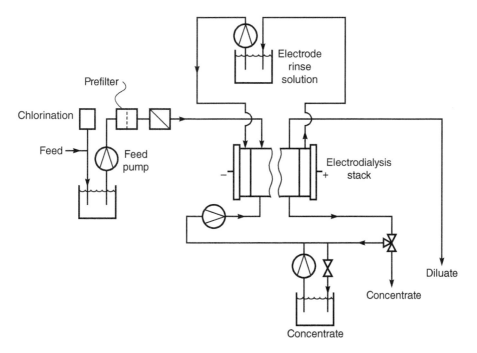

Figure 11.15   *Flow diagram of a typical small electrodialysis plant [10] / with permission of Elsevier.*

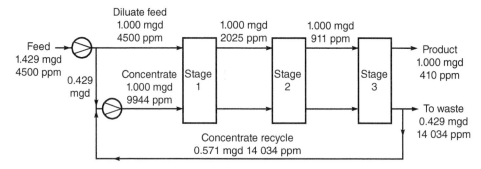

Figure 11.16   *Flow scheme of a three-stage electrodialysis plant. Source: Adapted from [26].*

velocity is controlled by the number of cell pairs through which the solution passes in each stack. The number of cell pairs used in series decreases from the first to the last stack; this is known as the taper of the system. The flow scheme of a three-stage design is shown in Figure 11.16. The use of optimized multi-stage electrodialysis designs has significantly reduced the energy consumption of electrodialysis. In 2000, seawater desalination used 6–7 kWh/m^3 of product, much better than the first systems, but still more than double the energy consumption of the competitive reverse osmosis process. This was a major reason why the market for seawater desalination by electrodialysis did not take off. Today, some process developers have reduced the energy consumption to 3.5 kWh/m^3 and further reductions may be possible [27, 28]. Nonetheless, reverse osmosis remains by far the more widely used process.

11.4.3.1   Feed Pretreatment

The type and complexity of the feed pretreatment system depends on the content of the water to be treated. As in reverse osmosis, most feed water is sterilized by chlorination to prevent bacterial growth on the membrane. Scaling on the membrane surface by precipitation of sparingly soluble salts such as calcium sulfate can be controlled by adding precipitation inhibitors, such as sodium hexametaphosphate. The pH may also be adjusted to maintain salts in their soluble range. Large, charged organic molecules or colloids, such as humic acid, are particularly troublesome impurities, because they are drawn by their charge to the membrane surface but are too large to permeate. They accumulate at the dilute solution side of the membrane and precipitate, causing an increase in membrane resistance. All of the electrodialysis plants installed in the 1950s through the 1960s were operated unidirectionally, that is, the polarity of the two electrodes, and hence the positions of the dilute and concentrated cells in the stack, was fixed. In this mode of operation, formation of scale on the membrane surface by precipitation of colloids and insoluble salts was often a problem. In the early 1970s, a breakthrough in system design, known as electrodialysis polarity reversal, was made by Ionics [9]. In this mode of operation, the polarity of the DC power applied to the membrane electrodes is reversed periodically. When the polarity is reversed, the desalted water and brine chambers are also reversed by automatic valves that control the flows in the stack. By switching cells and reversing current direction, freshly precipitated scale is flushed from the membrane before it can solidify. The direction of movement of colloidal particulates drawn to the membrane by the flow of current is also reversed, so colloids do not form a film on the membrane. Electrodialysis plants using the reverse polarity technique have been operating since 1970 and have proved more reliable than their fixed polarity predecessors.

11.4.3.2   Membrane Stack

After the pretreatment step, the feed water is pumped through the electrodialysis stack. This stack normally contains 100–200 membrane cell pairs, each with a membrane area between 1 and 2 m². Plastic mesh spacers form the channels through which the feed and concentrate solutions flow. Most manufacturers use one of the spacer designs shown in Figure 11.17. In the tortuous path cell design of Figure 11.17a, a solid spacer grid forms a long open channel through which the feed solution flows at relatively high velocity. The channel is not held open by netting, so the membranes must be thick and sturdy to prevent collapse of the channels. In the sheet flow design of Figure 11.17b, the gap between the membrane leaves is maintained by a polyolefin mesh spacer. The spacer is made as thin as possible without producing an excessive pressure drop.

Two membranes and two gasket spacers form a single cell pair. Holes in the gasket spacers are aligned with holes in the membrane sheet to form the manifold channels through which the dilute and concentrated solutions are introduced into each cell. The end plate of the stack is a rigid plastic frame containing the electrode compartment. The entire arrangement is compressed together with bolts between the two end flow plates. The perimeter gaskets of the gasket spacers are tightly pressed into the membranes to form the cells. A large electrodialysis stack has several hundred meters of fluid seals around each cell. Early units often developed small leaks over time, causing unsightly salt deposits on the outside of the stacks. These problems have now been largely solved. In principle, an electrodialysis stack can be disassembled and the membranes cleaned or replaced on-site. In practice, this operation is performed infrequently and almost never in the field.

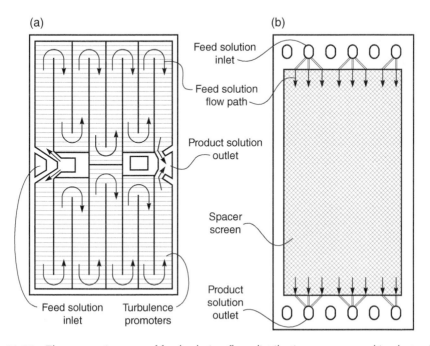

Figure 11.17   *The two main types of feed solution flow distribution spacers used in electrodialysis are (a) tortuous path, and (b) sheet flow.*

11.4.3.3   Power Supply and Process Control Unit

Electrodialysis plants use large amounts of direct current power; the rectifier required to convert AC to DC and to control the operation of the system represents a significant portion of the capital cost of a plant. A typical voltage across a single cell pair is 1–2 V, and the normal current flow is 40 mA/cm^2. For a 200-cell-pair stack containing 1 m^2 of membrane per cell, the total voltage is about 200–400 V, and the current about 400 A per stack. This is a considerable amount of electric power, and care must be used to ensure safe operation.

11.4.3.4   Solution Pumping System

A surprisingly large fraction of the total power used in electrodialysis systems is consumed by the water pumps required to circulate feed and concentrate solutions through the stacks. This fraction increases as the average salt concentration of the feed decreases and can become dominant in electrodialysis of low-concentration solutions (less than 500 ppm salt). The pressure drop per stack varies from 15 to 30 psi for sheet flow cells to as much as 70–90 psi for tortuous path cells. Depending on the separation required, the fluid will be pumped through two to four cells in series, requiring interstage pumps for each stack.

11.5   Electrodialysis Applications

11.5.1   Water Desalination

Desalination of brackish water containing less than about 5000 ppm salt is the largest application of electrodialysis. The competitive technologies are ion exchange for very dilute saline solutions, below 500 ppm, and reverse osmosis for concentrations above 2000 ppm. In the 500–2000 ppm range, power consumption is modest, less than 1 kWh/m^3 of product water, and electrodialysis is often the low-cost process. One advantage of electrodialysis is that a large fraction, typically 80–95% of the brackish feed, is recovered as product water. These high recoveries mean that the concentrated brine stream produced is 5–20 times more concentrated than the feed. The degree of water recovery is limited by precipitation of insoluble salts in the brine.

Several thousand electrodialysis plants have been installed around the world. Modern plants are generally fully automated and require only periodic operator attention. Typical plants have a production capacity of a few hundred to 1000 m^3/day of desalted water, and are used to make potable water, boiler feed water and industrial process water for the electronics manufacturing and pharmaceutical industries. A few larger municipal water plants with production rates of 10 000–50 000 m^3/day have also been installed.

11.5.2   Continuous Electrodeionization and Ultrapure Water

Electrodeionization systems were first suggested to remove small amounts of radioactive elements from contaminated waters [29], but the principal recent application is the preparation of ultrapure water for the electronics industries [30, 31]. Use of this process has grown rapidly with the growth of the industry, and it is now often used as a polishing step after the water has been pretreated by conventional electrodialysis or reverse osmosis.

In the production of ultrapure water, salt concentrations must be reduced to the ppb range. This is a problem with conventional electrodialysis, because the low conductivity of very dilute feed sets a practical limit on the amount of salt that can be removed. This means that the lower limit on salt content in the product water is generally 10 ppm or above. This

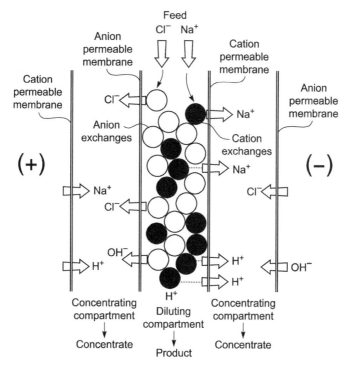

Figure 11.18 *Schematic of an electrodeionization process using a mixed-bed ion exchange resin to increase the conductivity of the dilute compartments of the electrodialysis stack. This type of process is often used in the production of ultrapure water.*

limitation can be overcome by filling the dilute chambers of the electrodialysis stack with fine mixed-bed ion exchange beads, as shown in Figure 11.18. The ions in the chamber partition into the beads, and are concentrated many times. As a result, ions and current flow through the beads in the resin bed, and the resistance of the cell is much lower than for a normal cell operating on the feed of the same dilution. An additional benefit is that, toward the bottom of the bed where the ion concentration is in the ppb range, a certain amount of water splitting occurs. This produces hydrogen and hydroxyl ions that also migrate to the membrane surface through the ion exchange beads. The presence of these ions maintains a high pH in the anion exchange beads and a low pH in the cation exchange beads. These extreme pHs enhance the ionization and removal of weakly ionized species, such as carbon dioxide and silica, that would otherwise be difficult to remove. Such electrodeionization systems can reduce most ionizable solutes to below ppb levels.

A process flow diagram of a plant to produce ultrapure boiler feed water by a combination of RO and electrodeionization is shown in Figure 11.19 [10]. The reverse osmosis unit removes the bulk of the feed water ions; the electrodeionization unit removes the rest.

In the alternative, the output from the reverse osmosis unit could be treated by non-membrane mixed-bed ion exchange. However, electrodeionization offers the advantages of a simpler, continuous process with no need for regeneration chemicals, as well as a lower overall cost.

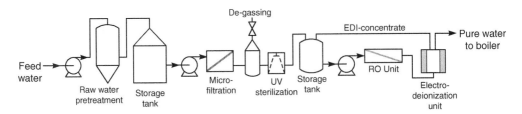

Figure 11.19 Ultrapure water production line used to produce boiler feed water. Source: Reproduced from [10] with permission from Elsevier. Copyright (2004).

11.5.3   Salt Recovery from Seawater

In the processes described above, the product is desalted water. Another major application is the production of table salts from seawater [8]. This process is only practiced in Japan, which has no other domestic salt supply, and is subsidized by the government. The plants are large, in total using more than 500 000 m^2 of membrane, and produce about 1.2 million tons/year of salt.

A flow scheme of one such plant is shown in Figure 11.20. A cogeneration unit produces the power required for the electrodialysis operation, which concentrates the salt to about 18–20 wt%. Steam from the power plant is then used as an energy source to evaporate water from the concentrated brine, yielding the crystallized salt product.

Seawater contains relatively high concentrations of sulfate (SO_4^{2-}), calcium (Ca^{2+}), magnesium (Mg^{2+}), and other multivalent ions that can precipitate in the concentrated salt compartments of the stack and cause severe scaling. This problem has been solved by applying a thin polyelectrolyte layer of opposite charge to the ion exchange membrane on the surface facing the seawater solution [8]. A cross-section of a coated anion exchange membrane is shown in Figure 11.21. Because the Donnan exclusion effect is much stronger for multivalent ions than for univalent ions, the polyelectrolyte layer rejects multivalent ions but allows the univalent ions to pass relatively unhindered. This method of enhancing the separation of univalent from multivalent ions is occasionally used in other electrodialysis applications.

11.5.4   Other Electrodialysis Applications

The water desalination applications described above represent the majority of the market for electrodialysis systems. Applications exist in the food industry to desalt whey, and to remove tannic acid from wine, and citric acid from fruit juice. Other applications exist in wastewater treatment, particularly regeneration of waste acids used in metal pickling operations and removal of heavy metals from electroplating rinse waters [10, 11]. These applications rely on the ability of electrodialysis membranes to separate electrolytes from nonelectrolytes and to separate multivalent from univalent ions.

The arrangement of membranes in these systems depends on the application. Figure 11.22a shows a stack comprised solely of cation exchange membranes to soften water; Figure 11.22b shows an all-anion exchange membrane stack to deacidify juice [26]. In the water-softening application, the objective is to exchange divalent cations such as calcium and magnesium for sodium ions. In the juice deacidification process, the anion stack is used to exchange citrate ions for hydroxyl ions. These are both ion exchange processes, and the salt concentration of the feed solution remains unchanged.

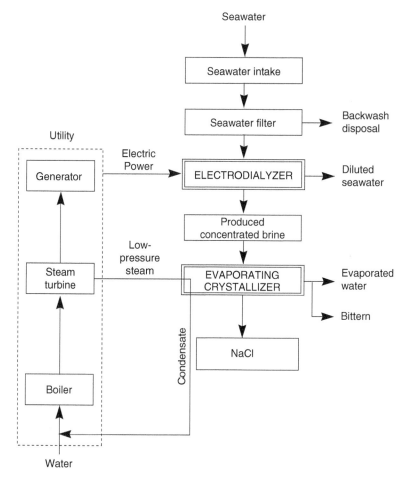

Figure 11.20 *Flow scheme of a process using electrodialysis to produce table salt from seawater [8] / John Wiley & Sons.*

11.6 Fuel Cells

The use of fuel cells in the space program in the 1960s took the technology out of the laboratory and made it a reliable, if expensive, way of generating power. The space program remained the principal application until the 1990s when there was a surge of interest in developing fuel cells as the power source for all sorts of electric-motor-driven cars, trucks, buses and boats. Fuel cells also found a place in non-interruptible power systems and as portable power sources to replace batteries for laptops, computers, and military electronic devices. The industry has gone through several boom-and-bust cycles – one in 2000–2001 and another in 2008–2009 – and there is currently a new surge of interest in fuel cells as part of the hydrogen economy [15, 32, 33].

The idea is to use hydrogen-powered cells as energy storage devices. Hydrogen will be produced by electrolysis and stored when excess renewable energy is available. When energy is needed, the fuel cell will perform a process that is essentially the opposite of electrolysis, using the stored hydrogen to combine with oxygen, thereby producing water and generating electric current. Fuel cells are too expensive at present for general use, but the cost of the power they

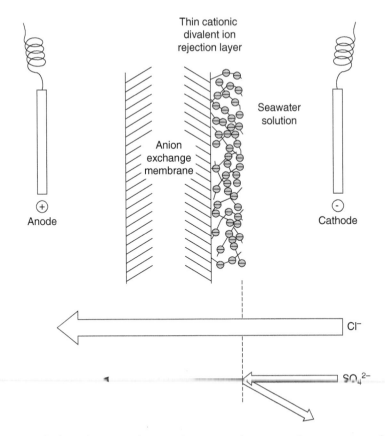

Figure 11.21 *Polyelectrolyte-coated ion exchange membranes used to separate multivalent and univalent ions in seawater salt concentration plants.*

produce is coming down and a few special uses have been found. There are about 50 000 small stationary units in operation worldwide, mainly in Japan, where the government is subsidizing their use as a source of power and heat for buildings.

Many different types of fuel cells exist, differing mostly in the nature of the barrier media separating the two electrodes. Membrane developers are mostly interested in polymer electrolyte membrane (PEM or proton exchange membrane) fuel cells of the type shown in Figure 11.23 [15]. In this device, the membrane has three functions:

it separates the anode and cathode to prevent an electrical short circuit;
it separates the hydrogen and oxygen fuels to prevent a chemical short circuit;
it selectively transports protons (H$^+$) from the anode to the cathode.

A hydrogen/oxygen fuel cell produces electricity by the reaction

$$2H_2 + O_2 \rightarrow 2H_2O \qquad (11.22)$$

Hydrogen is introduced into the cell at the anode, consisting of a porous carbon fabric, through which the gas diffuses to come into contact with one side of a cation exchange membrane. The dissociation of hydrogen to protons is normally slow, but can be catalyzed by a finely dispersed

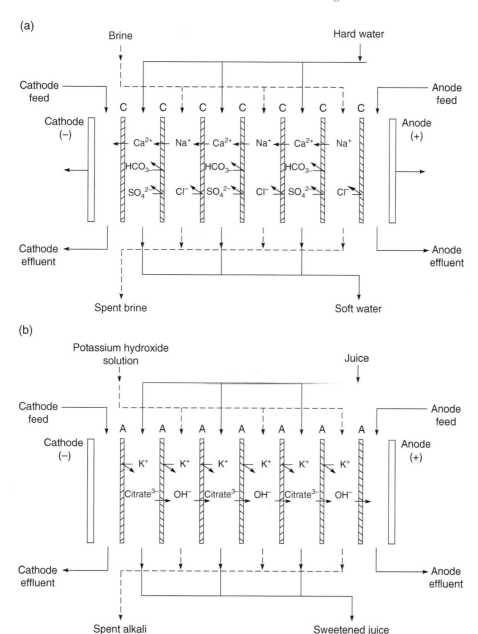

Figure 11.22 *Flow schematic of electrodialysis systems used to exchange target ions in the feed solution. (a) An all-cation exchange membrane stack to exchange sodium ions for calcium ions in water softening. (b) An all-anion exchange membrane stack to exchange hydroxyl ions for citrate ions in deacidification of fruit juice.*

layer of platinum at the membrane surface. In the presence of this catalyst, the hydrogen ionizes by the reaction

$$H_2 \rightarrow 2H^+ + 2e^- \tag{11.23}$$

Power is generated by the electrons flowing to the cathode through the external circuit. The protons diffuse through the negatively charged membrane. At the cathode, the protons combine with oxygen by the reaction

$$1/2O_2 + 2H^+ + 2e^- \rightarrow H_2O \tag{11.24}$$

Two types of fuel cells use polymer electrolyte membranes. The first is the proton exchange (or polymer electrolyte) membrane fuel cell (PEMFC) illustrated in Figure 11.23. The second is the direct methanol fuel cell (DMFC). The direct methanol fuel cell uses methanol instead of hydrogen at the anode of the cell. In the presence of platinum or platinum–ruthenium catalysts, methanol releases protons by the reaction

$$CH_3OH + H_2O \rightarrow CO_2 + 6H^+ + 6e^- \tag{11.25}$$

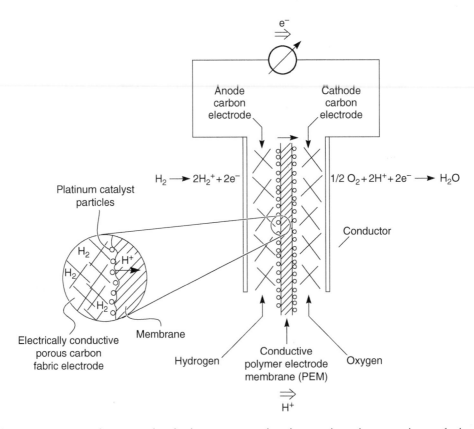

Figure 11.23 *A schematic of a hydrogen-powered polymer electrolyte membrane fuel cell (PEMFC). Hydrogen introduced at the anode is converted to protons and electrons. The electrons flow to the cathode, producing useful power. The protons diffuse through the proton-permeable (cationic) membrane to the cathode, where they react with incoming oxygen to produce water.*

Methanol is easier to transport and use than hydrogen and has a much higher volumetric energy density, which are significant advantages. The principal problem is that cation exchange membranes, as well as being permeable to protons, are also significantly permeable to methanol. Methanol permeation (crossover) from the anode to the cathode compartment leads to several unwanted effects, including a loss in cell voltage and the consumption of methanol and oxygen without electricity generation, lowering the fuel efficiency of the cell.

By far the most important PEM material is DuPont's Nafion perfluorosulfonic acid, for which the chemical structure was shown in Figure 11.9. The performances of Nafion membranes in a hydrogen-powered PEMFC cell and a methanol-powered DMFC cell are shown in Figure 11.24. The power generated by the fuel cell is the product of the cell voltage and current density. The data show that the performance of the hydrogen-powered PEM fuel cell gets better as the membrane becomes thinner. The only limitation is the mechanical weakness of very thin membranes. W.R. Gore has tried to circumvent this problem by impregnating Nafion into very thin microporous PTFE membranes. The PTFE provides the mechanical strength; the Nafion conducts the protons.

The data for the DMFC show that the power output is significantly less under comparable conditions than with the hydrogen-powered cell, and in contrast to the hydrogen-powered cell, the performance of the DMFC increases as the membrane becomes thicker. This is because of reduced methanol crossover to the cathode. Making the membrane thicker decreases the proton transport rate, but this is more than offset by the decrease in methanol crossover.

In the past decade, a significant effort has been spent developing better PEM membranes [13–15, 34]. However, as of today, Nafion remains the material to beat, despite its high cost.

11.7 Chlor-Alkali Processes

Caustic soda (sodium hydroxide) and chlorine are produced by the electrolysis of aqueous sodium chloride. The process has been carried out on an industrial scale since 1892. For many years, mercury cells were used, but environmental problems caused by leakage of mercury into the environment have meant these plants have been closed and replaced with membrane processes, first using asbestos diaphragms, and more recently using polymeric cation exchange membranes.

A schematic diagram of the chlor-alkali process is shown in Figure 11.25. Sodium chloride brine is sent to the anode compartment of the electrolysis cell. The brine is pretreated to bring calcium, aluminum, magnesium, and other impurities down to the ppb level. Make up water is sent to the cathode compartment of the cell. The cell compartments are separated by a cation exchange membrane, usually a Nafion or perfluorocarboxylic acid (Flemion®) membrane. When a voltage is generated between the electrodes, sodium ions pass through the membrane, which is an almost perfect barrier to chloride and hydroxyl ions. Chloride ions discharged at the anode create chlorine gas; hydroxide ions discharged at the cathode produce hydrogen and dissolved sodium hydroxide. The sodium hydroxide concentration builds up until the concentration reaches about 35 wt%. The sodium hydroxide is then removed as a product stream. The current efficiency of the process is very high, reaching up to 97%. In recent years, better membranes have allowed the concentration of the sodium hydroxide product produced in the cathode compartment to rise to as high as 50 wt%.

The main products of the chlor-alkali industry are chlorine – used directly in the pulp and paper industry and in the production of intermediates such as ethylene dichloride and vinyl

Hydrogen (PEMFC) Fuel Cell

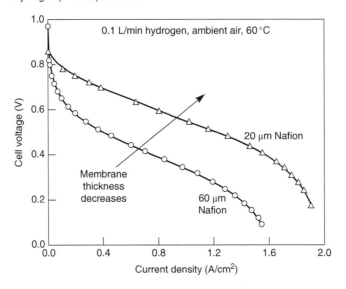

Methanol (DMFC) Fuel Cell

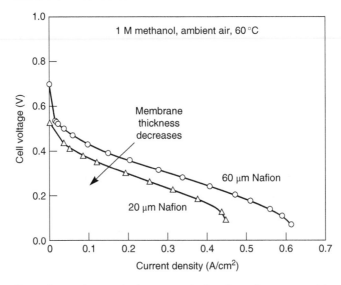

Figure 11.24 *Effect of membrane thickness on fuel cell performance with two commercial Nafion® membranes. Source: Reproduced from [14] with permission from John Wiley and Sons. Copyright (2007).*

chloride – and sodium hydroxide – used in a wide variety of neutralization reactions and in the production of hypochlorite for bleach. Hydrogen is also generated as a by-product, and is usually burned for fuel. In recent years, some plants have improved the economics of the process by using a porous cathode permeated by CO_2-free air. Oxygen in the air is reduced by hydrogen at the cathode and the potential of the oxygen reduction process reduces the

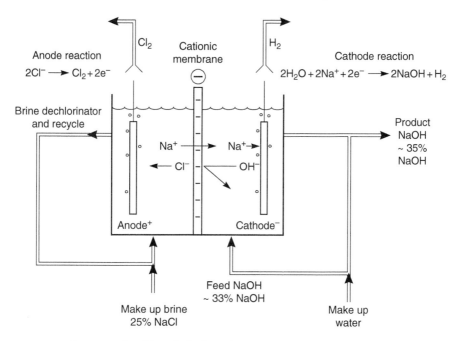

Figure 11.25 *Schematic of a chlor-alkali electrolysis cell.*

decomposition voltage needed for chlor-alkali electrolysis by as much as 30%. Since the consumption of electricity is a large contributor to the cost of the process, this savings is significant.

11.8   Other Electrochemical Processes

The use of membranes in chlor-alkali cells is by far the largest electrochemical application of ion exchange membranes. There are other smaller uses under development, two of which are described below: water splitting by means of bipolar membranes, and energy storage/power generation using flow-through redox batteries.

11.8.1   Water Splitting Using Bipolar Membranes

Bipolar membranes consist of an anion exchange membrane and a cation exchange membrane laminated together [10]. When the bipolar membrane is placed between two electrodes, as shown in Figure 11.26, the interface between the membranes becomes depleted of ions. The only way a current can then be carried is by the water splitting reaction, which liberates hydrogen ions that migrate to the cathode and hydroxyl ions that migrate to the anode. The mechanism of water splitting in these membranes has been discussed in detail by Strathmann et al. [10, 35] and others [36, 37]. Water splitting takes place in the very thin interfacial region shown in the figure, which is typically less than 1% of the total thickness of the laminate. All of the ions are removed from the region, which becomes the major electrical resistance across the cell. It follows that the nature of this region is critical to the performance of the overall unit. The splitting reaction can be catalyzed by metal hydroxides, typically

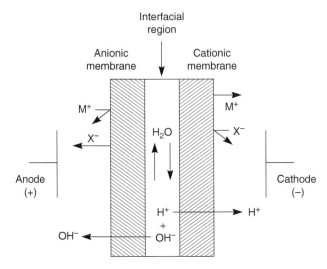

Figure 11.26 Schematic of a single bipolar membrane showing generation of hydroxyl and hydrogen ions by water splitting in the interior of the membrane. Electrolysis takes place in the very thin interfacial space between membranes.

$Cr(OH)_3$ or $Fe(OH)_3$, incorporated into the interfacial layer, or by $-NR_2$ and $-PO_3$ groups bonded to the membrane surface.

Early membranes were made by simply laminating preformed cation and anion exchange membranes together, but this created small gaps that led to high electrical resistance, as shown in Figure 11.27. Nowadays, one of the membranes is cast directly onto the other, which achieves a better result.

Bipolar membranes are utilized in an electrodialysis stack composed of a number of sets of three-chamber cells between two electrodes, as shown in Figure 11.28, to divide a neutral salt into the conjugate acid and base. Salt solution flows into the middle chamber; cations migrate to the chamber on the left and anions to the chamber on the right. Electrical neutrality is maintained in these chambers by hydroxyl and hydrogen ions provided by water splitting in the bipolar membranes that bound each set of three chambers. The process is limited to the generation of relatively dilute acid and base solutions, and the product acid and base are contaminated with 2–4% salt. Nevertheless, the process is significantly more energy efficient than the conventional electrolysis process because no gases are created at the cell electrodes. Total current efficiency is about 80%, and the system can often be integrated into the process generating the feed salt solution [16, 17].

11.8.2   Redox Flow Batteries

A flow battery is a type of rechargeable fuel cell. As in a conventional fuel cell, an electrochemical cell converts chemical energy into electricity. The difference is that the flow battery is configured such that the process can be reversed and electricity can be used to regenerate the fuel. The most well-developed example of the process uses the different valency states of aqueous vanadium solutions as a fuel source.

An illustration of a vanadium-based redox battery is shown in Figure 11.29a. Two carbon electrodes are separated by a cation exchange membrane. The membrane permeates protons

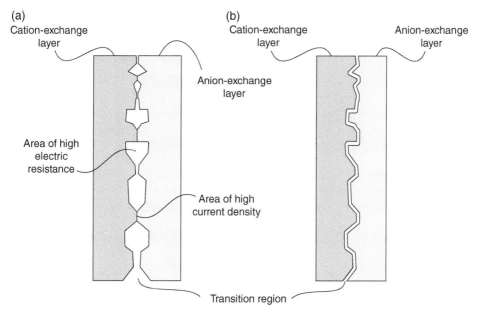

Figure 11.27 *Schematic diagram illustrating the transition region in a bipolar membrane prepared by (a) laminating two conventional ion-exchange membranes back-to-back by pressing and (b) casting liquid anion-exchange resin on a solid ion-exchange membrane. [10] / with permission of Elsevier.*

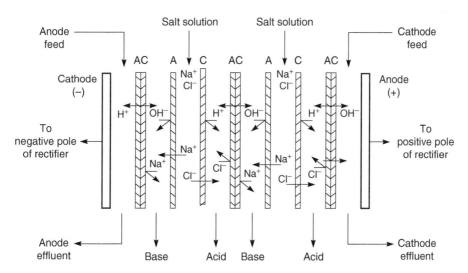

Figure 11.28 *Schematic of a bipolar membrane process to split sodium chloride into sodium hydroxide and hydrochloric acid.*

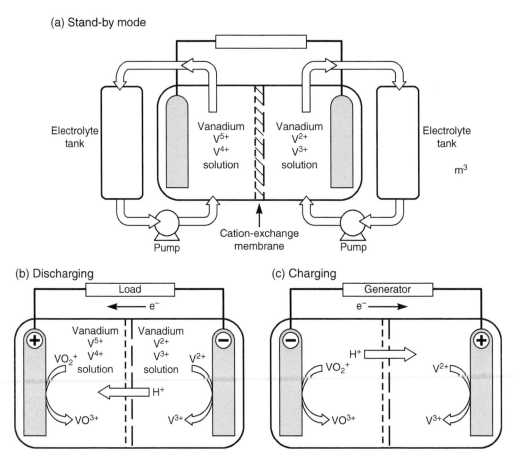

Figure 11.29 *Schematic drawing of (a) a vanadium redox flow battery (b) the redox reactions that occur during discharge, and (c) the reactions that occur during charging. Source: [38] / John Wiley & Sons.*

but has a very low permeability to all other cations. Different solutions are circulated through the two compartments from two large electrolyte storage tanks. Both tanks hold solutions containing high concentrations of vanadium ions but the ions are in different valency states. One solution contains vanadium in the V^{4+} and V^{5+} valency states, the other contains vanadium in the V^{2+} and V^{3+} valency states. The different solutions cause different reactions to occur at the electrodes. When the battery is off-line, no solution flows from the electrolyte tanks and no reactions occur. Figure 11.29b shows the operation of the battery in discharge mode. At the positive electrode, vanadium V^{5+} is reduced to V^{4+} by the reaction

$$2H^+ + VO_2^+ \rightarrow VO^{2+} + H_2O - e^-$$

And at the negative electrode, V^{2+} is oxidized to V^{3+} by

$$V^{2+} \rightarrow V^{3+} + e^-$$

As a consequence, protons move through the ion exchange membrane into the positive electrolyte solution and an equivalent flow of electrons generates an electric current through the exterior load. Power continues to be created as long as fresh vanadium solutions are circulated through the cell.

To recharge the batteries, an external electric current is applied in the direction indicated in Figure 11.29c. The polarity of the cell is reversed, and the reaction in the right-hand chamber now proceeds in the reverse direction, thereby returning the V^{3+} ions to the V^{2+} state. Likewise, in the left-hand chamber the V^{4+} vanadium is oxidized back to the V^{5+} state. The battery is then ready to be reused. Because vanadium solutions are used on both sides of the membranes, some leakage of vanadium ions through the membrane does not seriously affect the battery performance. This is a major advantage of vanadium redox batteries. In other redox batteries, the oxidation/reduction chemistries on either side of the membrane involve different metal ions. Long term stability then requires that permeation of all ions other than H^+ be completely prohibited, which is difficult to obtain.

Unlike most conventional batteries, the vanadium redox battery can go through discharge/recharge cycles several thousand times without the performance degrading. The power generated depends on the size of the cell, but the chemical energy that can be created, stored and used when needed depends only on the size of the electrolyte tanks, which can be very large. Batteries capable of storing the equivalent of up to 100 MWh of power have been built. The discharge/recharge time of these units is typically in the 2–10 hours range.

Large batteries have the same type of plate-and-frame form as the electrodialysis stack shown in Figure 11.3, with multiple single-cell batteries in the stack. The cells are electrically connected in series, but the electrolyte circulates through all the cells in parallel.

Although the vanadium redox chemistry is the most well-studied, a number of other chemistries are being developed and this area is the center of a considerable research effort [38, 39]. The main target application is stabilization of the grid when renewable wind and solar energy are used. Because power production from these sources can fluctuate quickly, it is hard to create a stable grid containing more than 30% renewable sources. Large flow-through batteries have quick response times and go a long way to solving this problem. Currently, storing and using electricity this way is too expensive for general use, but several demonstration systems have been built and costs are expected to come down.

11.8.3 Reverse Electrodialysis

Reverse electrodialysis has been suggested as a way of harvesting the energy of mixing created when the fresh water of a river mixes with the salty water of the sea. In a conventional electrodialysis process, salt water is circulated through a membrane stack formed of alternating anionic and cationic membranes. When a voltage is applied across the stack, a movement of ions occurs that results in a separation of the salt ions into concentrate and diluate solutions. In reverse electrodialysis, the process is reversed. A salt solution (seawater) and a dilute solution (river water) are circulated through the stack and ions move from the salt solution into the dilute solution. This movement of ions creates a voltage difference across the membrane stack and an electric current can then be produced. The process is illustrated in Figure 11.30 [40].

Reverse electrodialysis (RED) has much in common with pressure-retarded osmosis (PRO), described in Chapter 13. Both processes use the energy of mixing salt and fresh water streams to produce useful power. Both processes were first studied in a systematic fashion in the mid-1970s and more sporadically since then. PRO converts the potential energy of mixing into a hydrostatic pressure, that is then converted into electric power by a turbine generator. RED has the advantage

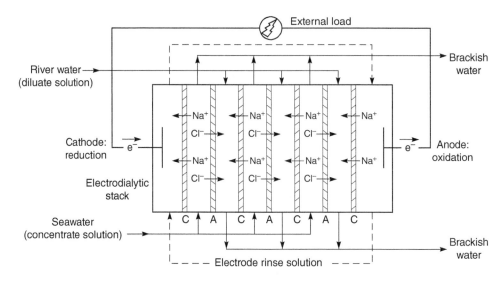

Figure 11.30 *Energy production using reverse electrodialysis. A is an anion exchange membrane, C is a cation exchange membrane. Source: Reproduced with permission from [40] Copyright (2009) Elsevier.*

that the process produces an electric current, so no generator is needed. Unfortunately, the power produced is only 5–10 watts/m^2 for PRO, and less than 1–2 watts/m^2 of membrane for most RED [41, 42]. Although RED continues to be the subject of occasional research papers, prospects for industrial application are dim. A number of other processes that use ion exchange membranes (diffusion dialysis, piezodialysis) are covered in Chapter 15.

11.9   Conclusions and Future Directions

Electrodialysis and the chlor-alkali process are by far the largest uses of ion exchange membranes. These two processes are both well established, and major technical innovations that will change the competitive position of ion exchange membranes do not appear likely. Some new applications of electrodialysis exist in the treatment of industrial process streams, food processing, and wastewater treatment systems, but the total market is small. Long-term major new applications for ion exchange membranes may develop in non-separation areas such as fuel cells, flow batteries, electrochemical reactions, and production of acids and alkalis with bipolar membranes. These processes have found some niche applications, but have yet to fulfill the hopes of their developers.

References

1. Ostwald, W. (1890). Elektrische Eigenschaften Halbdurchlässiger Scheidewande. *Z. Physik. Chem.* **6**: 71.
2. Donnan, F.G. (1911). Theory of membrane equilibria and membrane potentials in the presence of non-dialyzing electrolytes. *Z. Elektrochem.* **17**: 572.

3. Manegold, E. and Kalauch, C. (1939). Uber Kapillarsysteme, XXII Die Wirksamkeit Verschiedener Reinigungsmethoden (Filtration, Dialyse, Electrolyse und Ihre Kombinationen). *Kolloid Z.* **86**: 93.

4. Meyer, K.H. and Strauss, W. (1940). La Permeabilité des Membranes, VI, Sur la Passage du Courant Electrique a Travers des Membranes Selectives. *Helv. Chim. Acta* **23**: 795.

5. Kressman, T.R.E. (1950). Ion exchange resin membranes and resin-impregnated filter paper. *Nature* **165**: 568.

6. Murphy, E.A., Paton, F.J., and Ansell, J. (1941). Apparatus for the electrical treatment of colloidal dispersions. US Patent 2,331, 494, issued 12 October 1943.

7. Juda, W. and McRae, W.A. (1950). Coherent ion-exchange gels and membranes. *J. Am. Chem. Soc.* **72**: 1044.

8. Seko, M., Miyauchi, K., and Omura, J. (1983). Ion exchange membrane application for electrodialysis, electroreduction, and electrohydrodimerisation. In: *Ion Exchange Membranes* (ed. D.S. Flett), 179–193. Chichester: Ellis Horwood Ltd.

9. Strathmann, H. (1991). Electrodialysis. In: *Membrane Separation Systems: Recent Developments and Future Directions* (ed. R.W. Baker, E.L. Cussler, W. Eykamp, et al.), 396–448. Park Ridge, NJ: Noyes Data Corp.

10. Strathmann, H. (2004). *Ion Exchange Membrane Separation Processes*. Amsterdam: Elsevier.

11. Ran, J., Wu, Y., Yang, Z. et al. (2017). Ion exchange membranes: new developments and applications. *J. Membr. Sci.* **522**: 267.

12. Grotheer, M.P. (1992). Electrochemical processing (inorganic). In: *Kirk Other Encyclopedia of Chemical Technology*, 4e, vol. **9**, 124. New York: Wiley Interscience, Wiley.

13. Nunes, S.P. and Peinemann, K.-V. (2006). Membranes for fuel cells. In: *Membrane Technology in the Chemical Industry*, 2e (ed. S.P. Nunes and K.-V. Peinemann), 45–52. Wiley-VCH.

14. Pintauro, P.N. and Wycisk, R. (2008). Fuel cell membranes. In: (ed. N.N. Li, A.G. Fane, W.S.W. Ho, and T. Matsuura), 755–786. Hoboken, NJ: Wiley.

15. Barbir, F. (2005). *PEM Fuel Cells*. Amsterdam: Elsevier.

16. Nagasubramanian, K., Chlanda, F.P., and Liu, K.-J. (1977). Use of bipolar membranes for generation of acid and base – an engineering and economic analysis. *J. Membr. Sci.* **2**: 109.

17. Nagasubramanian, K., Chlanda, F.P., and Liu, K.-J. (1980). Bipolar membrane technology: an engineering and economic analysis. In: *Recent Advances in Separation Tech-II*, AIChE Symposium Series Number 1192, vol. **76**, 97. New York: AIChE.

18. Korngold, E. (1984). Electrodialysis – membranes and mass transport. In: *Synthetic Membrane Processes: Fundamentals and Water Applications* (ed. G. Belfort), 191–220. Orlando, FL: Academic Press.

19. Eisenberg, A. and Yeager, H.L. (1982). *Perfluorinated Ionomer Membranes*, ACS Symposium Series Number 180. Washington, DC: American Chemical Society.

20. Seko, M., Yomiyama, A., and Ogawa, S. (1983). Chlor-alkali electrolysis using perfluorocarboxylic acid membranes. In: *Ion Exchange Membranes* (ed. D.S. Flett), 121–135. Chichester: Ellis Horwood, Ltd.

21. Sata, T., Motani, K., and Ohaski, Y. (1983). Perfluorinated ion exchange membrane, Neosepta-F and its properties. In: *Ion Exchange Membranes* (ed. D.S. Flett), 137–150. Chichester: Ellis Horwood Ltd.

22. Krol, J.J., Wessling, M., and Strathmann, H. (1999). Concentration polarization with monopolar ion exchange membranes: current-voltage curves and water dissociation. *J. Membr. Sci.* **162**: 145.

23. Cowan, D. and Brown, J. (1959). Effect of turbulence on limiting current in electrodialysis cells. *Ind. Eng. Chem.* **51**: 1445.

24. Rautenbach, R. and Albrecht, R. (1989). *Membrane Processes*. Chichester: Wiley.

25. Balster, J., Yildirim, M.H., Stamatialis, D.F. et al. (2007). Morphology and microtopology of cation-exchange polymers and the origin of the overlimiting current. *J. Phys. Chem. B* **111**: 2152.

26. Rogers, A.N. (1984). Design and operation of desalting systems based on membrane processes. In: *Synthetic Membrane Processes* (ed. G. Belfort), 437–476. Orlando, FL: Academic Press.

27. Chehayeb, K.M., Nayer, K.G., and Lienhard, J.H. (2018). On the merits of using multi-stage and counterflow electrodialysis for reduced energy consumption. *Desalination* **439**: 1.

28. Doornbusch, G.J., Bel, M., Tedesco, M. et al. (2020). Effect of membrane area and membrane properties in multistage electrodialysis on seawater performance. *J. Membr. Sci.* **611**: 118303.

29. Walters, W.R., Weiser, D.W., and Marek, L.J. (1955). Concentration of radioactive aqueous wastes. Electromigration through ion-exchange membranes. *Ind. Eng. Chem.* **47**: 61.

30. Ganzi, G.C., Jha, A.D., DiMascio, F., and Wood, J.H. (1997). Electrodeionization: theory and practice of continuous electrodeionization. *Ultrapure Water* **14**: 64.

31. Dey, A. (2008). Ultrapure water by membranes. In: *Advanced Membrane Technology and Applications* (ed. N.N. Li, A.G. Fane, W.S.W. Ho, and T. Matsuura), 371–406. Hoboken, NJ: Wiley.

32. Meeks, N.D. and Baxley, S. (2016). The hydrogen economy. *Chem. Eng. Prog.* **112**: 34.

33. Satypal, S. and Vora, S. (2016). Establishing the fuel cell industry. *Chem. Eng. Prog.* **112**: 38.

34. Kreuer, K.D. (2001). On the development of proton conducting polymer membranes for hydrogen and methanol fuel cells. *J. Membr. Sci.* **185**: 29.

35. Strathmann, H., Krol, J.J., Rapp, H.-J., and Eigenberger, G. (1997). Limiting current density and water dissociation in bipolar membranes. *J. Membr. Sci.* **125**: 123.

36. Pärnamäe, R., Mareev, S., Nikonenko, N. et al. (2021). Bipolar membranes: a review on principles, latest developments and applications. *J. Membr. Sci.* **617**: 118538.

37. Kemperman, A.J.B. (ed.) (2000). *Handbook on Bipolar Membrane Technology*. Enschede, The Netherlands: Twente University Press.

38. Kear, G., Shah, A.A., and Walsh, F.C. (2012). Development of the all-vanadium redox flow battery for energy storage: a review of technological, financial and policy aspects. *Int. J. Energy Res.* **36**: 1105.

39. Ding, C., Zhang, H., Li, X. et al. (2013). Vanadium flow battery for energy storage: prospects and challenges. *J. Phys. Chem. Lett* **4**: 1281.

40. Post, J.W., Veermann, J., Hamelers, H.V.M. et al. (2009). Salinity gradient power: evaluation of pressure retarded osmosis and reverse electrodialysis. *J. Membr. Sci.* **327**: 136.

41. Dlugolecki, P., Neimeijer, K., Metz, S., and Wessling, M. (2008). Current status of ion exchange membranes for power generation from salinity gradients. *J. Membr. Sci.* **319**: 214.

42. Veerman, J., Saakes, M., Metz, S.J., and Harmsen, G.J. (2009). Reverse electrodialysis: performance of a stack with 50 cells on the mixing of sea and river water. *J. Membr. Sci.* **327**: 136.

12

Carrier Facilitated Transport

12.1 Introduction

Carrier transport uses a selectively reactive carrier incorporated within a membrane matrix. At the feed side, the carrier combines chemically with the desired permeant, and assists its transport across the membrane. Much early work on carrier transport employed liquid mixtures containing a dissolved carrier. The liquid was held by capillarity in the pores of a microporous membrane, so the process was originally often called liquid membrane transport. Nowadays many membranes contain carriers dissolved or dispersed in solid polymer films, and carrier transport is a better generic name for carrier-mediated processes.

Carrier transport has its origins with physiologists trying to understand biological processes, such as the transport of oxygen from the lung to distant muscle or other cells. We now know this is a two-stage process, as shown in Figure 12.1. In the lung, oxygen-containing air is separated from flowing arterial blood by a permeable film a few cells thick. Oxygen passes across the film and is absorbed by hemoglobin in the circulating blood, forming oxyhemoglobin. The oxygenated blood leaving the lung contains 98–99% of its saturation oxygen concentration. Blood

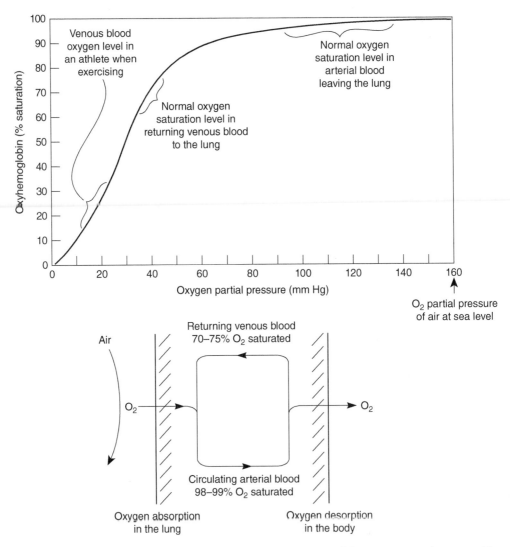

Figure 12.1 *Absorption of oxygen by hemoglobin as a function of the oxygen partial pressure. The diagram shows how oxygen from the lung is carried by blood to distant cells, such as muscle cells. The net process is similar to the operation of a two-stage membrane contactor.*

circulation carries this oxygen to distant parts of the body where the oxygen diffuses through the cell membrane and becomes available for metabolic processes. The partially oxygen-depleted blood, now containing about 70–75% of its saturation oxygen concentration, then returns to the lung. The total process has the form of a two-stage membrane contactor with hemoglobin carrying the oxygen from one membrane contactor to the other (see Chapter 13 for a discussion of membrane contactors).

In 1960, Scholander [1] showed that the biological process could be modeled in a single membrane by immobilizing a solution of hemoglobin in the pores of a thin microporous film. Scholander's membrane is illustrated in Figure 12.2a. Oxygen reacts with hemoglobin (HEM) at the high-pressure side of the membrane to form oxyhemoglobin (HEM O_2). Diffusion of the oxyhemoglobin transports oxygen across the membrane to the low-pressure side, where it is released, thereby re-forming hemoglobin, which diffuses back to the feed side. The carrier facilitated membrane produced by Scholander was stable only for a few hours, but during that time, the membrane had a selectivity for oxygen from air (nitrogen) far above that of any known polymeric membrane.

The two broad categories of carrier transport are shown in Figure 12.2. The first, exemplified by Scholander's work, Figure 12.2a, is called *facilitated transport*, and involves a reactive carrier that reacts selectively with a component in the feed and enhances its transport. Most potential applications of facilitated transport involve gas separations, such as oxygen from air or carbon dioxide from nitrogen or methane. The second category is called *coupled transport* and usually involves the transport of ions across a membrane. The example of a coupled transport process

(a)

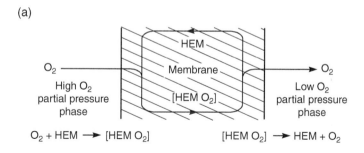

(b)

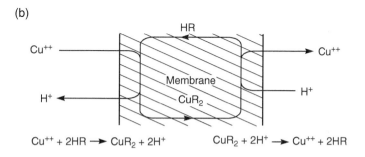

Figure 12.2 *Schematic examples of (a) facilitated transport and (b) coupled transport in a liquid membrane. The facilitated transport example shows permeation of oxygen across a membrane using hemoglobin as the carrier agent. The coupled transport example shows permeation of copper and hydrogen ions across a membrane using a reactive mobile oxime as the carrier agent.*

shown in Figure 12.2b is the separation of copper (Cu^{2+}) from an aqueous solution containing various metal ions. The carrier contained in the membrane is an oxime (HR) dissolved in kerosene. The oxime forms an organic soluble complex (CuR_2) selectively with copper ions. At the feed side of the membrane, two oxime carrier molecules pick up a copper ion, liberating two hydrogen ions to the feed solution.

$$Cu^{2+} + 2HR \rightleftarrows CuR_2 + 2H^+ \tag{12.1}$$

The complex is soluble in the hydrocarbon solution and transports the copper to the permeate surface of the membrane. An acid solution is circulated on the permeate side. The high concentration of hydrogen ions in the solution causes a reversal of the copper oxime reaction, re-forming copper ions and the protonated oxime.

$$CuR_2 + 2H^+ \rightleftarrows Cu^{2+} + 2HR \tag{12.2}$$

The difference between the two processes is that coupled transport requires movement of two ionic species across the membrane: Cu^{2+} and H^+ in the example shown. Electrical neutrality means that the flux of these components is inextricably linked; for every copper ion that permeates the membrane left to right, two protons must permeate right to left. One of the consequences of the linkage is that the flux of copper ions across the membrane is determined by the concentration differences of both copper and hydrogen ions across the membrane, hence the term coupled transport.

Before proceeding further, it seems only fair to warn the reader that, unlike the happy ending in all of the process chapters to this point, there are, to date, no commercial membrane processes that use either facilitated transport or coupled transport. This is not from lack of interest or effort by the membrane community. Carrier transport has been the subject of serious study for more than 50 years, both in academia and industry, as the timeline of Figure 12.3 shows. The driving force for this work is the spectacular results achievable under laboratory conditions. For example, using silver salts as the complexing agent, olefin/paraffin selectivities in the hundreds can be obtained. Unfortunately, converting these laboratory results into practical processes requires the solution of a number of tough technological problems.

The key problem hindering industrial use is membrane stability, both physical and chemical. The first generation of immobilized liquid membranes failed rapidly, because the liquid carrier was lost from the pores of the membrane into the strip or concentrate solutions in just a few days. This prompted efforts to develop better ways to contain the carrier by using dense, rather than microporous, polymer films, as the support matrix. As a result of these efforts, described in the next section, the problem of physical failure is on the way to being solved.

A more stubborn problem is that of chemical stability. The presence of even trace amounts of contaminants in the feed stream may poison the carrier agent by binding permanently with it or reducing it to non-ionic form. In the case of olefin/paraffin separation, such reduction seems to be an intrinsic effect, brought about by the olefin component. Similar types of problems affect oxygen/nitrogen separation. For carbon dioxide/nitrogen separations on the other hand, chemical stability seems to be less of a problem, and it is possible that this separation may be the first industrial application.

The prominence of stability problems in both facilitated and coupled transport processes has driven much of the development work. On the other hand, the theory and operational considerations for each process require different treatment. It seemed best, therefore, to divide the following discussion into separate facilitated transport and coupled transport sections.

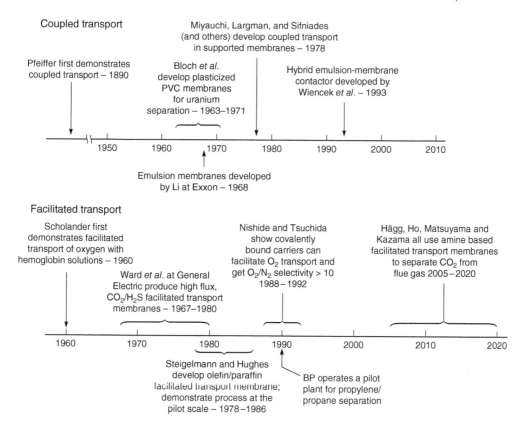

Figure 12.3   Milestones in the development of carrier transport.

12.2   Facilitated Transport

12.2.1   Membrane and Process Development

Following the publication of Scholander's work on hemoglobin-based membranes, various immobilized liquid membranes were developed. All suffered from mechanical and chemical instability and loss of the carrier to the surrounding media. Few achieved a lifetime of more than a week or two. Membranes could be regenerated periodically, but this was impractical in an industrial setting. Nonetheless, the separations obtained are often so extraordinary that carrier transport continues to be the subject of research interest. For example, in an important early paper, Ward and Robb at General Electric showed that by using immobilized aqueous sodium carbonate as a carrier, it was possible to completely separate carbon dioxide and hydrogen sulfide from other gases by the reversable reactions:

$$CO_2 + H_2O + CO_3^{2-} \rightleftarrows 2HCO_3 \tag{12.3}$$

and

$$H_2S + CO_3^{2-} \rightleftarrows HS^- + HCO_3 \tag{12.4}$$

The carrier was simple, low-cost, and chemically stable, but the aqueous-based membrane failed after a few days, even when pre-humidified gases were used [2].

At about the same time (from the mid-1970s to the early 1980s), Steigelman and Hughes at Standard Oil developed similar aqueous-based membranes containing dissolved silver salts [3]. Silver ions form coordination complexes with olefins and so are able to separate propylene/propane mixtures and ethylene/ethane mixtures. Propylene/propane selectivities of several hundred were obtained and the process was developed to the bench-test stage. The principal problem was mechanical degradation by evaporation and displacement of water from the microporous membrane, compounded by chemical instability of the silver salt solution. These problems lead to a steady decline in membrane flux over a period of 10–20 days.

A breakthrough occurred in the 1960s when Bloch and Vofsi, working on the related process of coupled transport, showed that carrier molecules dissolved or dispersed into a solid polymer matrix could still enhance permeation and were much more stable than immobilized liquid membranes [4, 5]. Their work is discussed in the coupled transport section. Such solid polymer membranes are now sometimes referred to as polymer inclusion membranes.

Since Bloch's work, the same approach has been used by many facilitated transport researchers, using a variety of polymers to contain the carrier material. Pinnau and Toy [6], Merkel et al. [7, 8] and Kang et al. [9] dispersed silver salts in Pebax® and related polar rubbers to make selective membranes for olefin/paraffin separations. Hägg [10, 11], Ho [12, 13], Duan et al. [14], and Matsuyama et al. [15] used amines dispersed in polyvinyl alcohol, or in various cross-linked hydrogels, to separate CO_2/N_2 mixtures. The carrier component is usually directly dissolved or dispersed in the rubbery polymer that forms the membrane matrix. In some cases, a hydrogel-forming polymer is used which, when hydrated, contains enough water to dissolve the carrier.

Development of solid polymer membranes containing dissolved carriers meant that the mechanism used to describe membrane transport had to be revised. When immobilized liquid membranes were originally used, it was thought that the complexed carrier diffused from the feed to the permeate side, where the carrier reaction was reversed, and that the carrier would then diffuse back to the feed side. This mechanism implies unrealistically high diffusion coefficients for a large, complexed molecule traversing a solid membrane. A better explanation is that, in this environment, the permeant complexes and decomplexes rapidly with the relatively immobile carrier molecule. At low carrier loadings, the carrier sites are far apart and do not assist permeation, but at higher carrier loadings, a point is reached where the permeating component can hop from one carrier site to another nearby site [16, 17]. Above this loading, transport is significantly enhanced. The loading level where permeation enhancement begins is called the percolation threshold.

Some results that nicely illustrate the percolation threshold are shown in Figure 12.4 from a paper by Merkel et al. [8] on the separation of ethylene from ethylene/ethane mixtures. The carrier was $AgBF_4$ dissolved in a solid membrane made from Pebax 2533, a rubbery polyamide-polyether block copolymer.

The effect of $AgBF_4$ loading on ethylene and ethane permeances is clear from the figure. At low $AgBF_4$ loadings, the permeance of both gases decreases with increasing silver loading. The silver acts as a filler for the non-facilitated gases so reducing the volume of the polymer phase through which permeation can occur. At loadings above about 30 wt% $AgBF_4$, the permeance of ethylene sharply increases with increasing $AgBF_4$ loading as facilitation begins. At an $AgBF_4$ loading of about 60 wt%, the ethylene/ethane selectivity, initially close to 1, has increased to almost 100. Alas, these very promising results could not be sustained due to carrier instability.

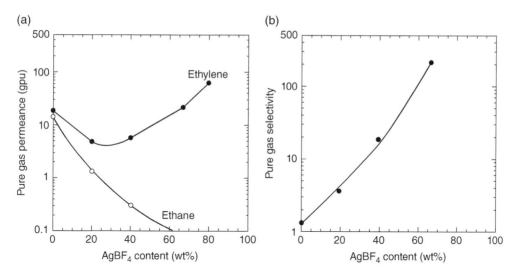

Figure 12.4 *Pure gas permeance (a) and selectivity (b) of ethylene and ethane in Pebax® 2533 membranes containing dispersed AgBF₄ as a function of AgBF₄ content. Enhanced permeation of ethylene occurs due to percolation when the AgBF₄ content is greater than about 40 wt%. Source: Reproduced with permission from [8]. Copyright (2013) Elsevier.*

Over a 20 day test run, the ethylene facilitation effect slowly dissipated as Ag^+ ions were converted to Ag^0 metal.

12.2.2   Theory

The simplest and most straightforward starting point to develop the equations for facilitated transport is to make two assumptions. The first is that the chemical reactions taking place at the feed and permeate side of the membrane are fast compared to diffusion through the membrane, so that chemical equilibrium is reached at the membrane interfaces; the second is that the majority of the target gas is transported across the membrane by the carrier; the amount transported by normal diffusion is so much smaller that it can be ignored.

The assumption that chemical equilibrium is reached at the membrane interfaces allows the coupled transport process to be modeled easily [7]. The carrier-species equilibrium in the membrane is

$$R + A \rightleftharpoons RA \tag{12.5}$$

where R is the carrier, A is the permeant transported by the carrier (usually a gas such as CO_2, O_2 or a light olefin), and RA is the permeant-carrier complex. Examples of reactions that have been performed in laboratory studies of facilitated transport processes are shown in Table 12.1. The process can be represented schematically as shown in Figure 12.5. The carrier–permeate reaction within the membrane is described by the equilibrium constant

$$K = \frac{[RA]_{(m)}}{[R]_{(m)}[A]_{(m)}} \tag{12.6}$$

Table 12.1 *Facilitated transport reactions.*

Permeant gas	Reaction
CO_2	$CO_2 + H_2O + Na_2CO_3 \rightleftarrows 2NaHCO_3$
O_2	$O_2 + CoSchiffs\ base \rightleftarrows CoSchiffs\ base\ (O_2)$
SO_2	$SO_2 + H_2O + Na_2SO_3 \rightleftarrows 2NaHSO_3$
H_2S	$H_2S + Na_2CO_3 \rightleftarrows NaHS + NaHCO_3$
CO	$CO + CuCl_2 \rightleftarrows CuCl_2(CO)$
C_2H_4	$C_2H_4 + AgNO_3 \rightleftarrows AgNO_3(C_2H_4)$

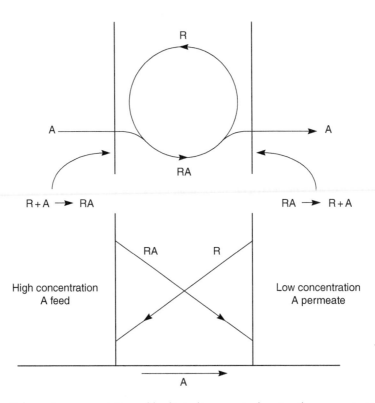

Figure 12.5 *Schematic representation of facilitated transport, showing the concentration gradients that form in the membrane.*

The concentration of permeant, $[A]_{(m)}$, within the membrane phase is linked to the concentration of permeant A in the adjacent gas phase, $[A]$, by the Henry's law expression

$$[A]_{(m)} = k[A] \tag{12.7}$$

Hence Eq. (12.6) can be written as

$$\frac{[RA]_{(m)}}{[R]_{(m)}[A]} = K \cdot k = K' \tag{12.8}$$

The components $[R]_{(m)}$ and $[RA]_{(m)}$ are linked by a simple mass balance expression to the total concentration of carrier $[R]_{(m)tot}$ within the membrane phase, so Eq. (12.8) can be rearranged to

$$[RA]_{(m)} = \frac{[R]_{(m)tot}}{1 + 1/[A]K'} \tag{12.9}$$

Equation (12.9) shows the fraction of the carrier that reacts to form a carrier complex. At very large values of the term $[A]K'$; that is, at high concentrations (partial pressures) of component A, all the carrier is complexed, and $[RA]_{(m)} \rightarrow [R]_{(m)tot}$. At low concentrations/partial pressures of A, $[A]K' \rightarrow 0$ and none of the carrier is complexed ($[RA]_{(m)} \rightarrow 0$). Using Eq. (12.9), the concentration of the carrier–permeant complex at each side of membrane is calculated in terms of the equilibrium constant between the carrier and the permeant, and the concentration partial pressure of permeant A in the adjacent feed and permeate gases. Transport through the membrane can then be calculated using Fick's law. The flux, j_{RA}, of RA through the membrane is given by

$$j_{RA} = \frac{D_{RA}\left([RA]_{o(m)} - [RA]_{\ell(m)}\right)}{\ell} \tag{12.10}$$

Substituting Eq. (12.9) into Eq. (12.10) yields

$$j_{RA} = \frac{D_{RA}[RA]_{(m)tot}}{\ell} \left[\frac{1}{1 + 1/[A]_o K'} - \frac{1}{1 + 1/[A]_\ell K'}\right] \tag{12.11}$$

To illustrate the dependence of the membrane flux, j_{RA}, on the equilibrium constant K' and the pressure gradient across the membrane, the partial pressure of permeant A on the permeate side can be set to zero, that is, $[A]_\ell = 0$, and Eq. (12.11) becomes

$$j_{RA} = \frac{D_{RA}[R]_{(m)tot}}{\ell} \left(\frac{1}{1 + 1/[A]_o K'}\right) \tag{12.12}$$

This expression is plotted in Figure 12.6 as flux as a function of feed pressure for different values of the equilibrium constant, K'. In this example, at an equilibrium constant K' of $0.01\ \text{atm}^{-1}$, very little of carrier R reacts with permeant A, even at a feed pressure of 10 atm, so the flux is low. As the equilibrium constant increases, the fraction of carrier reacting with permeant at the feed side of the membrane increases, so the flux increases. This result would suggest that, to achieve the maximum flux, a carrier with the highest possible equilibrium constant should be used. For example, the calculations illustrated in Figure 12.6 indicate a carrier with an equilibrium constant of $10\ \text{atm}^{-1}$ or more will provide the maximum flux.

The calculations illustrated in Figure 12.6 assume that a hard vacuum is maintained on the permeate side of the membrane. More realistically, processes are operated either with a compressed gas feed and a lower but still significant pressure on the permeate side, or with an ambient pressure feed gas and a vacuum of 0.1–0.2 bar on the permeate side. By substitution of specific values for the feed and permeate pressures into Eq. (12.11), the optimum values of the equilibrium constant can be calculated. A plot illustrating this calculation for two example cases

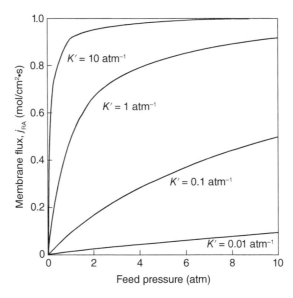

Figure 12.6 *Flux through a facilitated transport membrane calculated using Eq. (12.12) and setting the partial pressure of A on the permeate side at zero.*

is shown in Figure 12.7. One case used a feed pressure of 10 atm and a permeate pressure of 1 atm. The other case uses 1 atm feed pressure and 0.1 atm on the permeate side.

Under the assumptions of the calculations, the optimum equilibrium constant is 0.3 atm^{-1} for compression operation (feed pressure, 10 atm; permeate pressure, 1 atm), and 3 atm^{-1} for vacuum operation (feed pressure, 1 atm; permeate pressure, 0.1 atm). It follows that the feed and permeate pressures used must be matched to the equilibrium constant value to achieve the optimum membrane flux. However, operating pressures of industrial processes are often difficult and expensive to change. In these cases, an alternative approach sometimes used is to change the equilibrium constant by changing the carrier chemistry or the operating temperature of the process.

12.2.3 Membranes

The first generation facilitated transport membranes were liquid solutions held by capillarity in the pores of a microporous membrane [2, 3]. Evaporation of the aqueous carrier phase could be minimized by humidifying the feed gas, but even so, useful membrane lifetimes were only a few days. The ionic liquids development in the 1980s provided ultra-low vapor pressure solvents as substitute carrier solvents for water [18]. Solvent evaporation was diminished as a problem, but other problems remained, and the stability problem was reduced but not eliminated.

Despite the lack of success, there is still interest in the technology because of the spectacular separation that can be obtained in the short-term laboratory tests. Figure 12.8 shows some of the results that maintain this interest.

The carrier in this case, a cobalt(II) Schiff base that forms a reversible complex with oxygen in air, is held in an aqueous solution in the pores of a microporous membrane. With an oxygen/nitrogen selectivity of 20–30 and an oxygen permeability of 500 barrer, the performance of this membrane far surpassed that of any other membrane. If this membrane could have been

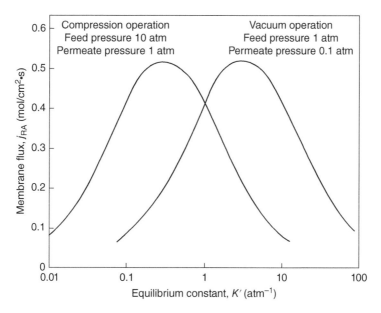

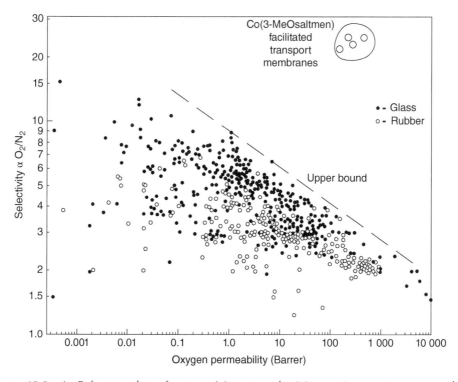

Figure 12.7 *Illustration of the effect of feed and permeate pressure on the optimum carrier equilibrium constant,* $D_{RA}[R]_{(m)tot}/\ell \approx 1$. *Vacuum operation: feed pressure 1 atm, permeate pressure 0.1 atm; compression operation: feed pressure 10 atm, permeate pressure 1 atm.*

Figure 12.8 *A Robeson plot of oxygen/nitrogen selectivity against oxygen permeability comparing numerous polymeric membranes with a supported liquid membrane containing a Co(3-MeOsaltmen) carrier. Source: Adapted from [19].*

developed to the industrial scale, it would have transformed the production and use of oxygen. This was not to be. Even in carefully controlled laboratory tests, the maximum lifetime of the membrane was only 1 or 2 weeks before a combination of membrane dewetting, solvent evaporation, and oxidation of the cobalt(II) carrier to cobalt(III) led to flux decline and loss of selectivity.

Since about 2000, polymer inclusion membranes, with the carrier dissolved or dispersed in a solid polymer matrix, have become the membranes of choice for most studies. None has reached industrial use, but the best membranes are getting closer. Most of the research effort for the last 20 years has focused on olefin/paraffin separations and removal of CO_2 from N_2, CH_4, and H_2. These two applications are covered below.

12.2.4   Applications

12.2.4.1   *Olefin/Paraffin Separations*

These separations are difficult because the molecules are similar in size and boiling point. To produce ethylene and propylene, two of the most important petrochemicals feedstocks, the separations of ethylene from ethane and propylene from propane are done on a massive scale. Currently, cryogenic distillation is the state-of-the-art technology, even though very large columns of more than 100 trays using reflux ratios of 10 or more are required. It has been known for many years that solutions of silver salts selectively absorb ethylene and propylene through π bond complexation with the olefin bond, and for a number of years, attempts were made to develop absorption processes based on silver salts [20, 21]. Silver cations of all silver salts will form this complex, but the nature of the anion affects the reaction. Silver tetrafluoroborate ($AgBF_4$) is significantly more reactive than the other salts and is usually the preferred silver salt used in facilitated transport membranes.

Beginning in about 2000, two groups, one in the United States [6–8] and the other in Korea [9, 22], developed solid polymer membranes for olefin separations. The carrier was generally $AgBF_4$, dissolved in polar polymeric polymers such as Pebax 2533 (a polyamide-polyether block copolymer) or POZ (poly 2-ethyl-2-oxazoline) or PVP (poly(vinyl pyrrolidone)). The membranes were made from casting solutions of a volatile solvent containing dissolved polymer and carrier. After casting and solvent evaporation, clear films containing 20–60 wt% carrier resulted.

Some pure gas permeation results are shown in Figure 12.9. Figure 12.9a repeats Figure 12.4. Figure 12.9b shows similar results for propylene and propane. Both figures are consistent with the theory that the carrier acts as a block for all permeants until the loading reaches the percolation threshold, at which point facilitated olefin transport takes off, yielding a sharp increase in olefin permeance and olefin/paraffin selectivities.

The stability of these membranes was determined with membranes of varying thicknesses in the 2–100 µm range. As with other membranes, thick films were more stable than thin membranes, but loss of carrier or carrier solvents was not a problem. The chemical stability of the carrier, however, remained an issue. Reducing gases such as hydrogen, a common contaminant of all olefin gas mixtures, converted the silver ions to silver metal via the reaction

$$Ag^+ + e^- \rightarrow Ag^0 \tag{12.13}$$

Exposure of the membranes to light had the same effect. Acetylene and hydrogen sulfide, two other trace contaminants found in industrial streams, also reacted with the $AgBF_4$ through the reactions

$$C_2H_2 + 2\,AgBF_4 \rightarrow Ag_2C_2 + 2HBF_4 \tag{12.14}$$

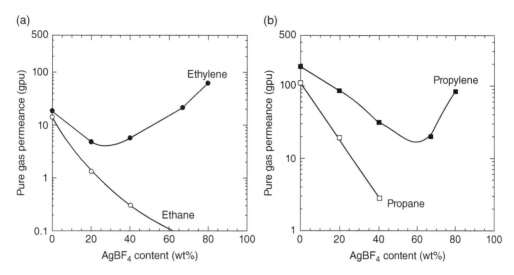

Figure 12.9 *Pure gas permeances of (a) ethylene and ethane, and (b) propylene and propane as a function of the AgBF₄ content in Pebax 2533 membranes (50 psig feed, 0 psig permeate, 25 °C). Source: [8] / With permission from Elsevier.*

$$H_2S + 2\,AgBF_4 \rightarrow Ag_2S + 2HBF_4 \tag{12.15}$$

Merkel et al. [8] showed that reduction of the silver carrier to metal or conversion to silver carbide could be reversed by exposure of the membrane to acidic hydrogen peroxide vapor mixtures for a few hours. Unfortunately, Merkel's results also showed that permeance decline occurred even with completely pure two-component olefin/paraffin mixtures. The rate of decline was much more noticeable in thin membranes and was proportional to the olefin content of the gas mixture being separated, as can be seen in Figure 12.10.

12.2.4.2 *CO₂ Separation*

In the last 15 years, recognition of the global warming problem has led to membrane development programs targeted on separating CO_2 from power plant and industrial process exhaust gases, which are composed primarily of nitrogen, carbon dioxide, and water vapor. The most advanced programs, using conventional thin polymer composite membranes, are discussed in Chapter 9. The best of these membranes have mixed gas CO_2/N_2 selectivity of about 30, and CO_2 permeances of more than 1000 gpu. However, for this separation, membranes with even higher permeances and higher selectivities would be useful. Building on the work of Ward from 40 years ago, a number of groups, led by Matsuyama [23] and Kazama [24] in Japan, Hägg [11, 25, 26] in Norway and Ho [13, 27, 28] in the United States, have developed facilitated transport membranes for this separation. Polymer inclusion membranes containing low molecular weight CO_2 carriers in a polymer matrix were tried at first, but it was soon discovered that better and much more stable membranes were possible if the polymer forming the membrane matrix is itself the carrier. Typical membranes of this type, generally known as fixed site carrier membranes, use poly(vinyl amine) as the carrier component. To improve the mechanical properties, a second polymer, commonly poly(vinyl alcohol), can form part of the membrane matrix. The facilitated transport mechanism is illustrated in Figure 12.11.

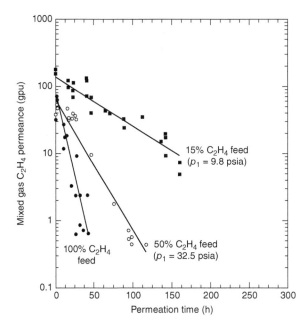

Figure 12.10 *The permeance of ethylene/ethane gas mixture through AgBF₄/Pebax 2533 membranes as a function of time. Source: Adapted from [8].*

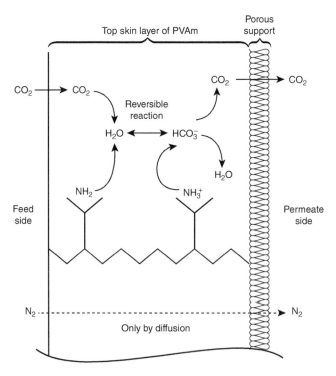

Figure 12.11 *An illustration of facilitated transport using fixed site amine carriers to transport CO_2 through a water swollen poly(vinyl amine) (PVAm) membrane. Source: [25] / John Wiley & Sons.*

It is believed that the CO_2 molecules travel through the membrane by hopping across the fixed amine groups utilizing the reaction

$$CO_2 + H_2O + NH_2 \rightleftharpoons HCO_3^- + NH_3^+ \tag{12.16}$$

The membrane performance depends on the carrier activity and can be improved by adding monomeric amines as supplementary mobile carriers. Mixed gas selectivities of greater than 100 are routinely reported, generally with carbon dioxide permeances in the 1000 gpu range.

But there are problems, perhaps the most important being that the above complexing reaction requires as many moles of water as of CO_2. The raw flue gas may be humid initially, but the membranes are extremely permeable to water, which means that water is rapidly lost with the outgoing permeate, and both the feed and permeate sides of the membrane must be humidified to maintain relative humidity therein at 80–100%. Some typical results showing the effect of relative humidity on the CO_2 permeance and the CO_2/N_2 selectivity of a poly(vinyl amine) fixed carrier membrane are shown in Figure 12.12. Maintaining this high level of humidity is difficult in an industrial process.

A second problem is the pressure dependence of membrane permeability and selectivity. At low partial pressures of CO_2, there is an excess of open carrier amine sites in the membrane, and both CO_2 flux and CO_2/N_2 selectivity are high. At higher feed carbon dioxide partial pressures, however, a point is reached where the carrier sites are saturated, in which case the carbon dioxide flux levels off, and increasing the feed pressure merely causes the selectivity to fall, as more nitrogen is driven through the membrane by regular diffusion. The saturation effect is

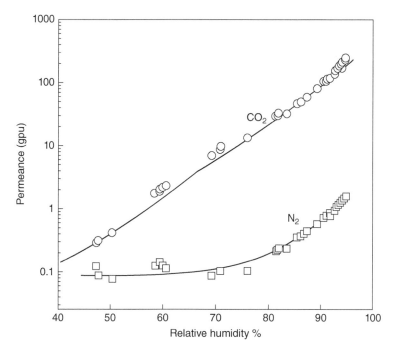

Figure 12.12 *The effect of feed gas humidity on the permeance of poly(vinyl amine) fixed site carrier membranes to CO_2/N_2 gas mixtures (2 bar feed, helium sweep gas, 10% CO_2, 90% N_2 containing varying amounts of H_2O). Source: [11] / With permission from Elsevier.*

very dependent on the carrier chemistry, the CO_2 partial pressure of the feed gas, and the membrane temperature. The membrane performance must be characterized and optimized for each application, and performance under one set of conditions cannot be easily extrapolated to another.

12.3   Coupled Transport

12.3.1   Membrane and Process Development

The main industrial operation that the developers of coupled transport membranes hope to displace is solvent extraction to separate solutions of metal ions by means of a mixer/setter process [29]. A conventional mixer/settler process and the potential carrier membrane alternative are shown in Figure 12.13.

The solvent extraction mixer/settler process in Figure 12.13a shows the separation of copper from dilute aqueous solutions containing a mixture of metal ions. Such solutions are generated when metal is extracted from copper ores using sulfuric acid; several hundred large extraction/recovery plants of this type are in operation around the world. The aqueous feed solution is mixed

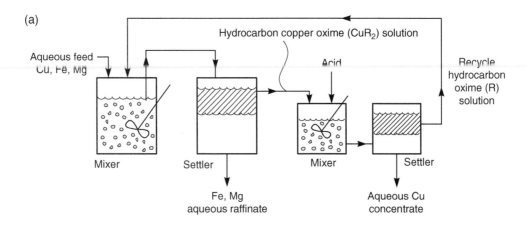

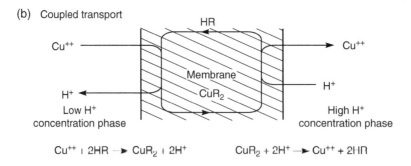

Figure 12.13   *Use of coupled transport to replace a mixer/settler. In the conventional scheme (a), an oxime carrier dissolved in kerosene separates copper ions selectively from an aqueous solution of metal ions. The membrane scheme (b) uses the same oxime-in-kerosene carrier solution held in a membrane matrix.*

with a hydrocarbon solvent containing an oxime carrier that reacts selectively with copper. The carrier extracts the copper into the hydrocarbon phase, simultaneously releasing hydrogen ions to the aqueous phase. The two-phase mixture is then sent to a large settling tank where the phases separate. The aqueous solution, now stripped of its copper content, is discharged; the organic phase, containing the copper carrier complex, is sent to a second, smaller mixing tank, where it is contacted with a strong acid solution. In the presence of a high concentration of hydrogen ions, the protonated oxime carrier re-forms, and the copper ion is released to the aqueous acid solution. The two-phase mixture is allowed to phase separate in a second settling tank. The aqueous acidic copper solution is removed for downstream treatment, and the hydrocarbon phase, containing the regenerated carrier, is recycled to the first mixed tank. Figure 12.13b shows the equivalent membrane process, which has already been described in Figure 12.2b.

The conventional process shown in Figure 12.13a uses an oxime kerosene mixture. Slow degradation of the carrier component occurs, so the recycling carrier solution must be replaced on a regular basis. The membrane process requires a much smaller inventory of carrier to treat the same amount of feed, which cuts costs and allows a wider range of more expensive carrier compounds to be used. The membrane system is also far more compact and potentially has a significantly lower cost.

Most work on carrier transport uses schemes of the type shown in Figures 12.2 and 12.13, where a single membrane performs the total separation. Such processes have many attractive features; nonetheless, none has reached the commercial stage because of the stability issues mentioned in the introduction. Conventional solvent extraction processes use very large amounts of carrier. In a membrane process, the amount of carrier used to perform the same separation is reduced more than 1000 fold because the loading and unloading of the carried species from the carrier occurs very much faster than in conventional operations. Unfortunately, however, this means that chemical degradation of the carrier to an inactive form, which might become noticeable in the conventional process after a year of operation, can become apparent after a few days in a carrier membrane.

Liquid membranes holding carriers for metal ions were developed by Miyauchi [30], Largman and Sifniades [31], and others [32]. The largest application was the recovery of metals from hydrometallurgical solutions, as described above. Efficient and stable carrier compounds, developed by others for solvent extraction processes, were available for use in membranes, so chemical stability was not an immediate problem, but the problem of loss of the liquid carrier phase from the support membrane was not solved. By the 1980s, it was clear that liquid membranes immobilized in microporous supports, despite their ability to perform spectacular separations, were going nowhere unless the membrane stability problem could be solved. The problem remains, although a number of possible solutions have been developed. The first, and still most promising, approach to improved membrane stability was pioneered by Bloch and Vofsi. They were using phosphate ester carriers [4, 5] for the coupled transport separation of uranium. Because phosphate esters plasticize poly(vinyl chloride) (PVC), Bloch and Vofsi were able to prepare immobilized carrier films by dispersing the liquid esters in a PVC matrix. A PVC/ester film, containing up to 60 wt% ester, was cast from the dispersion onto a paper support. Work continued until the early 1970s, when interest lagged, apparently because the fluxes obtained did not make the membranes competitive with conventional separation techniques. Nonetheless, membrane stability was much improved.

Dissolving the carrier in some form of polymer matrix remains the most widely used approach to solving the membrane stability problem. Ways to enhance the stability further have included dissolving the carrier in an ionic liquid and then dispersing the liquid mixture into the matrix. Because ionic liquids have no vapor pressure, loss of the liquid by evaporation is no longer

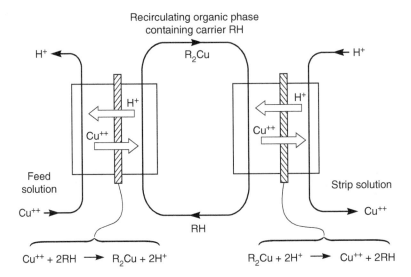

Figure 12.14 *Use of two contactors in a liquid membrane process.*

an issue. Dai et al. have reviewed this approach [33]. Others have chemically attached the carrier molecule to the polymer backbone [34]. In this way, migration and loss of the carrier is completely eliminated. However, chemical instability problems remain, and it is often difficult to incorporate enough carrier to exceed the percolation threshold.

One method of circumventing the stability problem of immobilized carrier membranes is to change the single membrane process of Figure 12.13 into a two-step membrane contactor process. The concept is illustrated in Figure 12.14.

Two membrane contactors are used: one to separate the organic carrier phase from the feed, the other to separate the organic carrier phase from the permeate. In the first contactor, metal ions from the feed solution diffuse across the microporous membrane and react with the carrier, liberating hydrogen counter ions. The organic carrier solution is then pumped from the first to the second membrane contactor, where the reaction is reversed. The metal ions are liberated to the permeate solution and hydrogen ions are picked up by the carrier. The re-formed carrier solution is pumped back to the first membrane contactor. Because a relatively large volume of carrier solution is used, the effect of loss of carrier by degradation or solvation into the feed and strip solutions is considerably ameliorated. Sirkar and his students [35] have used oxime carriers in this type of system to separate metal ions by coupled transport. A similar process was developed to the large pilot plant scale by Davis et al. at British Petroleum to separate ethylene/ethane mixtures by facilitated transport, using silver nitrate solution as a carrier for ethylene [36, 37]. Bessarabov et al. also pursued this approach [38].

Another variant related to the contactor processes described above uses liquid emulsion membranes. The technique, developed by Li at Exxon, employs a surfactant stabilized in an emulsion, and requires four discrete operations, as shown in Figure 12.15 [39–41]. The organic phase containing the carrier forms the wall of an emulsion droplet, separating the aqueous feed from the aqueous product solution. Thus, the droplet wall is the membrane, as shown in Figure 12.15a. The four-step process is shown in Figure 12.15b. Metal ions are removed from the aqueous feed and are concentrated in the interior of the droplets. Once sufficient metal has

(a) Emulsion liquid membrane

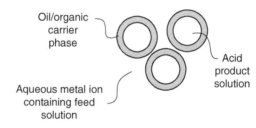

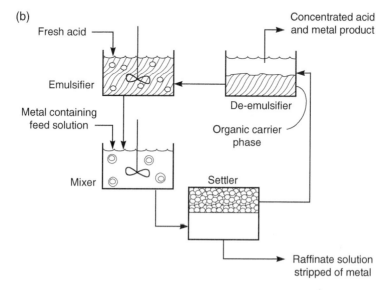

Figure 12.15 *(a) Structure of an emulsion liquid membrane. The walls of the emulsion droplets form the membrane. (b) Flow diagram of liquid emulsion membrane separation process.*

been extracted, the emulsion droplets are separated from the feed, and the emulsion is broken to liberate a concentrated product solution and an organic carrier phase.

The process has elements in common with the conventional solvent extraction/mixer-settler processes illustrated in Figure 12.13. First, fresh acid solution is emulsified in the liquid organic carrier membrane phase. The acid/oil emulsion then enters a second large mixer vessel where it is again emulsified into the feed solution to form a feed solution/oil carrier/acid emulsion. Metal ions in the feed solution permeate by coupled transport through the walls of the emulsion droplets, forming a product solution in the interior of the droplet. The mixture then passes to a settler tank, where the oil droplets separate from the metal-depleted raffinate solution. A single mixture-settler step is shown in Figure 12.15, but in practice, a series of mixer-settlers may be used to extract the metal completely. The emulsion concentrate then passes to a de-emulsifier where the emulsion is broken and the organic and concentrated product solutions are separated. The regenerated organic carrier solution is recycled to the first emulsifier.

Although emulsion degradation must be avoided in the mixer and settler tanks, complete and rapid breakdown is required in the de-emulsifier in which the product solution is separated from the organic complexing agent. Currently, electrostatic coalescers seem to be the best method of breaking these emulsions. Even then, some of the organic phase is lost with the feed raffinate.

The process was worked on by Exxon for more than a decade and brought to the pilot plant stage in 1979 before finally being abandoned.

One solution to the emulsion stability problem is to use the modified emulsion membrane contactor process developed by Wiencek et al. [42, 43] and shown in Figure 12.16. The contactor is fitted with a hydrophobic finely microporous membrane, the pores of which are filled with an organic solvent containing a dissolved carrier agent, as in Figure 12.16a. An aqueous feed solution containing a low concentration of heavy metal ions passes on one side of a membrane contactor. On the permeate side is more of organic solvent phase, now containing dispersed droplets of an acidic aqueous strip solution in addition to the dissolved carrier agent.

As the aqueous feed passes across the membrane, metal ions in the solution react with the carrier agent and displace hydrogen ions into the feed. The metal ion/carrier complex diffuses through the organic-liquid-filled membrane into the emulsified permeate solution, where it reacts with the dispersed acid droplets. The metal ion is extracted into these droplets and the carrier

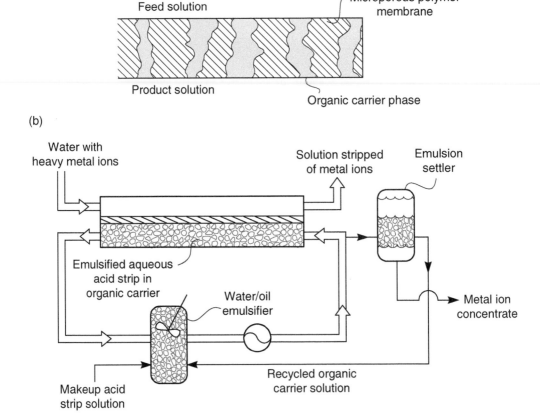

Figure 12.16   *(a) A supported liquid membrane. The carrier agent in an organic solvent is held in the pores of a finely microporous membrane. (b) A hybrid emulsion-membrane contactor system developed for coupled transport applications by Wiencek and his students [42]. The stability problems of liquid membranes and emulsion membranes are circumvented.*

agent is re-formed. The dispersed aqueous droplets in the permeate organic solution are much larger than the membrane pores, so mixing of the acid strip solution with the aqueous feed is completely prevented.

The dispersed droplets of strip solution are circulated counter-currently to the aqueous feed. In a continuous process, a portion of this strip solution is removed from the loop and allowed to phase separate. The organic phase is returned to the emulsifier unit and the aqueous phase is the concentrated metal ion product solution.

This device has the stability of the two contactor system shown in Figure 12.14, but uses only half the membrane area, and produces significantly higher membrane fluxes. Fouad and Bart [44] and Ho and coworkers [45] have applied this idea in the laboratory to a number of potential coupled transport applications.

Reviews of aspects of both facilitated and coupled transport have been given by Ho et al. [45, 46], Noble and Way, [47], O'Bryan et al. [48], and Nghiem [49].

12.3.2   Theory

A number of models of coupled transport have been proposed using the same basic assumptions as for facilitated transport, namely that the chemical reactions involved are fast, chemical equilibrium is reached at the membrane interfaces, and simple uncoupled diffusive transport can be ignored. The process is shown schematically in Figure 12.17, in which the reaction of the carrier (RH) with the metal (M^{n+}) and hydrogen ion (H^+) is given as

$$n\text{RH} + \text{M}^{n+} \rightleftharpoons \text{MR}_n + n\text{H}^+ \qquad (12.17)$$

This reaction is characterized by an equilibrium constant

$$K = \frac{[\text{MR}_n][\text{H}]^n}{[\text{RH}]^n[\text{M}]} \qquad (12.18)$$

where the terms in square brackets represent the molar concentrations of the particular chemical species. The equilibrium equation can be written for the organic phase or the aqueous phase. As in earlier chapters, the subscripts o and ℓ represent the position of the feed and permeate interfaces of the membrane. Thus, the term $[\text{MR}_n]_o$ represents the molar concentration of component MR in the aqueous solution at the feed/membrane interface. The subscript m is used to represent the membrane phase. The term $[\text{MR}_n]_{o(m)}$ is the molar concentration of component MR_n in the membrane at the feed interface (point o).

Only $[\text{MR}_n]$ and $[\text{RH}]$ are measurable in the organic phase, where $[\text{H}]$ and $[\text{M}]$ are negligibly small. Their concentration gradients and directions of flow are shown in the lower portion of the figure. Similarly, only $[\text{H}]$ and $[\text{M}]$ are measurable in the aqueous phase, where $[\text{MR}_n]$ and $[\text{RH}]$ are negligibly small. Equation (12.18) can, therefore, be written for the feed solution interface as

$$K' = \frac{[\text{MR}_n]_{o(m)}[\text{H}]_o^n}{[\text{RH}]_{o(m)}^n[\text{M}]_o} = \frac{k_m}{k_a} \cdot K \qquad (12.19)$$

where k_m and k_a are the partition coefficients of M and H between the aqueous and organic phases. This form of Eq. (12.18) is preferred because all the quantities are accessible experimentally.

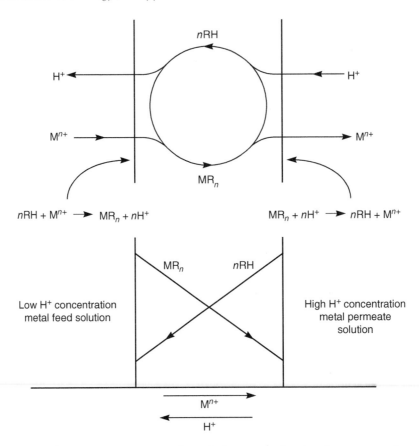

Figure 12.17 *Schematic representation of coupled transport showing the concentration gradients that form in coupled transport membranes.*

The term $[MR_n]_{o(m)}/[M]_o$ is easily recognizable as the distribution coefficient of metal between the organic and aqueous phases.

The same equilibrium applies at the permeate–solution interface, and Eq. (12.19) can be recast to

$$K' = \frac{[MR_n]_{\ell(m)}[H]_\ell^n}{[RH]_{\ell(m)}^n[M]_\ell} \tag{12.20}$$

Consider now the situation when a counter ion concentration gradient that exactly balances the metal ion concentration gradient is established, so no flux of either ion across the membrane occurs. Under this condition, $[MR_n]_{o(m)} - [MR_n]_{\ell(m)}$ and $[RH]_{o(m)}^n = [RH]_{\ell(m)}^n$, producing the expression

$$\frac{[M]_o}{[M]_\ell} = \left(\frac{[H]_o}{[H]_\ell}\right)^n \tag{12.21}$$

Thus, the maximum enrichment of metal ion that can be established across the membrane varies with the counter ion (hydrogen ion) concentration ratio (in the same direction) raised to the *n*th power, where *n* is the valence of the metal ion (M^{n+}).

This result says nothing about the metal ion flux across the membrane under non-equilibrium conditions; this is described by Fick's law. At steady state, the flux j_{MR_n}, in mol/cm^2·s, of metal complex MR_n across the liquid membrane is given by

$$j_{MR_n} = \frac{D_{MR_n}\left([MR_n]_{o(m)} - [MR_n]_{\ell(m)}\right)}{\ell} \tag{12.22}$$

where D_{MR_n} is the mean diffusion coefficient of the complex in the membrane of thickness ℓ. To put Eq. (12.22) into a more useful form, the terms in $[MR_n]$ are eliminated by introduction of Eq. (12.19). This results in a complex expression involving the desired quantities [M] and [H], but also involving [RH]. However, mass balance provides the following relationship

$$n[MR_n]_{(m)} + [RH]_m = [R]_{(m)tot} \tag{12.23}$$

where $[R]_{(m)tot}$ is the total concentration of R in the membrane.

Substitution of Eqs. (12.19) and (12.20) into (12.22) gives an expression for the metal ion flux in terms of the concentration of metal and counter ion in the aqueous solutions on the two sides of the membrane [32]. The solution is simple only for $n = 1$, in which case

$$j_{MR_n} = \frac{D_{MR_n}[R]_{(m)tot}}{\ell}\left[\left(\frac{1}{[H]_o/[M]_o K' + 1}\right) - \left(\frac{1}{[H]_\ell/[M]_\ell K' + 1}\right)\right] \tag{12.24}$$

This equation shows the coupling effect between the metal ion [M] and the hydrogen ion [H] because both appear in the concentration term of the Fick's law expression linked by the equilibrium reaction constant K'. Thus, there will be a positive 'uphill' flux of metal ion from the downstream to the upstream solution (that is, in the direction $\ell \rightarrow o$) as long as

$$\frac{[M]_o}{[H]_o} > \frac{[M]_\ell}{[H]_\ell} \tag{12.25}$$

When the inequality is opposite, the metal ion flux is in the conventional or 'downhill' direction. The maximum concentration factor, that is, the point at which metal ion flux ceases, can be determined in terms of the hydrogen ion concentration in the two aqueous phases

$$\frac{[M]_o}{[M]_\ell} = \frac{[H]_o}{[H]_\ell} \tag{12.26}$$

This expression is identical to Eq. (12.21) for the case of a monovalent metal ion.

12.3.3   Coupled Transport Membrane Characteristics

12.3.3.1   Concentration Effects

Equations (12.21)–(12.26) provide a basis for rationalizing the principal features of coupled transport membranes. It follows from Eq. (12.24) that coupled transport membranes can move metal ions from a dilute to a concentrated solution against the metal ion concentration gradient, provided the gradient in the second coupled ion concentration is sufficient. An experiment demonstrating this counterintuitive result is shown in Figure 12.18. The process is counter-transport of copper driven by hydrogen ions. In this particular experiment, a pH difference of 1.5 is maintained across the membrane. The initial product solution copper concentration is higher than the feed solution concentration. Nonetheless, copper diffuses against its concentration gradient from the feed to the product side of the membrane. The ratio of the counter hydrogen ions between the solutions on either side of the membrane is about 32 to 1 which, according to the appropriate form of Eq. (12.21), should give a copper concentration ratio of

$$\frac{\left[Cu^{2+}\right]_\ell}{\left[Cu^{2+}\right]_o} = \left(\frac{[H^+]_\ell}{[H^+]_o}\right)^2 = (32)^2 \approx 1000 \tag{12.27}$$

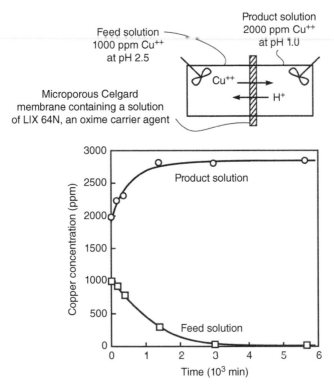

Figure 12.18   Demonstration of coupled transport. In a two-compartment cell, copper flows from the dilute (feed) solution into the concentrated (product) solution, driven by a gradient in hydrogen ion concentration [32]. Membrane, microporous Celgard 2400/LIX 64N; feed, pH 2.5; product, pH 1.0.

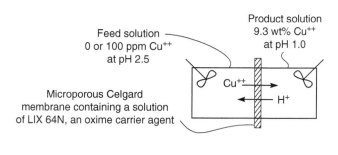

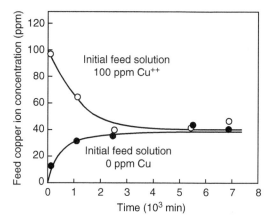

Figure 12.19 Experiments to demonstrate the maximum achievable concentration factor. Membrane, microporous Celgard 2400/LIX 64N; feed, pH 2.5, copper ion concentration, 0 or 100 ppm; product, pH 1.0, 9.3 wt% copper [32]. The concentration in the feed solution moves to a plateau value of 40 ppm at which the copper concentration gradient across the membrane is balanced by the hydrogen ion gradient in the other direction.

In the experiment shown in Figure 12.18, this means that the feed solution copper concentration should drop to just a few ppm, and this is the case.

A more convenient method of measuring the maximum copper concentration factor is to maintain the product solution at some high copper concentration and to allow the feed solution copper concentration to reach an easily measurable steady-state value. Figure 12.19 shows the feed copper concentration in such an experiment, in which the steady-state feed solution concentration was about 40 ppm. The feed solution was allowed to approach steady state from both directions, that is, with initial copper concentrations higher and lower than the predicted value for the given pH gradient. As Figure 12.19 shows, regardless of the starting point, the copper concentration factors measured by this method are in reasonable agreement with the predictions of Eq. (12.22).

12.3.3.2   Feed and Product Metal Ion Concentration Effects

A second characteristic of coupled transport membranes is that the membrane flux increases with increasing metal concentration in the feed solution, but is usually independent of the metal concentration in the product solution. This behavior follows from the flux Eqs. (12.22) and (12.24). In typical coupled transport experiments, the concentration of the driving ion (H^+) in the product solution is very high. For example, in coupled transport of copper, the driving ions are hydrogen ions, and 100 g/l sulfuric acid is often used as the product solution. As a result, on the product side of the membrane the carrier is in the protonated form, the term $[MR_n]_{\ell(m)}$ is very small compared to $[MR_n]_{o(m)}$, and Eq. (12.24) reduces to

$$j_{MR_n} = \frac{D_{MR_n}[R]_{(m)tot}}{\ell} \cdot \frac{1}{[H]_o/[M]_o K' + 1} \tag{12.28}$$

The permeate solution metal ion concentration, $[M]_\ell$, does not appear in the flux equation, which means that the membrane metal ion flux is independent of the concentration of metal on the permeate side. However, the flux does depend on the concentration of metal ions, $[M]_o$, on the feed solution side. At low values of $[M]_o$, the flux will increase linearly with $[M]_o$, but at higher concentrations the flux reaches a plateau value as the term $[H]_o/[M]_o K'$ becomes small compared to 1. At this point all of the available carrier molecules are complexed and no further increase in transport rate across the membrane is possible. The form of this dependence is illustrated for the feed and product solution metal ion concentrations in Figure 12.20. Both sets of experiments were carried out with the same membrane, with a feed solution pH of 2.5, and a sulfuric acid concentration of 100 g/l on the product side. The left-hand graph shows a flux increase with increasing copper concentration in the feed, up to a plateau value of about 14 $\mu g/cm^2/min$, whereas the right hand graph shows that the flux stays constant at about 12 $\mu g/cm^2/min$, despite a fivefold increase in the product solution copper concentration.

12.3.3.3 pH and Metal Ion Effects

It follows from flux Eq. (12.28) that the concentration of the counter hydrogen ion and the equilibrium coefficient K' for a particular metal ion will affect the metal ion flux. The effect of these factors can best be understood by looking at curves of metal ion extraction versus pH. Examples are shown in Figure 12.21 for copper and other metals using the carrier LIX 64N [16]. The counter ion is hydrogen, and the metal ions are extracted by reactions of the type shown at the beginning of Section 12.3.2.

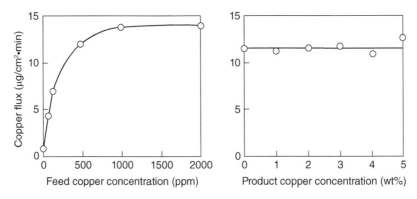

Figure 12.20 *Effect of metal concentration in the feed and product solution on flux. Membrane, microporous Celgard 2400/30% Kelex 100 in Kermac 470B; feed, pH 2.5; product, 100 g/l H_2SO_4 [32]. The metal ion flux increases to a plateau value with varying feed solution copper concentration in the feed solution (left-hand graph) but is independent of the copper concentration in the product (right-hand graph).*

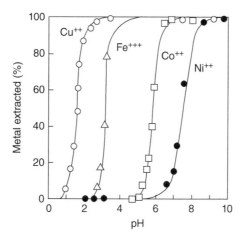

Figure 12.21 *Metal extraction curves for four metal ions, using LIX 64N as a carrier agent. The aqueous phase initially contained 1000 ppm metal as the sulfate salt. Source: [16] / Noyes Publications.*

The pH at which metal ions are extracted depends on the distribution coefficient for the particular metal and complexing agent. As a result, the pH at which the metal ions are extracted varies, as shown by the results in Figure 12.21. This behavior allows one metal to be separated from another. For example, consider the separation of copper and iron with LIX 64N. As Figure 12.21 shows, LIX 64N extracts copper at pH 1.5–2.0, but iron is not extracted until above pH 2.5. The separations obtained when 0.2% solutions of copper and iron are tested with a LIX 64N membrane at various pHs are shown in Figure 12.22. The copper flux is approximately 100 times higher than the iron flux at a feed pH of 2.5.

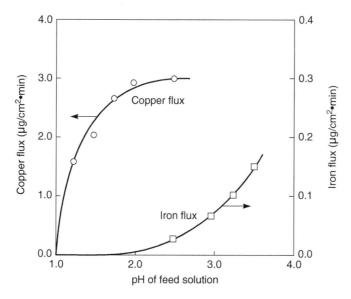

Figure 12.22 *Copper and iron fluxes as a function of feed pH [32]. Membrane, Celgard 2400/LIX 64N; feed, 0.2% copper and 02% iron; product, pH 1.0.*

12.3.3.4   *Carrier Agents and Membranes*

In Figures 12.20–12.22, the carriers are two commercially available oxime products, LIX 64N and Kelex 100. These are but two of a huge number of known complexing agents. Table 12.2 shows a few widely used examples.

For both facilitated and coupled transport, supported liquid membranes, formed as hollow fibers, are easy to make and operate, and the initial fluxes would often be sufficient for an economically feasible process if they could be sustained. But as discussed above, decline in flux, related to loss of the complexing agent from the support membrane, is an insoluble problem for liquid membranes [50–52]. Figure 12.23 shows some typical results from long-term coupled transport tests [50]. Although the flux can be restored by reloading the membrane with fresh complexing agent, this is not practical in a commercial system. As with facilitated transport, solid polymer support membranes are now used for many research studies.

Table 12.2   Commonly used metal ion carrier agents.

Trade name and Manufacturers	Chemical formula	Complexed ion
LIX 84® Henkel Crop. Minneapolis, MN	OH NOH C_9H_{19} CH$_3$ LIX 84®	Cu^{2+}, Ni^{2+}, Co^{2+}
LIX 64N® (a mixture of a few % LIX 63 in LIX 65N®) Henkel Corp.	C_9H_{19} C N OH OH LIX 64®	Cu^{2+}, Ni^{2+}
Kelex 100® Sherex Chemical Co. Dublin, OH	OH $C_{12}H_{25}$ N Kelex 100®	Cu^{2+}, Co^{2+}, Ni^{2+}
Alamine 336® Henkel Crop.	R N — R R Alamine 336® R = C_8H_{17} or $C_{10}H_{21}$	$UO_2(SO_4)_3^{4-}$, $Cr_2O_7^{2}$
Crown ethers various suppliers	n = 1: 18 • Crown • 6 (18C6) n = 2: 21 • Crown • 7 (21C7) Macrocyclic crown ethers	Li^+, K^+, Cs^+, rare earth lanthanides

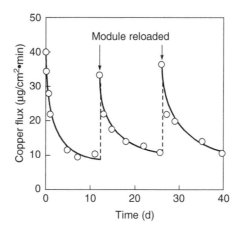

Figure 12.23 *The effect of replenishing a hollow fiber coupled transport module with fresh complexing agent. Membrane, polysulfone, hollow-fiber/Kelex 100; feed, 0.2% copper, pH 2.5; product, 2% copper, 100 g/l H₂SO₄ [50].*

12.3.4 Applications

The application that has received the most attention is the recovery of copper from dilute hydrometallurgical process streams. Such streams are produced by extraction of low-grade copper ores with dilute sulfuric acid. Typically, the leach stream contains 500–5000 ppm copper and various amounts of other metal ions, principally iron. Currently, copper is removed from these streams by precipitation with metallic iron or by solvent extraction. A scheme for recovering the copper by coupled transport is shown in Figure 12.24. The dilute copper solution from the dump leach stream forms the feed solution; concentrated sulfuric acid from the electrowinning operation forms the product solution. Copper from the feed solution permeates the membrane, producing an acidic raffinate solution containing 50–100 ppm copper, which is returned to the dump. The product solution, which contains 2–5% copper, is sent to the electrowinning tank house. Many papers have described this application of coupled transport with supported and emulsion membranes. Membrane stability is the problem and, although the economics appeared promising, the advantages have not proven sufficient to encourage industrial adoption.

12.4 Conclusions and Future Directions

Carrier transport membranes have been the subject of serious study for half a century, but no commercial process has resulted. Nevertheless, interest remains strong. The membranes are a popular topic with researchers because spectacular separations can be achieved with simple laboratory equipment. The difficulty lies in converting these performances into viable industrial processes. This is not a question of issues of cost or scale up, as with some other membrane types. Unfortunately, the problems revolve around membrane stability, and have so far proved intractable.

The stability problems are both physical and chemical. The first generation of immobilized liquid membranes rarely had a useful lifetime of more than a few days. The development of polymer inclusion membranes and more recently polymer membranes using fixed site carriers

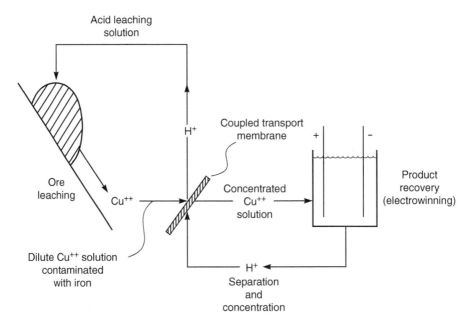

Figure 12.24 *Schematic of copper recovery by coupled transport from dump leach streams. The concentrated copper solution produced by coupled transport separation of the dump leach liquid is sent to an electrolysis cell where copper sulfate is electrolyzed to copper metal and sulfuric acid.*

has gone a long way to solving the physical stability of the membrane. The chemical stability of the carrier material remains an issue, depending on the particular separation. For olefin/paraffin and oxygen/air separations the chemical stability of all the carriers used to date has been poor. Some problems are caused by trace contaminants in the feed, but others seem to be intrinsic to the components under separation. One bright spot is the separation of carbon dioxide from nitrogen, where chemical instability seems to be less of a problem. This separation is currently receiving much attention because of its relevance to the amelioration of global warming caused by emissions from burning of fossil fuels. It is possible that this separation may be the first industrial application. We will see.

References

1. Scholander, P.F. (1960). Oxygen transport through hemoglobin solutions. *Science* **131**: 585.
2. Ward, W.J. III and Robb, W.L. (1967). Carbon dioxide-oxygen separation: facilitated transport of carbon dioxide across a liquid film. *Science* **156**: 1481.
3. Hughes, R.D., Mahoney, J.A., and Steigelmann, E.F. (1986). Olefin separation by facilitated transport membranes. In: *Recent Developments in Separation Science* (ed. N. N. Li and J.M. Calo), 173–196. Boca Raton, FL: CRC Press.
4. Kedem, O., Vofsi, D., and Bloch, R. (1963). Ion specific polymer membrane. *Nature* **199**: 802.
5. Bloch, R. (1970). Hydrometallurgical separations by solvent membranes. In: *Proceedings of Membrane Science and Technology Symposium* (ed. J.E. Flinn), 171–187. New York: Colombus Laboratories of Battelle Memorial Institute/Plenum Press.

6. Pinnau, I. and Toy, L.G. (2001). Solid polymer electrolyte composite membranes for olefin/paraffin separation. *J. Membr. Sci.* **184**: 39.

7. Morrisato, A., He, Z., Pinnau, I., and Merkel, T.C. (2002). Transport properties of PAI2-PTMO/AgBF$_4$ solid polymer electrolyte membranes for olefin/paraffin separation. *Desalination* **145**: 347.

8. Merkel, T.C., Blanc, R., Ciobanu, I. et al. (2013). Silver salt facilitated transport membranes for olefin/paraffin separations: carrier instability and novel regeneration method. *J. Membr. Sci.* **447**: 177.

9. Kim, J., Kang, S., and Kang, Y.S. (2012). Threshold silver concentration for facilitated transport in polymer/silver salt membranes. *J. Polym. Sci.* **19**: 9753.

10. Deng, L., Kim, T.-J., and Hägg, M.B. (2009). Facilitated transport of CO$_2$ in novel PVAm, PUA blend membranes. *J. Membr. Sci.* **350**: 154.

11. Deng, L. and Hägg, M.B. (2010). Swelling behavior and gas permeation performance of PVAm/PVA blend FSC membrane. *J. Membr. Sci.* **363**: 295.

12. Zou, J. and Ho, W.S.W. (2006). CO$_2$-selective polymer membranes containing amines in crosslinked poly(vinyl alcohol). *J. Membr. Sci.* **286**: 310.

13. Huang, J., Zou, J., and Ho, W.S.W. (2008). Carbon dioxide capture using a CO$_2$-selective facilitated transport membrane. *Ind. Eng. Chem. Res.* **47**: 1261.

14. Duan, S., Taniguchi, I., Kai, T., and Kazama, S. (2012). Poly(amidoamine) dendrimer/poly(vinyl alcohol) hybrid membranes for CO$_2$ capture. *J. Membr. Sci.* **423**: 107.

15. Matsuyama, H., Terada, A., Nakagawara, T. et al. (1999). Facilitated transport of CO$_2$ through polyethylenimine/poly(vinyl alcohol) blend membrane. *J. Membr. Sci.* **163**: 221.

16. Baker, R.W. and Blume, I. (1990). Coupled transport membranes. In: *Handbook of Industrial Membrane Technology* (ed. M.C. Porter), 511–558. Park Ridge: NJ Noyes Publications.

17. Cussler, E.L. (1991). Facilitated transport. In: *Membrane Separation Systems* (ed. R.W. Baker, E.L. Cussler, W. Eykamp, et al.), 242–275. Park Ridge, NJ: Noyes Data Corp.

18. Bara, J.E., Lessmann, S., Gabriel, C.J. et al. (2007). Synthesis and performance of polymerizable room temperature ionic liquids as gas separation membranes. *Ind. Eng. Chem. Res.* **46**: 5397.

19. Johnson, B.M., Baker, R.W., Matson, S.L. et al. (1987). Liquid membranes for the production of oxygen-enriched air: II. facilitated-transport membranes. *J. Membr. Sci.* **31**: 31.

20. Quinn, H.W. (1971). Hydrocarbon separations with silver(I) systems. In: *Progress in Separation and Purification*, vol. **4** (ed. E.S. Perry). New York: Interscience.

21. Safarik, D.J. and Eldridge, R.B. (1998). Olefin/paraffin separations by reactive absorption: a review. *Ind. Eng. Chem. Res.* **37**: 2571.

22. Kim, J.H., Won, J., and Kang, Y.S. (2004). Olefin-induced dissolution of silver salts physically dispersed in inert polymers and their application to olefin/paraffin separation. *J. Membr. Sci.* **241**: 403.

23. Yegani, R., Hirozawa, H., Teramoto, M. et al. (2007). Selective separation of CO$_2$ by using novel facilitated transport membranes at elevated temperatures and pressures. *J. Membr. Sci.* **291**: 157.

24. Kai, T., Kouketsu, T., Duan, S. et al. (2008). Development of commercial-sized dendrimer composite membrane modules for CO$_2$ removal from flue. *Gas, Sep. Purif. Tech* **63**: 524.

25. Kim, T.-J., Li, B., Kim, T.-J. et al. (2004). Novel fixed site carrier polyvinylamine membrane for carbon dioxide capture. *J. Polym. Sci., Part B* **42**: 4326.

26. Kim, T.-J., Vrålstad, H.K., Sandra, M., and Hägg, M.B. (2013). Separation performance of PVAm composite membrane for CO$_2$ capture at various pH levels. *J. Membr. Sci.* **428**: 218.

27. Chen, Y., Hao, L.Z., Wang, B. et al. (2016). Amine containing polymer/zeolite composite membranes for CO_2/N_2 separation. *J. Membr. Sci.* **497**: 21.
28. Wu, D., Han, Y., Hao, L.Z. et al. (2018). Scale-up of Zeolite-Y/polyethersulfone substrate for composite membrane fabrication in CO_2 separation. *J. Membr. Sci* **562**: 56.
29. Seader, J.D. and Henley, E.J. (2006). *Separation Process Principles*, 2e. New Jersey: Wiley.
30. Miyauchi, T. (1974). Liquid-liquid extraction process of metals. US Patent 4,051,230, issued 27 September 1977.
31. Largman, T. and Sifniades, S. (1978). Recovery of copper(II) from aqueous solutions by means of supported liquid membranes. *Hydrometallurgy* **3**: 153.
32. Baker, R.W., Tuttle, M.E., Kelly, D.J., and Lonsdale, H.K. (1977). Coupled transport membranes: I. Copper separations. *J. Membr. Sci.* **2**: 213.
33. Dai, Z., Noble, R.D., Gin, D.L. et al. (2016). Combination of ionic liquids with membrane technology: a new approach for CO_2 separation. *J. Membr. Sci.* **497**: 1.
34. Suzuki, T., Yasuda, H., Nishide, H. et al. (1996). Electrochemical measurement of facilitated oxygen transport through a polymer membrane containing cobaltporphyrin as a fixed carrier. *J. Membr. Sci.* **112**: 155.
35. Majumdar, S. and Sirkar, K.K. (1992). Hollow fiber contained liquid membrane. In: *Membrane Handbook* (ed. W.S.W. Ho and K.K. Sirkar), 764. New York: Van Nostrand Reinhold.
36. Davis, J.C., Valus, R.J., Eshraghi, R., and Velikoff, A.E. (1993). Facilitated transport membrane hybrid systems for olefin purification. *Sep. Sci. Technol.* **28** (1–3): 463.
37. Tsou, D.T., Blachman, M.W.B., and Davies, J.C. (1994). Silver-facilitated olefin/paraffin separation in a liquid membrane contactor system. *Ind. Eng. Chem. Res.* **33**: 3209.
38. Bessarabov, D.G., Sanderson, R.D., Jacobs, E.P., and Beckman, I.N. (1995). High-efficiency separation of an ethylene/ethane mixture by a large-scale liquid-membrane contactor containing flat-sheet nonporous polymeric gas-separation membranes and a selective flowing-liquid absorbent. *Ind. Eng. Chem. Res.* **34**: 1769.
39. Li, N.N. (1971). Permeation through liquid surfactant membranes. *AIChE J.* **17**: 459.
40. Matulevicius, E.S. and Li, N.N. (1975). Facilitated transport through liquid membranes. *Sep. Purif. Rev.* **4**: 73.
41. Li, N.N., Cahn, R.P., Naden, D., and Lai, R.W.M. (1983). Liquid membrane processes for copper extraction. *Hydrometallurgy* **9**: 277.
42. Raghuraman, B. and Wiencek, J. (1993). Extraction with emulsion liquid membranes in a hollow fiber contactor. *AIChE J.* **39**: 1885.
43. Hu, S.-Y. and Wiencek, J. (1998). Emulsion-liquid-membrane extraction of copper using a hollow fiber contactor. *AIChE J.* **44**: 570.
44. Fouad, E.A. and Bart, H.-J. (2008). Emulsion liquid membrane extraction of zinc by a hollow fiber contactor. *J. Membr. Sci.* **307**: 156.
45. Zou, J., Huang, J., and Ho, W.S.W. (2008). Facilitated transport membranes for environmental, energy, and biochemical applications. In: *Advanced Membrane Technology and Applications* (ed. N.N. Li, A.G. Fane, W.S.W. Ho, and T. Matsuura), 721–754. Hoboken, NJ: Wiley.
46. Han, Y. and Ho, W.S.W. (2021). Polymeric membranes for CO_2 separation and capture. *J. Membr. Sci.* **628**: 119244.
47. Noble, R.D. and Way, J.D. (1987). *Liquid Membranes, ACS Symposium Series*, vol. **347**. Washington, DC: American Chemical Society.

48. O'Bryan, Y., Truong, Y.B., Cattrall, R.W. et al. (2017). A new generation of highly stable and permeable polymer inclusion membranes (PIMs) with their carrier immobilized in a crosslinked semi-interpenetrating polymer network. application to the transport of thiocyanate. *J. Membr. Sci.* **529**: 55.
49. Nghiem, L.D., Mornane, P., Potter, I.D. et al. (2006). Extraction transport of metal ions and small organic compounds using polymer inclusion membranes. *J. Membr. Sci.* **281**: 7.
50. Babcock, W.C., Baker, R.W., Kelly, D.J., et al. (1979). Coupled Transport Membranes for Metal Separations. *Phase IV Final Report, US Bureau of Mines Technical Report*, Springfield, VA.
51. Kemperman, A.J.B., Bargeman, D., van den Boomgaard, T., and Strathmann, H. (1996). Stability of supported liquid membranes: state-of-the-art. *Sep. Sci. Technol.* **31**: 2733.
52. Zha, F.F., Fane, A.G., and Fell, C.J.D. (1995). Instability mechanisms of supported liquid membranes in phenol transport process. *J. Membr. Sci.* **107**: 59.

13

Membrane Contactors

13.1   Introduction

The majority of the processes discussed in this book use a membrane as a one-way selective separation device, as illustrated in Figure 13.1a. A fluid, gas, or liquid flows across the feed side of the membrane. A portion of the fluid permeates the membrane and is collected from the permeate side. One stream enters the module and two streams leave. In a membrane contactor, fluids are introduced to both sides of the membrane, as shown in Figure 13.1b. Two streams enter

---

(a) Membrane filtration device

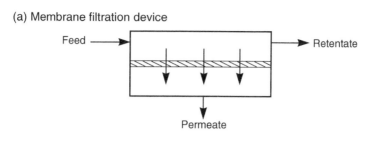

(b) Membrane contactor

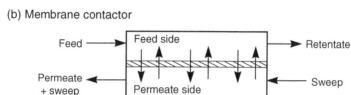

Figure 13.1 *(a) Most membrane processes filter a portion of the feed flow across the membrane surface to form the membrane permeate. (b) In membrane contactors, fluid flows on both sides of the membrane. Components can permeate from feed-to-permeate and permeate-to-feed.*

the module and two streams leave. Often the fluids on both sides are at a similar pressure, and the driving force to generate permeation is a difference in concentration or temperature. In contactor applications, the principal flow of material is by definition feed-to-permeate, but passage of some components from permeate to feed can also occur. In many membrane contactors, the most important function of the membrane is to provide a permeable interface to separate the feed and permeate fluids so that no direct mixing occurs. In some applications, this is the only function of the membrane. The membrane must be permeable but does not need to be selective. In other applications, membrane selectivity is also required.

A number of membrane contactor applications are described elsewhere in this book. For example, in Chapter 14, Medical Applications of Membranes, the use of blood oxygenators for open-heart surgery is covered. In these devices, blood is circulated on one side of the membrane, while a mixture of oxygen and carbon dioxide is circulated on the other. Oxygen permeates the membrane into the blood; carbon dioxide permeates from the blood into the gas. The permeation rates of the gases are controlled by adjusting their concentrations in the circulating gas. The membrane is a non-wetting, finely microporous material and is completely non-selective; its only function is to provide a large surface area for gas exchange in a small volume device. Similar membranes are used in many other contactors. In blood oxygenation, the contactor separates a liquid and a gas phase and so falls into the category of liquid/gas contactors. Membrane contactors can also be used to separate two immiscible liquids (liquid/liquid contactors), two miscible liquids (usually called membrane distillation), or to selectively absorb one component from a gas mixture into a liquid (gas/liquid contactors). Gas/gas contactors are also known.

In most contactor applications, the principal function of the membrane is to provide an interface between different phases. The membrane then performs the same function as the packing of a gas/liquid absorber or a liquid/liquid extractor. The advantage of the contactor is the high contact area per unit volume of the membrane module. The contact area of membrane contactors compared to traditional absorber columns is shown in Table 13.1. A membrane contactor can provide a 10-fold higher contact area than a packed tower of equivalent size. This makes contactors small and light. The first generation of blood oxygenators were small packed columns in which blood flowed down and a mixture of oxygen and carbon dioxide flowed up. As an exchange device, the towers were efficient, but blood oxygenators were not widely used for

Table 13.1   *The contact area per unit volume of different devices used as contactors between two phases [1, 2].*

Contactor	Surface area per volume (m^2/m^3)
Free dispersion columns	3–30
Packed/tray columns	30–300
Mechanically agitated columns	200–500
Membranes	1000–2000

open-heart surgery until the much more compact membrane blood oxygenator was developed, reducing the volume of blood required to operate the device to a manageable level.

A second advantage of membrane contactors, also important in blood oxygenators, is physical separation of the counter-flowing phases. The relative flow rates of the two phases are then independent, so large flow rate differences can be used without producing channeling, flooding, or poor phase contact. This enables maximum advantage to be taken of the ability of counter-flow to separate and concentrate the components crossing the membrane, and allows small volumes of high-cost extractants to be used to treat large volumes of low-value feed. Separation of the two phases also eliminates entrainment of one phase by the other, as well as foaming. Also, unlike traditional contactors, fluids of equal density can be used for the two phases. A final advantage is that the membrane can optionally be selective for the permeating components. In many applications this is not necessary and high permeability is key, but in some, such as pressure-retarded osmosis (PRO) and some gas/gas processes, good selectivity is also important.

The main disadvantages of contactors are related to the nature of the membrane interface. The membrane acts as a barrier to permeation, which may lead to an undesirable slowing of the rate of transport between the two phases. Over time, the membranes can foul, reducing the permeation rate further, or develop leaks, allowing direct mixing of the two phases. Also, the membranes are necessarily thin (to maximize their permeation rate) and consequently cannot withstand large pressure differences or exposure to harsh solvents and chemicals. In industrial settings, this lack of robustness may limit the use of membrane contactors.

13.2   Contactor Modules

All of the main module configurations described in Chapter 4 can be adapted for use as membrane contactors. The types of modifications required are illustrated in Figure 13.2. Many variations to module designs of the type shown in the figure are used to reduce the pressure required, to circulate the feed and sweep fluids efficiently on both sides of the membrane, or to reduce concentration polarization either side of the membrane.

Capillary hollow fiber modules generally require the least modification and are compact and easy to make, so they are a commonly used contactor form. Good control of concentration polarization is possible on the bore side of the membrane, but is harder on the shell side. Flow-directing baffles are sometimes used on the shell side to minimize polarization effects.

In spiral-wound modules, sweep flow on the permeate side of the membrane can be achieved by blocking the perforated product pipe at its midpoint. Sweep fluid is pumped into the pipe at one end then enters the permeate envelope, flows across the membrane, and exits at the second end of the pipe. As with hollow fiber modules, baffles may be used to minimize the creation of stagnant areas on the permeate side of the module.

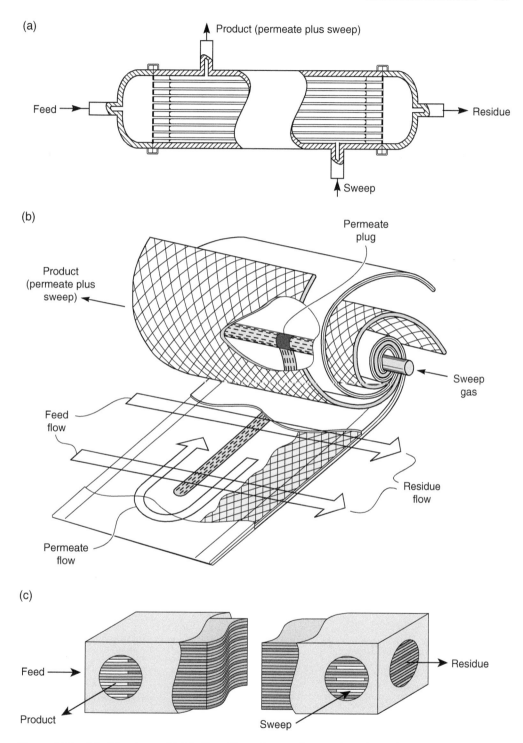

Figure 13.2   *Modifications of (a) hollow fiber, (b) spiral-wound, and (c) plate-and-frame module configurations to adapt them to use as membrane contactors.*

Plate-and-frame modules, because of the simple flow path on both sides of the membrane, generally give good control of fluid boundary layers. However, edge seals are required on every membrane sheet to separate the feed and permeate spaces, which means careful construction is required to produce leak-free modules. Estay et al. have created a useful tabulation with drawings of some of the more common membrane contactor module designs [3].

The relative flow direction of the feed fluid to the permeate sweep fluid is an important module characteristic. All of the contactors illustrated in Figure 13.2 are designed to achieve at least a partial counter-flow effect. That is, the feed flow and the permeate flow directions are counter to each other. As in other membrane processes, counter-flow is generally better than cross-flow, where the feed and permeate flow directions are at right angles. Cross-flow, in turn, is generally better than co-flow, where the feed and permeate flow directions are in the same direction. These different flow modes are described in Chapter 5, Concentration Polarization, and will not be covered again here.

13.3   Applications of Membrane Contactors

Some examples of different membrane contactor applications are shown in Figure 13.3. In all of these separations, the membrane has a finely microporous structure. The membrane is completely non-selective to the components in the feed and permeate fluids but still performs

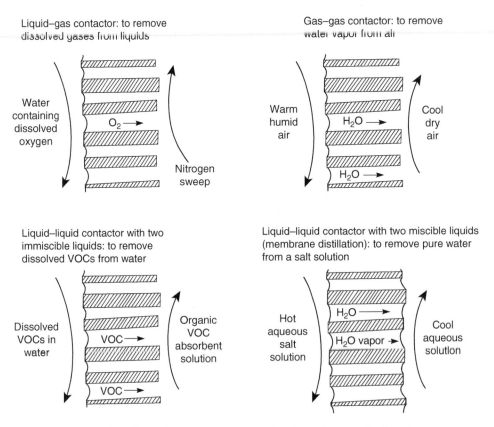

Figure 13.3 *Examples of membrane contactor applications that use finely microporous, non-wetting membranes to create an air barrier between the feed and the permeate phases.*

Table 13.2   *Applications of membrane contactors.*

	Chapter	References
Liquid/liquid contactors		
Hemodialysis	Chapter 14	[4]
Dialysis	This chapter	[5–7]
Membrane distillation	This chapter	[8–10]
Forward osmosis/Pressure retarded osmosis	This chapter	[11–13]
Carrier transport	Chapter 12	[14, 15]
Electrodialysis	Chapter 11	[16]
Gas/liquid–liquid/gas contactors		
Blood oxygenators	Chapter 14	[17]
Liquid deaeration	This chapter	[18, 19]
Membrane gas absorber	This chapter	[20, 21]
Gas/gas contactors		
Gas dehumidification	This chapter	[22, 23]
CO_2 selective recycle	This chapter	[24, 25]

a useful separation because of the driving force difference between the fluids on either side of the membrane.

A list of the more important membrane contactor applications is given in Table 13.2. The applications are broken into three categories depending on the fluid flows on either side of the membrane: liquid/liquid, liquid/gas and gas/gas applications. A few of these applications are covered in other chapters; we will cover the remainder here. We have chosen to describe the technology in terms of applications because they are so varied and each has a different set of advantages and problems. However, some authors have tried a more integrated approach; examples can be found in the book of Drioli, Criscuoli, and Curcio [26] and a review by Sirkar [27].

13.3.1   Liquid/Liquid Contactor Applications

13.3.1.1   Membrane Distillation

Membrane distillation has been widely studied [8–10, 28]. The process is shown schematically in Figure 13.4. A warm salt solution is circulated on one side of the membrane and cool pure water is circulated on the other. The membrane is made of an extremely hydrophobic microporous material that is not wetted by the aqueous solutions, and so forms an air gap between the two solutions. Because the solutions are at different temperatures, their vapor pressures are different and water vapor diffuses across the membrane from the salty side to the pure water side. Figure 13.4 shows the gradients driving the process in simple schematic form. All the non-volatile solutes in the feed solution are left behind; a separation is achieved between the salt components of the feed and the condensed water vapor permeate.

The latent heat required to vaporize the water is removed from the feed solution and is carried to the permeate solution with the water vapor. Consequently, the feed solution cools at the feed membrane interface and warms at the permeate membrane interface. This flow of heat produces temperature gradients in the fluid on both sides of the membrane. The heat required to vaporize the liquid at the feed side comes from the bulk feed, and the heat released when the vapor is condensed on the permeate side has to be removed into the distillate. As a consequence,

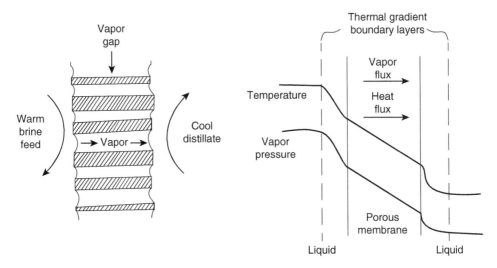

Figure 13.4   *A schematic illustration of the membrane distillation process showing temperature and water vapor pressure gradients that drive the process.*

temperature gradients in the membrane boundary layer significantly reduce the vapor pressure difference across the membrane and hence the membrane flux. Membrane fluxes in module tests are often as little as half the value of laboratory tests using membrane stamps in well-stirred test cells [28, 29].

Because of the exponential rise in vapor pressure with temperature, flux increases sharply as the temperature difference across the membrane is increased. Some typical results illustrating the dependence of flux on the temperature and vapor pressure difference across a membrane are shown in Figure 13.5. Dissolved salts in the feed solution decrease the vapor pressure driving force, but this effect is small unless the salt concentration is very high.

Membrane distillation offers a number of advantages over alternative pressure-driven processes such as reverse osmosis. Because the process is driven by a temperature gradient, low-grade heat can be used and expensive high-pressure pumps are not required. Fluxes are usually less than reverse osmosis fluxes, but not excessively less. Also, the process is still effective with slightly reduced fluxes, even for concentrated salt solutions. This is an advantage over reverse osmosis, in which the feed solution osmotic pressure places a practical upper limit on the acceptable salt concentration in the solution to be processed. The temperature difference across the membrane is generally about 20–40 °C so typical fluxes are 5–20 kg/m²/h.

The process shown in Figure 13.4, now called direct contact membrane distillation, (DCMD), was first described in a 1967 patent and was the subject of a few papers at that time. It was then largely abandoned for 20 years, until it was picked up in the 1980s by a research group at Enka, then a division of Akzo, who developed membrane distillation to the commercial scale using microporous polypropylene membrane modules. Their process is shown in Figure 13.6 [29]. The incoming salt solution is heated to close to 100 °C and circulated on one side of the membrane. An approximately equal flow of cooler distillate solution is circulated countercurrently on the other side of the membrane. As water vapor passes from feed to permeate,

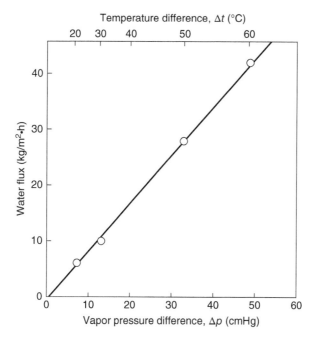

Figure 13.5 *Water flux across a microporous membrane as a function of temperature and vapor pressure difference (distillate temperature 18–38 °C; feed solution temperature 50–90 °C). Source: Data from Schneider et al. [29].*

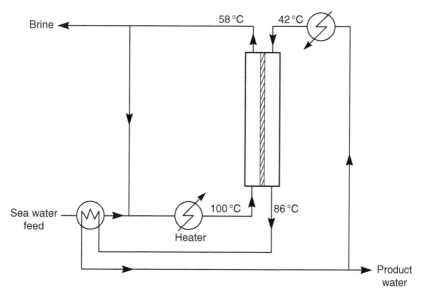

Figure 13.6 *Flow scheme of a membrane distillation process developed by Enka for the production of water from seawater and other high concentration salt solutions. The distillate is almost salt-free and the brine solution can be concentrated up to 15–20% NaCl. Source: [29] / Reproduced with permission from Elsevier.*

the feed solution cools and the permeate (distillate) solution warms by an equivalent amount. In the unit illustrated, the temperature of the brine solution cools from 100 to 58 °C, and the distillate warms from 42 to 86 °C. Heat recovery is practiced by running the warm distillate counter-current against the cooled incoming salt solution. The remaining energy required is provided by a supplementary heater that warms the circulating brine solution to 100 °C. In this unit, about 10% of the water in the salt solution passes through the membrane each time the solution is circulated through the device. This means the salt solution is recirculated several times before a fraction is removed as concentrated brine. The condensed distillate produced by the process is almost salt free. Essentially all the power used in the process is provided as heat.

Despite the technical success of the Enka device, a significant market did not develop. For large applications such as seawater desalination, the capillary membrane contactor modules were too expensive compared to low-cost, reliable reverse osmosis modules. Also the amount of low-grade heat required was high and far above the theoretical value because of heat lost from the feed by simple conduction through the membrane to the distillate. Wetting and fouling by the membranes over time were also issues.

The most widely studied process remains the direct contact (DCMD) configuration used by Enka and shown in Figure 13.7a. This is the simplest, lowest cost configuration, but heat losses caused by conduction are high and thermal efficiency is low. Developers of the technology often suggest waste heat or heat from solar collectors can be used to drive the process, but large amounts of waste heat are not a common commodity and may not be available where the water is needed. Also the use of solar collectors implies the process is limited to daytime use in sunny climates.

In direct contact devices, the condensed distillate is in direct contact with the membrane. The air gap (AGMD) configuration shown in Figure 13.7b was developed to reduce conductive heat loss by creating a gap of 1–2 mm between the membrane and the cooled condensing distillate. The cooling required to condense the distillate is provided through a conducting surface by a flow of cold water. By using an air gap, temperature polarization on the permeate side of the membrane and thermal conduction losses are much reduced, but this improvement is achieved at the expense of a more complex and costly membrane module.

Since the Enka work, a number of other groups have tried to bring the process to the industrial scale. In the early 2000s, Memstill, a spin-off of TNO in the Netherlands, built a number of pilot

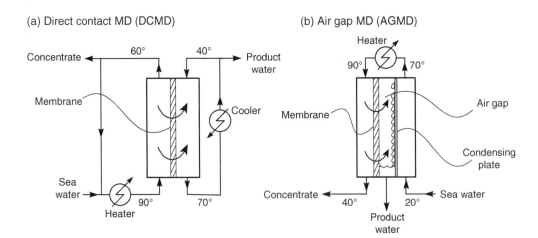

Figure 13.7　*Two of the most widely used membrane distillation process designs. (a) Direct contact membrane distillation. (b) Air gap membrane distillation. Heat integration is not shown for clarity.*

systems for seawater desalination. But RO proved to be a hard technology to beat. To become a commercial process, membrane distillation has to find a significant application where RO fails. The best hope seems to be in dewatering high salt concentration solutions. The brine concentrate produced by seawater RO plants typically contains about 5% salt. The osmotic pressure of this solution makes further concentrations by RO uneconomic. In principle, membrane distillation could bring the salt concentration to a higher level, reducing brine disposal costs, while simultaneously producing additional desalted product water. Similar high salt concentration applications exist in the treatment of water pumped up in the production of oil and natural gas, the so-called produced water. Unfortunately, the expected cost of salt concentration by membrane distillation is in the US$10–20/m³ range, not better than the competitive technologies of evaporation or reinjection [30]. Although the subject of a continuing research effort, the prospects for significant industrial use of membrane distillation are bleak.

13.3.1.2   *Pressure-Retarded Osmosis (PRO)*

Pressure-retarded osmosis (PRO) and forward osmosis (FO) are related processes in which the driving force is a difference in concentration and pressure. The connection between the two processes and the more well-known process of reverse osmosis (RO) is illustrated in Figure 13.8.

In the processes shown in Figure 13.8, a membrane permeable to water but impermeable to salt is used to separate a salt solution (in this example seawater) from pure water. When no pressure is applied across the membrane ($\Delta p = 0$) water permeates from the fresh water into the seawater. This is FO, and the driving force is the osmotic pressure difference created by the difference in the salt concentrations on either side of the membrane. The osmotic pressure difference between seawater and pure water is approximately 25 bar. If a pressure is applied to the seawater side of the process, the flow of water into the seawater is slowed (retarded) in proportion to the pressure applied. This is PRO. When the applied pressure is about 25 bar ($\Delta\pi - \Delta p = 0$), the water flux into the seawater is reduced to zero. If the pressure on the seawater side is increased above 25 bar ($\Delta p > \Delta\pi$), the osmotic flow of water is reversed and water flows from the seawater side to the pure water side. This is RO. Each of the processes described above has found applications. RO is a well-established desalination technique and was described in Chapter 6. FO and PRO are developing technologies and are described below.

The objective of PRO is to capture some of the vast amount of energy of mixing dissipated when fresh river water mixes with the sea. In principle, this energy could be captured by using a suitable osmotic membrane system to convert the osmotic pressure that would otherwise be dissipated by seawater dilution into hydrostatic pressure that drives a turbine and generates electricity. Such a scheme was proposed by Norman [31] and independently by Jellinek [32] and Loeb [11]. The process experienced a brief vogue after the first oil shock in the mid-1970s, but was later largely abandoned. However, in the early 2000s, workers at SINTEF Norway [12] revived interest and brought a small demonstration plant online between 2009 and 2011. The plant used a very large area of membrane to produce a few kW/h of electricity. The plant showed that, although PRO was theoretically capable of producing electricity, very large improvements to the process would be required to make it economically viable. Plans for construction of a larger demonstration project were put on hold.

The scheme used in the SINTEF plant is illustrated in Figure 13.9. Low-pressure freshwater from a river is circulated on one side of a permselective membrane; seawater is circulated under an applied pressure of typically 10–20 bar on the other side. Because the osmotic pressure difference of about 25 bar between freshwater and seawater is greater than the applied hydrostatic pressure, a flow of water takes place from the freshwater into the seawater. This flow will continue as long as the applied pressure on the seawater side is less than the osmotic pressure.

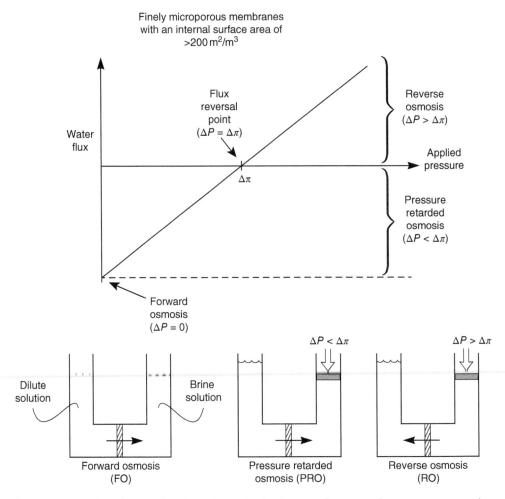

Figure 13.8   *A plot of water flux through a salt selective membrane used to separate water from seawater under conditions of forward osmosis, pressure-retarded osmosis and reverse osmosis.*

The pressurized salt solution is removed from the seawater side of the membrane. This solution has a larger volume than the seawater feed so a portion of the high-pressure solution is sent to a pressure exchanger/seawater pump to provide the energy needed to pressurize the incoming seawater. The remaining high-pressure solution is sent to a turbine to generate electricity. The water flux through the membrane from the fresh water to the salt solution is given by Eq. (13.1) (Eq. (2.44) from Chapter 2), namely:

$$J_w = A(\Delta\pi - \Delta p) \qquad (13.1)$$

where A is the water permeation coefficient, $\Delta\pi$ is the osmotic pressure difference across the membrane, and Δp is the hydrostatic pressure difference across the membrane.[1]

---

[1] This relationship is conventionally written $J_w = A(\Delta p - \Delta\pi)$ where J_w is the water flux from the salt solution to the fresh water side of the membrane. We have chosen the form shown in Figure 13.1 so that the water flux in PRO is positive.

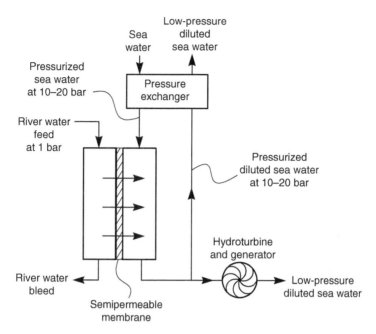

Figure 13.9 *A block diagram illustrating the use of a PRO process to recover the energy of mixing when a river meets the sea.*

The flow of low-pressure sea water into the pressurized permeate side of the membrane is powered by an equal flow of brackish water from the permeate side through a pressure exchange device. These devices have high efficiencies of up to 95%. So to a good approximation, the power per unit membrane area, or the power density (watts/m^2), W, that can be generated in PRO is equal to the product of the water flux across the membrane and the hydrostatic pressure of the salt solution:

$$W = J_w \Delta p = A(\Delta \pi - \Delta p) \cdot \Delta p \tag{13.2}$$

Differentiating Eq. (13.2) with respect to Δp shows that the work per unit area W, reaches its maximum value when $\Delta p = \dfrac{\Delta \pi}{2}$. Substituting this value for Δp in Eq. (13.2) yields

$$W_{max} = A\Delta \pi^2 / 4 \tag{13.3}$$

Equation (13.3) shows that the maximum power in a PRO system is directly proportional to the water permeability coefficient, A, and thus high flux membranes are preferred. The maximum power is also proportional to the square of the osmotic pressure difference. This arises because increasing the osmotic pressure of the salt solution increases both the optimum pressure at which the system operates and the water flux through the membrane at that pressure.

In calculating the maximum power per unit area using Equation (13.3), the osmotic pressure difference $\Delta \pi$ has been assumed to be constant at every point along the membrane. However, significant brine dilution has to be expected in any real system. In optimized systems, the

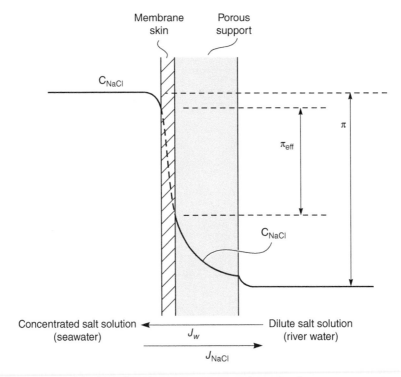

Figure 13.10   *Diagram showing the salt concentration gradients that form in a composite membrane during pressure-retarded osmosis. Diffusion of small amounts of salt into the porous support layer significantly reduces the effective osmotic pressure.*

available osmotic pressure will be reduced by a significant factor (for example, 25%) if a factor of 2 dilution is accepted over the length of the membrane [12].

Concentration polarization is the major problem of PRO [12, 33]. The problem is illustrated in Figure 13.10. External concentration polarization in the liquid boundary layers on either side of the membrane can be easily controlled in properly designed modules. However, even with very selective membranes, a small amount of salt will diffuse across the membrane. This salt becomes trapped in the porous support layer of the membrane. The salt can only escape by diffusing through the support layer into the freshwater solution against a convective flow of water in the opposite direction. Internal concentration polarization sharply reduces the effective osmotic pressure across the membrane and hence the fluxes obtained [33].

The problem of internal concentration polarization can be mitigated by membranes that combine extremely water-permeable and very selective skin layers with highly permeable porous substructures. The measure of membrane performance is usually taken to be W_{max}, the power produced per square meter of membrane. The first generation of PRO membranes had a power density of 3–5 watts/m^2 when used with seawater feeds [12, 34]. Over time, better membranes have been produced and power densities of 20 watts/m^2 have been reported in laboratory measurements using fresh water and concentrated brine [35, 36]. A PRO plant fitted with such membranes could produce 170 kW/h electricity/year·m^2 of membrane, probably still not good enough for a commercial process, but closer.

13.3.1.3   *Forward Osmosis (FO)*

FO was first characterized by Traube and Pfeffer in systematic measurements of the osmotic pressure of sugar solutions in the 1870s. These measurements were later used by van't Hoff as the basis of his theory of solutions. The process is being considered as a method of concentrating already concentrated feed solutions or as a method of desalinating water. A potential process design is shown in Figure 13.11 [37].

Seawater containing about 3.5% sodium chloride flows on one side of a selective membrane contactor; a much more concentrated ammonium carbonate solution $(NH_4)_2CO_3$, usually called the draw solution, flows counter-current on the other side. There is no pressure difference across the membrane but there is a large concentration difference, and hence a large osmotic pressure difference, so water from the seawater side of the membrane permeates to the ammonium carbonate solution. The seawater leaves the membrane unit as concentrated brine. A portion of the draw solution is removed, sent to a boiler, and heated to 80–100 °C. At these temperatures, the ammonium carbonate solution decomposes, producing a mixture of ammonia, carbon dioxide, and water. The water is removed as a condensed product; the ammonia and carbon dioxide are recycled to the draw solution reservoir where they react to produce fresh concentrated ammonium carbonate. The outcome of the process is to separate seawater into brine and potable water using thermal energy to drive the process. Ammonium carbonate has been the most widely studied draw solution but a variety of alternative regenerable draw solutions have been described.

FO, like PRO, is struggling to get out of the laboratory. Many potential applications have been suggested, but the process economics do not look favorable. In some applications, the ability of the process to concentrate a high brine salt concentration stream is the driver, for example, concentrating salty (produced) water generated in the production of oil. In other applications, the ability to produce a more concentrated product from a difficult-to-dispose-of feed by

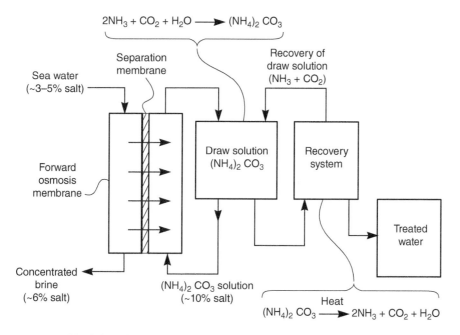

Figure 13.11   *A block flow diagram of a forward osmosis process that, in principle, could be used to produce desalted seawater.*

removing clean water is the driver. For feed solutions in the 1–3% concentration range, RO is the competitor. For more concentrated solutions, the competitor is simple evaporation. One problem leading to high cost is the relatively low flux caused by internal concentration polarization in the support layer of the membrane [37, 38]. The measured flux is often less than half of the expected value because of this effect. Also, the thermal energy needed to regenerate the draw solution is expensive. The calculated theoretical energy need is often quite low, but in actual practice the energy required may be several-fold higher.

13.3.1.4 Dialysis

A discussion of liquid/liquid contactors would not be complete without at least a mention of dialysis. By far the most important dialysis process, and in fact one of the most important of all of membrane processes, is hemodialysis. This is set aside for the moment for a full discussion in the next chapter, which covers medical applications of membranes. Simple dialysis is also used to draw some but not all components from a multicomponent solution into water by using a selective membrane as contactor. This was a widely used laboratory technique until the 1970s. The process was used to separate biological mixtures, and apparatuses able to separate several liters of biological solutions were developed [6].

Simple dialysis was the first process to use membranes on an industrial scale, with the development of the Cerini dialyzer in Italy [5]. The production of rayon from cellulose expanded rapidly in the 1930s, resulting in a need for technology to recover sodium hydroxide from large volumes of hemicellulose/sodium hydroxide by-product solutions. The hemicellulose was of little value, but the 17–18 wt% sodium hydroxide, if separated, could be directly reused in the process. Hemicellulose has a much higher molecular weight than sodium hydroxide, so parchmentized woven fabric or impregnated cotton cloth made an adequate dialysis membrane. The Cerini plate-and-frame dialyzer, illustrated in Figure 13.12, consisted of a large tank containing 50 membrane bags. Feed liquid passed through the tank and the dialysate solution passed counter-currently through each bag in parallel. The product dialysate solution typically contained 7.5–9.5% sodium hydroxide and was essentially free of hemicellulose. About 90% of the sodium hydroxide in the original feed solution was recovered. The economics of the process were very good, and the Cerini dialyzer was widely adopted. Later, improved membranes and improved dialyzer designs, mostly of the plate-and-frame type, were produced. A description of these early industrial dialyzers is given in Tuwiner's book [39].

Aside from its use in artificial kidneys, dialysis has now essentially been abandoned as a separation technique because it relies on diffusion, which is usually relatively unselective and inherently slow, to achieve a separation. Most potential dialysis separations are better handled by ultrafiltration or electrodialysis, in which an outside force and more selective membrane provide better, faster separations.

13.3.2 Liquid/Gas and Gas/Liquid Contactors

13.3.2.1 Liquid Degassing

The use of liquid/gas contactors in blood oxygenation is described in the next chapter. However, a number of similar liquid/gas applications exist, for example, deoxygenation of ultrapure water in the electronics industry, oxygen removal from boiler feed water in power plants, and the adjustment of carbonation levels in beverages [18, 19].

The performance of an industrial scale oxygen removal system treating boiler feed water is shown in Figure 13.13. The unit consists of a capillary hollow fiber module containing about $130\,\text{m}^2$ of membrane. Oxygen-containing feed water flows on the shell side of the device.

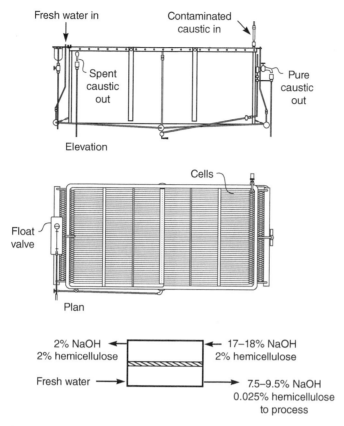

Figure 13.12 *Elevation and plan drawings showing the flow scheme of the Cerini dialyzer. The device was used to recover sodium hydroxide from waste streams from rayon production [5], and represented the first successful use of membranes in an industrial process.*

A vacuum can be drawn on the bore side of the fibers, but more commonly a sweep flow of nitrogen is used. As Figure 13.13 shows, a relatively small device can remove essentially all of the dissolved oxygen from a large water flow (50 gal/min). Oxygen removal is required to control corrosion. Although air-saturated water contains only 8 ppm oxygen, the partial pressure of the dissolved oxygen on the feed side is high (21% of atmospheric pressure), whereas it is close to zero on the sweep side. The result is a large partial pressure driving force for oxygen permeation and a highly effective separation. The aqueous phase is circulated on the outer, shell side of the fibers to avoid the excessive pressure required to circulate fluid at a high velocity down the fiber bore. The major resistance to mass transfer is in the liquid boundary on the liquid shell side of the fiber, so a baffled hollow fiber membrane module design is used to cause radial-flow of the fluid across the membrane from a central fluid distribution tube.

13.3.2.2 *Membrane Absorber–Strippers*

A number of efforts have been made to use similar but much larger liquid/gas contactors to replace the packed towers used in various gas/liquid absorption processes. An example of this type, described in Chapter 12, is the separation of gaseous olefin/paraffin mixtures by absorption of the olefin into silver nitrate solution, which was brought to the pilot scale by British Petroleum [15].

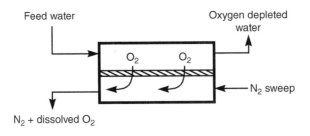

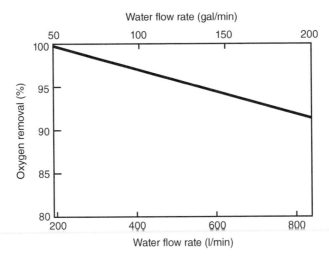

Figure 13.13   *Oxygen removal from boiler feed water using a liquid/gas contactor with nitrogen as the sweep gas on the permeate side [40].*

A more widely studied process has been the separation of carbon dioxide from natural gas or power plant flue gas as a replacement to conventional amine absorber-stripper operations. The first developer was Kvaerner, who were trying to adapt amine absorption systems for use on off-shore natural gas production platforms [20]. Kvaerner's feasibility program showed membrane contactors could reduce the equipment size by 70% and the weight by 60%. The process was later adapted to remove CO_2 from turbine power plant exhaust gas and brought to the pilot plant scale before work was halted in the late 1990s. In the last 10 years, there has been renewed interest in using membrane contactors to separate CO_2 from power plant flue gas [21, 41–43]. The reduced weight and footprint of contactors could produce process cost savings, but more importantly, contactors can significantly reduce the environmental issues of solvent emissions with the CO_2-stripped flue gas brought about by amine volatility and aerosol generation. A block flow diagram of the contactor process is shown in Figure 13.14.

13.3.2.3   Membrane Structured Packing

A final application of gas/liquid membrane contactors was described in a number of papers by Cussler and his students [44, 45]. The idea is to replace the structured packing of a conventional distillation column with a hollow fiber gas/liquid contactor. A schematic illustration of the two processes is shown in Figure 13.15. The vapor phase flows up the column on the bore side of the membrane; the liquid phase flows down on the shell side. Depending on the diameter of the

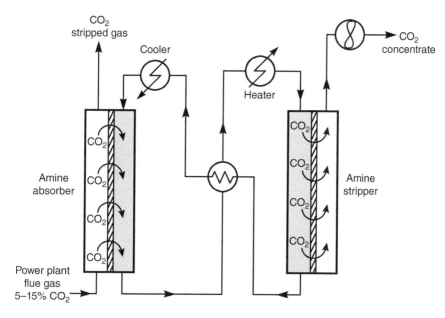

Figure 13.14 *Block diagram of absorber-stripper membrane contactors using aqueous amine solution to remove CO$_2$ from power plant flue gas.*

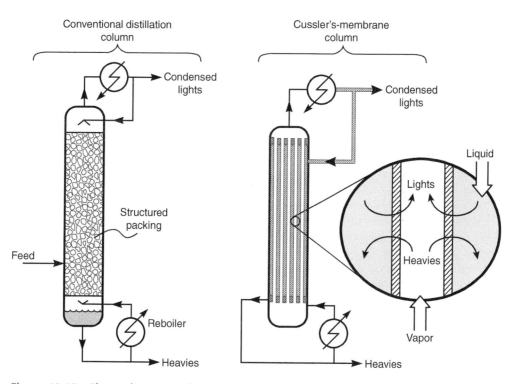

Figure 13.15 *Flow schematics of a conventional distillation column and Cussler's proposed membrane column.*

hollow fiber membranes, a large increase in liquid–gas contact area is achieved so the efficiency of the column as a separating device should be much improved. If the membrane contactor also has some selectivity between the heavy and light components even greater enhancements in separation might be possible. After the idea was proposed, a number of attempts were made to actually reduce the device to practice. Alas, like Leonardo's helicopter, there were mechanical problems that were hard to solve. A working model has yet to be made.

13.3.3   Gas/Gas Membrane Contactors

The two most developed applications of gas/gas contactors are water vapor recovery in hydrogen fuel cells and dehumidification in building air conditioners. Both applications use water vapor-permeable membranes to separate water vapor from air (oxygen and nitrogen). Many hydrophilic materials have water vapor/air selectivities of 100–1000 and the water permeability of these materials can be in the 10^4–10^5 Barrer range. When these materials are fabricated into thin membranes, it is usually found that most of the resistance to permeation is in the gas boundary layers on either side of the membrane. This means that once the selective layer membrane is less than about 5 μm thick, reducing the membrane thickness does not produce an increase in water vapor permeance, but does begin to degrade membrane selectivity. Nonetheless, even these relatively thick membranes have high enough permeances for useful devices as described below.

13.3.3.1   *Water Recovery in Hydrogen Fuel Cells*

Hydrogen fuel cells are beginning to be used to power buses, trucks, and some cars. In such a fuel cell, gaseous hydrogen contacts a catalyst-impregnated anode, where it disassociates into electrons and protons. The electrons flow through an external circuit, thereby providing electric power for the vehicle. The protons diffuse across the electrolyte layer to the cathode compartment, where they react with the electrons and with oxygen from the air to produce water.

$$2H^+ + O_2 + 2e^- \rightarrow 2H_2O$$

The water vapor produced is then removed from the cathode air compartment with the used air. Water needed to maintain the electrolyte solution is carried out with the produced water, and unless the incoming air is humid, the cell may dry out and cease to function. A membrane contactor solves this problem, as shown in Figure 13.16.

The incoming air flows on one side of the membrane and humid oxygen-depleted air leaving the fuel cell flows counter-current on the other side. Water vapor permeates from the humid exhaust air to the incoming air. The amount of water vapor recycled to the fuel cell this way changes with the humidity of the incoming air. The used air is always close to saturation with water vapor, so the partial pressure of water vapor on the exhaust side remains roughly constant. On the air intake side, however, the partial pressure varies with the water content of the atmosphere. In dry conditions the driving force for water permeation is higher than when the incoming air is more humid. Thus, the device is self-regulating in terms of the fraction of the water vapor from the used air that is recycled. In this way, the water content of the cathode compartment is controlled and flooding or desiccation of the cathode chamber is avoided.

13.3.3.2   *Membrane Energy Recovery Ventilation Systems (ERVs)*

A much bigger application of gas/gas contactors that permeate water vapor is in building air conditioning systems. Air-conditioning systems already consume a huge amount of energy and their use is growing rapidly as the tropical regions of the world in Southeast Asia, Africa,

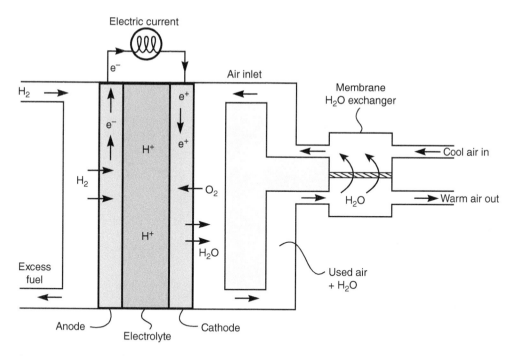

Figure 13.16 *A membrane contactor used to control the water balance of a hydrogen fuel cell. The contactor transfers a portion of the water vapor generated by the fuel cell into the incoming air stream. This maintains the water level in the cell in the required operating range.*

and South America become more urban and developed. Mechanical vapor compression, the main air-conditioning technology used in large buildings, was commercialized in the 1850s and has been the subject of optimization and refinement ever since. Nonetheless, new technology is required to reduce the energy costs, especially in the developing world. A variety of membrane processes are being worked on; currently the most advanced approach is the Energy Recovery Ventilation (ERV) unit illustrated in Figure 13.17. The operating principle of this device is similar to that described for the fuel cell application above.

Hot humid air from the outside enters on one side of the membrane contactor, and ventilation air leaving the building flows counter-current on the other side. Ventilation of stale building air is required to remove odors and CO_2 generated in the building. Building ventilation rates depend on the use and size, as well as the number of occupants, but normally the air volume of a building will turn over every 1–3 hours. This means an equivalent flow of hot humid air enters and the heat and water vapor content of this volume of outside air must be removed.

The membrane permeates water from the outside air into exhaust air. As with fuel cell dehumidification membranes, a high permeance to water vapor is required. About two-thirds of the air cooling that takes place is due to the latent heat removed from the incoming air by permeation of water; the remaining one-third is sensible heat removed by conduction through the membrane. A separate refrigeration unit then further cools the incoming dry air as needed, but the size of this unit is much reduced.

The water permeability of some of the materials developed for this application is extraordinarily high. Akhtar et al., for example, have produced polyethylene glycol-based copolymer materials with water permeabilities of 5 to 10×10^4 Barrer [46] and self-assembled block copolymer

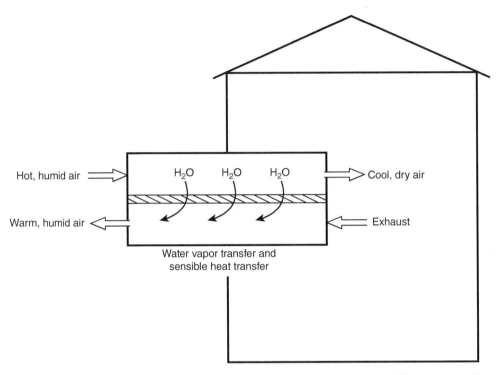

Figure 13.17 *The operating principle of a membrane ERV to cool and dehumidify incoming air for a building air conditioner. Cool dry exhaust air from the building is used to pick up much of the water coming from fresh but warm and humid outside air*

membranes have been reported with permeabilities of 5×10^5 Barrer [47]. As well as being very permeable to water, the membranes must also be impermeable to CO_2 and odor-causing VOCs to prevent these components permeating from the exhaust air back into the building intake stream. The reported selectivity of membranes in current use is about 1000 for H_2O/CO_2 and 100 for H_2O/VOC. A detailed analysis of these devices is given by Engarnevis et al. [23].

The efficiency of humidity exchange devices is limited by concentration polarization. A number of studies have shown that the boundary layer on the incoming air side is responsible for the bulk of the resistance to heat transfer through the membrane. Boundary layers also affect water vapor (mass transfer) through the membrane. Increasing the membrane permeance beyond a certain value by using thinner or more permeable membranes does not change the boundary layer resistance, which becomes the dominant resistance to permeation. For this reason, the selective layers of the membranes used are relatively thick at 2–5 μm. An illustration of the water vapor concentration gradient through an operating ERV membrane is shown in Figure 13.18. Boundary layers on either side of the membrane and within the porous support layer produce more than half of the resistance to transport [23]. Nonetheless, a selective barrier is required to prevent CO_2 and odors permeating with the water. Membrane ERVs are commercial products, and their use is growing, but current use is still only a small fraction of the total potential market. Improvements to reduce cost and increase efficiency are under way and the use of ERVs is likely to grow. Although ERVs are the most developed devices, a number of related cooling and dehumidification designs are under active development. A useful introduction to the technology is given in the review of Woods [22].

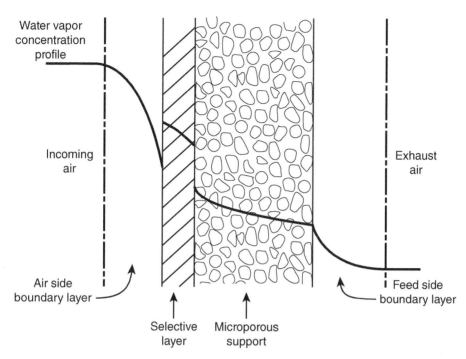

Figure 13.18 *An illustration (not to scale) of the water vapor concentration profile in an operating ERV.*

13.3.3.3 Selective Exhaust Gas Recycle

The capture of carbon dioxide from the exhaust gas of fossil fuel combustion is an important developing membrane application. The objective is to prevent CO_2 emissions entering the atmosphere and thereby to mitigate global warming. Some of the CO_2 capture technologies being developed incorporate a membrane contactor in their design. The type of process being considered is illustrated in Figure 13.19 [24, 25, 48].

In the process illustrated, combustion of fossil fuel with air produces heat to make steam, which drives a turbine. The same concept can be used in cement or steel plants, for example, which produce similar exhaust gas streams containing 15–20% CO_2. The exhaust gas is usually treated with a primary CO_2 separation unit. This unit could be a membrane unit as described in Chapter 9, or another separation technology, such as adsorption or amine absorption. The primary separation step will generally remove 70–80% of the CO_2 from the exhaust gas, producing a CO_2 concentrate suitable for sequestration. The primary unit can be designed to remove 90–95% of the CO_2 but capturing the last few percent of CO_2 is often disproportionately expensive. A membrane contactor offers a low energy, low-cost way of capturing this last fraction of CO_2.

When a membrane contactor is used, the treated flue gas from the primary capture step, still containing a few percent CO_2, is passed over one side of a CO_2-permeable membrane that is relatively impermeable to oxygen and nitrogen. Air to be used in the combustion step is passed over the other side of the membrane. There is no pressure difference across the membrane, but there is a concentration difference. As a result, CO_2 in the exhaust gas passes into the air stream and is selectively recycled. A portion of the oxygen in the air passes into the exhaust gas. Because

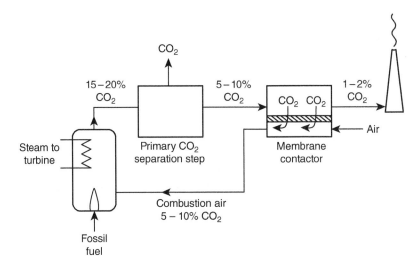

Figure 13.19   *Flow diagram showing the application of a membrane contactor to capture CO_2 in the exhaust gas from a fossil fuel combustion process.*

the membrane is selective for CO_2 over oxygen, oxygen loss is small. The net result is to remove the last fraction of CO_2 in the exhaust gas and recycle it back to the incineration step so it can be captured by the primary separation unit. Contactors of this type are being evaluated in pilot-scale systems.

13.4   Conclusions and Future Directions

Many of the membrane contactor devices and processes described in this chapter are struggling to leave the laboratory. A common problem is concentration polarization. The driving force in many applications is a concentration gradient, so concentration gradients that develop in the feed and permeate solutions adjacent to the membrane become a significant contributor to the total resistance to permeation. The performance of many devices would be improved if good counter-current flow between the fluids on either side of the membrane could be achieved. Flow maldistribution and channeling are additional common module design problems.

Major improvements to current technology will be required to bring membrane distillation, FO or PRO to the large scale. Prospects for gas/liquid contactors to replace conventional absorption towers are better. Use of gas/gas contactors for air conditioning and CO_2 capture applications is also likely to grow.

References

1.   Reed, B.W., Semmens, M.J., and Cussler, E.L. (1995). Membrane contactors. In: *Membrane Separation Technology: Principles and Applications* (ed. R.D. Noble and S.A. Stern), 467–498. Amsterdam: Elsevier Science.
2.   Qi, Z. and Cussler, E.L. (1985). Microporous hollow fibers for gas absorption: I. Mass transfer in the liquid. *J. Membr. Sci.* **23**: 321.

3. Estay, H., Troncoso, E., Ruby-Figueroa, R., and Romero, J. (2018). Performance evaluation of mass transfer correlations in the GFMA process: a review with perspectives to the design. *J. Membr. Sci.* **554**: 140.

4. Keen, M.L. and Grotch, F.A. (1991). Dialysers and delivery systems. In: *Introduction to Dialysis* (ed. M.C. Cogan, P. Schoenfield, and F.A. Grotch). New York: Churchill Livingstone.

5. Cerini, L. (1926). Apparatus for the purification of impure solutions of caustic soda and the like, on osmotic principles. US Patent 1,719,754, issued July 1929.

6. Craig, L.C. (1964). Differential dialysis. *Science* **144**: 1093.

7. Wallace, R.M. (1967). Concentration and separation of ions by Donnan membrane equilibrium. *Ind. Eng. Chem. Process Des. Dev.* **6**: 423.

8. El-Bourawi, M.S., Ding, Z., Ma, R., and Khayat, M. (2006). A framework for better understanding membrane distillation separation process. *J. Membr. Sci.* **285**: 204.

9. Lawson, K.V. and Lloyd, D.R. (1997). Membrane distillation. *J. Membr. Sci.* **124**: 1.

10. Srisurichan, S., Jiraratananon, R., and Fane, A.R. (2006). Mass transfer mechanisms and transport resistance in direct contact distillation process. *J. Membr. Sci.* **277**: 186.

11. Loeb, S. (1976). Production of energy from concentrated brines by pressure-retarded osmosis: I. Preliminary technical and economic correlations. *J. Membr. Sci.* **1**: 49–63.

12. Thorsen, T. and Holt, T. (2009). The potential for power production from salinity gradients by pressure retarded osmosis. *J. Membr. Sci.* **335**: 103–110.

13. McCutcheon, J., McGinnis, R.L., and Elimelech, M. (2006). Desalination by ammonia-carbon dioxide forward osmosis: influence of draw and feed solution concentration on process performance. *J. Membr. Sci.* **278**: 114.

14. Baker, R.W., Tuttle, M.E., Kelly, D.J., and Lonsdale, H.K. (1977). Coupled transport membranes: I. Copper separations. *J. Membr. Sci.* **2**: 213.

15. Davis, J.C., Valus, R.J., Eshraghi, R., and Velikoff, A.E. (1993). Facilitated transport membrane hybrid systems for olefin purification. *Sep. Sci. Technol.* **28** (1–3): 463.

16. Strathmann, H. (2004). *Ion-exchange Membrane Separation Processes*. Amsterdam: Elsevier.

17. Wiese, F. (2008). Membranes for artificial lungs. In: *Membranes for Life Sciences* (ed. K.V. Peinemann and S.P. Nunes), 49–68. Weinheim: Wiley-VCH GmbH.

18. Page J.K.R. and Kalhod, D.G. (1996). Control of dissolved gases in liquids. US Patent 5,565,149, issued October 1996.

19. Honda, K and Yamashita, M. (1994). Method for deaerating liquid products. US Patent 5,522,917, issued 4 June 1996.

20. Falk-Pedersen, O. and Dannstrom, H. (1997). Method for removing carbon dioxide from gases. US Patent 6,228,145, issued 6 May 2001.

21. Zhao, S., Ferow, P.M.P., Deng, L. et al. (2016). Status and progress of membrane contactors in post-combustion carbon capture: a state-of-the-art review of new developments. *J. Membr. Sci.* **511**: 180.

22. Woods, J. (2014). Membrane processes for heating and air-conditioning. *Renew. Sustainable Energy Rev.* **33**: 290.

23. Engarnevis, A., Huizing, R., Green, S., and Rogak, S. (2018). Heat and mass transfer modeling in enthalpy exchangers using asymmetric composite membranes. *J. Membr. Sci.* **556**: 248.

24. Merkel, T.C., Lin, H., Wei, X., and Baker, R.W. (2010). Power plant post combustion carbon dioxide capture: an opportunity for membranes. *J. Membr. Sci.* **359**: 126.

25. Hao, P., Wijmans, J.G., Kniep, J., and Baker, R.W. (2014). Gas/gas membrane contactors – an emerging membrane unit operation. *J. Membr. Sci.* **4623**: 131.

26. Drioli, E., Criscuoli, A., and Curcio, E. (2006). *Membrane Contactors: Fundamentals, Applications and Potentialities*. Amsterdam: Elsevier.

27. Sirkar, K.K. (2008). Membrane contactors. In: *Advanced Membrane Technology and Applications* (ed. N.N. Li, A.G. Fane, W.S.W. Ho, and T. Matsuura), 687–702. New York: Wiley.

28. Anvari, A., Yancheshme, A.A., Kekre, K.M., and Ronen, A. (2020). State-of-the-art methods for overcoming temperature polarization in membrane distillation processes: a review. *J. Membr. Sci.* **616**: 118413.

29. Schneider, K., Hölz, W., Wollbeck, R., and Ripperger, S. (1988). Membranes and modules for transmembrane distillation. *J. Membr. Sci.* **39**: 25.

30. Bartholomew, T.V., Duochenko, A.V., Siefert, N.S., and Mutter, M.S. (2020). Cost optimization of high recovery single gap membrane distillation. *J. Membr. Sci.* **611**: 118370.

31. Norman, R.S. (1974). Water salination: a source of energy. *Science* **186**: 350.

32. Jellinek, H.H. (1975). Osmotic Work I. Energy Production from Osmosis on Fresh Water Systems, Kagaku Kogy **19**, 87.

33. Lee, K.L., Baker, R.W., and Lonsdale, H.K. (1981). Membranes for power generation by pressure-retarded osmosis. *J. Membr. Sci.* **8**: 141–171.

34. Post, J.W., Veerman, J., Hamelers, H.V.M. et al. (2007). Salinity-gradient power: evaluation of pressure-retarded osmosis and reverse electrodialysis. *J. Membr. Sci.* **288**: 218–230.

35. Chan, S., Wang, R., and Fane, A.G. (2013). Robust and high-performance hollow fiber membranes for energy harvesting from salinity gradients by pressure retarded osmosis. *J. Membr. Sci.* **448**: 44.

36. Wan, C.F., Yang, T., Gai, W. et al. (2018). Thin film composite hollow fiber membrane with inorganic salt additives for high mechanical strength and high-power pressure retarded osmosis. *J. Membr. Sci* **555**: 388.

37. McCutcheon, J.R. and Elimelech, M. (2006). Influence of concentration and dilutive internal concentration polarization on flux in forward osmosis. *J. Membr. Sci.* **284**: 237.

38. Suh, C. and Lee, S. (2013). Modeling reverse draw solute flux in forward osmosis with external concentration polarization in both sides of the draw and feed solution. *J. Membr. Sci.* **427**: 365.

39. Tuwiner, S.B. (1962). *Diffusion and Membrane Technology*, 1e. New York: Reinhold Publishing Corp.

40. Wiesler, F. and Sodaro, R. (1996). Deaeration: degasification of water using novel membrane technology. *Ultrapure Water* **35**: 53.

41. Rahbari-Sisakht, M., Ismail, A.F., Rana, D., and Matsuura, T. (2013). Carbon dioxide from diethanolamine solution through porous surface modified PVDF hollow fiber membrane contactor. *J. Membr. Sci.* **427**: 270.

42. Cai, J.J., Hawboldt, K., and Abdi, M.A. (2016). Improving gas absorption efficiency using a novel dual membrane contactor. *J. Membr. Sci.* **510**: 249.

43. Ansaloni, L., Rennemo, R., Knuutila, H.K., and Deng, L. (2017). development of membrane contactors using volatile amine based absorbents for CO_2 capture: amine permeation through the membranes. *J. Membr. Sci.* **537**: 272.

44. Zhang, Z. and Cussler, E.L. (2003). Hollow fiber as structured distillation packaging. *J. Membr. Sci.* **215**: 185.

45. Chung, J.B. (2005). Distillation with nanopores or coated hollow fibers. *J. Membr. Sci.* **257**: 3.

46. Akhtar, F.H., Kumar, M., Vovusha, H. et al. (2019). Scalable synthesis of amphiphilic copolymers for CO_2-and-water-selective membranes: effect of copolymer composition and chain length. *Macromolecules* **52**: 6213.
47. Upadhyaya, L., Gebreyohannes, A.Y., Akhtar, F.H. et al. (2020). NEXAR™-coated hollow fibers for air dehumidification. *J. Membr. Sci.* **118**: 450.
48. Baker, R.W., Freeman, B., Kniep, J. et al. (2017). CO_2 capture from natural gas power plants using selective exhaust gas recycle membrane designs. *Int. J. Greenhouse Gas Control* **66**: 35–47.

14

Medical Applications of Membranes

14.1 Introduction

In this chapter, the use of membranes in medical devices is reviewed. In terms of total membrane area involved, medical applications are equivalent to all industrial membrane applications combined. In terms of dollars, the medical market is by far the larger. In spite of this, little communication between the two areas has occurred over the years. Medical and industrial membrane developers each have their own journals, societies and meetings, and rarely look over the fence to see what the other is doing. This book cannot reverse 50 years of history, but every industrial membrane technologist should at least be aware of the main medical applications. Therefore, in this chapter, the three most important – hemodialysis (the artificial kidney), blood oxygenation (the artificial lung), and controlled release drug delivery – are briefly discussed.

Membrane Technology and Applications, Fourth Edition. Richard W. Baker.
© 2024 John Wiley & Sons Ltd. Published 2024 by John Wiley & Sons Ltd.

14.2 Hemodialysis

The kidney is a key component of the body's waste disposal and acid–base regulation mechanisms. Each year approximately one person in 10 000 suffers irreversible kidney failure. Before 1960, this condition was universally fatal, but now a number of treatment methods can maintain patients in good or at least reasonable health. Hemodialysis is by far the most important therapy, and approximately 2.5 million patients worldwide benefit from the process. Each patient is treated two to three times per week with a dialyzer containing about 1 m^2 of membrane area; the devices are generally discarded after one or two uses.

The operation of the human kidney is illustrated in Figure 14.1 [1]. The process begins in the glomerulus, a network of tiny capillaries surrounding spaces called Bowman's capsules. Blood flowing through these capillaries is at a higher pressure than the fluid in the capsules, and the walls of the capillaries are finely microporous. As a result, water, salts, and other microsolutes

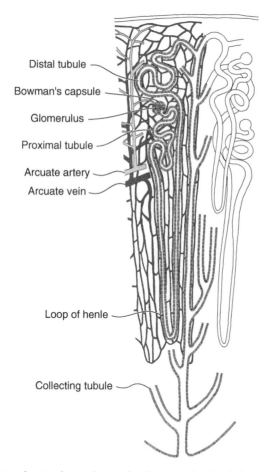

Figure 14.1 Schematic of a single nephron, the functional unit of the kidney. Microsolutes are filtered from blood cells in Bowman's capsules. As the filtrate passes toward the collection tubule, most of the microsolutes and water are reabsorbed. The fluid finally entering the collecting tubule contains the nitrogenous wastes from the body and is excreted as urine. There are about 1 million nephrons in the normal kidney. Source: [1] / Springer Nature.

in the blood are ultrafiltered into the capsule; blood cells stay behind. Each Bowman's capsule is connected by a relatively long, thin duct to the collecting tubule. The average kidney has approximately one million tubules and many Bowman's capsules are connected to each tubule. As the fluid from the Bowman's capsules travels down the collection duct to the central tubule, more than 99% of the water and almost all of the salts, sugars, and proteins are reabsorbed into the blood by a process similar to facilitated transport. The remaining concentrated fluid, rich in urea and creatinine, forms the urine, which exits the body via the ureters, bladder, and urethra. This is the principal method by which nitrogen-containing metabolites are discharged from the body. The acid–base balance is also controlled by the bicarbonate level of urine, and many drugs and toxins are excreted this way.

The first successful hemodialyzer was constructed by Kolff and Berk in The Netherlands in 1945 [2, 3]. Kolff's device used dialysis to remove urea and other waste products directly from blood. A flat cellophane (cellulose) tube formed the dialysis membrane; the tube was wound around a drum that rotated in a bath of saline. As blood was pumped through the tube, urea and other low molecular weight metabolites diffused across the membrane under a concentration gradient, and were picked up in the dialysate solution. The cellophane tubing did not allow diffusion of larger components in the blood, such as proteins or blood cells. By maintaining the salt, potassium, and calcium levels in the dialysate solution at the same levels as in the blood, loss of these components from the blood was prevented.

Kolff's early devices were used for patients who had suffered acute kidney failure as a result of trauma or poisoning and needed dialysis only a few times. Such emergency treatment was the main application of hemodialysis until the early 1960s, because patients suffering from chronic late-stage kidney disease require dialysis two to three times per week indefinitely, which was not practical with these early devices. Long-term hemodialysis for such patients was made possible by improvements in the dialyzer design and the development of a plastic shunt that could be permanently implanted to allow easy access to the patient's vascular system. This shunt, developed by Scribner and co-workers and first used in 1960 [4], allowed dialysis without the need for surgery to connect blood vessels to the dialysis machine for each treatment.

Kolff's first tubular dialyzer, shown in Figure 14.2, required several liters of blood to prime the system, a major operational problem. Tubular dialyzers were subsequently replaced with much smaller spiral devices, also developed by Kolff and coworkers, that did not require the large dialysate bath or rotating drum. This system was the basis for the first disposable dialyzer produced commercially in the early 1960s. The blood volume required to prime the device was still excessive, however. Plate-and-frame and hollow fiber devices were developed as better alternatives, and by 1975, about 65% of all dialyzers were coil, 20% hollow fiber systems, and 15% plate-and-frame. Within 10 years, the coil dialyzer had essentially disappeared, and today hollow fiber dialyzers of the type shown in Figure 14.3 are used exclusively.

Hollow fiber dialyzers typically contain $1-2 \text{ m}^2$ of membrane in the form of fibers 0.1–0.2 mm in diameter. A typical dialyzer module may contain several thousand fibers housed in a 2-in.-diameter tube, 10–15 in. long. Approximately 200 million hemodialysis procedures are performed annually worldwide, creating a demand for modules on a scale unknown in any other membrane application. Because hollow fiber dialyzers are produced in such large numbers, prices are very low. Today a $1-2 \text{ m}^2$ hollow fiber dialyzer is produced for about US$15, well below the module costs of any other membrane technology. These low costs have been achieved by the use of high-speed machines able to spin several hundred fibers simultaneously around the clock. The entire spinning, cutting, module potting and testing process is automated.

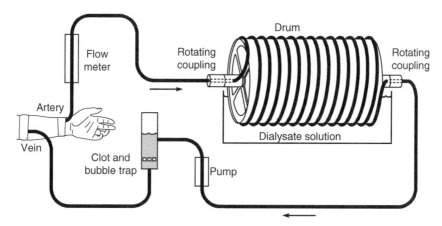

Figure 14.2 *Schematic of an early tubular hemodialyzer based on the design of Kolff's original device. The device required several liters of blood to fill the tubing and minor surgery to connect to the patient. Nonetheless, it saved the lives of patients suffering from acute kidney failure. Source: Adapted from [2].*

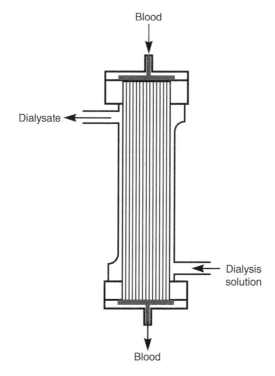

Figure 14.3 *Today, a hollow fiber artificial kidney contains several thousand hollow fibers with a total membrane surface area of 1–2 m².*

In a hollow fiber dialyzer, the blood flows down the bore of the fiber, providing good fluid flow hydrodynamics. An advantage of the hollow fiber design is that only 60–100 ml of blood is required to fill the dialyzer. At the end of a dialysis procedure, the module can be drained, flushed with sterilizing agent, and reused. Dialyzer reuse was once widely practiced, in part for economic reasons, but currently, many dialyzers are discarded after a single use.

The regenerated cellulose membranes used in Kolff's first dialyzer were still in use until the 1990s. Cellulose membranes are isotropic hydrogels generally about 10 µm thick and, although very water swollen, they have a high wet strength. The membrane has a molecular weight cut-off of about 2000 Da.

Although cellulose was used successfully in hemodialyzers for many years, the relatively low permeability and the ability of free hydroxyl groups on the membrane surface to activate the blood clotting process were problems. (When cellulose-based dialyzers are reused, blood compatibility improves, because a coating of protein has formed on the membrane surface.) Beginning in the 1980s, synthetic polymers began to replace cellulose as the membrane material. Initially, substituted cellulose derivatives, principally cellulose acetate, were used; polyacrylonitrile, polysulfone, and poly(methyl methacrylate) are now the materials of choice. The membranes are generally microporous, with a finely microporous skin layer on the inside (blood-contacting) surface of the fiber. The permeability of these fibers is up to 10 times that of cellulose membranes, and they can be tailored to achieve a range of molecular weight cutoffs by using different preparation procedures. The blood compatibility is good. Synthetic polymers, particularly polysulfone, have more than 90% of the current market.

An attractive feature of newer membranes is their ability to remove middle molecular weight metabolites from blood. This improvement in performance is illustrated by Figure 14.4. Cellulose membranes remove the major metabolites, urea, and creatinine, efficiently, but metabolites with molecular weights between 1000 and 10 000 Da are removed poorly. Patients on long-term dialysis are believed to accumulate these metabolites, which are associated with a number of health issues. Synthetic polymer membranes appear to simulate the function of the normal kidney more closely than cellulose membranes.

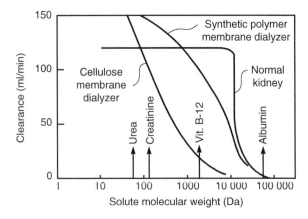

Figure 14.4 *Clearance, a measure of membrane permeability, as a function of molecular weight for hemodialyzers and the normal kidney. Source: Reproduced with permission from [5] Copyright (1991) Elsevier.*

The artificial kidney is a life-saving device, but patients are still attached to a machine two or three times a week and seldom feel completely well. This has led to programs to develop small devices that can be packaged as wearable or even implantable units. A number of prototype wearable devices are being evaluated in clinical trials. The benefits will be real, but safety and costs are significant issues.

14.3 Plasma Fractionation

A comparatively recent application that is related to hemodialysis, in that it separates blood components, is membrane plasma fractionation, which is used to separate high molecular weight toxic components from blood in the treatment of a number of diseases. The procedure is called by several names, including plasma fractionation, therapeutic hemapheresis, therapeutic plasma-pheresis, or therapeutic ampheresis. The process uses two filtration steps in series, as shown in Figure 14.5, taken from the review of Wiese [6].

The first step uses a cross-flow plasma filtration membrane with a nominal pore size of about 0.2 μm. The function of this step is to produce a permeate of blood plasma (cell-free blood).

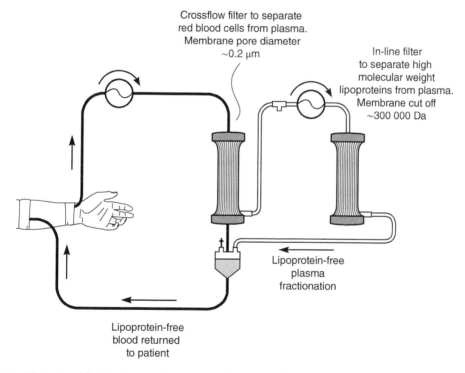

Figure 14.5 *Double filtration used to remove high molecular weight lipoproteins and triglycerides in the treatment of* myasthenia gravis, *some forms of lupus and hypercholesterolemia [6]. The first membrane, with pore diameter of ~0.2 μm, separates blood cells from the plasma. The plasma permeate is then treated with a second tighter membrane (MW cutoff of 300 000), which removes high molecular weight lipoproteins. The twice-filtered plasma is returned to the patient. Source: Reproduced with permission from [6]. Copyright (2011) John Wiley and Sons.*

The cellular materials are not lost, but are retained by the membrane and recirculated as shown to the patient. The plasma permeate is sent to a dead-end filter with a molecular weight cut-off of about 300 000 Da. The second filtration step retains very high molecular weight lipoproteins and triglycerides, while passing albumin, β-globulins, IgG-globulin, and all other plasma components. The filtered plasma is then recirculated as shown. The net result of the two steps is that only the harmful lipoproteins and fats are removed; the remainder of the blood is recovered undamaged and returned to the vascular system of the patient. The use of these devices, while not commonplace, is growing. Currently, it is estimated that about 600 000 procedures per year are performed worldwide.

14.4   Blood Oxygenators

Blood oxygenators are used during surgery when the patient's lungs cannot function normally. Pioneering work on these devices was carried out in the 1930s and 1940s by J.H. Gibbon [7, 8], leading to the first successful open-heart surgery on a human patient in 1953. Gibbon's heart–lung machine used a small tower filled with stainless steel screens to contact blood with counter-flowing oxygen. Direct oxygenation of the blood was used in all such devices until the early 1980s. Screen oxygenators of the type devised by Gibbon were first replaced with a disk oxygenator, which consisted of 20–100 rotating disks in a closed cylinder containing 1–2 l of blood. Later, bubble oxygenators, in which blood flowed through a packed plastic tower, were developed. Because these direct-contact oxygenators required large volumes of blood to prime the device and, more importantly, damaged some of the blood components, they were used in only a few thousand operations per year. The introduction of indirect-contact membrane oxygenators resulted in significantly less blood damage and lower blood priming volumes. The first devices tested in the laboratory used thin dense silicone rubber films, the low permeance of which resulted in bulky devices. The development of polyolefin hollow fiber membranes was a major breakthrough. Devices incorporating the new membranes were introduced in the early 1980s and rapidly accepted; by 1985, more than half of the oxygenators in use were membrane-based. This percentage had risen to 70% by 1990, and today membrane oxygenators are the only type used, and the number of procedures using blood oxygenators has risen to more than 1 million per year worldwide.

A membrane blood oxygenator is shown schematically in Figure 14.6. In the human lung, the total exchange membrane area between the blood capillaries and the inhaled/exhaled air is about $80 \, m^2$, and the membrane is estimated to be about 1 μm thick. The total exchange capacity is far larger than is required for ordinary activities, and people with impaired lung capacity can lead relatively normal lives. To maintain adequate blood oxygenation, a successful heart–lung machine must deliver about $250 \, cm^3(STP)/min$ oxygen and remove about $200 \, cm^3(STP)/min$ carbon dioxide [9]. Because of the limited solubility of these gases in the blood, large blood flows through the device are required, typically 2–4 l/min (approximately 10 times the blood flow through a kidney dialyzer). To maintain good mass transfer with minimal pressure drop through the device, blood is generally circulated on the outside of the fibers and high oxygen concentration, low carbon dioxide concentration gas is circulated down the lumens. Concentration polarization in the blood-side liquid boundary layer significantly affects gas transport. The vast majority of blood oxygenators are used in surgical procedures. However, improvements in the equipment now allow longer term use in a procedure called extracorporeal membrane oxygenation (ECMO). Patients can be supported for days, or even weeks, on ECMO, until their condition improves sufficiently

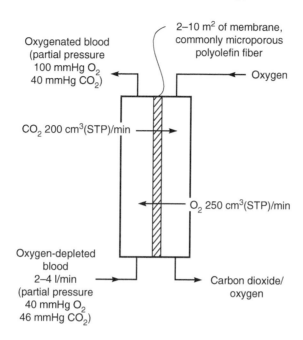

Figure 14.6 *Flow schematic of a membrane blood oxygenator.*

for their own lungs to function. About 100 000 patients per year are treated this way. A detailed review of blood oxygenators has been written by Wiese [6].

14.5   Controlled Drug Delivery

In controlled drug delivery systems, a membrane is used to moderate the delivery rate of a therapeutic agent to the body. In some devices, the membrane controls permeation of the drug from a reservoir. Other devices use the osmotic pressure produced by diffusion of water across a membrane to power miniature pumps. In yet other devices, the drug is impregnated into the membrane material, which then slowly dissolves or degrades in the body. Drug delivery is then controlled by a combination of diffusion and biodegradation.

The objective of all of these devices is a delivery rate predetermined by the device design and independent of the changing environment of the body. In conventional medications, only the total mass of drug delivered to a patient is controlled. In controlled drug delivery, both the mass and dosing rate can be controlled, providing three important therapeutic benefits:

1. The drug is metered to the body slowly over a long period; the problems of overdosing and underdosing associated with conventional periodic delivery of medication are ameliorated or avoided.
2. The drug can often be given locally, ideally to the affected organ directly. Localized delivery results in high concentrations of the drug at the site of action, and improves the therapeutic effect, using a fraction of the drug dose needed for conventional therapy.
3. Localized drug delivery results in low concentrations of drug in the body away from the delivery site, with correspondingly lower systemic side effects.

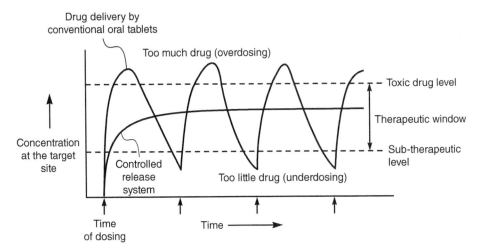

Figure 14.7　*Drug concentrations at the target site for drug action achieved by repeated doses of a conventional medication (tablet, injection) and a single, long-lasting controlled release system.*

The concept of controlled delivery is not limited to drugs. Similar principles are used to control the delivery of agrochemicals, fertilizers, pesticides, and a number of household products. However, most of the technology development in the past 30 years has focused on drug delivery; only this aspect of the topic is covered here.

Some benefits of a controlled release system are illustrated in Figure 14.7. The figure shows drug concentration at the site of drug action as a function of time. Conventional tablets or injections produce highly fluctuating concentration levels. After each drug dose, the drug concentration rises to a peak value and then declines. With drugs having a narrow therapeutic window, it is easy to produce excessively high drug concentrations, leading to adverse side effects, or excessively low drug concentrations, leading to inadequate therapy. The controlled release system meters the drug so that concentration is maintained in the therapeutic window.

The origins of controlled release drug delivery go back a long way. For example, Rose and Nelson [10] described the first miniature osmotic pump in 1955. A key early publication was the 1964 paper of Folkman and Long [11] describing the use of silicone rubber membranes to control the release of anesthetics and cardiovascular drugs. Concurrent discoveries in the field of hormone regulation of female fertility quickly led to the development of controlled delivery systems to release steroids for contraception [12, 13]. The founding of Alza Corporation by Alex Zaffaroni in the late 1960s gave the entire technology a decisive thrust. Alza was dedicated to developing novel controlled release drug delivery systems [14]. The products developed by Alza during the subsequent 25 years stimulated the entire pharmaceutical industry. The first pharmaceutical product in which the drug registration document specified both the total amount of drug in the device and the delivery rate was an Alza product, the Ocusert®, launched in 1974 [15]. This device, shown in Figure 14.8, consists of a three-layer laminate with the drug sandwiched between two rate-controlling polymer membranes. The device, an ellipse about 1 mm thick and 1 cm in diameter, is placed in the cul-de-sac of the eye where it delivers pilocarpine at a constant rate for 7 days, after which it is removed and replaced. The Ocusert was a technical tour de force, although only a limited marketing success. Alza later developed a number of more widely used products, including transdermal

Release rate curve

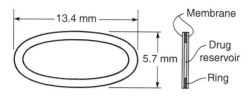

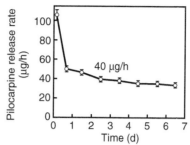

Pilocarpine concentration in the aqueous humor of the eye

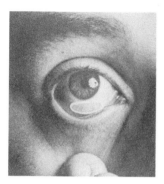

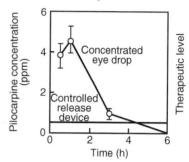

Figure 14.8 *The Ocusert pilocarpine system is a thin multilayer membrane device. The central sandwich consists of a core containing the drug pilocarpine. The device is placed in the eye, where it releases the drug at a continuous rate for 7 days. Devices with release rates of 20 or 40 µg/h are used. Controlled release of the drug eliminates the over- and under-dosing observed with conventional eye drop formulations, which must be delivered every few hours. Source: With permission from [15]. Copyright (1975) Elsevier.*

patches designed to deliver drugs through the skin [16] and osmotic devices for oral drug delivery. Many imitators followed Alza's success, and an entire sub-industry has grown up that produces controlled release systems for a wide variety of drugs.

14.5.1 Membrane Diffusion-Controlled Systems

In membrane diffusion-controlled systems, a drug is released from a device by permeation from an interior reservoir to the surrounding medium. The rate of diffusion of the drug through the membrane governs its rate of release. The reservoir device illustrated in Figure 14.9 is the simplest diffusion-controlled system. An inert membrane encloses the drug to be released; the drug diffuses through the membrane at a finite, controllable rate. If the concentration (or thermodynamic activity) of the material in equilibrium with the inner surface of the enclosing membrane is constant, then the concentration gradient, the driving force for diffusional release of the drug, is constant. This occurs when the inner reservoir contains a saturated solution of the material, providing a constant release rate for as long as excess solid drug is maintained in the solution. This is called zero-order release. If, however, the active drug within the device is initially present as an unsaturated solution, its concentration falls as it is released. The release rate declines exponentially, producing a first-order release profile.

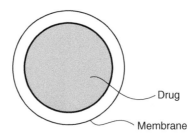

Figure 14.9　*Reservoir device.*

For a device containing a saturated solution of drug, and excess solid drug, Fick's law

$$J = -DK\frac{dc_s}{dx} \tag{14.1}$$

can be restated for a slab or sandwich geometry as

$$\frac{dM_t}{dt} = \frac{AJ}{\ell} = \frac{ADKc_s}{\ell} \tag{14.2}$$

where M_t is the mass of drug released at any time t, and hence dM_t/dt is the steady-state release rate at time t; A is the total surface area of the device (edge effects being ignored); c_s is the saturation solubility of the drug in the reservoir layer; J is the flux of drug through the membrane; and D and K are the familiar diffusion and sorption coefficients.

The Ocusert system illustrated in Figure 14.8 is one example of a diffusion-controlled reservoir device. Another is the steroid-releasing intrauterine device (IUD) shown in Figure 14.10 [17]. Inert IUDs of various shapes were widely used for birth control in the 1950s and 1960s. The contraceptive effect of these IUDs was based on physical irritation of the uterus. The devices resulting in the lowest pregnancy rate were often associated with unacceptable levels of pain and bleeding, whereas more comfortable devices were associated with unacceptably high pregnancy rates. Researchers tried a large number of different IUD shapes in an attempt to produce a device that combined a low pregnancy rate with minimal pain and bleeding, but without real success.

Steroid-releasing IUDs, in which the contraceptive effect of the device comes largely from the steroid, enable less irritating IUD forms to be used. Scommegna et al. performed the first clinical trials to test this concept [13]. The commercial embodiment of these ideas is shown in Figure 14.10, together with the drug release rate curve [17]. This curve shows an initial high drug release during the first 30–40 days, representing drug that has migrated into the polymer during storage of the device and which is released as an initial burst. Thereafter, the device maintains an almost constant drug release rate until it is exhausted at about 400 days. Later versions of this device incorporated synthetic steroids that were more biologically active. These devices contained sufficient steroid to last 2 years or more.

A more familiar type of diffusion-controlled device is the transdermal patch shown in Figure 14.11. A variety of patch designs are used, but a typical patch has an area of 3–30 cm². The drug is contained in a liquid, gel, or polymer reservoir layer and a membrane is used to control delivery of drug to the skin. In some devices, the membrane is a separate layer; in others, the membrane may be the adhesive layer that sticks the device to the skin.

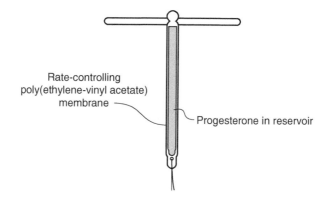

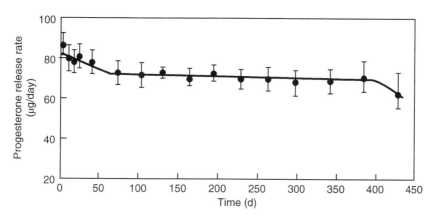

Figure 14.10 *Progestasert® intrauterine device (IUD) designed to deliver progesterone for contraception at 75 μg/day for 1 year. Source: Adapted from [17].*

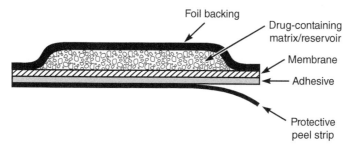

Figure 14.11 *Schematic of a typical transdermal patch. Depending on the drug and the drug delivery rate to be achieved, these devices can vary from 3 to 30 cm² in area.*

Human skin is a remarkably efficient barrier designed to keep 'our insides in and the outsides out!' For this reason, transdermal delivery is limited to potent drugs with efficacy when delivered at a few micrograms to milligrams per day. Only a handful of such drugs has been found that are able to penetrate the skin at a therapeutically useful rate; examples include scopolamine, nitroglycerine, nicotine, clonidine, fentanyl, estradiol, testosterone, lidocaine, and oxybutynin.

Nonetheless, taken altogether the current US market for transdermal patches is over US$3 billion per year.

Despite the best efforts of the manufacturers, drug delivery is often controlled more by the skin barrier than the device membrane. Many attempts have been made to increase the number of drugs that can permeate the skin at a useful rate. Compounds such as DMSO (dimethyl sulfoxide), Azone (1-dodecylazacycloheptane-2-one), and alcohols have been used to enhance the permeability of the skin and so allow the drug to permeate, but success to date has been very limited [18]. Another approach has been to use an applied DC electric current to produce an increased flux of ionic drugs through the skin (iontophoresis), by application of short high voltage (100–1000 V) pulses of electricity to disrupt the skin and increase its permeability (electroporation). Neither of these approaches has become commercial [19].

14.5.2   Monolithic Systems

The controlled release products described above are all examples of membrane-controlled reservoir systems. A second category of diffusion-controlled device is the monolithic system, in which the agent to be released is dispersed uniformly throughout the rate-controlling polymer medium, as shown in Figure 14.12. The release profile is determined by the loading of dispersed agent, the nature of the components, and the geometry of the device. Depending on the exact details of these parameters, drug release tends to decline over time, following first-order kinetics, as shown in Figure 14.13. Thin spots, pinholes, and other similar defects, which can be problems with reservoir systems, do not substantially alter the release rate from monolithic devices. This, together with the ease with which dispersions can be compounded (by milling and extruding, for example), results in low production costs. These advantages can often outweigh the declining release rate with time.

14.5.3   Biodegradable Systems

The diffusion-controlled devices outlined so far are permanent, in that the membrane or matrix of the device remains intact after its delivery role is completed. In some applications, particularly if the device is implanted, this is undesirable; such applications require a device that degrades during or subsequent to its delivery role.

Many polymer-based devices that slowly biodegrade when implanted in the body have been developed; the most important are based on polylactic acid, polyglycolic acid, and their copolymers. In principle, the release of an active agent can be programmed by dispersing the material within a polymer matrix, with erosion of the polymer effecting release of the agent [21, 22]. One class of biodegradable devices relies on *surface eroding* of the polymer material. As the name suggests, the surface area from which drug is delivered diminishes

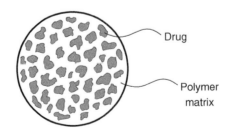

Figure 14.12   Monolithic dispersion device.

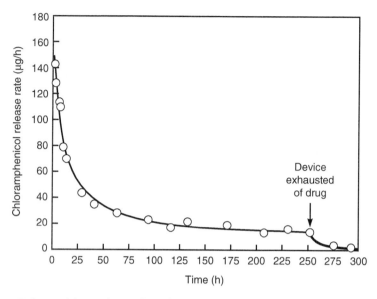

Figure 14.13   *Release of the antibiotic drug chloramphenicol dispersed in a matrix of poly(ethylene-vinyl acetate) [20] / John Wiley & Sons.*

with time, resulting in a decreasing release rate, unless the geometry of the device is manipulated or the drug loading is higher in the interior than in the surface layers. In a more common class of biodegradable polymer, the initial period of degradation occurs very slowly, after which the degradation rate increases rapidly. The bulk of the polymer then erodes over a comparatively short period. In the initial period of exposure to the body, the polymer chains are being cleaved, but the molecular weight is still high, so the mechanical properties of the device are not seriously affected. As chain cleavage continues, a point is reached at which the polymer fragments become swollen or soluble in water. At this point, the polymer begins to dissolve. This type of polymer can be used to make reservoir or monolithic diffusion-controlled systems that degrade after their delivery role is over. A final category of polymer has the active agent covalently attached by a labile bond to the backbone of a matrix polymer. When the device is placed at the site of action, the labile bonds slowly degrade, releasing the active agent and forming a soluble polymer. The methods by which these concepts can be formulated into practical systems are illustrated in Figure 14.14.

14.5.4   Osmotic Systems

Osmotic effects are often a problem in diffusion-controlled systems because imbibition of water swells the device or dilutes the drug. However, several devices have been developed that use osmotic effects to control the release of drugs. These devices, called osmotic pumps, use the osmotic pressure developed by diffusion of water across a semipermeable membrane into a salt solution to push a solution of the active agent from the device. Osmotic pumps of various designs are widely applied in the pharmaceutical area, particularly in oral tablet formulations [23].

The forerunner of modern osmotic devices, developed around 1955, was the Rose–Nelson pump. Rose and Nelson were two Australian physiologists interested in the delivery of drugs

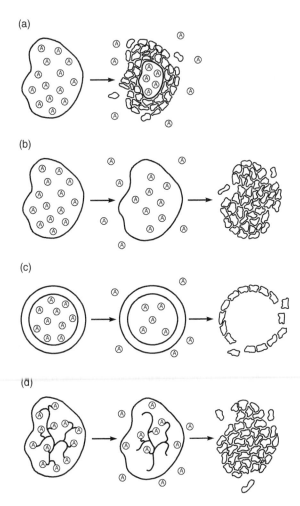

Figure 14.14 *Methods of using biodegradable polymers in controlled release implantable devices to release the active agent, A. (a) Degradation-controlled monolithic system. (b) Diffusion-controlled monolithic system. (c) Diffusion-controlled reservoir system. (d) Erodable polyagent system.*

to the gut of sheep and cattle [10]. Their pump, illustrated in Figure 14.15, consists of three chambers: a drug chamber, a salt chamber containing excess solid salt, and a water chamber. The salt and water chambers are separated by a rigid semipermeable membrane. The difference in osmotic pressure across the membrane moves water from the water chamber into the salt chamber. The volume of the salt chamber increases because of this water flow, which distends the latex diaphragm separating the salt and drug chambers, thereby pumping drug out of the device.

The osmotic pressure of the saturated salt solution is high, on the order of tens of atmospheres, and the small pressure required to pump the suspension of active agent is insignificant in comparison. Therefore, the rate of water permeation across the semipermeable membrane remains constant as long as sufficient solid salt is present in the salt chamber to maintain a saturated solution and hence a constant osmotic pressure driving force. One big disadvantage of the

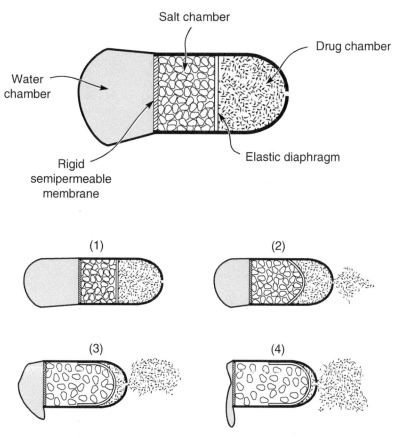

Salt chamber

Drug chamber

Water chamber

Rigid semipermeable membrane

Elastic diaphragm

(1)   (2)

(3)   (4)

Figure 14.15 *Principle of the three-chamber Rose–Nelson osmotic pump first described in 1955. Source: Adapted from [10].*

Rose–Nelson pump is that the device will start pumping out drug as soon as the water chamber contains water, so the chamber can only be filled immediately before use.

The pump designs developed by Alza in the 1970s represent a series of simplifications of the Rose–Nelson concept. The Higuchi–Leeper pump [24] shown in Figure 14.16 has no water chamber; the device is activated by water imbibed from the surrounding environment. The drug-laden pump can be prepared and then stored for weeks or months prior to use, as it is only activated when it is swallowed or implanted in the body. Higuchi–Leeper pumps have a rigid housing, and the semipermeable membrane is supported on a perforated frame. This type of pump usually has a salt chamber containing a fluid solution with excess solid salt. The target application was the delivery of antibiotics and growth hormones to cattle because repeated delivery of oral medications to animals is difficult. The problem is solved by these devices, which are designed to be swallowed by the cow and then to reside in the rumen, delivering a full course of medication over a period of days to weeks.

Theeuwes and Higuchi developed even simpler variants of the Rose–Nelson pump [25, 26]. One such device is illustrated in Figure 14.17. As with the Higuchi–Leeper pump, water to

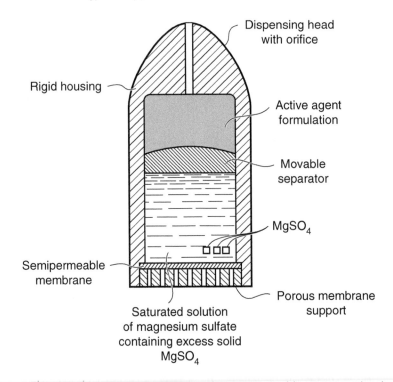

Figure 14.16 *The Higuchi–Leeper osmotic pump design [24] / Public Domain. This device has no water chamber and can be stored in a sealed foil pouch indefinitely. However, once removed from the pouch and placed in an aqueous environment, for example, by an animal swallowing the device, the pumping action begins. The active agent is pumped at a constant rate.*

activate the osmotic action of the pump comes from the surrounding environment. The Theeuwes–Higuchi device, however, has no rigid housing – the membrane acts as the outer casing of the pump. This membrane is sturdy and strong enough to withstand the pumping pressure developed inside the device. The device is loaded with the desired drug prior to use. When the device is placed in an aqueous environment, release of the drug follows a time course set by the salt used in the salt chamber and the permeability of the outer membrane casing.

The principal application of these small osmotic pumps has been as implantable systems for experimental studies. The devices are made with volumes of 0.2–2 ml. Figure 14.17 shows one such device suitable for implantation in a laboratory rat. The delivery pattern obtained with the device is constant and independent of the site of implantation, as shown by the data in Figure 14.18.

The development that made osmotic delivery a major method of achieving controlled drug release was the invention of the elementary osmotic pump by Theeuwes in 1974 [27]. The concept behind this invention is illustrated in Figure 14.19. The device is a further simplification of the Theeuwes–Higuchi pump, and eliminates the separate salt chamber by using the drug itself as the osmotic agent. The device is formed by compressing a drug having a suitable osmotic

(a)

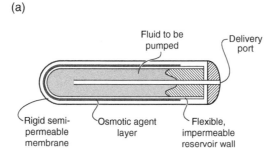

(b)

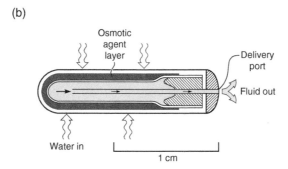

Figure 14.17 *The Theeuwes–Higuchi osmotic pump has been widely used in drug delivery tests in laboratory animals. The device is small enough to be implanted under the skin of a rat and delivers up to a milliliter of drug solution over a period of 3–4 days. (a) Pump filled and assembled. (b) Pump in use. Source: Reprinted from [26] under CC-BY 4.0 license. Copyright (1976) Springer New York.*

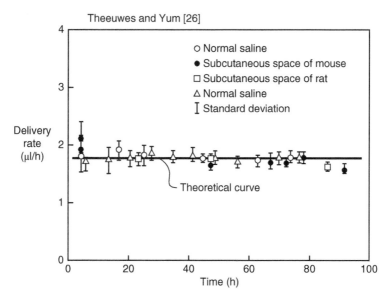

Figure 14.18 *Drug delivery data obtained with an implantable osmotic pump. Source: [26] / Reproduced with permission of Springer Nature.*

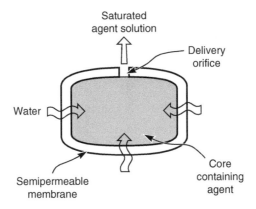

Figure 14.19   *The Theeuwes elementary osmotic pump.*

pressure into a tablet using a tableting machine. The tablet is then coated with a semipermeable membrane, usually cellulose acetate, and a small hole is drilled through the membrane coating. When the tablet is placed in an aqueous environment, the osmotic pressure of the soluble drug inside the tablet draws water through the semipermeable coating, forming a saturated aqueous solution inside the device. The membrane does not expand, so the increase in volume caused by the imbibition of water raises the hydrostatic pressure inside the tablet slightly. This pressure is relieved by a flow of saturated drug solution out of the device through the small orifice. Thus, the tablet acts as a small chemical pump, in which water is drawn osmotically into the tablet through the membrane wall and then leaves as a saturated drug solution through the orifice. This process continues at a constant rate until all the solid drug inside the tablet has been dissolved and only a solution-filled shell remains. This residual dissolved drug continues to be delivered, but at a declining rate, until the osmotic pressures inside and outside the tablet are equal. The driving force that draws water into the device is the difference in osmotic pressure between the outside environment and a saturated drug solution. Therefore, the osmotic pressure of the dissolved drug solution has to be relatively high to overcome the osmotic pressure of the body, but for drugs with solubilities greater than 5–10 wt%, these devices function very well. Later variations on the simple osmotic tablet design have been made to overcome the solubility limitation. The elementary osmotic pump was developed by Alza under the name OROS®, and is commercially available for a number of drugs.

The four types of osmotic pumps described above are interesting examples of how true innovation is sometimes achieved by leaving things out. The first osmotic pump produced by Rose and Nelson contained six critical components, had a volume of 80 cm^3, and was little more than a research tool. In the early 1980s, Felix Theeuwes and others progressively simplified and refined the concept, leading in the end to the elementary osmotic pump, a device that looks almost trivially simple. It has been described as a tablet with a hole, but is, in fact, a truly elegant invention having a volume of less than 1 cm^3, containing only two components, achieving almost constant drug delivery, and allowing manufacture on an enormous scale at minimal cost. Figure 14.20 shows examples of the four main types of osmotic pumps taken from the patent drawings. The pumps are drawn to scale to illustrate the progression that occurred as the design was simplified.

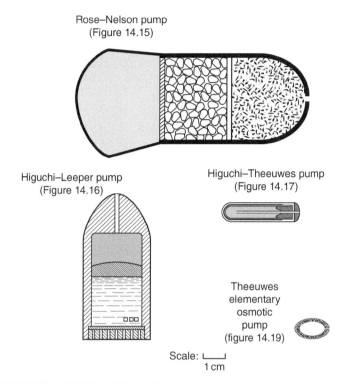

Rose–Nelson pump
(Figure 14.15)

Higuchi–Leeper pump
(Figure 14.16)

Higuchi–Theeuwes pump
(Figure 14.17)

Theeuwes
elementary
osmotic
pump
(figure 14.19)

Scale: ⌐___⌐
1 cm

	Rose–Nelson pump	Higuchi–Leeper pump	Theeuwes–Higuchi pump	Theeuwes elementary osmotic pump
Approximate volume (cm³)	80	35	3	< 1
Components	Rigid housing Water chamber Salt chamber Drug chamber Elastic diaphragm Membrane	Rigid housing Salt chamber Drug chamber Elastic diaphragm Membrane	Salt chamber Drug chamber Elastic diaphragm Membrane	Drug chamber Membrane
Number of components	6	5	4	2

Figure 14.20 The main types of osmotic pump drawn to scale.

References

1. Merrill, J.P. (1961). The artificial kidney. *Sci. Am.* **205**: 56.
2. Kolff, W.J. and Berk, H.T. (1944). The artificial kidney: a dialyzer with great area. *Acta Med. Scand.* **117**: 121.
3. Kolff, W.J. (1947). *New Ways of Treating Uremia: The Artificial Kidney, Peritoneal Lavage, Intestinal Lavage*. London: J & A Churchill, Ltd.

4. Quinton, W., Dilpard, D., and Scribner, B.H. (1960). Cannulation of blood vessels for prolonged hemodialysis. *Trans. Am. Soc. Artif. Inter. Organs* **6**: 104.
5. Keen, M.L. and Gotch, F.A. (1991). Dialyzers and delivery systems. In: *Introduction to Dialysis* (ed. M.C. Cogan, P. Schoenfeld, and F.A. Gotch), 1–7. New York: Churchill Livingston.
6. Wiese, F. (2008). Membranes for artificial lungs. In: *Membranes for Life Sciences* (ed. K.-V. Peinemann and S.P. Nunes), 49–68. Weinheim, Germany: Wiley-VCH.
7. Gibbon, J.H. Jr. (1937). Artificial maintenance of circulation during experimental occlusions of pulmonary artery. *Arch. Surg. Chicago* **34**: 1105.
8. Gibbon, J.H. Jr. (1954). Application of a mechanical heart and lung apparatus to cardiac surgery. *Minn. Med.* **37**: 171.
9. Galletti, P.M. and Brecher, G.A. (1962). *Heart-Lung Bypass*. New York: Grun and Stratton.
10. Rose, S. and Nelson, J.F. (1955). A continuous long-term injector. *Aust. J. Exp. Biol.* **33**: 415.
11. Folkman, J.M. and Long, D.M. (1964). The use of silicone rubber as a carrier for prolonged drug therapy. *J. Surg. Res.* **4**: 139.
12. Croxatto, H.B., Diaz, S., Vera, R. et al. (1969). Fertility control in women with a progestagen released in microquantities from subcutaneous capsules. *Am. Obstet. Gynecol.* **105**: 1135.
13. Scommegna, A., Pandya, G.N., Christ, M. et al. (1970). Intrauterine administration of progesterone by a slow releasing device. *Fert. Steril.* **21**: 201.
14. Zaffaroni, A. (1981). Applications of polymers in rate-controlled drug delivery. *Polym. Sci. Tech.* **14**: 293.
15. Sendelbeck, L., Moore, D., and Urquhart, J. (1975). Comparative distribution of pilocarpine in ocular tissues of the rabbit during administration of eyedrops or by membrane controlled delivery systems. *Am J. Opthalmol* **80**: 274.
16. Shaw, J. (1983). Development of transdermal therapeutic systems. *Drug Dev. Ind. Pharm.* **9**: 579.
17. Pharriss, B.B., Erickson, R., Bashaw, J. et al. (1974). Progestasert: a uterine therapeutic system for long-term contraception. *Fert. Steril.* **25**: 915.
18. Williams, A.C. and Barry, B.W. (2004). Penetration enhancers. *Adv. Drug Delivery Rev.* **56**: 603–618.
19. Stamatialis, D.F. (2008). Drug delivery through skin: overcoming the ultimate biological membrane. In: *Membranes for Life Sciences* (ed. K.-V. Peinemann and S.P. Nunes), 191–221. Weinheim, Germany: Wiley-VCH.
20. Baker, R.W. (1987). *Controlled Release of Biologically Active Agents*. New York: Wiley.
21. Heller, J. (1980). Controlled release of biologically active compounds from bioerodible polymers. *Biomaterials* **1**: 51.
22. Pitt, C.G. and Schindler, A. (1980). The design of controlled drug delivery systems based on biodegradable polymers. In: *Biodegradables and Delivery Systems for Contraception* (ed. E.S.E. Hafez and W.A.A. van Os), 17–46. Lancaster: MTP Press.
23. Santus, G. and Baker, R.W. (1995). Osmotic drug delivery: a review of the patent literature. *J. Controlled Release* **35**: 1.
24. Higuchi, T. and Leeper, H.M. (1971). Improved osmotic dispenser employing magnesium sulfate and magnesium chloride. US Patent 3,760,804, issued 25 September 1973.

25. Theeuwes, F. and Higuchi, T. (1972). Osmotic dispensing agent for releasing beneficial agent. US Patent 3,845,770, filed 5 June 1972 and issued 5 November 1974.
26. Theeuwes, F. and Yum, S.I. (1976). Principles of the design and operation of generic osmotic pumps for the delivery of semisolid or liquid drug formulations. *Ann. Biomed. Eng.* **4**: 343.
27. Theeuwes, F. (1975). Elementary osmotic pump. *J. Pharm. Sci.* **64**: 1987.

15

Other Membrane Processes

15.1 Introduction

Any book must leave something out, and this one has left out a good deal; it does not cover membranes used in packaging materials, sensors, ion-selective electrodes, battery separators, electrophoresis, affinity membranes, and thermal diffusion. In this chapter, we describe the preparation, properties, and potential uses of three very different membranes that come under the general heading of 'other'

- Palladium metal membranes
- Perovskite ion transport membranes
- Piezodialysis ion-exchange membranes

The permeation properties of these membranes are very different from each other, as well as the other membranes described in this book. The likelihood of their future commercial use is low. Nonetheless, the membrane energumen who has traveled this far may find them of interest.

15.2 Metal Membranes

Metal membranes, particularly palladium-based membranes, have been considered for hydrogen separation for a long time. In 1866 Thomas Graham, then Master of the Mint, used a palladium membrane to separate hydrogen from coal gas. One hundred years later, Union Carbide built and

Membrane Technology and Applications, Fourth Edition. Richard W. Baker.
© 2024 John Wiley & Sons Ltd. Published 2024 by John Wiley & Sons Ltd.

operated a palladium membrane plant to separate hydrogen from a refinery off-gas stream [1]. The plant produced 99.9% pure hydrogen in a single pass through 25-μm-thick palladium membranes. However, problems with membrane stability were hard to solve and the plant soon closed.

Palladium absorbs large amounts of hydrogen and forms an alloy which can exist in two possible phases (α and β) in the H_2/Pd phase diagram. In the presence of low concentrations of hydrogen, palladium exists in the α phase, but depending on the amount of hydrogen present and the temperature, a portion of the metal can convert to the β phase. Repeated transitions from the α to β phase and back again cause the metal to crack and deform, a phenomenon called hydrogen embrittlement. At temperatures above about 300 °C, the β phase no longer forms and embrittlement is not a problem. This means palladium membranes are normally operated at 350–450 °C. It also means plant start-up and shutdown must be done carefully to avoid entering a hydrogen composition/temperature zone where the β phase can form.

A further problem with palladium membranes is cost: a 5-μm-thick palladium membrane requires approximately 60 g of palladium/m^2 of membrane. At current palladium costs of US \$60/g, the metal cost alone is US\$3600/m^2 of membrane. This is 50 times the membrane material cost of even the most expensive polymeric membrane. However, the permeance of palladium membranes is extremely high and the membrane is almost infinitely permeable to hydrogen over all other components, properties that in some applications offset the high cost.

Nonetheless, if noble metal membranes are ever to be used on a large scale, their cost must be reduced. One approach [2, 3] is to make the membranes much thinner by coating a 1000–5000 Å film of the metal on a microporous metal or polymer support. Electrolytic deposition or sputter coating can be used. Another approach is to coat a thin palladium layer on a dense tantalum or vanadium support film. Tantalum and vanadium are quite permeable to hydrogen and much less expensive than palladium, but cannot be used alone because they tend to form impermeable oxide surface films. However, protected by a thin palladium layer, such membranes have good permeability at high temperatures. Buxbaum [4] and Edlund [5, 6] pursued this approach. A detailed discussion of hydrogen permeation in metals is given in the book by Alefeld and Völkl [7].

The high selectivity and permeance of metal membranes is caused by an unusual transport mechanism believed to follow the multistep process illustrated in Figure 15.1 [8]. Hydrogen molecules from the feed gas are sorbed on the membrane surface, where they dissociate into hydrogen atoms. Each individual hydrogen atom loses its electron to the metal lattice and diffuses through the lattice as an ion. Hydrogen atoms emerging at the permeate side reassociate to form hydrogen molecules, then desorb, completing the permeation process. Only hydrogen is transported through the membrane by this mechanism; all other gases are excluded.

If the sorption and dissociation are rapid, then the hydrogen atoms on the membrane surface are in equilibrium with the gas phase. The concentration, c, of hydrogen atoms on the metal surface is given by Sievert's law:

$$c = Kp^{1/2} \qquad (15.1)$$

where K is Sievert's constant, and p is the hydrogen pressure in the gas phase. At high temperatures (>300 °C), the surface sorption and dissociation steps are fast, and the rate-controlling step is diffusion of atomic hydrogen through the metal lattice. This idea is supported by data such as that of Ma et al. (Figure 15.2) [2], who have shown that the hydrogen flux through the metal membrane is proportional to the difference of the square roots of the hydrogen pressures on either

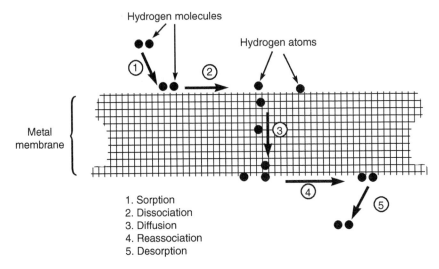

Figure 15.1   *Mechanism of hydrogen permeation through metal membranes.*

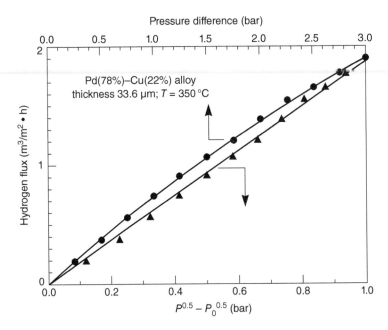

Figure 15.2   *Sievert's plot for a Pd–Cu/porous stainless steel composite membrane at 350 °C. Source: [2] / John Wiley & Sons.*

side of the membrane. At lower temperatures, however, the sorption and dissociation of hydrogen on the membrane surface become the rate-controlling steps, and permeation deviates from Sievert's law predictions.

Despite their extraordinary properties, metal membranes have found very limited industrial application. In the 1970s and early 1980s, Johnson Matthey built a number of systems to produce on-site hydrogen from hydrogen/carbon dioxide mixtures made by reforming methanol [9].

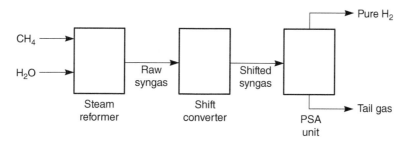

Figure 15.3   *Hydrogen production by steam methane reforming in a conventional process.*

This was not a commercial success, but the company and others still produce small systems, using palladium-silver alloy membranes, to generate ultrapure hydrogen from 99.9% hydrogen for the electronics industry and as feed gas to fuel cells.

In the early 2000s, the US Department of Energy (DOE) sponsored a number of research programs to develop hydrogen-separating metal membranes. DOE's support was driven in part by its interest in developing fuel cell technology, but also by the potential of metal membranes to improve the classic steam methane reforming process, by which most industrial hydrogen is made. Steam reforming, shown schematically in Figure 15.3, reacts methane with water at high temperature to produce hydrogen via the reaction

$$CH_4 + 2H_2O \rightleftharpoons 4H_2 + CO_2 \tag{15.2}$$

The product, known as syngas, is a mixture of hydrogen and CO_2. Not all of the methane is fully oxidized to CO_2; a significant fraction is converted to CO via the reaction

$$CH_4 + H_2O \rightleftharpoons 3H_2 + CO \tag{15.3}$$

To obtain more complete conversion, the raw syngas, consisting of hydrogen and CO_2 together with CO and some unconverted methane, is reacted a second time in a catalytic shift converter. Most of the remaining CO is converted in this reactor, but the shifted syngas still contains CO_2 and some unreacted CO and methane, and it is passed to a PSA unit for purification.

A proposed design for treatment of the raw syngas using metal membranes incorporated into the shift converter is shown in Figure 15.4. A palladium membrane is used to form a part of the wall of the shift converter, and is in direct contact with the reaction mixture in the chamber. The membrane reactor thus formed replaces both the conventional shift unit and the PSA purification step. The shift reaction now takes place on the feed side of the membrane and continuous withdrawal of hydrogen from the permeate side allows the reaction to be driven to completion while producing ultrapure hydrogen. About 15% of the hydrogen produced in the conversion process of Figure 15.3 is also lost in the fuel gas from the PSA unit. In the membrane reactor process, this hydrogen loss is eliminated.

The goal of much of this work was to allow the distributed production of hydrogen at refueling stations for hydrogen fuel cell trucks and buses. A side benefit of the process is that the membrane residue stream is at about 40 bar, and can be cooled to produce liquid CO_2 for sequestration. Though work on such membranes and designs continued for a number of years, the membranes were plagued by poisoning, scale-up, and stability issues and none of the projects progressed

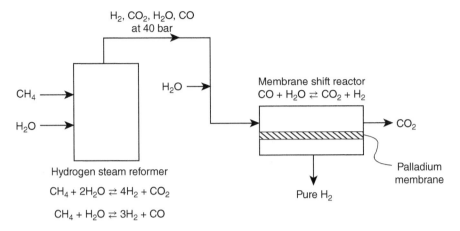

Figure 15.4 *Block diagram of a hydrogen steam reformer followed by a membrane shift reactor.*

Table 15.1 *Petrochemical reactions that yield a hydrogen product amenable to purification by metal membranes.*

$C_4H_{10} \rightleftarrows C_4H_8 + H_2$
$C_6H_{11}CH_3 \rightleftarrows C_6H_5CH_3 + 3H_2$
$C_6H_{12} \rightleftarrows C_6H_6 + 3H_2$
$H_2S \rightleftarrows H_2 + S$

beyond bench-scale tests. If such problems could be solved, however, the same technology could be applied to other petrochemical dehydrogenation processes, some of which are shown in Table 15.1.

A laboratory example of the improvement in conversion that is possible is shown in Figure 15.5 [10]. In this figure, the fractional conversion of methylcyclohexane to toluene in a conventional tube reactor is compared to that in a similar reactor having walls made from palladium/silver alloy, which serve as a hydrogen-selective membrane. Without the membrane, the degree of conversion is limited to the equilibrium value of the reaction. By continuously removing the hydrogen, much higher degrees of conversion can be achieved.

15.3 Ion-Conducting Membranes

The first papers on ion-conducting membranes were published in the 1980s by Teraoka [11, 12]. He showed that some oxygen-deficient perovskite metal oxides with the general formula $ABO_{3-\delta}$ were extraordinarily permeable to oxygen at temperatures above 500 °C. In these structures, the A component is a mixture of divalent lanthanides or alkaline earths, such as calcium, barium, or strontium. The B component is commonly a mixture of trivalent iron and cobalt. Teraoka used $SrCo_{0.8}Fe_{0.2}O_{3-\delta}$. The oxygen vacancies in the perovskite structure provide a pathway for ionic diffusion of oxygen ions $(O^=)$ from one empty site to another. Electrical neutrality is maintained by a flow of electrons through the membrane in the other direction. Oxygen permeation is a three-step process, as illustrated in Figure 15.6. On the feed side of

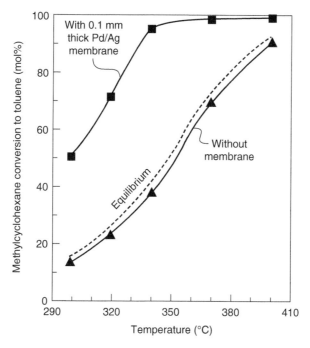

Figure 15.5 *Methylcyclohexane conversion to toluene as a function of reactor temperature in a membrane and a nonmembrane reactor. Source: Reproduced with permission from [10]. Copyright (1995) American Chemical Society.*

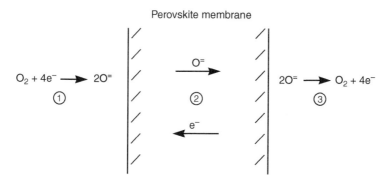

Figure 15.6 *Oxygen transport across a perovskite membrane.*

the membrane at temperatures above 500 °C, a portion of oxygen can be ionized to form $O^=$ (step 1). The $O^=$ ions then diffuse through the membrane to the permeate side of the membrane (step 2) where the ions recombine to reform oxygen molecules and produce the electrons needed for electrical neutrality. (Step 3) Permeation in these perovskites can be extraordinarily fast; permeabilities of 10 000 Barrer have been reported. For a thin membrane, permeance can be so high that the ionization and recombination reactions on the feed and permeate sides become the rate-limiting steps in the process. The membrane thickness at which half of the resistance to permeation is caused by the membrane and half is caused by the slower surface reactions is sometimes called the critical thickness. Making membranes thinner than the critical thickness, therefore, does not increase the permeance significantly. Critical thickness varies with the particular perovskite used, but is generally in the range of 5–10 μm. Nonetheless, even membranes this thick have useful overall permeances.

Membrane stability has proved to be a major problem with perovskites. The alkaline earth components in many perovskites react with carbon dioxide (unfortunately often present in the gas to be treated), to form carbonate. Also the perovskite structure is only thermodynamically stable at high oxygen concentrations and high temperatures. At other conditions, a phase transition to the generally more stable Brownmillerite structure can occur, which seriously reduces the oxygen permeance.

Many variants of Teraoka's first perovskite membranes have been made; most changes that improve stability seem to inevitably reduce permeability. One class of perovskite developed by Shao [13], $Ba_{0.5}Sr_{0.5}Co_{0.8}Fe_{0.2}O_{3-\delta}$, has been used as a starting point for many studies. Many potential applications for ion-conducting membranes have been suggested. The most important application (and the reason for U.S. Department of Energy support for more than 15 years), is the generation of electricity from coal or natural gas by an oxycombustion process that produces a flue gas of pure carbon dioxide for sequestration [14–16].

A simplified flow diagram of the process is shown in Figure 15.7. An ion-conducting membrane is used to produce a high purity oxygen stream from air. The oxygen is then sent to a boiler using coal or natural gas as fuel. The steam produced is used to generate electricity in a steam turbine. The boiler flue gas is mostly carbon dioxide, with 5–10% unreacted oxygen and a mix of contaminants coming from the coal. After a purification step to remove these contaminants, the stream is split into two fractions. One portion is cooled and compressed to make liquid carbon dioxide for sequestration. The other, the larger fraction, is recirculated as a sweep gas on the permeate side of the membrane. Two large U.S consortia, one headed by Praxair [17], the other by Air Products [18], worked on the development of membranes for this and other processes until about 2010. Later, there were other membrane development programs in Europe, China, and Japan.

Air Products took the technology to a 5 ton/day oxygen plant, using rigid ceramic discs coated on both sides with the perovskite membrane. The discs were stacked one on top of the other around a ceramic hollow core, as shown in Figure 15.8. This all-ceramic module design was used

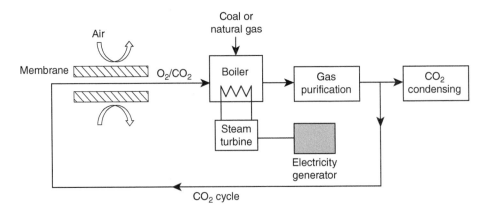

Figure 15.7 *Scheme for the membrane integrated oxy-fuel combustion process. Source: Adapted from [16].*

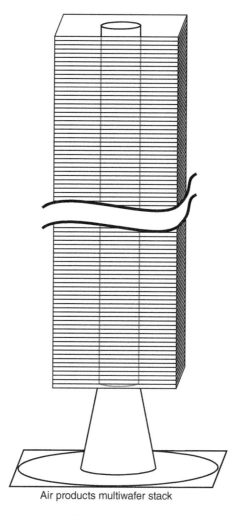

Air products multiwafer stack

Figure 15.8 *A stack module developed by Air Products consisting of multiple perovskite-coated ceramic discs.*

to avoid the problems of cracking caused by differential expansion when the brittle ceramic membrane assembly was heated to high temperatures. Praxair and others used tubular coated ceramic membranes closed at one end and suspended in a furnace. Solutions to many of the membrane stability problems were found, but operating the membranes was never going to be easy or cheap, and since 2010, interest from government and industrial participants has waned. Incremental improvements to the perovskites continue to be made by academic groups, but the transformational improvements required to make the process feasible have not been found.

One of the new approaches, initiated by Kang Li at Imperial College, is the development of ceramic perovskite capillary fibers [15, 19]. The fibers are prepared by the conventional spinning process described in Chapter 4, with the difference being that the polymer spinning solution contains a large amount of dispersed perovskite particles. After spinning, the green capillary fibers are sintered in a high-temperature furnace that burns off the polymer binder, leaving a ceramic perovskite skeleton behind. Scale-up of these fibers to the size of equipment needed for industrial

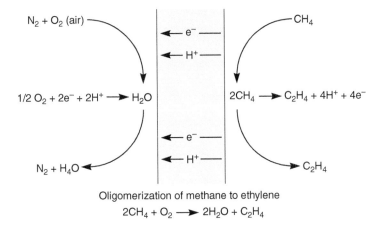

Oligomerization of methane to ethylene

$$2CH_4 + O_2 \longrightarrow 2H_2O + C_2H_4$$

Figure 15.9   Oligomerization of methane to ethylene by a process incorporating an ion-conducting membrane.

processes will not be easy, but the resulting product is likely to be closer to economic feasibility than the tubular or stacked discs used to date.

The bulk of ion-conducting membrane research has focused on developing oxygen-permeating membranes, although membranes that selectively permeate hydrogen can also be made. The most commonly suggested materials have the $ABO_{3-\delta}$ perovskite structure, although the chemistries are different; $Sr_{0.95}Yb_{0.05}CeO_{3-\delta}$ and $Ca_{0.1}Zr_{0.9}InO_{3-\delta}$, for example, have been used. As with oxygen, the membranes are only significantly permeable at temperatures above 500 °C. Figure 15.9 shows a potential application, using a hydrogen ion (proton) selective membrane in the oligomerization of methane to ethylene. The mechanism of proton transport through these materials is complex, but involves the same structural oxygen vacancy involved in oxygen transport.

15.4   Charge Mosaic Membranes and Piezodialysis

One of the characteristic features of ion-exchange membranes is that the membranes are quite impermeable to salts. Although counter-ions to the fixed charge groups in the membrane can easily permeate the membrane, ions with the same charge as the fixed charge groups are excluded and do not permeate. If a pressure driving force is applied across the membrane, the permeable ion creates a charge difference from the feed to the permeate side that counterbalances the pressure driving force. Sollner [20] proposed that, if ion exchange membranes consisting of separated small domains of anionic and cationic membranes could be made, they would be permeable to both anions and cations. These membranes are now called charge mosaic membranes; the concept is illustrated in Figure 15.10. Cations permeate the cationic membrane domain; anions permeate the anionic domain.

Charge mosaic membranes can preferentially permeate salts from water. This is because the principle of electroneutrality requires that the counter ion concentration inside the ion exchange regions be at least as great as the fixed charge density. Because the fixed charge density of ion exchange membranes is typically greater than 1 M, dilute counter ions present in the feed solution are concentrated 10- to 100-fold in the membrane phase. The large concentration

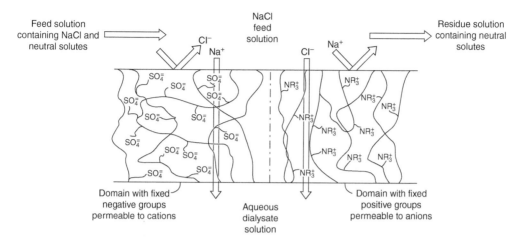

Figure 15.10 *A charge mosaic membrane, consisting of separate finely dispersed domains, one domain containing fixed negative charges, the other containing fixed positive charges [20].*

gradient that forms in the membrane leads to high ion permeabilities. Water and neutral solutes are not concentrated in the membrane and permeate at low rates. When used as dialysis membranes, therefore, these charged mosaic membranes are permeable to salts but relatively impermeable to non-ionized solutes.

For charge mosaic membranes to work most efficiently, the cationic and anionic domains in the membrane must be close together to minimize charge separation effects [21, 22]. The first charge mosaic membranes were made by distributing very small ion exchange beads in an impermeable support matrix of silicone rubber [23, 24]. A second approach, used by Platt and Schindler [25], was to use the mutual incompatibility that occurs when a solution containing a mixture of two different polymers is evaporated. Figure 15.11 shows a photomicrograph of a film cast from poly(styrene-*co*-butadiene) and poly(2-vinyl pyridine-*co*-butadiene). The *co*-butadiene fraction makes these two polymers mutually soluble in tetrahydrofuran but, on evaporation of the solvent, a two-phase-domain structure extending completely through the membrane layer forms. Once formed, the poly(2-vinyl pyridine-*co*-butadiene) portion of the membrane is quaternerized to form fixed positive groups, and the poly(styrene-*co*-butadiene) portion of the membrane is sulfonated to form fixed negative groups.

Itou et al. [26] have obtained an even finer distribution of fixed charge groups by casting films from multicomponent block copolymers such as poly(isoprene-*b*-styrene-*b*-butadiene-*b*-(4-vinyl benzyl)dimethylamine-*b*-isoprene). These films show a very regular domain structure with a 200–500 Å spacing. After casting the polymer film, the (4-vinyl benzyl)dimethylamine blocks were quaternerized with methyl iodide vapor, and the styrene blocks were sulfonated with chlorosulfonic acid.

Using the block copolymer membranes described above, significant selectivities for electrolytes over non-electrolytes have been observed. Some data reported by Hirahara et al. [27] are shown in Table 15.2. The ionizable electrolytes were 100 times more permeable than non-ionized solutes such as glucose and sucrose, suggesting a number of potential applications in which deionization of mixtures of neutral solutes and dissolved salts is desirable. The permeabilities of salts in these membranes are also orders of magnitude higher than values measured for normal ion exchange membranes. In principle then, these membranes can be used in deionization processes, for example, to remove salts from sucrose solutions in the sugar industry.

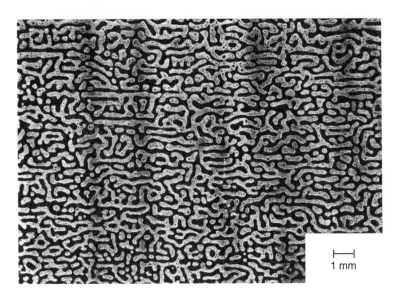

Figure 15.11 *Film cast from a 1:2 mixture of poly(styrene-co-butadiene) and poly(2-vinyl pyridine-co-butadiene) with about 15 mol% butadiene content (10 wt% solution of the copolymers in tetrahydrofuran). Dark areas: poly(styrene-co-butadiene); light areas: poly(2-vinyl pyridine-co-butadiene). Source: Reproduced with permission from [25]. Copyright (1971) John Wiley and Sons.*

Table 15.2 *Solute flux measured in well-stirred dialysis cells at 25 °C using 0.1 M feed solutions [27].*

Solute	Flux (10^{-8} mol/cm^2 • s)
Sodium chloride	7.5
Potassium chloride	9
Hydrochloric acid	18
Sodium hydroxide	10
Glucose	0.08
Sucrose	0.04

A second potential application is pressure-driven desalination. When a pressure difference is applied across the membrane, the concentrated ionic groups in the ion exchange domains are swept through the membrane, producing a salt-enriched permeate on the low-pressure side. This process, usually called piezodialysis, has a number of conceptual advantages over the alternative, conventional reverse osmosis, because the minor component (salt), not the major component (water), permeates the membrane.

Charge mosaic membranes and piezodialysis continue to be the subject of sporadic research [28], but so far, this has met with little commercial interest. It was originally hoped that the flow of water and salt through charge mosaic membranes would be strongly coupled. If this were the case, the 100-fold enrichment of ions within the charged regions of the membrane would provide substantial enrichment of salt in the permeate solution. In practice, the enrichment obtained is relatively small, and the salt fluxes are low even at high pressures. The salt enrichment also

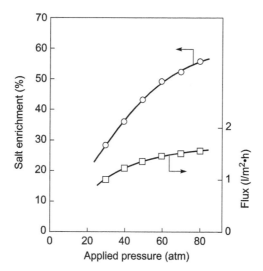

Figure 15.12 *Piezodialysis of 0.02 M potassium chloride solution with block copolymer charge mosaic membranes. Enrichment is calculated using the expression:*
$$\text{enrichment} = 100\left(\frac{\text{concentration permeate}}{\text{concentration feed}} - 1\right).$$
Source: Reproduced with permission from [26]. Copyright (1988) American Chemical Society.

decreases substantially as the salt concentration in the feed increases, limiting the potential applications of the process to desalination of low concentration solutions. Some results of piezodialysis experiments with block copolymer membranes and a potassium chloride solution are shown in Figure 15.12 [26].

References

1. McBride, R.B. and McKinley, D.L. (1965). A new hydrogen recovery route. *Chem. Eng. Prog.* **61**: 81.
2. Ma, Y.H. (2008). Hydrogen separation membranes. In: *Advanced Membrane Technology and Applications* (ed. N.N. Li, A.G. Fane, W.S.W. Ho, and T. Matsuura), 451. Hoboken, NJ: Wiley.
3. Athayde, A.L., Baker, R.W., and Nguyen, P. (1994). Metal composite membranes for hydrogen separation. *J. Membr. Sci.* **94**: 299.
4. Buxbaum, R.E. and Marker, T.L. (1993). Hydrogen transport through non-porous membranes of palladium-coated niobium, tantalum, and vanadium. *J. Membr. Sci.* **85**: 29.
5. Edlund, D.J. and McCarthy, J. (1995). The relationship between intermetallic diffusion and flux decline in composite metal membranes. *J. Membr. Sci.* **107**: 147–153.
6. Edlund, D.J., Friesen, D.T., Johnson, B., and Pledger, W. (1994). Hydrogen-permeable metal membranes for high-temperature gas separations. *Gas. Sep. Purif.* **8**: 13.
7. Alefeld, G. and Völkl, J. (ed.) (1978). *Hydrogen in Metals I: Basic Properties*. Berlin: Springer-Verlag.
8. Hunter, J.B. (1963). Ultrapure hydrogen by diffusion through palladium alloys. *Disv. Pet. Chem. Prepr.* **8**: 4.

9. Philpott, J.E. (1985). Hydrogen diffusion technology, commercial applications of palladium membrane. *Platinum Met. Rev.* **29**: 12.
10. Ali, J.K. and Rippin, D.W.T. (1995). Comparing mono and bimetallic noble-metal catalysts in a catalytic membrane reactor for methylcyclohexane dehydrogenation. *Ind. Eng. Chem. Res.* **35**: 733.
11. Teraoka, Y., Zhang, H., Furukawa, S., and Yamazoe, N. (1985). Oxygen permeation through perovskite-type oxides. *Chem. Lett.* **14**: 1743.
12. Teraoka, Y., Nobunaga, T., and Yamazoe, N. (1988). The effect of cation substitution on the oxygen semipermeability of perovskite-type oxides. *Chem. Lett.* **17**: 503.
13. Shao, Z., Yang, W., Cong, Y. et al. (2000). Investigation of the permeation behavior and stability of a $Ba_{0.5}Sr_{0.5}Co_{0.8}Fe_{0.2}O_{3-\delta}$ oxygen membrane. *J. Membr. Sci.* **172**: 177–188.
14. Chen, W., Chen, C.-S., Bouwmeester, H.J.M. et al. (2014). Oxygen-selective membranes integrated with oxycombustion. *J. Membr. Sci* **463**: 166–172.
15. Tan, X., Wang, Z., Meng, B. et al. (2010). Pilot-scale production of oxygen from air using perovskite hollow fibre membranes. *J. Membr. Sci.* **352**: 189–196.
16. Kneer, R., Toporov, D., Forster, M. et al. (2010). Oxycoal-AC: toward and integrated coal-fired power plant process with ion transport membrane-based oxygen supply. *Energy Enviro. Sci.* **3**: 198.
17. Mazanec, T.J., Cable, T.L., Frye, J.G, Jr., and Kliewer, W.R. (1997). Solid-component membranes electrochemical reactor components electrochemical reactors use of membranes reactor components and reactor for oxidation reactions. US Patent 5,591, 315 (1997); 5,306, 411 (1994); 5,648, 304 (1993) and many others.
18. Thorogood, R.M., Srinivasan, R., Yee, T.F., and Drake, M.P. (1993). Composite mixed conductor membranes for producing oxygen. US Patent 5,240,480 (1993); 5,240,473 (1993); 5,261,932 (1993); 5,269,822 (1993) and many others.
19. Li, K. (2007). *Ceramic Membranes for Separation and Reaction*. Hoboken, NJ: Wiley.
20. Sollner, K. (1932). Über Mosaik Membranen. *Biochem. Z.* **244**: 370.
21. Leitz, F.B. (1976). Piezodialysis. In: *Membrane Separation Processes* (ed. P. Meares), 261–294. Amsterdam: Elsevier Santi Publishing Company.
22. Gardner, C.R., Weinstein, J.N., and Caplan, S.R. (1973). Transport properties of charge-mosaic membranes III. Piezodialysis. *Desalination* **12**: 19.
23. Weinstein, J.N. and Caplan, S.R. (1970). Charge-mosaic membranes: dialytic separation of electrolytes from nonelectrolytes and amino acids. *Science* **169**: 296.
24. Caplan, S.R. (1971). Transport in natural and synthetic membranes in membrane processes. In: *Industry and Biomedicine* (ed. M. Bier), 1–22. New York: Plenum Press.
25. Platt, K.L. and Schindler, A. (1971). Ionic membranes for water desalination. *Ang. Makromol. Chem.* **19**: 135.
26. Itou, H., Toda, M., Ohkoshi, K. et al. (1988). Artificial membranes from multiblock copolymers. 6. Water and salt transports through a charge-mosaic membrane. *Ind. Eng. Chem. Res.* **27**: 983–987.
27. Hirahara, K., Takahashi, S., Iwata, M. et al. (1986). Artificial membranes from multiblock copolymers. 5. Transport behaviors of organic and inorganic solutes through a charge-mosaic membrane. *Ind. Eng. Chem. Prod. Res. Dev.* **25**: 305–313.
28. Higa, M., Masuda, D., Kobayashi, E. et al. (2008). Charge mosaic membranes prepared from laminated structures of PVA-based charged layers: 1. Preparation and transport properties of charges mosaic membranes. *J. Membr. Sci.* **310**: 466.

Appendix

Membrane Technology and Applications, Fourth Edition. Richard W. Baker.
© 2024 John Wiley & Sons Ltd. Published 2024 by John Wiley & Sons Ltd.

Table A1   *Constants.*

Mathematical
   e = 2.71828
   ln 10 = 2.30259
   π = 3.14159
Gas law constant, R
   1.987 cal/g-mol K
   82.05 cm^3 atm/g-mol K
   8.314×10^7 g cm^2/s^2 g-mol K
   8.314×10^3 kgm^2/s^{2}kg-molK
Standard acceleration of gravity
   980.665 cm/s^2
   32.1740 ft./s^2
Avogadro's number
   6.023×10^{23} molecules/g-mol
Faraday's constant, F
   9.652×10^4 abs-coulombs/g-equivalent
Standard temperature and pressure (STP)
   273.15 K and 1 atm pressure
Volume of 1 mol of ideal gas at STP = 22.41 l

Table A2   Conversion factors for weight and volume.

Given a quantity in these units	Multiply by	To convert quantity to these units
Pounds	453.59	Grams
Kilograms	2.2046	Pounds
Ton, short (US)	2000	Pounds
Ton, long (UK)	2240	Pounds
Ton, metric	1000	Kilograms
Gallons (US)	3.7853	Liters
Gallons (US)	231.00	Cubic inches
Gallons (US)	0.13368	Cubic feet
Cubic feet	28.316	Liters
Cubic meters	264.17	Gallons (US)

Table A3   Conversion factors – other.

Given a quantity in these units	Multiply by	To convert quantity to these units
Inches	2.54	Centimeters
Meters	39.37	Inches
Mils	25.4	Microns
Square meters	10.764	Square feet
Dynes	1	g cm/s
Centipoises	10^{-3}	kg/m s

Table A4 Conversion factors for pressure.

Given a quantity in these units	Multiply by value below to convert to corresponding units					
	Atmosphere (atm)	mmHg (torr)	lb/in² (psi)	kg/cm²	Kilopascal (kPa)	Bar
Atmosphere (atm)	1	760	14.696	1.0332	101.325	1.01325
mmHg (torr)	1.3158×10^{-3}	1	1.9337×10^{-2}	1.3595×10^{-3}	0.13332	1.3332×10^{-3}
lb/in² (psi)	6.8046×10^{-2}	51.715	1	7.0305×10^{-2}	6.8948	6.8948×10^{-2}
kg/cm²	0.96787	735.58	14.224	1	98.069	0.98069
Kilopascal (kPa)	9.8692×10^{-3}	7.5008	0.14504	10.197×10^{-3}	1	0.01
Bar	0.98692	750.08	14.504	1.0197	100	1

Table A5 Conversion factors for energy.

Given a quantity in these units	Multiply by value below to convert to corresponding units					
	g cm²/s² (ergs)	kg m²/s² (joules)	Calories (cal)	British thermal units (Btu)	Horsepower hour (hp h)	Kilowatt hour (kWh)
g cm²/s² (ergs)	1	10^{-7}	2.3901×10^{-8}	9.4783×10^{-11}	3.7251×10^{-14}	2.7778×10^{-14}
kg m²/s² (joule)	10^7	1	2.3901×10^{-1}	9.4783×10^{-4}	3.7251×10^{-7}	2.7778×10^{-7}
Calories (cal)	4.1840×10^7	4.1840	1	3.9657×10^{-3}	1.5586×10^{-6}	1.1622×10^{-6}
British thermal units (Btu)	1.0550×10^{10}	1.0550×10^3	2.5216×10^2	1	3.9301×10^{-4}	2.9307×10^{-4}
Horsepower-hour (hp h)	2.6845×10^{13}	2.6845×10^6	6.4162×10^5	2.5445×10^3	1	7.4570×10^{-1}
Kilowatt hour (kWh)	3.6000×10^{13}	3.6000×10^6	8.6042×10^5	8.4122×10^3	1.3410	1

Table A6 *Conversion factors for liquid flux.*

Given a quantity in these units	Multiply by value below to convert to corresponding units					
	l/m^2 h	gal (US)/ft^2 day (gfd)	cm^3/cm^2 s	cm^3/cm^2 min	m^3/m^2 day	l/m^2 day
l/m^2 h	1	0.59	2.78×10^{-5}	1.6667×10^{-3}	2.40×10^{-2}	24.0
gal (US)/ft^2 day (gfd)	1.70	1	4.72×10^{-5}	2.832×10^{-3}	4.07×10^{-2}	40.73
cm^3/cm^2 s	3.60×10^4	2.12×10^4	1	60	864	8.64×10^5
cm^3/cm^2 min	600	353	0.667	1	14.4	1.44×10^3
m^3/m^2 day	41.67	24.55	1.6×10^{-3}	6.944×10^{-2}	1	10^3
l/m^2 day	4.17×10^{-2}	2.46×10^{-2}	1.6×10^{-6}	6.944×10^{-4}	1×10^{-3}	1

Table A7 *Conversion factors for liquid permeance.*

Given a quantity in these units	Multiply by value below to convert to corresponding units				
	l/m^2hbar	l/m^2hMegPa	l/m^2sPa	gal(US)/ft^2 day psi	m^3/m^2 day bar
l/m^2 h bar	1	10	2.777×10^{-9}	4.064×10^{-2}	2.4×10^{-2}
l/m^2 h MegPa	0.1	1	2.777×10^{-10}	4.064×10^{-3}	2.4×10^{-3}
l/m^2 s Pa	3.600×10^8	3.600×10^9	1	1.463×10^7	8.642×10^6
gal (US)/ft^2 day psi	24.61	2.461×10^2	6.837×10^{-8}	1	0.5906
m^3/m^2 day bar	41.67	4.167×10^2	1.57×10^{-7}	1.693	1

Table A8 Conversion factors for gas flux.

Given a quantity in these units	Multiply by value below to convert to corresponding units						
	$cm^3(STP)/cm^2 s$	$L(STP)/cm^2 s$	$m^3(STP)/m^2 day$	$ft^3(STP)/ft^2 h$	$mol/m^2 s$	$mol/cm^2 s$	$\mu mol/m^2 s$
$cm^3(STP)/cm^2 s$	1	1×10^{-3}	864	118.1	0.4462	0.4462×10^{-4}	0.4462×10^{6}
$L(STP)/cm^2 s$	1000	1	8.64×10^{5}	1.181×10^{5}	0.4462×10^{3}	0.4462×10^{-1}	0.4462×10^{9}
$m^3(STP)/m^2$ day	1.157×10^{-3}	1.157×10^{-6}	1	0.1366	0.5162×10^{-3}	0.5162×10^{-7}	0.5162×10^{3}
$ft^3(STP)/ft^2$ h	8.467×10^{-3}	8.467×10^{-6}	7.315	1	3.778×10^{-3}	3.778×10^{-7}	3.778×10^{3}
$mol/m^2 s$	2.241	2.241×10^{-3}	1.936×10^{3}	2.647×10^{2}	1	1×10^{-4}	1×10^{6}
$mol/cm^2 s$	2.241×10^{4}	22.41	1.936×10^{7}	2.647×10^{6}	1×10^{4}	1	1×10^{10}
$\mu mol/m^2 s$	2.241×10^{-6}	2.241×10^{-9}	1.936×10^{-3}	2.647×10^{-4}	1×10^{-6}	1×10^{-10}	1

Table A9 Conversion factors for gas permeance.

Given a quantity in these units	Multiply by value below to convert to corresponding units						
	1×10^{-6} $cm^3(STP)/cm^2 \cdot s \cdot$ cmHg (gpu)	$L(STP)/$ $m^2 \cdot h \cdot bar$	$cm^3(STP)/$ $cm^2 \cdot min \cdot bar$	$mol/m^2 \cdot s \cdot Pa$	$m^3(STP)/$ $m^2 \cdot s \cdot bar$	$m^3(STP)/$ $m^2 \cdot s \cdot Pa$	$mol/cm^2 \cdot$ $s \cdot kPa$
1×10^{-6} $cm^3(STP)/cm^2 \cdot s \cdot$ cmHg (gpu)	1	2.700	4.501×10^{-3}	3.347×10^{-10}	7.501×10^{-7}	7.501×10^{-12}	3.347×10^{-11}
$L(STP)/m^2 \cdot h \cdot bar$	0.3703	1	1.666×10^{-3}	1.239×10^{-10}	2.778×10^{-7}	2.778×10^{-12}	1.241×10^{-11}
$cm^3(STP)/cm^2 \cdot min \cdot bar$	0.2222×10^{3}	0.6000×10^{3}	1	0.7438×10^{-7}	1.666×10^{-4}	1.666×10^{-9}	0.7437×10^{-8}
$mol/m^2 \cdot s \cdot Pa$	0.2988×10^{10}	0.8065×10^{10}	1.344×10^{7}	1	2.241×10^{3}	2.241×10^{-2}	0.1
$m^3(STP)/m^2 \cdot s \cdot bar$	1.333×10^{6}	3.6000×10^{6}	6.000×10^{3}	4.462×10^{-4}	1	1×10^{-5}	4.463×10^{-5}
$m^3(STP)/m^2 \cdot s \cdot Pa$	1.333×10^{11}	3.6000×10^{11}	6.000×10^{8}	44.62	1×10^{5}	1	4.463
$mol/cm^2 \cdot s \cdot kPa$	2.988×10^{10}	8.065×10^{10}	1.344×10^{8}	10	2.241×10^{4}	2.241×10^{-1}	1

A 1-μm-thick membrane having a permeability of 1 Barrer has a permeance of 1 gpu.

Table A10 Conversion factors for gas permeability.

Given a quantity in these units	Multiply by value below to convert to corresponding units						
	1×10^{-10} cm³(STP)· cm/cm²·s·cmHg (Barrer)	cm³(STP)· cm/cm²·s·bar	cm³(STP)· cm/cm²·s·Pa	mol·m/m²· s·Pa	m³(STP)·m/m²· s·bar	m³(STP)·m/m²· s·Pa	mol·cm/cm²· s·kPa
1×10^{-10} cm³(STP)·cm/cm²· s·cmHg (Barrer)	1	7.501×10^{-9}	7.501×10^{-14}	3.347×10^{-16}	7.501×10^{-13}	7.501×10^{-18}	3.347×10^{-15}
cm³(STP)·cm/cm²·s·bar	1.333×10^{8}	1	1×10^{-5}	4.462×10^{-8}	1×10^{-4}	1×10^{-9}	4.462×10^{-7}
cm³(STP)·cm/cm²·s·Pa	1.333×10^{13}	1×10^{5}	1	4.462×10^{-3}	10	1×10^{-4}	4.462×10^{-2}
mol·m/m²·s·Pa	2.988×10^{15}	2.241×10^{7}	224.1	1	2.241×10^{3}	2.241×10^{-2}	10
m³(STP)·m/m²·s·bar	1.333×10^{12}	1×10^{4}	0.1	4.462×10^{-4}	1	1×10^{-5}	4.462×10^{-3}
m³(STP)·m/m²·s·Pa	1.333×10^{17}	1×10^{9}	1×10^{4}	4.462×10^{-5}	1×10^{5}	1	4.462×10^{-4}
mol·cm/cm²·s·kPa	2.988×10^{14}	2.241×10^{6}	22.41	0.1	224.1	2.241×10^{-3}	1

Table A11 *Vapor pressure of water and ice.*

Temp (°C)	Pressure (mmHg)	Temp (°C)	Pressure (mmHg)	Temp (°C)	Pressure (mmHg)	Temp (°C)	Pressure (mmHg)	Temp (°C)	Pressure (mmHg)
−30	0.285	0	4.58	30	31.8	60	149	90	526
−29	0.317	1	4.93	31	33.7	61	156	91	546
−28	0.351	2	5.29	32	35.7	62	164	92	567
−27	0.389	3	5.69	33	37.7	63	171	93	589
−26	0.43	4	6.1	34	39.9	64	179	94	611
−25	0.476	5	6.54	35	41.2	65	188	95	634
−24	0.526	6	7.01	36	44.6	66	196	96	658
−23	0.58	7	7.51	37	47.1	67	205	97	682
−22	0.64	8	8.05	38	49.7	68	214	98	707
−21	0.705	9	8.61	39	52.4	69	224	99	733
−20	0.776	10	9.21	40	55.3	70	234	100	760
−19	0.854	11	9.84	41	58.3	71	244	101	788
−18	0.939	12	10.5	42	61.5	72	254	102	816
−17	1.03	13	11.2	43	64.8	73	266	103	845
−16	1.13	14	12	44	68.3	74	277	104	875
−15	1.24	15	12.8	45	71.9	75	289	105	906
−14	1.36	16	13.6	46	75.7	76	301	106	938
−13	1.49	17	14.5	47	79.6	77	314	107	971
−12	1.63	18	15.5	48	83.7	78	327	108	1004
−11	1.79	19	16.5	49	88	79	341	109	1039
−10	1.95	20	17.5	50	92.5	80	355	110	1075
−9	2.13	21	18.7	51	97.2	81	370	111	1111
−8	2.33	22	19.8	52	102	82	385	112	1149
−7	2.54	23	21.1	53	107	83	401	113	1187
−6	2.77	24	22.4	54	113	84	417	114	1228
−5	3.01	25	23.8	55	118	85	434	115	1267
−4	3.28	26	25.2	56	124	86	451	116	1310
−3	3.57	27	26.7	57	130	87	469	117	1353
−2	3.88	28	28.3	58	136	88	487	118	1397
−1	4.22	29	30	59	143	89	506	119	1443
0	4.58	30	31.8	60	149	90	526	120	1489

Table A12 *Composition of air.*

Component	Concentration (vol%)	Concentration (wt%)
Nitrogen	78.09	75.52
Oxygen	20.95	23.15
Argon	0.933	1.28
Carbon dioxide	0.043	0.066
Neon	0.0018	0.0012
Helium	0.0005	0.00007
Krypton	0.0001	0.0003
Hydrogen	0.0005	0.00003
Xenon	0.000003	0.00004

Table A13 *Typical osmotic pressures at 25 °C.*

Compound	Concentration (mg/l)	Concentration (moles/l)	Osmotic pressure (psi) (bar)	
NaCl	35 000	0.60	398	27.4
Seawater	32 000	—	340	23.4
NaCl	2000	0.0342	22.8	1.57
Brackish water	2000–5000	—	15–40	1–2.6
$NaHCO_3$	1000	0.0119	12.8	0.88
Na_2SO_4	1000	0.00705	6.0	0.41
$MgSO_4$	1000	0.00831	3.6	0.25
$MgCl_2$	1000	0.0105	9.7	0.67
$CaCl_2$	1000	0.009	8.3	0.57
Sucrose	1000	0.00292	1.05	0.071
Dextrose	1000	0.0055	2.0	0.14

Table A14 *Mean free path of gases (25 °C).*

Gas	λ (Å)
Argon	1017
Hydrogen	1775
Helium	2809
Nitrogen	947
Neon	2005
Oxygen	1039
UF_6	279

Table A15 *Estimated diameter of common gas molecules.*

Gas molecule	Kinetic diameter (Å)	Lennard-Jones diameter (Å)
Helium	2.60	2.55
Neon	2.75	2.82
Hydrogen	2.89	2.83
Water	2.65	2.72
Nitrous oxide	3.17	3.49
Carbon dioxide	3.30	3.94
Acetylene	3.30	4.03
Argon	3.40	3.54
Oxygen	3.46	3.47
Nitrogen	3.64	3.80
Carbon monoxide	3.76	3.69
Methane	3.80	3.76
Ethylene	3.90	4.16
Propane	4.30	5.12
Propylene	4.50	4.68

Gas diameters can be determined as kinetic diameter based on molecular sieve measurements or estimated as Lennard-Jones diameters based on viscosity measurements. The absolute magnitude of the estimated diameters is not important, but the ratio of diameters can give a good estimate of the relative diffusion coefficients of different gas pairs (see Eq. (8.4)). On this basis, the kinetic diameters do a better job of predicting the relative diffusion coefficients of carbon dioxide/methane (always greater than 1 and often as high as 5–10 in glassy polymers). However, the Lennard-Jones diameter does a better job of predicting the relative diffusion coefficients of propylene/propane (always greater than 1 and often as high as 5 in glassy polymers).
Å = 0.1 nm.

Table A16 *Experimental diffusion coefficient of water in organic liquids at 20–25 °C at infinite dilution.*

Liquid	Temperature (°C)	Viscosity	cm^2/s $\times 10^5$
Methanol	20	—	2.2
Ethanol	25	1.15	1.2
1-Propanol	20	—	0.5
2-Propanol	20	—	0.5
1-Butanol	25	2.60	0.56
Isobutanol	20	—	0.36
Benzyl alcohol	20	6.5	0.37
Ethylene glycol	25	—	0.24
Triethylene glycol	30	30	0.19
Propane-1,2-diol	20	56	0.075
2-Ethylhexane-1,3-diol	20	320	0.019
Glycerol	20	1500	0.008
Acetone	25	0.33	4.6
Furfuraldehyde	20	1.64	0.90
Ethyl acetate	20	0.47	3.20
Aniline	20	4.4	0.70
n-Hexadecane	20	3.45	3.8
n-Butyl acetate	25	0.67	2.9
n-Butyric acid	25	1.41	0.79
Toluene	25	0.55	6.2
Methylene chloride	25	0.41	6.5
1,1,1-Trichloroethylene	25	0.78	4.6
Trichloroethylene	25	0.55	8.8
1,1,2,2-Tetrachloroethane	25	1.63	3.8
2-Bromo-2-chloro-1,1,1-trifluoroethane	25	0.61	8.9
Nitrobenzene	25	1.84	2.8
Pyridine	25	0.88	2.7

Source: F.P. Lees and P. Sarram, *J. Chem. Eng. Data* **16**, 41 (1971).

Table A17　*Diffusion coefficient of salts in water at 25 °C at infinite dilution.*

Salt	Diffusion coefficient $(cm^2/s \times 10^5)$
NH_4Cl	1.99
$BaCl_2$	1.39
$CaCl_2$	1.34
$Ca(NO_3)_2$	1.10
$CuSO_4$	0.63
$LiCl$	1.37
$LiNO_3$	1.34
$MgCl_2$	1.25
$Mg(NO_3)_2$	1.60
$MgSO_4$	0.85
KCl	1.99
KNO_3	1.89
K_2SO_4	1.95
Glycerol	0.94
$NaCl$	0.61
$NaNO_3$	1.57
Na_2SO_4	1.23
Sucrose	0.52
Urea	1.38

Source: Data correlated by Sourirajan from various sources in
Reverse Osmosis, Academic Press, New York (1970)

Table A18　*Interdiffusion of gases and vapors into air at 20 °C.*

Gas or vapor	Diffusion coefficient (cm^2/s)
O_2-air	0.18
CO_2-air	0.14
H_2-air	0.61
H_2O-air	0.22
n-Propyl alcohol	0.085
Ethyl acetate	0.072
Toluene	0.071
n-Octane	0.051

Source: Selected values from *International Critical Tables*,
W.P. Boynton and W.H. Brattain.

Table A19   *Interdiffusion of vapors into air, carbon dioxide, or hydrogen.*

Gas/vapor	Diffusion coefficient (cm^2/s)		
	Air	CO_2	H_2
Oxygen	0.18	0.14	0.70
Water	0.22	0.14	0.75
Ethyl acetate	0.072	0.049	0.27
n-Propyl alcohol	0.085	0.058	0.32
Propyl butyrate	0.053	0.036	0.21

Index

Membrane Technology and Applications, Fourth Edition. Richard W. Baker.
© 2024 John Wiley & Sons Ltd. Published 2024 by John Wiley & Sons Ltd.